S0-BKE-418

Advances in Thermodynamics

Advances in Thermodynamics

Edited by

G. Ali Mansoori
Department of Chemical Engineering
University of Illinois
P.O. Box 4348, Chicago, IL 60680

Editorial Board

Larry G. Chorn
Mobile Research & Development Corp.
Dallas Research Laboratory
P.O. Box 819047, Dallas, TX 75381

Enrico Matteoli
Instituto di Chimica Quantistica ed
Energetica Molecolare del C.N.R.
Via Risorgimento 35, 56100 Pisa, Italy

Peter Salamon
Department of Mathematical Sciences
San Diego State University
San Diego, CA 92182-0314

Stanislaw Sieniutycz
Institute of Chemical Engineering
Warsaw Technical University
1 Warynskiego St., 00-645 Warsaw, Poland

Volume 1: **C_{7+} Fraction Characterization**

Volume 2: **Fluctuation Theory of Mixtures**

Volume 3: **Nonequilibrium Theory and Extremum Principles**

Volume 4: **Finite-Time Thermodynamics and Thermoeconomics**

Volume 5: **Analytic Thermodynamics: Origins, Methods, Limits, and Validity**

Volume 6: **Flow, Diffusion, and Rate Processes**

ADVANCES IN THERMODYNAMICS

VOLUME 6

Flow, Diffusion, and Rate Processes

EDITED BY

Stanislaw Sieniutycz and Peter Salamon

Taylor & Francis

New York • Philadelphia • Washington, D.C. • London

PHYSICS
o4448303

USA	Publishing Office:	Taylor & Francis New York Inc. 79 Madison Ave., New York, NY 10016-7892
	Sales Office:	Taylor & Francis Inc. 1900 Frost Road, Bristol, PA 19007-1598
UK		Taylor & Francis Ltd. 4 John St., London WC1N 2ET

FLOW, DIFFUSION, AND RATE PROCESSES

Copyright © 1992 Taylor & Francis New York Inc. All rights reserved. No part of this publication may be reproduced, stored in a retrieval system, or transmitted, in any form or by any means, electronic, electrostatic, magnetic tape, mechanical, photocopying, recording or otherwise, without the prior permission of the copyright owner.

1 2 3 4 5 6 7 8 9 0 BRBR 9 8 7 6 5 4 3 2 1

Cover design by Berg Design.
A CIP catalog record for this book is available from the British Library.

Library of Congress Cataloging-in-Publication Data

Flow, diffusion, and rate processes / edited by Stanislaw Sieniutycz
and Peter Salamon.
p. cm.—(Advances in thermodynamics ; v. 6)
Includes bibliographical references and index.

1. Rate processes. 2. Diffusion. 3. Thermodynamics.
I. Sieniutycz, Stanislaw. II. Salamon, Peter, date.
III. Series.
QC174.17.R37F56 1992
536'.7—dc20 91-36609
ISBN 0-8448-1692-2 CIP

QC 174.17
R37 F56
1992
PHYS

Contents

Contributors

A. Altenberger
Department of Chemistry, University of Minnesota, Minneapolis, MN

B. Baranowski
Institute of Physical Chemistry PAN, Warsaw, Poland

D. Bedeaux
Department of Physical and Macromolecular Chemistry, University of Leiden, Netherlands

X. L. Chu
Department of Physics, U.N.E.D., Madrid, Spain

J. Dahler
Department of Chemistry, University of Minnesota, Minneapolis, MN

H. Farkas
Institute of Physics, Technical University of Budapest, Hungary

K. S. Førland
Division of Inorganic Chemistry, The Norwegian Institute of Technology, Trondheim, Norway

T. Førland
Division of Inorganic Chemistry, The Norwegian Institute of Technology, Trondheim, Norway

A. Guy
University of Florida, Gainesville, FL

J. Karkheck
Physics Science and Mathematics Department, GMI Engineering and Management Institute, Flint, MI

S. Lengyel
Central Research Institute for Chemistry, Budapest, Hungary

A. V. Myshlyavtsev
Institute of Catalysis, Novosibirsk, USSR

Z. Noszticzius
Institute of Physics, Technical University of Budapest, Hungary

S. K. Ratkje
Division of Physical Chemistry, The Norwegian Institute of Technology, Trondheim, Norway

R. L. Rowley
Bingham Young University, Provo, Utah

P. Salamon
Department of Mathematical Sciences, San Diego State University, CA

J. S. Shiner
Physiologishes Institut, Universitat Bern, Switzerland

S. Sieniutycz
Department of Chemical Engineering, Warsaw, Poland

G. C. Sih
Institute of Fracture and Solid Mechanics, Lehigh University, Bethlehem, PA

M. G. Velarde
Department of Physics, U.N.E.D., Madrid, Spain

B. Vujanović
Faculty of Technical Sciences, University of Novi Sad, Yugoslavia

S. Wiśniewski
Institute of Heat and Refrigeration Engineering, Technical University of Lodz, Poland

G. S. Yablonski
Institute of Catalysis, Novosibirsk, USSR

Introduction

In this multiauthored volume recent results obtained for the nonequilibrium thermodynamics of transport and rate processes are reviewed. Kinetic equations, conservation laws, and transport coefficients are obtained for multicomponent mixtures. Thermodynamic principles are used in the design of experiments predicting heat and mass transport coefficients. Highly nonstationary conditions are analyzed in the context of transient heat transfer, nonlocal diffusion in stress fields and thermohydrodynamic oscillatory instabilities.

Unification of the dynamics of chemical systems with other sorts of processes (e.g. mechanical) is given. Thermodynamics of reacting surfaces is developed. Admissible reaction paths are studied and a consistency of chemical kinetics with thermodynamics is shown. Oscillatory reactions are analysed in a unifying approach showing explosive, conservation or damped behavior. A comprehensive review of transport processes in electrolytes and membranes is given. Applications of thermodynamics to thermoelectric systems and ionized gas (plasma) systems are reviewed.

Thermodynamics of Transport and Rate Processes

S. Sieniutycz
Department of Chemical Engineering
Warsaw Techical University
00-645 Warsaw, 1 Warynskiego St., Poland

P. Salamon
Department of Mathematics
San Diego State University, San Diego, CA 92182, USA

ABSTRACT
The role of irreversible thermodynamics and kinetic theory is discussed in the context of various transport phenomena and rate problems. Statistical mechanical approaches to the phenomenological description of nonequilibrium are briefly characterized. The role of the theory for treating highly nonstationary transients is discussed. The significance of the problem of consistency of chemical kinetics and thermodynamics and the unification of the chemical dynamics with other processes is pointed out. The relative merits of these approaches are discussed. The contents of the papers appearing in the later part of this volume are reviewed.

1. INTRODUCTION

The investigation of nonequilibrium behavior based on general statistical mechanical and stochastic properties was discussed in vol. 3 of this series (see especially the papers by Lavenda and Lindenberg in that volume). Here we outline an alternative approach based on kinetic theory. This approach is exceptionally suitable for dealing with transport and rate phenomena. A kinetic justification of the principles of nonequilibrium thermodynamics has been mostly limited to dilute gases and metals. However, substantial progress has recently been made extending Enskog's method (Chapman and Cowling 1952), Grad's moment method (Grad 1958; Jou at al 1989), and other methods of kinetic theory to dense and real gases (Schmidt at al 1981).

The short discussion of the basic properties of kinetic equations (section 2) is followed by a brief review of the progress made in the theory of transport and rate phenomena (section 3). The discussion touches on the various approaches presented in volumes 3, 4, 6 and 7 of the present series. In section 4, the papers appearing in the later part of this volume are discussed.

2. THE ROLE OF KINETIC EQUATIONS

The typical ingredients of kinetic theories are the distribution function [functions f_i ($\mathbf{r}$, $\mathbf{c}_i$, t) for each component of a multicomponent mixture, i= 1,2.....n] and the governing kinetic equation (equations) describing its time evolution in phase space. Here $\mathbf{r}$ is the radius vector and $\mathbf{c}_i$ is the molecular velocity of i-th species at time t. The best known example of such kinetic equations is the basic integro-differential equation of Boltzmann. Their use is usually preceded by a summary of the macroscopic properties that should result from any physically acceptable kinetic equation. It has to be shown that an H-theorem holds, the macroscopic expression for the entropy source must follow, and that the kinetic equations lead to the Onsager reciprocity relations in the linear case. Actually, these ingredients reflect the fact that the irreversibility is already built into the fundamental kinetic equation whether this involves the Boltzmann integro-differential equation or its extensions. For the standard derivation of the Boltzmann equation based on the assumption of the molecular chaos, the reader is referred to many texbooks on kinetic theory and nonequilibrium statistical mechanics (Chapman and Cowling 1952; Huang 1967, Keizer 1987; for example). For a onecomponent system, the Boltzmann equation can be written in the form:

$$\frac{\partial f}{\partial t} + \mathbf{c}.\nabla_r f + \mathbf{F}.\nabla_c f = \int \sigma_T g[f'f'_1 - ff_1]dc_1^3 \qquad (1)$$

where the simplified notation is introduced f ($\mathbf{r}$, $\mathbf{c}'$, t) = f', f($\mathbf{r}$, $\mathbf{c}'_1$, t)= f'_1, σ_T for a linear operator corresponding to the differential scattering cross section $\sigma(\Omega, g)$, where Ω is the solid angle and g the constant magnitude of a relative velocity. Gradient operators involve derivatives with respect to position and velocity (Keizer 1987). Since the collision term of eq. (1) tends toward thermal equilibrium, the relaxation approximation is suggestive

$$\frac{\partial f}{\partial t} + \mathbf{c}.\nabla_r f + \mathbf{F}.\nabla_c f = -\frac{f - f^{eq}}{\tau} \qquad (2)$$

This approximation, containing the relaxation time τ, yields frequently good estimates of the transport coefficients (Warner 1966).

To investigate nonequilibrium phenomena, one must obtain the time evolution of the distribution function by solving the kinetic equation with given initial conditions. This is an involved problem (Enskog 1917; Chapman and Cowling 1952; Cercignani 1975). However, some rigorous properties of any solution to the Boltzmann equation follow from the fact that in any molecular collision there are some conserved dynamical quantities. Hence the macroscopic conservation laws follow for mass, energy, and momentum (Huang 1967, Kreuzer 1981). The conservation laws appear in kinetic theory as the result of collisional invariants for mass, momentum and kinetic energy of molecules. When combined with the equations describing the irreversible fluxes of mass, heat and momentum, the so-called equations of change are obtained which describe the hydrodynamic fields of velocity, temperature, and concentrations. The formalism is very useful for describing a variety of reacting and nonreacting systems on the hydrodynamic level (Bird at al 1960).

The kinetic theory leads to definitions of the temperature, pressure (scalar and tensor), internal energy, heat flux density, diffusion fluxes, entropy flux and entropy source in terms of definite integrals of the distribution function with respect to the molecular velocities. The inequality $\sigma>0$, which express the fact that the entropy source strength must be positive in any irreversible process, constitutes, within the framework of the kinetic theory, a derivation of the second law. This is known as Boltzmann's H-theorem. One of the most important virtues of kinetic theories is that they provide us with expressions for the transport coefficients.

In the limit of infinite time and provided the boundary conditions are compatible with equilibrium, the entropy source vanishes identically, detailed balance follows, and logarithms of the distribution functions become summational invariants. They must therefore be equal to a linear combination of the microscopic quantities m_i, $m_i\mathbf{c}_i$ and $(1/2)m_i\mathbf{c}_i^2$. This allows one to recover the Maxwell equilibrium distribution, the definition of hydrostatic pressure, and the ideal gas law. It also shows that the equilibrium value of the kinetic entropy density coincides with the thermodynamic one. It follows that the diffusion fluxes and the nondiagonal elements of the pressure tensor be non-zero only if the entropy source σ does not vanish. In particular, we get that the diffusional part of the entropy flux vanishes at equilibrium.

An important issue is that, in disequilibrium, the classical phenomenological form of the expressions for the irreversible entropy flux and entropy source (involving the products of fluxes and forces) follows from the approximate solution of the Boltzmann kinetic equation. This corresponds to the linear phenomenological laws. and can be shown by using the Enskog iterational method (de Groot and Mazur 1984). More precisely: if one evaluates the fluxes occuring in the hydrodynamic equations by retaining in the distribution functions f_i only the first order perturbations $\Phi^{(1)}{}_i$, then the diffusion fluxes $\mathbf{j}_i$, the heat flux $\mathbf{q}$ and the off-diagonal elements of the pressure tensor $\mathbf{P}$ become linear functions of the gradients of the macroscopic functions μ_i, T and $\mathbf{u}$. Onsager's symmetry results from the kinetic theory in a straightforward manner under the essential condition of microscopic reversibility (de Groot and Mazur 1984). Indeed, the first approximation of Enskog corresponds to the linear laws of the phenomenological theory. On the other hand, if terms quadratic in $\Phi^{(1)}{}_i$ or linear in $\Phi^{(2)}{}_i$ are retained in the entropy density expansion $\rho_s = \rho_s{}^0 + \rho_s{}^1 + \rho_s{}^2 ....$ this entropy will become a function of the macroscopic gradients in the Enskog formalism. Consequently, the nonlinear terms in ρ_s correspond to nonequilibrium entropies.

The Enskog method only enables one to calculate the deviations from the local Maxwellian distribution which are related to the spatial nonuniformities in the system. This is not the most general solution of the Boltzmann equation. Higher order hydrodynamic theories (Burnett 1935) and/or analyzes of short-time effects require more detailed approaches. It can be shown that the homogeneous (spatially uniform) perturbations have a relaxation time, τ, of the order of a few collision times. Only for times larger than τ can the true solution of the Boltzmann equation be approximated by the Enskog solution.

The relaxation times can be evaluated in Grad's moment approach (Grad 1958). He introduces velocity tensors of order n, $v^n{}_{ijk....}$, and the corresponding moments given by integrals of the products of

$v^n_{ijk...}$ and fdv. The Boltzmann equation is multiplied successively by each of the $v^n_{ijk....}$and integrated over the velocities. Successive equations for the moments are obtained. The first two moment equations and the contraction of the third are the usual equations of change describing the density mass-average velocity and the temperature. For these equations, the collision terms vanish due to properties of the summational invariants. In general, the equation of each moment involves the moment of higher order explicitly as well as the entire distribution function. Such a set of moment equations is equivalent to the original Boltzmann equation. The method of solution is based on the trial distribution functions; the Maxwellian distribution multiplied by a finite series of multidimensional Hermite polynomials with a number of arbitrary parameters which may then be written in terms of the first scalar moments (Grad 1958). This method justifies the first approximation of Enskog and permits any level of description ranging from the hydrodynamic description to a the complete description in terms of f_i ($\mathbf{r}$, $\mathbf{c}_i$, t). Grad's treatment is valuable for the study of gases at densities that are sufficiently low that the gas ceases to behave as a continuous medium. In Grad's formalism the quadratic correction terms contain the diffusive fluxes rather than gradients; this is exploited in extended thermodynamic theories involving nonequilibrium entropies which contain such fluxes.

Our discussion of kinetic theory has thus far been limited to nonreacting mixtures of simple molecules. It has been assumed that the molecules are spherical and have no internal degrees of freedom. A nice analysis of systems with internal degrees of freedom (polyatomic gases) is available (Resibois and Leener 1977). It may be shown that the integral term of the Boltzmann equation represents the net number of molecules of a particular kind gained by a "group" of molecules in phase space because of collision processes (Hirschfelder at al 1966). Consequently, if the collisions result in a chemical reaction, the right side of the Boltzmann equation has to be changed. In general, a modification of the Boltzmann equation affects the integral properties used in obtaining the equations of change. Modifications of the kinetic approach for chemically reacting systems and extensions to monoatomic gases are not discussed here. See for example Hirschfelder at al (1966) where these problems are analysed extensively in the context of quantum theory, reaction paths, saddle points and chemical kinetics. However, mass, momentum and energy are conserved even in a collision which produces a chemical reaction.

3. TRANSPORT AND RATE PHENOMENA

Important understanding of irreversibility can be achieved by embedding the kinetic theory relationships into a maximum entropy formalism. Maximization of entropy has been employed for obtaining kinetic equations for both dilute and dense fluids. These approaches constitute significant progress; they should be distinguished from the previous ones where the entropy was maximized under constraints imposed on quantities at the hydrodynamic level, e.g., the macroscopic energy (Lewis 1967). It was shown only recently that one must go beyond hydrodynamic quantities in order to obtain expressions for the transport coefficients (Karkheck 1986, 1989). These results indicate that when the constraints involve the one-particle distribution function as well as the potential energy density then the transport coefficients appear naturally as the outcome of the formalism. Consequently, both conservation laws and transport coefficient expressions can be obtained via the generalized maximum entropy approach. Such an approach is analysed in Karkheck's review in this volume.

A knowledge of the numerical values of the transport coefficients (Waldmann 1958) as well as their qualitative behavior as the thermodynamic state changes is crucial for understanding many macroscopic effects in continua. These include nonlinear effects and the dynamics of fluctuations. Standard approaches based on kinetic equations are the most popular. However, entropy-free and/or entropy source-free approaches have also been proposed (Schofield 1966). Transport phenomena in multicomponent mixtures can be investigated via a statistical mechanical method for the constitutive relations that does not require the introduction of an entropy source and that allows one to establish with certainty the molecular definitions and symmetry properties of the Onsager coefficients (see the paper by Altenberger and Dahler in this volume).

Thermodynamic and kinetic principles can be used to predict values of heat-mass transport coefficients in mixtures (Rowley and Gubler 1988) and to design experiments measuring such values. This approach is never trivial but it is important for the design of many engineering devices, e.g., heat and mass exchangers. It has the advantage of providing both micro and macroscopic information, and allows comparison with results from molecular dynamics. Onsager's reciprocity has been experimentally verified (Miller 1974; Rowley and Horne 1978). With the help of the moment theories of Grad's type, the relaxation times have been evaluated and highly nonstationary conditions may be analyzed. It has been shown that the kinetic theory approach leads to relaxation terms in the phenomenological equations implying damped-wave diffusion rather than the standard parabolic behavior (Jou at al 1989). Furthermore, it turns out that the natural variable of the extended Gibb's equation is neither heat flux nor entropy flux but a momentum variable called thermal momentum (Sieniutycz and Berry 1989). This formulation allows one to define the nonequilibrium temperatures and pressures unambiguously and to develop the theory of nonequilibrium thermodynamic potentials in the field representation of thermodynamics (Sieniutycz and Berry 1991). In addition, in a macroscopic approach, certain variational principles of Hamilton's type for transient heat transfer and mass diffusion have been formulated and their approximate solutions found by direct variational methods (Vujanovic and Jones 1989, Sieniutycz 1979, 1983).

In the last two decades many articles have appeared treating non-Fickian diffusion and, in particular, the transport of mass or energy described in terms of (damped) wave equations rather than classical parabolic equations (see Sieniutycz's review in this volume). The topics pursued in the literature range from theoretical developments (based usually on the phenomenological non-Onsagerian thermodynamics or nonequilibrium statistical mechanics), to a growing number of applications in fluid mechanics, polymer science, environmental engineering and chemical engineering.

It has been recognized that non-Fickian theories of diffusive mass transport provide a more realistic description of transfer phenomena than classical theories. In particular, such descriptions avoid the paradox of infinite propagation speed by incorporating inertial effects resulting from the finite mass of the diffusing particles. For concentrational disturbances, the paradox can be removed in a classical way (Baranowski 1974; Fitts 1962, de Groot and Mazur 1984). For thermal inertia, however, the situation is more subtle, and calls for the Boltzmann equation or certain extended quantum approaches (Sieniutycz 1991). The nonequilibrium energy obtained from the relaxation time approximation of the Boltzmann equation has lead to an extended Gibbs equation involving irreversible fluxes. The corresponding thermodynamic field theory is based on nonequilibrium field potentials which include the diffusion fluxes and hydrodynamic velocity as state variables (Sieniutycz and Berry 1989, 1991).

The nonclassical effects are particularly strong in diffusion of aerosols containing heavy particles and in polymers characterized by large molecular weight. Dispersed systems, common in chemical engineering, exhibit enhanced nonclassical effects which have been experimentally confirmed in drying processes. Other experiments show that the non-Fickian models of diffusion of mass and energy are more appropriate in the case of highly unsteady-state transients, occuring, for example, when laser pulses of high frequency are acting on the surfaces of solids or in the case of high frequency acoustic dispersion. Diffusion in polymers shows, as a rule, severe nonclassical effects related to the complex structure of polymeric media. The phenomenological thermodynamics of diffusion links some of the nonclassical effects with the absence of local equilibrium in the system. An extension of the chemical potential to stress fields has been given and nonlocal diffusion effects were experimentally verified (see Baranowski's paper in this volume).

A unifying picture of thermohydrodynamic oscillatory instabilities has been presented for Newtonian and viscoelastic fluids (Velarde and Chu 1990). It has been shown that the competition of dissipation and damping with restoring forces leads to oscillations and limit cycles. A thermomechanical theory, based on the isoenergy density (Sih 1985), can be used to describe irreversible continua.

A novel Lagrangian formulation of chemical dynamics yielding mass action kinetics has been developed (Shiner 1984, 1987; see also the Shiner's synthesis in this volume). It represents a unification of the dynamics of chemical systems with other processes (e.g. mechanical) and simplifies the treatment of complex couplings. The unification is based on a concept of a nonlinear chemical resistance (Shiner 1990, Sieniutycz 1987). New symmetries in chemical systems have been revealed corresponding to the nonlinear properties of these systems (Shiner 1987, Lengyel and Gyarmati 1986). The thermodynamics of the reacting surfaces has been developed and basic relations on surfaces have been formulated (Bedeaux 1986). Thermodynamically allowed reaction paths in open catalytic systems have also been studied.

Bykov, Gorban and Mirkes (1990) have worked out a new qualitative method for the thermodynamic analysis of the equations of chemical kinetics. An estimate of the thermodynamic function called the convexity margin has been introduced for the ideal chemical system. An algorithm has been proposed for defining regions of accessible composition based on a given reaction mechanism, location of equilibrium and the original composition. Allowable reaction paths have been studied and open catalytic systems have been investigated. The method suggests a new procedure for planning unsteady-state experiments. Also, it can solve the problem of boundary equilibria, and evaluate the region of admissibility for linear estimation. The relation between the times of relaxation to equilibrium and equilibrium flows can be found.

New procedures for planning steady and nonsteady-state kinetic experiments have been proposed. An analysis of the reacting systems displaced from equilibrium has been made showing that, in general, the total dissipation is not minimized at steady states (see Månsson and Lindgren review in vol. 3). By integrating the kinetic mass action law into nonequilibrium thermodynamics, a consistency of chemical kinetics with thermodynamics has been demonstrated in the framework of the so-called governing principle of dissipative processes (Lengyel 1988). However, as shown in Lengyel's review in this volume, this takes place at the expense of rejecting the standard affinity.

Oscillatory chemical reactions have been analyzed in a collection of treatments with sequentially enhanced degrees of unification. A synthetic review is presented in this volume by Farkas and Noszticzius for the three basic groups of chemical systems showing explosive, conservative or damped behavior (depending on parameters of a kinetic model). A matrix method for calculating surface reactions has been developed allowing effective calculations on lattice gases (see Myshlyavtsev's and Yablonski's review in this volume). A simple comprehensive treatment of transport processes in electrolytes and membranes have been given in a monograph by Forland et al.. Analysis of thermo-electric coefficients in metal couples have been presented with a new insight into the contact potential. Applications of thermodynamics to the ionized gas (plasma) systems have been summarized in the context of plasma formation, thermionic diodes and MHD generators (see Wisniewski's review in this volume).

A synthesis of the statistical thermodynamics of Onsager and the kinetic molecular theory of Boltzmann has recently been achieved

(Keizer 1987). This theory provides a mechanistic foundation for thermodynamics. It is based on the idea of elementary molecular processes (indexed by i)

$$(n_{i1}^{+}, n_{i2}^{+},.....) == (n_{i1}^{-}, n_{i2}^{-},.....) \tag{3}$$

with the extensive thermodynamic variables represented by the vector **n**= (n_1, n_2, ..,n_k). The superscripts + and - refer to the forward and backward steps in generalized reactions. It can be shown that these processes provide a natural description of bimolecular collision dynamics and chemical kinetics. The transition rate of an elementary process, that is the number of times per second that it occurs in the forward or reverse direction, is given by the canonical form

$$V_i^{+} = \Omega_i^{+}\exp(-\sum_l Y_l^{+} n_{il}/k),$$

and likewise (4)

$$V_i^{-} = \Omega_i^{-}\exp(-\sum_l Y_l n_{il}/k).$$

The constants Ω_i^{+-} are the intrinsic rates of the forward and reverse steps of the elementary process (i) and are the basic transport coefficients in the canonical theory. Y_l are intensive variables which are functions of the extensive variables; they are the derivatives of a local equilibrium entropy S(**n**). Using the canonical representation for the rates of the elementary processes, one can develop the statistical thermodynamics of molecular processes in a form applicable to systems close or far from equilibrium. Both transport and rate processes can be treated in a unified way; an extension of the idea already pursued in Eyring's works (Eyring 1962, Eyring and Jhon 1969) imbeded in a rather different formalism. The same basic approach may be used for molecular collisions, hydrodynamics and chemical or electrochemical processes. Nonlinear molecular mechanism can be shown to give rise to critical points, bifurcations, oscillations and chaotic behavior.

4. DISCUSSION

Below the papers appearing in the later part of the volume are briefly reviewed. The first group of papers pertains to transport and flow phenomena. The second group pertains to rate processes occuring in both homogeneous and heterogeneous systems with possible electric effects. The latter approaches, discussed in the final part of the volume, are exceptionally important for their applications in chromatography, membrane technology and catalysis.

J. Karkheck develops the Boltzmann equation and other kinetic equations in the context of the maximum entropy principle and the laws of hydrodynamics. He uses a general approach based on the maximum entropy principle, and provides a specific example with a constraint imposed on the one particle distribution function. Hydrodynamics emerges from the observation that a vast amount of detailed microscopic information is macroscopically irrelevant. His interpretation of minimum information (maximum entropy) is not the conventional interpretation of information which is missing but rather information which is irrelevant to a chosen level of observing the behavior of a system. His approach yields generalizations of Boltzmann and Enskog theories. The generalizations are applicable to real fluids and dense gases. An interparticle potential is used which can be adapted to real fluids by using perturbation theory of equilibrium statistical mechanics (Mansoori and Canfield 1969). A sequence of kinetic equations is generated in the maximum entropy approach by imposing different constraints to yield different forms of closure. Examples of closed kinetic theories for simple liquids are given, and a correspondence with Boltzmann and Vlasov ensembles is shown when the interparticle potential is neglected. Then, from the moment approach, the conservation laws of mass, momentum and energy follow, with very general flux expressions. By expansion of the fluxes in powers of the gradients of density, velocity and temperature, the transport coefficients are obtained. Existence of two distinct temperatures out of equilibrium is shown. A modified Chapman-Enskog procedure (to include extra degrees of freedom) is proposed for solving the kinetic equations. A comparison with molecular dynamics is presented; the agreement is good. This paper should be read by everyone who is interested in the origin and nature of irreversibility.

A. Altenberger and J. S. Dahler propose an entropy source-free method which allows one to answer unequivocally questions concerning the symmetry of the kinetic coefficients. It is based on nonequilibrium statistical mechanics and it involves a linear functional generalization of Onsagerian thermodynamics. It extends some earlier attempts (Schofield, 1966; Shter 1973) to generalize fluctuation dynamics and Onsager's relations using convolution integrals to link thermodynamic fluxes and forces. By using the Mori operator identity (Mori 1965), applied to the initial values of the currents of mass and heat, constitutive equations linking the thermodynamic fluxes and forces are obtained. Correspondence of the results of the functional theory with the classical one is verified by the use of Fourier representations for the fields and for the functional derivatives. This approach provides a consistent description of transport processes with temporal and spatial dimensions and yet enables one to recover the classical results in the long-wave limit, where the gradients in the system are small. The possibility of extending this linear theory to situations far from equilibrium is postulated. The extension is based on the probability densities of the phase-space variables determined for stationary nonequilibrium reference states. The roles of external fields and various reference frames are investigated, with the conclusion that the laboratory frame description is as a rule the most suitable for the theory. Transformations of the thermodynamic forces are thoroughly analyzed, focusing on violations of the Onsagerian symmetries.

R. Rowley reviews an approach using thermodynamic principles to predict values of heat-mass transport coefficients in mixtures and to design experiments measuring such values. The approach is a combined experimental and theoretical examination of diffusion, thermal conductivity and heat of transport in nonelectrolyte liquid mixtures. Mutual diffusion coefficients are measured in a variety of liquids. In an effort to develop a predictive technique for such diffusion coefficients, the mixtures are modelled as Lenard-Jones fluids and the values of the mutual diffusion coefficients are obtained from molecular dynamics simulations. Onsager coefficients, heats of transport, and diffusion coefficients are found from experiments near liquid-liquid critical point, and Onsager reciprocity is verified. The thermal effect of diffussion is measured and developed into an accurate experimental technique for measurements of heat of transport. A new predictive method free of adjustable parameters is developed for the thermal conductivity. Conclusions of a methological nature are drawn predicting a lasting role for molecular dynamic simulations.

M. G. Velarde's paper presents a unifying picture of various thermo-hydrodynamic oscillatory instabilities in both Newtonian and viscoelastic fluids. It shows that the competition of dissipation and damping with restoring forces can lead to oscillations and limit cycles. It explains how nonlinearity helps sustain these oscillations in the form of limit cycles. First, oscillatory Rayleigh-Benard convection driven by buoyancy forces is discussed. Then, interfacial oscillators are investigated. Starting with the perturbation equations, various effects are analyzed, such as, Marangoni instabilities, the onset of oscillatory motion, transverse gravity-capillary oscillations, neutral stability conditions, longitudinal waves, interfacial electrohydrodynamics, etc. Solitons excited by Marangoni effect are introduced. It is shown how the Korteweg-de Vries soliton description can be used as a building block for some nonlinear traveling wave phenomena. The role played by dissipation in triggering and eventually sustaining nonlinear motions is shown. This acquints the reader with the beauty of the methodological unification of various nonlinear phenomena in hydrodynamics.

S. Sieniutycz's tutorial review discusses the so-called paradox of infinite speed of propagation of thermal and concentrational disturbances. This paradox has influenced the construction and development of the important thermodynamic theory now called extended irreversible thermodynamics (EIT). (See volume 7 of this series for a comprehensive presentation of recent developments in that theory). To acquaint the reader with a historical development of attempts to eliminate the paradox, solutions of the parabolic and the hyperbolic partial differential equation of heat are compared. A plausible concept of relaxation time is proposed; a simple evaluation of relaxation times for heat, mass and momentum in ideal gases with a single propagation speed is presented. Extending these ideas to nonideal systems, the simplest extended theory of coupled heat and mass transfer is constructed based on the nonequilibrium entropy. The entropy source analysis yields a matrix formula for the relaxation matrix, τ. This matrix can be applied to coupled transfer and corresponds to the well known values of relaxation times of pure heat transfer and isothermal diffusion. Various simple forms of wave equations for coupled heat and mass transfer are discussed. The simplest extensions of the second differential of the entropy and the excess entropy production are given. They allow one to prove the stability of the coupled heat and mass transfer equations by the second method of Liapounov. Dissipative variational formulations are presented, leading to aproximate solutions by a direct variational method. Applications of the hyperbolic equations to a description of short-time effects and high frequency behavior are outlined. Some related ideas dealing with variational formulations for hyperbolic equations of heat can be found in the paper by Vujanovic in this volume.

B. Baranowski's paper explores the influence of stress fields on diffusion in isotropic elastic media. The thermodynamics of solids with stress fields is developed. An extension of the chemical potential to stress fields is given under (experimentally proved) assumption neglecting the role of off-diagonal elements in the stress tensor. Attention is directed to stresses developed by a diffusing component, causing displacements, and resulting stresses, in the solid. Plastic deformations are neglected i.e. an ideal elastic medium is assumed. Diffusion equations are derived in a one-dimensional case. It is shown that the resulting equation of change for concentration is a nonlinear integro-differential equation. Nonlocal diffusion effects are verified by an original experiment. Experimental results of hydrogen diffusion in a PdPt alloy are compared with the theory. The general importance of the stress field to diffusion in solids is outlined. Effect of stresses on the entropy production is discussed with the conclusion that a stress-free steady-state obeys the minimum entropy production principle as a stationarity criterion. See also Wisniewski (1984) for an analysis of the role of diffusion rather than mechanical stresses in a general multicomponent system.

B. D. Vujanovic develops a variational theory related to the thermodynamics of transport phenomena with a finite propagation rate. The theory is presented in the context of wave equations of heat conduction but it can be generalized to diffusion problems involving both isothermal and coupled diffusion. The variational approach is intended to obtain approximate solutions for problems of classical (Fourier's) transient heat conduction. The approximation is obtained by a limiting procedure in which the relaxation time tends to zero (variational principle with a vanishing parameter). The essential value of author's approach lies in its ability to handle a variety of nonlinear problems with temperature dependent thermal conductivities and heat capacities. A nice representation of the theory in terms of Biot's heat vector (Biot 1970) can also treat some nonlinearities. Related aspects of the theory and its modifications are discussed. Many illustrative examples using the approximate "method of partial integration" are presented. This method is based essentially on the Ritz-Kantorovich direct method in variational calculus (Gelfand and Fomin 1963) and works effectively for complex boundary value problems of radiation, melting, heating of slabs, etc..

G. C. Sih proposes a new approach to nonequilibrium thermomechanics based on the so-called isoenergy density theory. A variety of examples is discussed in terms of the predictions of the theory which emphasises the exchange of volume and surface energy. Nonequilibrium behavior is characterized by the rate at which a process can no longer be described in terms of physical parameters representing the system as a whole. Thermal changes are synchronized with the motion of mass elements. The isoenergy density theory is employed for characterizing irreversible behavior in gases, liquids, solids and phase transformations. It can take into account both laminar and turbulent flows as well as the transition between them. The conditions for a fluid to separate from an immersed object are predicted along with the position of the separation. Nonuniform condensation of a gas in a confined region, fluid-solid separation, and heating of an uniaxial specimen are other examples analyzed. A critical analysis is presented concerning the application of concepts and notions established from equilibrium theories to nonequilibrium phenomena.

J. S. Shiner presents a novel Lagrangian formulation of chemical dynamics consistent with mass action kinetics. It is based on the idea of a nonlinear chemical resistance and a corresponding Rayleigh dissipation function. These can be applied to chemical processes in a fashion which is formally similar to the use of the well known resistances and dissipation functions of physical processes. The author's approach yields a unification of the dynamics of chemical systems with other sorts of processes (e.g. mechanical) in the framework of Lagrangian dynamics and yet simplifies the treatment of complex couplings. A major virtue of the work is a physical Lagrangian based on the Gibbs energy. The author's formulation stresses a natural algebraic symmetry in the chemical reaction system at stationary states far from equilibrium. This symmetry can be viewed as a more general case of Onsager's reciprocity, and can explain the complex nature of the symmetry properties in chemical systems. The formalism leads naturally to the network representation of chemical systems; this allows a treatment using standard methods of network thermodynamics. When the inertial effects are negligible, the Lagrangian dynamics can be set in terms of a variational formulation which simplifies near equilibrium to the theory of minimum power dissipation.

S. Lengyel analyses the consistency between chemical kinetics and thermodynamics. His approach involves expressions for the entropy production as a bilinear form in the stoichiometrically dependent and independent elementary chemical reactions. For independent reaction systems, the generalized reciprocity relations are satisfied

(Gyarmati-Li relations, Li 1958). Dissipation potentials are introduced and the kinetic law of mass action is shown to result from the "governing principle of dissipative processes". This is an important result since chemical reaction kinetics is frequently considered as a phenomenon which only reluctantly obeys a standard thermodynamic formalism. However, the traditional concept of chemical affinity has to be rejected; rather the partial Gibbs free energies of the reactants should be used. The corresponding flux potential is a function of both the partial Gibbs free energy of the reactants and of the products. It is shown that the theory can solve some difficult problems of chemical kinetics which were unsolved by other methods. This approach offers several surprises, but its consistency is firmly established by the worked out examples.

H. Farkas and Z. Noszticzius present a unified treatment of chemical oscillators treated as thermodynamic systems far from equilibrium. They start with the theory of the Belousov-Zhabotynski (BZ) reaction pointing out the role of positive and negative feedback. Then, they show that consecutive autocatalytic reaction systems of the Lotka-Volterra type play an essential role in the BZ reaction and other oscillatory reactions. Consequently, they introduce and then analyze certain (two-dimensional) generalized Lotka-Volterra models. They show that these models can be conservative, dissipative, or explosive, depending of the value of a parameter. The transition from dissipative to explosive behavior occurs via a critical Hopf bifurcation. The system is conservative at certain critical values of the parameter. The stability properties can be examined by testing the sign of a Liapounov function selected as an integral of the conservative system. By adding a limiting reaction to an explosive network, a limit cycle oscillator surrounding an unstable singular point is obtained. Some general thermodynamic aspects are discussed pointing out the distinct role of oscillatory dynamics as an inherently far-from-equilibrium phenomenon. The existence of the attractors is recognized as a form of the second law. This paper gives a pedagogical perspective on the very broad class of problems involving the time order in homogeneous chemical systems.

K. S. Forland, T. Forland and S. K. Ratkje present a new, simple, approach to the thermodynamics of irreversible transport processes in electrolytes and membranes. The description is lumped (i.e. discontinuous) and preserves many features of the classical Onsagerian formalism as, e.g., local equilibrium, bilinearity of the entropy production, reciprocity relations, etc. On the other hand, the classical chemical affinity is not applied in the formalism (see also Lengyel's work giving up the affinity, in this volume). In such systems scalar-vector couplings can take place. Only vectorial fluxes and forces (gradients of the chemical potentials) are used. The treatment yields valuable insight concerning some important processes, as, e.g., active transport, or muscle contraction. Many examples are discussed including glass electrodes and cells with liquid junctions, cation transport in water in a cation exchange membrane, and transport-reaction interaction in a battery and in muscle contraction. Extensions for nonisothermal systems are reviewed. This minimum-mathematics approach provides a good introduction to the irreversible thermodynamic problems of physiology, biophysics and physical chemistry of electrolytes.

S. Wisniewski introduces the reader to the thermodynamics of solutions of ionized gases. When a large fraction of the species present is ionized, such solutions are called plasmas. Plasma formation is analysed from the view point of terms like: ionization potential, conditions of ionized gas equilibrium, Saha equation, nonideal plasmas, equilibrium composition, and degree of ionization. Thermionic generators, i.e. devices which convert internal energy into electric energy, are analysed with the thermionic diode serving as the example. Energy balance is given, entropy source obtained, and the phenomenological equations postulated for the flows of energy and electric current. The thermal efficiency of thermionic energy conversion treated as a heat engine is determined. Equations of plasma in an electrodynamic field are given along with Onsager-Casimir relations. The Hall effect in a plasma is quantitatively described. Finally, the main types of magnetohydrodynamic generators are discussed taking the Hall effect into account. This paper is a nice, tutorial-style, presentation of nonequilibrium problems in plasma systems.

A. G. Guy analyses thermoelectric effects in metal couples. His analysis points out the necessity to distinguish the electrochemical potential of the electrons from the electrochemical potential itself. On this basis, he discusses in sequence: thermoelectric effects in a single metal, equilibrium in an isothermal couple, and thermoelectric effects in a metal-metal couple. His analysis links the Seebeck coefficient with the electron enthalpy in an unconventional way, describes quantitatively equilibrium conditions in metal couples, and specifies some thermoelectric characteristics. His conclusion is that thermoelectric phenomena in thermocouples are determined by the thermoelectric behavior of single isolated metals. The contact potential has no influence on the thermoelectric behavior of metal-metal couples and hence the Seebeck and Peltier coefficients for a couple are postulated to depend in a simple way on the corresponding coefficients of single metals.

D. Bedeaux formulates rigorous nonequilibrium thermodynamics of surfaces. He considers a boundary layer between two phases, which may change its position and curvature. The analysis requires time dependent curvilinear coordinates. At the surface the field properties, densities, fluxes and sources exhibit discontinuities which are quantitatively described by appropriate excess functions. The excesses of the densities and currents are described as singular interfacial contributions to the distributions of the densities and currents. After formulating detailed balances of mass, momentum and energy, the interface contributions are singled out from global equations. Finally, the entropy production for interfaces, σ^s, is found and the phenomenological equations postulated. Onsager relations are formulated on surfaces. The general expression for σ^s involves a variety of fluxes and forces and the author presents reasonable simplifications. Applications to membrane transport, evaporation, surfactant physics, and chemistry are briefly discussed. The virtue of the relatively simple analysis achieved in this paper is a unified approach to bulk and interface, i.e. to continuous and discontinuous properties.

A. V. Myshlyavtsev and G. S. Yablonski offer the reader extensive studies in the thermodynamics and kinetics of surface reactions. Their method is called "method of transfer matrix". It allows one to evaluate the thermostatistical and kinetic properties of surfaces. These properties are essential in the theories of catalysis, grain boundaries, and adsorption. They are also of value to a general understanding of the thermodynamics of surfaces. The transfer matrix method turns out to be the most effective for calculating with models of lattice gases. It allows one to determine the statistical thermodynamic properties from the grand partition function fitting a one-site, which is treated as a stage along a chain. Every site along the chain is found in one of m possible states where m is any non-negative integer. It follows that the largest eigenvalue of the transfer matrix is equal to the grand partition function of a one-site along the chain. The probability that the site is in a definite state can be found. Kinetic rate constants and coefficients of surface diffusion can be determined. The authors give sample applications of the method in variety of involved cases: monomolecular adsorption, desorption spectra, surface diffusion coefficients, and desorption from a reconstructed surface. They also predict applications to more complicated lattices, with several kinds of centers. This paper should be studied by everyone who wants to apply thermodynamics to practical problems of surfaces.

ACKNOWLEDGEMENT.
One of the authors (S. S.) acknowledges the kind hospitality of the Department of Mathematical Sciences at San Diego State University. Both authors gratefully acknowledge partial support from the Office of Naval Research, grant number N00014-88-K-0491.

REFERENCES

Bedeaux, D. 1986. Nonequilibrium Thermodynamics and Statistical Physics. In Advances in Chemical Physics. Vol. 64, p. 67. Eds. I. Prigogine, and S. A. Rice. New York:Wiley.

Biot, M.A. 1970. Variational Principles in Heat Transfer. Oxford: Clarendon Press.

Bird, R.B., W.E. Steward and E.N. Lightfoot. 1960. Transport Phenomena. New York: J. Wiley.

Burnett, D. 1935. The distribution of the molecular velocities and the mean motion in a non-uniform gas. Proc. Lond. Math. Soc. 40: 382- 435.

Bykov, V.I., A. N. Gorban and E. M. Mirkes. 1990. Chemical Dynamics and Equilibrium. In preparation. Computational Center, Krasnoyarsk, USSR.

Callen, H. 1988. Thermodynamics and an Introduction to Thermostatistics. New York: J. Wiley.

Cercignani, C. 1975. Theory and Application of the Boltzmann Equation. Edinburgh: Scottish Academic Press.

Chapman, S., and T.G. Cowling. 1952. The mathematical Theory of Nonuniform Gases. Cambridge.

Enskog, D. 1917. Kinetic Theory of Processes in Moderately Dense Gases. Dissertation. Upsala University.

Eyring, H. 1962. The transmission coefficient in reaction rate theory. Rev. Mod. Phys. 34: 616-619.

Eyring, H., and Jhon, M.S. 1969. Significant Liquid Structures. New York: Wiley.

Fitts, D.D.1962. Nonequilibrium Thermodynamics. New York: Mc Graw-Hill.

Forland, K.S., T. Forland and S.K. Ratkje. 1988. Irreversible Thermodynamics. Theory and Applications. Chichester:Wiley.
Gelfand, J.M., and S.W. Fomin. 1970. Variational Calculus. Warsaw: PWN.
Glansdorff, P., and I. Prigogine. 1971. Thermodynamic Theory of Structure Stability and Fluctuations. New York: J.Wiley.
Grad, H.1958. Principles of the theory of gases. In Handbook der Physik 12, ed. S. Flugge. Berlin: Springer.
de Groot, S.R., and P. Mazur. 1984. Nonequilibrium Thermodynamics. New York: Dover.
Huang, K. 1967. Statistical Mechanics. New York: J.Wiley.
Karkheck, J. 1986. Kinetic theories and ensembles of maximum entropy. Kinam 7A: 191-208.
Karkheck, J. 1989. Kinetic theories and ensembles of maximum entropy. In: Maximum Entropy and Bayesian Methods, ed. J.Skilling, 491-6. Kluwer Academic Publishers.
Keizer, J. 1987. Statistical Thermodynamics of Nonequilibrium Processes. New York: Springer.
Kestin, J. 1979. Foundations of Non-Equilibrium Thermodynamics. Domingos at al, eds. London: MacMillan Press.
Kreuzer, H.J. 1981. Nonequilibrium Thermodynamics and its Statistical Foundations. Oxford: Clarendon Press.
Landau, L., and E. Lifshitz. 1974. Mechanics of Continua. London: Pergamon.
Lebon, G., and P. Mathieu. 1983. Comparison of the diverse theories of nonequilibrium thermodynamics. Intern. Chem. Engng. 23: 651- 662.
Lengyel, S. 1988. Deduction of the Guldberg-Waage mass action law from Gyarmai's governing principle of dissipative processes. J. Chem. Phys. 88: 1617-1621.
Lengyel, S. and Gyarmati. 1986. Constitutive equations and reciprocity relations of nonideal homogeneous closed chemical systems. Acta Chem. Hung.122: 7-17.
Lewis, R.M. 1967. A unifying principle in statistical mechanics. J. Math. Phys. 8: 1448-1459.
Li, J.C.M. 1958. Thermodynamics of nonisothermal system. The classical formulation. J. Chem. Phys. 29: 747-754.
Mansoori, G. A., Canfield, F. B., 1969. Variational approach to the equilibrium thermodynamic properties of simple fluids. J. Chem. Phys. 51: 4958.-67.
Miller, D.G. 1974. The Onsager relation: experimental evidence. In Foundations of Continuum Mechanics. Eds. J.J. Delgado Domingos, M.N. Nina, and J.H. Whitelaw. London: Mac Millan

Mori, H. 1965. Transport, collective motion and Brownian motion. Progr. Theor. Phys. 33: 423-455.

Müller, I. 1985. Thermodynamics. London: Mc Millan

Jou, D., J. Casas-Vazquez, and G. Lebon. 1989. Extended Irreversible Thermodynamics. Rep. Progr. Phys. 51: 1105-1115.

Prigogine, I. 1949. On the domain of validity of the local equilibrium hypothesis. Physica 15:1942.

Resibois, P., and de Leener. 1977. Classical Kinetic Theory of Fluids. New York: Wiley.

Rowley, R.L., and F.H. Horne. 1978. The Dufour effect: Experimental confirmation of the Onsager heat-mass reciprocity relation for a binary liquid mixture. J. Chem. Phys. 68: 325-326.

Rowley, R.L., and V. Gruber. 1978. Thermal conductivities in seven ternary mixtures at 40^{0} C and 1 atm. J. Chem Eng. Data 33: 5-8.

Schmidt, G., W. Kohler and S. Hess 1981. (A treatment of a kinetic model of a dense fluid). Z. Naturforsch. 36a:545

Schofield, P. 1966. Wavelength-dependent fluctuations in classical fluids. Proc. Phys. Soc. 88: 149-170.

Shter, I.M. 1973. The generalized Onsager principle and its application. Inzh. Fiz. Zh. 25: 736-742.

Sieniutycz, S. 1979. The wave equations for simultaneous heat and mass transfer in moving media- structure testing, space-time transformations and variational approach. Intern. J. Heat Mass Transfer 22: 585-599.

Sieniutycz, S. 1983. The inertial relaxation terms and the variational principles of least action type for nonstationary energy and mass diffusion. Intern. J. Heat Mass Transfer 26: 55-63.

Sieniutycz, S. 1987. From a least action principle to mass action law and extended affinity. Chem. Engng Sci. 42: 2697-2711.

Sieniutycz, S. 1990/91. Thermal Momentum, Heat Inertia and a Macroscopic Extension of the de Broglie Micro-thermodynamics I and II. In Advances in Thermodynamics 3 and 7, eds. S. Sieniutycz and P. Salamon. New York: Taylor and Francis.

Sieniutycz, S., and R.S. Berry. 1989. Conservation laws from Hamilton's principle for nonlocal thermodynamic equilibrium fluids with heat flow. Phys. Rev. A. 40: 348-361.

Sieniutycz, S., and R.S. Berry. 1991. Field thermodynamic potentials and geometric thermodynamics with heat transfer. Phys. Rev. A. 43: 2807-2818.

Shiner, J.S. 1984. A dissipative Lagrangian formulation of the network thermodynamics of pseudo- first order reaction-diffusion systems. J. Chem. Phys. 81: 1455-1465.

Shiner, J.S. 1987. Algebraic symmetry in chemical reaction systems at stationary states arbitrarily far from thermodynamic equilibrium. J. Chem. Phys. 87: 1089-1094.

Shiner, J.S. 1991. A Lagrangian formulation of chemical reaction dynamics far from equilibrium. Adv. in Thermodyn. vol. 6., S. Sieniutycz and P. Salamon eds.

Sih, G.C.1985. Mechanics and physics of energy density and rate of change of volume with surface. Journ. of Theor. and Applied Fract. Mech. 4: 157-173.

Truesdell, C. 1962. Mechanical basis of diffusion. J. Chem. Phys. 37: 2236-2344.

Truesdell, C. 1969. Rational Thermodynamics. New York: Mc Graw-Hill.

Velarde, M.G., and X.L. Chu. 1990. Interfacial Instabilities. London:World Scientific.

Vujanovic, B.D., and S.D. Jones. 1989. Variational Methods in Nonconservative Phenomena. Boston: Academic Press.

Waldmann, L. 1958. Transporterscheinungen in Gase von mittlerem Druck. In Hanbuch der Physik, vol12, Thermodynamics of Gases. Ed. S. Flugge. Berlin: Springer.

Warner, G.H. 1966. Statistical Physics. New York: Wiley.

Wisniewski, S. 1984. Balance equations of extensive quantities for a multicomponent fluid taking into account diffusion stresses. Archiw. Termod. 5: 201- 213.

Irreversible Processes, Kinetic Equations, and Ensembles of Maximum Entropy

John Karkheck
Department of Science and Mathematics
GMI Engineering and Management Institute
Flint, MI 48504-4898, USA

ABSTRACT

Maximization of entropy is explored as a technique for producing explicit nonequilibrium ensembles that can be used to describe kinetic and transport processes in dense fluids. When applied to an appropriate subset of the BBGKY hierarchy equations, these ensembles yield closed kinetic theories. The nonequilibrium ensemble correctly reproduces the data about which the ensemble is constructed. For the one-particle distribution function and the potential energy density as data, an Enskog-like kinetic theory is obtained that subsumes or eclipses many previous theories that have been applied to simple liquids. Novel features of the new theory include complete representation of the hydrodynamic equations, including energy conservation, a strong H-theorem and correct equilibrium solution, and the presence of two distinct nonequilibrium temperature scales, one associated with kinetic energy and the other with potential energy. The temperatures couple through conservation of energy. Their distinction is shown to affect the value of the bulk viscosity. The new transport coefficients are "bare" quantities for which an adaptation is described that yields an accurate theory when compared to simulation and experimental data. More general ensembles that consider the full two-particle distribution function are also discussed.

INTRODUCTION

The great contributions made by Boltzmann and Enskog toward the development of a general kinetic theory of fluids were the products of brilliant intuition. Both Boltzmann and Enskog theories have contributed immeasurably to our understanding of transport processes in dilute and dense fluids, respectively. Boltzmann's H-theorem is a landmark in demonstrating a relation between microscopic dynamics and irreversible processes. These kinetic equations serve as a bridge between the microscopic domain and the realm of macroscopic irreversible processes by providing a framework that encompasses and explains hydrodynamics in terms of interparticle dynamics. However, their link to the microscopic domain is obscure since there the fundamental equations are reversible.

Irreversibility is a widely observed but incompletely understood phenomenon. At issue is what conditions engender irreversibility and how to reconcile these conditions with an underlying microscopic reversibility that, for simple classical fluids, is inherent in Newton's second law, or the Liouville equation for ensembles. Here we describe an approach to these problems that is based upon maximization of entropy. The goal of this program is to elucidate the origin and nature of irreversibility by producing generalizations of Boltzmann and Enskog theories that are applicable to real liquids and dense gases. The procedure is illustrated with construction of a family of closed kinetic theories for simple liquids. The results obtained here are equivalent to, or eclipse, those obtained by different approaches to the development of kinetic equations, such as linear response theory (Lebowitz et.al. 1969), graphical expansions (van Beijeren and Ernst 1973), the Mori-Zwanzig formalism (Sung and Dahler 1984), short-time principles (Davis et.al. 1961), and many-body methods (Konijnendijk and van Leeuwen 1973). The maximum entropy method is systematic and seems more amenable to the production of more general kinetic theories.

As a prototypical example of irreversibility, hydrodynamics teaches us that a broad range of nonequilibrium states can be specified by just the mass density, velocity, and energy density of the fluid, reflecting those same quantities that are conserved during collisions amongst the fluid particles. The hydrodynamic equations then comprise a set of five coupled equations that accurately portray a wide range of relaxation processes. The decay processes associated with the relaxation of nonequilibrium states to equilibrium are characterized by phenomenological relations, such as Fourier's law of heat conduction, and the decay rates are governed by the transport coefficients. These can be obtained from measurements. Ready accessibility of this information has permitted hydrodynamics to become a well developed subject without recourse to the methods of statistical mechanics.

The important lesson of hydrodynamics is that a small amount of reproducible information is sufficient for accurate description of a wide range of nonequilibrium states and their temporal evolution. Since the transport coefficients must be supplied from experiment, hydrodynamics is not closed in a strict sense. But its profundity lies in the demonstration that a vast amount of information associated with the detailed microscopic state is irrelevant to the macroscopic picture. What information is essential to that description is the compelling question.

Linear response theory yields formulas for the transport coefficients in terms of integrals over time correlation functions. These many-body expressions are obtained by solving the Liouville equation to linear order in driving forces, hence appear to be reversible in form. For two-body interparticle forces, these time correlation functions depend upon the one and two particle distribution functions. The latter are rendered from some approximate kinetic equations in practice. Computer simulations, however, employ these expressions directly (Hoheisel and Vogelsang 1988) to produce "exact" results. These provide a basis for quantitative assessment of the correctness of the kinetic equations. Moreover, simulations show that these expressions can render transport coefficients for real fluids using reasonable potential models such as the Lennard-Jones.

The Boltzmann equation describes the temporal evolution of the one particle distribution function:

$$\left(\frac{\partial}{\partial t} + \vec{v}_1 \circ \nabla_1\right) f_1(\vec{r}_1,\vec{v}_1,t) = \int d\vec{v}_2 \int d\Omega\ \sigma(g,\Omega) g \times$$

$$\{f_1(\vec{r}_1,\vec{v}_1^{*},t) f_1(\vec{r}_1,\vec{v}_2^{*},t) - f_1(\vec{r}_1,\vec{v}_1,t) f_1(\vec{r}_1,\vec{v}_2,t)\}.$$

Among its distinctive features are closure, support of the hydrodynamic equations and expressions for the transport coefficients, and of course, irreversibility. In the collision integral, which describes binary encounters, a factored form for the two particle distribution, f_2, is employed, in essence, $f_2^B(\vec{r}_1,\vec{v}_1,\vec{r}_2,\vec{v}_2,t) = f_1(\vec{r}_1,\vec{v}_1,t) f_1(\vec{r}_1,\vec{v}_2,t)$. Boltzmann's f_2^B exhibits velocity chaos and localizes collisions, that is, ignores spatial separation between colliding particles. The locality condition $\vec{r}_1 = \vec{r}_2$ applies in a dilute gas since the interaction length is much smaller than the mean free path, the characteristic length for transport processes. Lewis showed (1967) that velocity chaos is inherent in the ensemble that results by maximizing entropy subject to f_1 as a constraint.

In striking contrast, the Vlasov equation

$$\left(\frac{\partial}{\partial t} + \vec{v}_1 \circ \nabla_1\right) f_1(\vec{r}_1,\vec{v}_1,t) = \frac{1}{m} \int d\vec{r}_2 d\vec{v}_2 \times$$

$$\nabla_1 V(|\vec{r}_1 - \vec{r}_2|) \circ \frac{\partial}{\partial \vec{v}_1} f_1(\vec{r}_1,\vec{v}_1,t) f_1(\vec{r}_2,\vec{v}_2,t)$$

is reversible, even though the interparticle potential V yields the scattering cross section σ and the same velocity chaos is employed. The nonlocality here is not the reason for reversibility, as will be shown later, but then neither is velocity chaos alone the origin of the irreversibility in the former. That is an effect of the combined operation of the statistical assumption and the collision events.

Since hydrodynamics applies to dilute and dense fluids, the contracted description exhibits universality; the f_2^B does not. Even at equilibrium, it fails to represent the spatial structure of the dense fluid. One of Enskog's principal contributions (1922) was demonstration of the significance of the fluid structure in the transport coefficients of the dense hard-sphere fluid. Yet, it remains an open question whether a tractable representation can be given to the nonequilibrium f_2 that applies to real fluids of arbitrary density. The maximization of entropy seems to be an appropriate tool for studying this question.

Maximization of entropy has been widely applied under the rubric of information theory and the notion of missing information (Levine and Tribus 1978). Our perspective here is not information that is missing but rather information that is irrelevant to a chosen level of discussion of fluid behavior. Recently, alternate interpretations for the utility of the maximum entropy formalism (Tikochinsky et.al. 1984, Shalitin 1986) have been proposed which seem closer to our purpose, namely to objectively evaluate what input information an ensemble must bear to permit accurate inferences about behavior and properties of the many-body system. The information here is not what is known to the observer, but rather what information within the system drives its temporal behavior. Clearly, what is pertinent information is commensurate with the questions asked. Anything less than the full many-body information is only an approximation, thus, is of limited applicability. Hydrodynamics cannot explain the phenomenological relations, nor the transport coefficients, and it is limited to spatially and temporally slow relaxation processes. Its accuracy within that regime shows that a small amount of particular information is sufficient to permit reliable theoretical description of experimental results at that level. Going further, via f_2^B, the one particle distribution function is sufficient to describe its temporal evolution and the transport properties of the dilute gas. A posteriori, one observes that Boltzmann's f_2 is not comprehensive, for it fails to describe postcollisional correlations, but the information it lacks is unimportant to the purpose to which it is applied. So too, the ensemble which supports velocity chaos lacks correlation information which must be present in the "exact" ensemble, but this appears to be unimportant to a discussion of transport phenomena in the dilute gas.

In the next section, the kinetic theory is developed, using a model interparticle potential that can be adapted to real fluids through equilibrium statistical mechanical perturbation theory. A sequence of kinetic equations is generated by imposing different constraints in the maximum entropy procedure to produce different forms of closure. The third section is devoted to probing the structural and predictive properties of the theories, including comparison of transport coefficient predictions to simulation and experimental results. The concluding section includes discussion of generalizations of the theory and some comments.

DEVELOPMENT OF THE KINETIC THEORY

The essential ingredients in Boltzmann's kinetic theory are statistical independence of a pair of particles about to collide and a distinct collision time scale. These conditions prevail in a dilute gas of neutral particles because collisions are sufficiently rare that particles stream freely over distances that are large compared to the interaction length. The well separated collisional and free streaming time scales lead to simplification of the description of the collisional dynamics. Also, a given colliding pair is highly likely to have originated from different regions of the gas. This enhances the degree of mixing and diminishes the likelihood of precollisional correlations.

Neither of these conditions prevails in the dense gas or liquid, which can be characterized such that the mean separation between neighboring particles does not exceed the interaction length. Thus, each particle is in continual interaction with several neighbors. This obscures the notion of collision event, but at the same time may result in a smearing out of the net long-range interaction acting on any particle into a weak mean-field attraction, leaving in effect a well-defined short ranged repulsion.

Model Potentials

In this spirit, we employ the following model interparticle

potential, where θ is the Heaviside function,

$$V(r) = \begin{cases} \infty & r < d \\ \phi(r)\theta(R-r) & r > d\,. \end{cases} \tag{2.1}$$

For application to real liquids, a smooth function ϕ is used that contains a short-ranged repulsion and a long ranged attraction. The d and R are chosen, respectively, such that the repulsive hard sphere core is interposed over the smooth repulsion and the cutoff is made on the smooth attraction. These features are not required by the structure of the theory below, which could just as well be formulated for charged hard spheres and many other fluids. In fact, the form (2.1) lends itself to many limiting cases: pure hard spheres when $\phi = 0$, the full potential ϕ when $d = 0$ and $R \to \infty$, the square well attraction when ϕ is a negative constant, and the square well repulsion when ϕ is a positive constant. The discontinuities at d and R yield instantaneous transfer of momentum and energy (collisions) which ensures a strong mixing of kinetic and potential energy. They also establish a vanishingly small time scale over which the smooth parts of the potential produce vanishingly small transfer of momentum and energy between particles. This decouples collisional and streaming effects.

The discontinuity in V at R, henceforth referred to as "well edge", permits bound states, if $\varepsilon = -\phi(R^-) > 0$. In this case, four distinct types of collision can occur: i) hard-core repulsion, ii) entering the well, iii) leaving the well, and iv) a bound-state collision at the inside well edge, which appears as a hard-core repulsion. The respective postcollisional velocities for particles of mass m and initial velocities $\vec{v}_1$ and $\vec{v}_2$ follow from conservation of energy and momentum:

i and iv $$\vec{v}_1' - \vec{v}_1 = \hat{\sigma} g_\sigma = \vec{v}_2 - \vec{v}_2' \tag{2.2a}$$

ii $$\vec{v}_1^{+} - \vec{v}_1 = \frac{\hat{\sigma}}{2} \left[g_\sigma - (g_\sigma^2 + v_e^2)^{1/2} \right] \tag{2.2b}$$

iii $$\vec{v}_1^{-} - \vec{v}_1 = \frac{\hat{\sigma}}{2} \left[g_\sigma - (g_\sigma^2 - v_e^2)^{1/2} \right] . \tag{2.2c}$$

Since they emanate from a Hamiltonian, the postcollisional values also serve as initial values for restituting collisions. The escape speed $v_e = (4\varepsilon/m)^{1/2}$, and the unit vector $\hat{\sigma}$, which points from the center of particle 2 to the center of particle 1, is oriented so that the component of the relative velocity $\vec{g} = \vec{v}_2 - \vec{v}_1$, namely $g_\sigma = \hat{\sigma}\circ\vec{g}$, is positive to ensure a collision. These formulas can be readily generalized to mixtures (Castillo et.al. 1989).

The spherical symmetry of the model potential precludes consideration of rotational degrees of freedom. Generalizations of the hard-sphere potential, including loaded spheres, ellipsoids (Dahler and Theodosopulu 1975), spherocylinders (Kirkpatrick 1988), and rough spheres (Cercignani and Lampis 1988) are amenable to the techniques described here. Rough spheres are defined such that collisions are perfectly elastic and the relative velocity at contact of a colliding pair is completely reversed. However, since rough spheres are not a Hamiltonian system, the basic dynamical equation cannot be elicited from the Liouville equation in the manner to be described. The postcollisional velocities do not also serve as restituting velocities which then must be inserted by hand.

Dynamical Equations

A kinetic theory for simple liquids and dense gases must have sufficient structure to support hydrodynamical equations, which describe the time evolution of the mass density, velocity, and energy density of the fluid. These are given in terms of the one (f_1) and two (f_2) particle distribution functions, respectively:

$$\rho(\vec{r},t) = m\int d\vec{v}\ f_1(\vec{r},\vec{v},t) \tag{2.3a}$$

$$\vec{u}(\vec{r},t) = \int d\vec{v}\; \vec{v} f_1 \tag{2.3b}$$

$$e(\vec{r},t) = e_k + e_p = \int d\vec{v}\; \frac{1}{2} mv^2 f_1 + \\ + \frac{1}{2} \times \int_{d<r_{12}} dx_1 dx_2\; V(r_{12}) f_2(x_1,x_2,t)\,\delta(\vec{r}_1-\vec{r}) . \tag{2.3c}$$

The $x = (\vec{r},\vec{v})$, $r_{12} = |\vec{r}_1-\vec{r}_2|$ and δ is the Dirac delta. To obtain expressions for the transport coefficients requires going deeper than the hydrodynamic equations; for this a "minimal" kinetic theory is developed that describes the time evolution of f_1 and e_p, the potential energy density. To construct such a highly contracted description, one starts with the BBGKY hierarchy.

For specific distribution functions, F_s, which are normalized $\int d^s x\; F_s(x^s,t) = \mathcal{V}^s$, where $x^s = (x_1,x_2,\ldots,x_s)$ and $\mathcal{V}$ is the volume of the fluid, the BBGKY hierarchy is (Uhlenbeck and Ford 1963)

$$\frac{\partial}{\partial t} F_s + \{H_s,F_s\} = n\int dx_{s+1} \sum_{i=1}^{s} \nabla_i V(r_{is+1}) \circ \frac{\partial}{\partial \vec{p}_i} F_{s+1} , \tag{2.4}$$

where $N = n\mathcal{V}$ is the number of particles in the fluid, the $\{\ ,\ \}$ is the Poisson bracket, and H_s is the Hamiltonian for a subset of s particles. This set of equations is equivalent to the Liouville equation, hence is reversible. Neither are solvable for a potential such as (2.1). To make progress requires closing the hierarchy at some s to form a self-contained subset of these equations. This involves expressing F_{s+1} in terms of the lower order functions.

The earlier comparison between the Boltzmann and Vlasov equations pointed out the significance of collisions. The integrodifferential equation form is not as useful for constructing collision integrals as is the solution (Lewis 1961) in terms of streaming operators:

$$F_s(x^s,\ t+\tau) =$$

$$\sum_{k=0}^{\infty} n^k \int dx_{s+1} \ldots dx_{s+k} \sum_{j=0}^{k} \frac{(-1)^{k-j}}{j!(k-j)!} T_{-\tau}^{(j+s)} F_{k+s}. \tag{2.5}$$

The T operation is defined in terms of conventional streaming operators $S_{-\tau}^{(j)} = \exp\tau\{H_j,\ \}$ by $T_{-\tau}^{(j+s)} F_{k+s} = F_{k+s}(S_{-\tau}^{(j+s)} x^{j+s},$ $S_{-\tau}^{(1)} x_{j+s+1}, \ldots,\ S_{-\tau}^{(1)} x_{j+k}, t)$. Equation (2.4) is recovered immediately from (2.5) for smooth potentials in the limit $\tau \to 0$. For the discontinuous parts of V(r), the streaming operators must be handled carefully (Ernst et.al. 1969).

Equation (2.5) demonstrates that there are two parameters at our disposal, n and τ, that solution for F_s requires knowledge of all higher distribution functions at a time τ earlier, and that the j+s-particle interactions must be followed over that time interval. The importance of F_{k+s}, $k > 1$, is tempered, however, by n^k and by τ. For dilute fluids, only leading order in n need be retained; for dense fluids, a priori, all orders must be retained. To illustrate, obtain for s = 1

$$\begin{aligned} F_1(x_1,t+\tau) - T_{-\tau}^{(1)} F_1(x_1,t) &= n\int dx_2 \left[T_{-\tau}^{(2)} F_2(x_1,x_2,t) - \right. \\ & \left. T_{-\tau}^{(1)} F_2(x_1,x_2,t) \right] + n^2 \int dx_2 dx_3 \left[\tfrac{1}{2} T_{-\tau}^{(3)} F_3(x^3,t) - \right. \\ & \left. T_{-\tau}^{(2)} F_3(x^3,t) + \tfrac{1}{2} T_{-\tau}^{(1)} F_3(x^3,t) \right] + \ldots \end{aligned} \tag{2.6}$$

Inasmuch as the smooth part of (2.1) results in the form (2.4) for the time evolution of F_s, only the contributions from the discontinuities, that is, the collision terms, need to be constructed. Due to symmetry of F_s under interchange of particles, each integrand at each power of density exhibits a cluster property such that the integrand vanishes unless each of the s+1 to s+k particles is in a position such that it could interact within time interval τ with at least one of the s particles. This limits the relevant volume (Karkheck and Stell 1982) that each of the former could occupy in relation to the latter to a spherocylinder, placed appropriately at d or at R, whose height is proportional to τ and whose axis is oriented according to the relative velocity of the colliding pair. Thus, the k+s-particle term is weighted by $(n\tau)^k$. The desired equation is obtained by dividing (2.5), or (2.6), by τ and taking the limit $\tau \to 0$. The remaining equation exhibits linear n dependence, regardless of density, dependence only on F_s and F_{s+1}, and only binary collisions.

In summary, both the Boltzmann equation and kinetic equations for dense fluids can be developed from (2.6) (Karkheck and Stell 1982) utilizing the features afforded by each density regime. For a dense fluid governed by (2.1), (2.6) yields, where the generic function $f_s = n^s F_s$ has replaced F_s and the thermodynamic limit is understood (Karkheck et.al. 1988),

$$
\begin{aligned}
\left(\frac{\partial}{\partial t} + \vec{v}_1 \circ \nabla_1\right) f_1(x_1,t) &= \frac{1}{m} \times \int_{d<r_{12}<R} dx_2\, \nabla_1\phi \circ \frac{\partial}{\partial \vec{v}_1} f_2(x_1,x_2,t) \; + \\
&+ \int d\vec{v}_2 \int d\hat{\sigma}\; \theta(g_\sigma) g_\sigma \Big\{ d^2 \big[f_2(x_1', \vec{r}_1 + \vec{d}^+, \vec{v}_2', t) - \\
&\qquad f_2(x_1, \vec{r}_1 - \vec{d}^+, \vec{v}_2, t) \big] \; + \\
&+ R^2 \big[f_2(x_1^+, \vec{r}_1 + \vec{R}^-, \vec{v}_2^+, t) - f_2(x_1, \vec{r}_1 - \vec{R}^+, \vec{v}_2, t) \; + \\
&+ \theta(g_\sigma - v_e) \big(f_2(x_1^-, \vec{r}_1 - \vec{R}^+, \vec{v}_2^-, t) - f_2(x_1, \vec{r}_1 + \vec{R}^-, \vec{v}_2, t) \big) \; + \\
&+ \theta(v_e - g_\sigma) \big(f_2(x_1', \vec{r}_1 - \vec{R}^-, \vec{v}_2', t) - f_2(x_1, \vec{r}_1 + \vec{R}^-, \vec{v}_2, t) \big) \big] \Big\} \; .
\end{aligned}
\tag{2.7}
$$

The R^+, R^- refer to values just outside, inside the well edge. The companion equation for e_p is obtained similarly. The collisional part is constructed here explicitly to illustrate the procedure (Karkheck 1986). By definition,

$$\frac{\partial e_p}{\partial t} = \lim_{\tau\to 0^+} \frac{1}{2} \times \int_{d<r_{12}} dx_1 dx_2 \; \delta(\vec{r}-\vec{r}_1) V(r_{12}) \frac{f_2(x^2,t+\tau)-f_2(x^2,t)}{\tau} \; .$$

Applying $\int d\vec{v}_1 d\vec{v}_2$ to the expression for $f_2(x^2,t+\tau)$ given by (2.5) yields for $\partial e_p/\partial t$

$$\frac{\partial e_p}{\partial t} = \lim_{\tau\to 0^+} \frac{1}{2} \times \int_{d<r_{12}} dx_1 dx_2 \; \delta(\vec{r}-\vec{r}_1) V(r_{12}) \times \frac{f_2(S_{-\tau}^{(2)}x^2,t)-f_2(x^2,t)}{\tau} \; . \tag{2.8}$$

The first term in (2.8) is transformed, using the canonical transformation $(\vec{\imath}_1,\vec{\mathfrak{v}}_1,\vec{\imath}_2,\vec{\mathfrak{v}}_2) = S_\tau^{(2)}(x_1,x_2)$ which has unit Jacobian, to obtain

$$\frac{\partial e_p}{\partial t} = \lim_{\tau\to 0^+} \frac{1}{2} \times \int_{d<r_{12}} dx_1 dx_2 \; f_2 \; \frac{V(\imath_{12})\delta(\vec{r}-\vec{\imath}_1)-V(r_{12})\delta(\vec{r}-\vec{r}_1)}{\tau} \; .$$

To extract the collision term, as above, $d\vec{r}_2 = O(\tau)$ and the identity is used:

$$\frac{V(\imath_{12})\delta(\vec{r}-\vec{\imath}_1)-V(r_{12})\delta(\vec{r}-\vec{r}_1)}{\tau} = V(\imath_{12})\frac{\delta(\vec{r}-\vec{\imath}_1)-\delta(\vec{r}-\vec{r}_1)}{\tau} + \frac{V(\imath_{12})-V(r_{12})}{\tau}\,\delta(\vec{r}-\vec{r}_1) \, .$$

The collision term emanates from the second term on the right hand side which contributes only across the well edge when $\tau \to 0$, because $d\vec{r}_2$ is also $O(\tau)$. For small, but finite, τ, such as time steps used in molecular dynamics simulations, the second term would contribute over the whole potential. The equation for e_p is obtained,

$$\frac{\partial}{\partial t} e_p(\vec{r},t) + \nabla\circ\left[e_p\vec{u}(\vec{r},t) + \vec{J}_\phi(\vec{r},t)\right] =$$

$$\frac{1}{2} \times \int_{d<r_{12}<R} dx_1 dx_2\, \delta(\vec{r}-\vec{r}_1)\, f_2\, \vec{g}\circ\nabla_2\phi +$$

$$+ \frac{1}{2}\, \varepsilon R^2 \int d\vec{v}_1 d\vec{v}_2 \int d\hat{\sigma}\; g_\sigma \left[\theta(g_\sigma - v_e) f_2(x_1, \vec{r}+\vec{R}^-, \vec{v}_2, t) - \theta(g_\sigma) f_2(x_1, \vec{r}-\vec{R}^+, \vec{v}_2, t)\right]. \qquad (2.9)$$

The flux $\vec{J}_\phi = \frac{1}{2}\int dx_1 dx_2 (\vec{v}_1 - \vec{u}) V(r_{12}) \delta(\vec{r}-\vec{r}_1) f_2$. The coupled equations (2.7) and (2.9) contain the unknown f_2, with no provision to obtain it. Thus this set is not selfcontained. Closure involving expressing f_2 in terms of f_1 and e_p is pursued next.

The Closure Principle

The equilibrium two particle distribution function ℓ_2 factorizes

$$\ell_2 = \ell_1(v_1)\ell_1(v_2)g_2(r_{12})$$

where ℓ_1 is the Maxwellian distribution and g_2 is the radial distribution function (rdf). The same form is used to define the nonequilibrium pair distribution function G,

$$G(x_1, x_2, t) = f_2(x_1, x_2, t)/f_1(x_1, t) f_1(x_2, t). \qquad (2.10)$$

For pure hard spheres, only G at contact, $r_{12} = d$, is needed in the kinetic theory. The form used by Enskog is that of the radial distribution function evaluated at the density found at the point of contact of the spheres. There is not a unique generalization of this notion for mixtures and whatever choice is used yields (Barajas et.al.1973) a theory that is inconsistent with irreversible thermodynamics, especially the Onsager reciprocal relations. A revised version, obtained independently by diagrammatic methods (van Beijeren and Ernst 1973a), linear response (Lebowitz et.al. 1969), and many-body methods (Konijnendijk and van Leeuwen 1973), equates G to the pair distribution function for a nonuniform fluid at equilibrium. This depends on the density at all points in the fluid. This form does generalize (van Beijeren and Ernst 1973b) to mixtures in conformity with irreversible thermodynamics. For the square-well fluid, DRS (Davis et.al. 1961) used a variant of the Enskog G that also included temperature at the points of collision. This bears similar shortcomings. In all these cases, G is velocity independent, that is, exhibits velocity chaos.

Accurate description of hydrodynamics and the transport coefficients requires only a part of the information contained in the "exact" G. Thus, our goal is to construct an approximation to G that yields accurate representation of the transport coefficients and closure of (2.7) and (2.9) such that they evolve as a self-contained set. In this context, closure involves expressing G in terms of f_1 and e_p, possibly in a nonlocal or nonMarkovian form.

The maximum entropy procedure produces closure by generating an ensemble distribution function that depends on the information held as constraints (Lewis 1967; Karkheck and Stell 1982). The ensemble accurately reproduces that information and also imposes dependence of the higher order distribution functions on that information, which is the desired closure. The procedure is exhibited for a fixed number, N, of particles to simplify notation. Formulation for an open system is straightforward; the additional degree of freedom, the number of particles, participates in the variational operation. The entropy

$$S[W_N] = -k\int d^N x\, W_N(x^N,t)\,\ell n(W_N/\Gamma) \tag{2.11}$$

is maximized subject to constraints. The W_N is the exact ensemble distribution function, and Γ is a measure on the phase space whose value is not pertinent here. The constraining functions are related to W_N:

$$h_i^{(\alpha,\nu)}(\vec{r}_1,\ldots,\vec{r}_\alpha,\vec{v}_1,\ldots,\vec{v}_\nu,t) = \int d^N x\, H_i(x^N,t)W_N\ . \tag{2.12}$$

The H_i, $i = 1, 2,\ldots$ are functions associated with the microscopic properties of the fluid. The constraints are built into the maximization procedure by using Lagrange multipliers, $\lambda_i(\vec{r}_1,\ldots,\vec{r}_\alpha,\vec{v}_1,\ldots,\vec{v}_\nu,t)$. Here, three constraints are imposed on W_N: i) normalization, for which $h_1^{(0,0)} = 1$, $H_1 = 1$, and the Lagrange multiplier $\lambda_1 = \gamma(t)$, ii) f_1, for which $h_2^{(1,1)} = f_1(\vec{r},\vec{v},t)$, $H_2 = \sum_{i=1}^{N} \delta(\vec{r}-\vec{r}_i)\delta(\vec{v}-\vec{v}_i)$, and $\lambda_2 = \lambda(\vec{r},\vec{v},t)$, iii) e_p, for which $h_3^{(1,0)} = e_p(\vec{r},t)$, $H_3 = \frac{1}{2}\sum_{i,j=1}^{N}{}' V(r_{ij})\delta(\vec{r}-\vec{r}_i)$, $\lambda_3 = \beta(\vec{r},t)$.

The variation is performed on the trial functional

$$I[W_N] = S[W_N] + k[\int dx\, \lambda(x,t)f_1 - \sum_{i=1}^{N}\int d^N x\, \lambda(x_i,t)W_N] +$$

$$k\gamma[1-\int d^N x\, W_N] + k[\int d\vec{r}\, \beta(\vec{r},t)e_p - \sum_{i<j}\int d^N x\, \beta_{ij}V(r_{ij})W_N].$$

The $\beta_{ij} = \frac{1}{2}[\beta(\vec{r}_i,t)+\beta(\vec{r}_j,t)]$. The variational condition $\frac{\delta I}{\delta W_N} = 0$ yields the maximizing distribution W_N^m,

$$W_N^m(x^N,t) = \frac{1}{Z}\prod_{i=1}^{N}\exp\{-\lambda(x_i,t)\}\prod_{j<k}^{N}\exp\{-\beta_{ij}V(r_{ij})\} \tag{2.13}$$

such that $Z(t)$ is the normalizing factor. This ensemble yields for G:

$$G = g_2(\vec{r}_1,\vec{r}_2|n,\beta) \tag{2.14}$$

which has a formal graphical expansion (Stell 1963) like the rdf except that each vertex is weighted by the density field $n(\vec{r},t) = \int d\vec{v}\ f_1$ and each Mayer bond $f^M_{ij} = \exp\{-\beta_{ij}V(r_{ij})\}-1$. In the rdf, both n and β are constant. The insertion here of space and time dependent quantities is analogous to the procedure used to construct generalized hydrodynamics from conventional hydrodynamics (Boon and Yip 1980). Both f_1 and e_p are functionals of the Lagrange multipliers λ and β. Formally, λ and β may be obtained as functionals of f_1 and e_p, so W^m_N is determined by the latter. Thus, inserting (2.10) and (2.14) into (2.7) and (2.9) yields a closed kinetic theory, the coupled equations for f_1 and e_p.

This theory is the third member of a family of kinetic variational theories (Stell et.al. 1983) all of which are Enskog-like by exhibiting velocity chaos. The first member, KVT I (Karkheck and Stell 1981), utilizes only the repulsive part of V in the trial function. This builds overlap exclusion of the hard cores into the W^m_N. The corresponding G is identical to that of the revised Enskog theory (van Beijeren and Ernst 1973a). The associated kinetic theory yields an approximate equation of state, commensurate with a first order perturbation expansion in the tail part of the potential. The KVT I for mixtures is compatible with irreversible thermodynamics only in the Kac limit (Karkheck et.al. 1982). If the tail is suppressed altogether, the KVT I is just the revised Enskog kinetic theory.

The second member, KVT II (Castillo et.al. 1989), utilizes the full potential energy $E_p(t) = \int d\vec{r} e_p$ which engenders a conjugate Lagrange parameter $\beta(t)$. This theory corresponds to that of Sung and Dahler (1984). At equilibrium, this theory yields an exact equation of state, but the lack of position dependence in β prohibits the formulation of a correct local pressure for the near equilibrium state. This theory linearizes (Castillo et.al. 1989) about absolute equilibrium such that, in the limit $R \to \infty$, the one particle equation is identical to that of LPS (Lebowitz 1969), which is accurate for short times.

The KVT III bears none of these deficiencies. Applied to the square-well potential (Karkheck et.al. 1985), ϕ = negative constant in (2.1), it eclipses DRS theory (Davis et.al. 1961) in two ways - by accounting for energy conservation at well-edge collisions via (2.9) and by functional dependence of G on both a density and a temperature field.

In closing this section, we note that by setting V = 0 in (2.13), the ensemble appropriate for the Boltzmann and Vlasov equations is obtained. Whether W_N^m bears enough information about the fluid to ensure that it aptly describes the time evolution of the fluid, particularly for long times, at least at the hydrodynamic level, is examined below.

STRUCTURAL PROPERTIES AND QUANTITATIVE PREDICTIONS

All members of the KVT family exhibit (Karkheck 1989) an H-theorem and support expressions for the transport coefficients. The KVT III is clearly the most interesting, because it demonstrates the role of well-edge collisions in the relaxation process, and allows for the existence of two temperatures (van Beijeren et.al. 1988), one associated with f_1 and kinetic energy, the other, $1/k\beta$, associated with G and potential energy. In general, these temperatures need not be equal.

H-Theorem

For the ensemble (2.13), the entropy (2.11) takes the form

$$\frac{1}{k}S^m(t) = \ell n\mathcal{Z} + \int dx_1 f_1 \lambda + \int d\vec{r}_1 e_p \beta \ .$$

We study dS^m/dt, where the time evolution of f_1 and e_p is governed by (2.7,9,10,14). Via normalization of the W_N^m, there occurs $\frac{d}{dt}\ell n\mathcal{Z}$ $+ \int dx_1 f_1 \frac{\partial \lambda}{\partial t} + \int d\vec{r}_1 e_p \frac{\partial \beta}{\partial t} = 0$ so that

$$\frac{1}{k}\frac{dS^m}{dt} = \int dx_1 \lambda \frac{\partial f_1}{\partial t} + \int d\vec{r}_1 \beta \frac{\partial e_p}{\partial t} . \tag{3.1}$$

Applying the kinetic equations to (3.1) and performing some manipulations yields the H-theorem (Karkheck et.al. 1985)

$$\frac{dS^m}{dt} = \tfrac{1}{2}k\int dx_1 dx_2 f_2(x_1,x_2,t)\Big\{\vec{v}_{12}\circ\hat{r}_{12}\big[\delta(r_{12}-d^+) + \delta(r_{12}-R^+) -$$

$$\delta(r_{12}-R^-)\big] + |\vec{v}_{12}\circ\hat{r}_{12}|\sum_{\ell=1}^{4} \delta(r_{12}-R_\ell^*)\theta_\ell(\vec{v}_{12}\circ\hat{r}_{12}) \times$$

$$\ln\frac{f_1(x_1,t)f_1(x_2,t)\exp(\beta_{12}E_\ell)}{b^{(\ell)}(\hat{r}_{12})f_1(x_1,t)f_1(x_2,t)}\Big\} \geq 0 .$$

The abbreviations are used for ℓ =1, 2, 3, 4 respectively: $R_\ell^* = d^+, R^+, R^-$, and R^-; $E_\ell = 0, -\varepsilon, +\varepsilon, 0$; θ_ℓ are the θ factors in the collision integrals of (2.7); $b^{(\ell)}$ are operators which effect the transformations (2.2). The inequality follows after applying $\ln(y/z) \geq 1-z/y$, where y and z are numerator and denominator, respectively, in the argument of the logarithm. This implies equality holds for distribution functions f_1^0 that satisfy the independent conditions:

$$\delta(r_{12}-R_\ell)\theta_\ell(\vec{v}_{12}\circ\hat{r}_{12})\big[\exp(\beta_{12}E_\ell) -$$

$$b^{(\ell)}(\hat{r}_{12})\big]f_1^0(x_1,t)f_1^0(x_2,t) = 0 .$$

That at the hard core, $\ell = 1$ and $R_\ell = d$, yields

$$f_1^0(x,t) = n(\vec{r},t)\left\{\frac{m}{2\pi kT(t)}\right\}^{3/2}\exp\left\{-\frac{m[\vec{v}-\vec{u}(t)]^2}{2kT(t)}\right\} . \tag{3.2}$$

Those at the well edge, $\ell = 2,3,4$ and $R_\ell = R$, yield

$$kT(t) = 1/\beta(t). \tag{3.3}$$

Thus, the parameters n, T, $\vec{u}$ completely determine the ensemble W_N^0 which satisfies $dS^m/dt = 0$. At complete equilibrium, n, T, $\vec{u}$ do not depend on time (by Galilean invariance a constant $\vec{u}$ can be transformed away), and n then satisfies

$$\nabla_1 \ln n(\vec{r}_1) = -\beta \times \int_{d<r_{12}<R} d\vec{r}_2 \, \nabla_1\phi \, n(\vec{r}_2) g_2(\vec{r}_1,\vec{r}_2) +$$

$$\int d\vec{r}_2 \hat{r}_{12} [\delta(r_{12}-d^+) + (1-e^{\beta\varepsilon})\delta(r_{12}-R^+)] n(\vec{r}_2) g_2(\vec{r}_1,\vec{r}_2).$$

This is the first equation of the YBG hierarchy, an equilibrium result. Therefore, at equilibrium the usual canonical ensemble $W_N = \frac{1}{Z}\exp\{-\beta H_N\}$ is obtained. The theory can be formulated in the presence of an external field, yielding similar results.

This H-theorem shows that it is possible to construct an irreversible kinetic theory for strongly interacting particles which can exhibit a true liquid state and in which nonlocality is a dominant feature. The nonlocality plays a special role in that the f_1^0, which satisfies $dS^m/dt = 0$, does not also make the collision integrals vanish. In the Boltzmann theory, the analog to f_1^0, the local Maxwellian, makes both vanish simultaneously. In the hard sphere limit, it has been shown (Karkheck and Stell 1982) that the collision integral can be decomposed into "reversible" and "irreversible" parts, in the sense of time reversal symmetry. (The former vanishes identically in Boltzmann theory.) There it was shown that the f_1^0 made the irreversible part of the hard sphere collision integral vanish, but not the reversible part, due to the nonlocality of the collision and the position dependence of n. Thus, the presence of a density field is sufficient to maintain collisional relaxation. Therefore, relaxation processes in the dense fluid are more complicated than can be surmised from Boltzmann theory. In particular, $dS^m/dt = 0$ is a necessary, but not sufficient, condition for the achievement of equilibrium. The initial stages of relaxation do not produce a form like f_1^0, with subsequent relaxation proceeding about that form. A similar conclusion was reached by Grad (1965) in regard to Boltzmann theory.

As a statement of global stability, the H-theorem ensures the existence of well-defined transport coefficients. However, the H-theorem reveals that the mean-field terms in (2.7) and (2.9) do not contribute to the irreversible behavior. Thus, the transport coefficients do not depend on the smooth part of V dynamically, but only statically, that is, through the equilibrium factors in the formulas.

The irreversibility is a consequence of statistics - the form (2.14) is crucial to the analysis - and of collisional dynamics in which transfer of energy and momentum occur. The minimal kinetic theory is sufficient to relax a class of nonequilibrium states to equilibrium. Included in this class are states shown by Jaynes (1971) to fall outside the realm of Boltzmann theory, namely, states for which initial potential energy is not insignificant.

Hydrodynamics and Transport Coefficients

To obtain expressions for the transport coefficients, it is first necessary to construct expressions for the fluxes of mass, momentum, and energy. These are deduced from the laws of conservation of mass, momentum, and energy that follow from (2.7) and (2.9) by taking moments of (2.7) with the mass, momentum, and kinetic energy of a particle and combining acccordingly. The conservation laws obtained are

(mass) $$\frac{d\rho}{dt} + \rho\nabla\cdot\vec{u} = 0 \qquad (3.4a)$$

(momentum) $$\rho\frac{d\vec{u}}{dt} + \nabla\cdot\overleftrightarrow{\mathbb{P}} = 0 \qquad (3.4b)$$

(energy) $$\frac{\partial e}{\partial t} + \nabla\cdot\left[e\vec{u} + \vec{J} + \vec{u}\cdot\overleftrightarrow{\mathbb{P}}\right] = 0. \qquad (3.4c)$$

The momentum flux, $\overleftrightarrow{\mathbb{P}}$, is the sum of three distinct contributions, a streaming term, $\overleftrightarrow{\mathbb{P}}_s$, corresponding to free motion of the particles, a contribution from the smooth tail, $\overleftrightarrow{\mathbb{P}}_t$, and the collisional contribution, $\overleftrightarrow{\mathbb{P}}_c$. The energy flux, $\vec{J}$, consists of four terms, the fourth corresponding to convection, $\vec{J}_\phi$, shown below (2.9). The one component kinetic theory cannot provide an expression for the mass flux, which measures relative migration of a species. The self diffusion coefficient can be inferred from a mixture version of the kinetic theory, for which the species are ultimately made identical. Generalization of the theory to mixtures is straightforward (Castillo et.al. 1989).

The flux expressions are:

$$\overleftrightarrow{\mathbb{P}}_s(\vec{r},t) = \int d\vec{v}\; m(\vec{v}-\vec{u})(\vec{v}-\vec{u})\, f_1(\vec{r},\vec{v},t) \qquad (3.5a)$$

$$\overleftrightarrow{\mathbb{P}}_t = -\frac{1}{2}\int_0^1 d\lambda \int_{d<s<R} d\vec{s}\; \hat{s}\vec{s}\frac{d\phi}{ds}\; n_2(\vec{r}+\lambda\vec{s},\vec{r}+(\lambda-1)\vec{s},t) \qquad (3.5b)$$

$$\overleftrightarrow{\mathbb{P}}_c = \int dM\; \hat{\sigma}\hat{\sigma}H \qquad (3.5c)$$

$$\vec{J}_s = \int d\vec{v}\; \frac{1}{2}\, m(\vec{v}-\vec{u})^2(\vec{v}-\vec{u})\, f_1 \qquad (3.6a)$$

$$\vec{J}_t = -\frac{1}{2}\int_0^1 d\lambda \int_{d<s<R} d\vec{s}\int d\vec{v}_1 d\vec{v}_2\; \frac{d\phi}{ds}\; \vec{s}\hat{s}\circ\left[\vec{v}_1-\vec{u}(\vec{r},t)\right] \times f_2(\vec{r}+\lambda\vec{s},\vec{v}_1,\vec{r}+(\lambda-1)\vec{s},\vec{v}_2,t) \qquad (3.6b)$$

$$\vec{J}_c = \int dM\hat{\sigma}\hat{\sigma}\circ\vec{G}H \qquad (3.6c)$$

$$\vec{J}_\phi = \frac{1}{2}\int d\vec{v}_1 dx_2(\vec{v}_1-\vec{u})Vf_2 \qquad (3.6d)$$

where $\int dM\ldots = \frac{m}{2}\int_0^1 d\lambda\int d\vec{v}_1 d\vec{v}_2\int d\hat{\sigma}\;\theta(g_\sigma)g_\sigma\ldots$, $\vec{G} = (\vec{v}_1+\vec{v}_2)/2$, $n_2 = \int d\vec{v}_1 d\vec{v}_2 f_2$, and

$$H = \hat{\sigma}\circ(\vec{v}_1'-\vec{v}_1)d^3 f_2(\vec{r}+(1-\lambda)\vec{d}^+,\vec{v}_1,\vec{r}-\lambda\vec{d}^+,\vec{v}_2,t) -$$

$$R^3[\theta(v_e-g_\sigma)\hat{\sigma}\circ(\vec{v}_1'-\vec{v}_1)f_2(\vec{r}+(\lambda-1)\vec{R}^-,\vec{v}_1,\vec{r}+\lambda\vec{R}^-,\vec{v}_2,t) +$$

$$+\theta(g_\sigma-v_e)\hat{\sigma}\circ(\vec{v}_1^--\vec{v}_1)f_2(\vec{r}+(\lambda-1)\vec{R}^-,\vec{v}_1,\vec{r}+\lambda\vec{R}^-,\vec{v}_2,t)$$

$$-\hat{\sigma}\circ(\vec{v}_1^+-\vec{v}_1)f_2(\vec{r}+(1-\lambda)\vec{R}^+,\vec{v}_1,\vec{r}-\lambda\vec{R}^+,\vec{v}_2,t)].$$

These expressions are exact, i.e., independent of any particular form for f_2, and are valid to all orders in deviation from equilibrium. The ordinary transport coefficients are the coefficients of the linear terms in the expansion of the fluxes in powers of the gradients of density, velocity, and temperature of the fluid. In order to achieve these expansions, the distribution functions must be obtained by solving the kinetic equations to this order in gradients.

The original Chapman-Enskog solution procedure (Chapman and Cowling 1970) was formulated to solve the Boltzmann equation. For both the dilute gas, and the dense hard-sphere gas to which the procedure also applies, the energy is purely kinetic. In the KVT III, the potential energy is not trivial so the procedure must be modified to accomodate the extra degree of freedom provided by β. Here (van Beijeren et.al. 1988), both f_1 and e are expanded to linear order in gradients such that their leading orders, $f_1^{(0)}$ and $e^{(0)}$ fully describe the local number density, velocity, and energy density of the fluid. The $f_1^{(0)}$ is the local Maxwellian; $e^{(0)}$ is given by

$$e^{(0)}(\vec{r},t) = \int d\vec{v}\ \tfrac{1}{2}\ m(\vec{v}-\vec{u})^2 f_1^{(0)} + \qquad (3.7)$$

$$\tfrac{1}{2}\times\int_{d<r_{12}} dx_1 dx_2 V(r_{12}) f_2^{(0)}\delta(\vec{r}_1-\vec{r}).$$

The $f_1^{(1)} = f_1^{(0)}\Phi(\vec{v})$, and Φ, which is linear in the gradients, is determined subject to the conditions that its moments associated with n and $\vec{u}$ vanish. Also, the condition $e^{(1)} = 0$ is imposed to ensure that energy density is fully determined by $e^{(0)}$. The β is expanded similarly,

$$\beta(\vec{r},t) = \beta^{(0)}(\vec{r},t) + \beta^{(1)} \tag{3.8}$$

where $\beta^{(0)} = 1/kT^{(0)}(\vec{r},t)$ follows from the linearized collision integral, and $T^{(0)}$ is the temperature contained in $f_1^{(0)}$. The next terms, $T^{(1)}$ and $\beta^{(1)}$, are proportional, so β and T are not reciprocals and the fluid exhibits two nonequilibrium temperatures. The $\beta^{(1)}$ couples to $\nabla\circ\vec{u}$ through the friction term $\mathbb{P}^{(0)}\nabla\circ\vec{u}$ appearing in the linearized version of (3.4c): $\beta^{(1)} = \alpha\nabla\circ\vec{u}$, where α is also to be determined. This produces a distortion in the fluid structure proportional to $\nabla\circ\vec{u}$. Only bulk viscosity is affected by this distortion.

By construction, $e^{(0)}$ is endowed with the same functional dependence on density and temperature as its equilibrium counterpart, but its temporal relaxation is governed by the gradients. The linearized version of (2.7) provides one relation between Φ and α; $e^{(1)} = 0$ provides the second. Thereby, unique solutions for f_1 and β are obtained and so for the transport coefficients.

For the square-well potential, the KVT III expressions for shear viscosity, thermal conductivity, and self diffusion coefficients are identical to those of DRS (Davis et.al. 1961), but the bulk viscosity is different (van Beijeren et.al. 1988):

$$\kappa^{III} = \kappa^{DRS} + \kappa^{new} \tag{3.9}$$

where

$$\kappa^{DRS} = \frac{64\pi}{45}\,\eta_0 d^2[Y_4(d) + Y_4(R)\Xi(1/T^*)] \tag{3.9a}$$

and

$$\kappa^{new} = \frac{8\eta_0 d^2}{5Y_2(R)}\Big[\frac{\pi}{3}Y_3(R)\left(1+2\pi^{-1/2}\int_0^\infty dx\, e^{-x}\,(x + 1/T^*)^{1/2}\right) + T^*\left(\beta\mathcal{P} - \frac{3}{2}n\left(\frac{\partial\mathcal{P}}{\partial e}\right)_n\right)\Big]^2. \tag{3.9b}$$

The η_0 is the hard-sphere dilute gas shear viscosity, $Y_\ell(x) = n^2x^\ell g_2(x^+)$, $T^* = kT/\varepsilon$, $\mathcal{P}$ and e are the equilibrium pressure and energy density, and $\Xi(2x) = e^{2x} - x[1+e^x K_1(x)]$, where K_1 is a modified Bessel function. A comparison between molecular dynamics results (Joslin et. al. 1990) and the theory is given in Table 1. The molecular dynamics results (MD) are based upon the time correlation function expression. The DRS, and KVT III results (III), were evaluated using equilibrium data generated during the MD runs, the latter analysis effected differentiation with a best fit curve between pressure and energy density. The table entries portray the reduced bulk viscosity $\kappa^* = \kappa d^2(mkT)^{-1/2}$. For the two lower densities, the estimated error for the MD results is about 5%, for the higher densities, 7% and 10% respectively. As shown, the full theory gives good agreement with the MD results at intermediate densities, $n^* = nd^3 \simeq 0.3$-0.4, a condition found also for shear viscosity and thermal conductivity (Michels and Trappeniers 1980, 1981) and for self-diffusion (Wilbertz et.al. 1988). At lower densities, particles are less likely to experience the partial collisions (2.2) as uncorrelated events, which is as (2.14) builds them into the kinetic theory. At higher densities, mode coupling and other dynamic effects become important.

A large amount of simulation and experimental data is available for the shear viscosity and thermal conductivity of simple liquids. For appropriate theory, the limit $R \to \infty$ is imposed to obtain the respective expressions (Karkheck et.al.1988)

$$\eta = \frac{\eta_0}{g_2(d^+)} \left[1 + 0.8h_3 + 0.7615(h_3)^2\right] \tag{3.10}$$

$$\lambda = \frac{\lambda_0}{g_2(d^+)} \left[1 + 1.2h_3 + 0.7575(h_3)^2\right] \tag{3.11}$$

in which $h_3 = \frac{2}{3}n^* g_2(d^+)$. These formulas are identical in form to those of the Enskog hard-sphere theory. In the KVT III, the g_2 depends upon the whole potential, not just the repulsive core as it does in Enskog theory and the KVT I (Karkheck and Stell 1981).

Table 1. Bulk Viscosity of the Square-Well Fluid.

T^*	n^*	0.3	0.5	0.7	0.9
4	MD	0.146	0.51	1.14	
	DRS	0.130	0.45	1.25	
	III	0.143	0.46	1.32	
3	MD	0.164	0.58	1.18	4.4
	DRS	0.144	0.48	1.28	3.3
	III	0.165	0.48	1.33	3.3
2	MD	0.221	0.82	1.30	4.3
	DRS	0.181	0.55	1.36	3.46
	III	0.231	0.55	1.39	3.54
1.5	MD	0.321	1.23	1.57	4.5
	DRS	0.233	0.64	1.49	3.6
	III	0.322	0.65	1.5	3.8
1	MD			2.33	5.5
	DRS			1.83	4.1
	III			1.85	4.6

The unphysical hard-sphere core poses a problem in applying these formulas to real fluids. In approaches known as modified Enskog theory, pioneered by Enskog and supported in form by KVT I, the d and g_2 at contact are obtained by equating two properties of the hard-sphere fluid to those of the real fluid, in essence making the diameter state dependent. These approaches actually engender two different hard-sphere diameters within the same formula (Karkheck and Stell 1981). They are useful as correlation tools (Dipippo et.al. 1977), but not as explanatory measures.

Several routes by which to adapt the hard-sphere fluid to the real fluid have been developed in the framework of thermodynamic perturbation theory (Weeks et.al. 1971; Mansoori and Canfield 1969; Rasaiah and Stell 1970; Barker and Henderson 1967). The aim here is to accurately reproduce some real-fluid equilibrium properties by forcing the hard-sphere fluid to fit those properties. This makes the diameter state dependent, but in a way that can be used consistently throughout the transport coefficient formulas and with no adjustable parameters. For fluids that can be accurately represented by the Lennard-Jones model, an extensive comparison among these diameter choices in the framework of level I theory (Karkheck and Stell 1981) showed that the shear viscosity was better represented by the WCA diameter (Weeks et.al. 1971), and thermal conductivity by the MC/RS diameter (Mansoori and Canfield 1969; Rasaiah and Stell 1970) which is smaller than the former.

Also revealed in that study is that (3.10) serves as a bare transport coefficient, that is, acts a reference for more detailed theory such as mode-coupling or ring kinetic theory. Computer simulations with hard spheres (Alder et.al. 1970) give a shear viscosity that is significantly larger than that predicted by theory at higher densities. These discrepancies are attributed to the use of velocity chaos in G for hard spheres; presumably, such discrepancy carries into (2.14). Assuming the discrepancies to be similar, the analytic correction factor constructed (Dymond 1976) for the hard-sphere fluid, that makes Enskog theory agree with simulation, is applied to (3.10) in the following way. As it applies to shear viscosity, the exact G can be expanded $G = g_2^{(0)} + \overleftrightarrow{\delta g_2}:\nabla\vec{u}$. The unknown $\overleftrightarrow{\delta g_2}$ affects Φ linearly as it is the solution to the linearized integral equation. Since shear viscosity depends linearly upon Φ, the perturbation $\overleftrightarrow{\delta g_2}$ contributes additively. The theory that ensues is called kinetic reference theory (KRT), and the KRT III is described by (Karkheck et.al. 1989)

$$\eta^{III} = \eta + \eta^{HS} - \eta^{E}. \tag{3.12}$$

The η^{E} is just (3.10) taken in the hard-sphere limit, and η^{HS} is the exact result. The interpretation is that η is a bare transport coefficient, and $\eta^{HS} - \eta^{E}$ is the contribution to shear viscosity of the hard-sphere fluid due to velocity correlations. Both KRT I and KRT III predictions differ significantly at higher densities from the corresponding KVT values. No corrections seem to be needed for thermal conductivity. The level III transport theory improves upon level I in two ways (Karkheck et.al. 1989). Both shear viscosity and thermal conductivity are described well by using the WCA diameter. In general, the former is superior in terms of quantitative agreement with both simulation and experimental results. Along the saturated liquid line for argon, the theory reproduces the data to within 10% for thermal conductivity, and 5% for shear viscosity. For illustration, Table 2 shows a comparison between theory and experimental (E) or simulation (S) results for argon, which is modeled with a Lennard-Jones potential in which σ_{LJ} = 0.3405nm and ε_{LJ}/k = 119.8K. Here, the $n^{*} = n\sigma_{LJ}^{3}$, $T^{*} = kT/\varepsilon_{LJ}$. For references, see (Karkheck et.al. 1988).

The Gibbs relation

$$\frac{ds}{dt} = \frac{1}{T}\frac{de}{dt} - \frac{\mu}{T}\frac{dn}{dt} \tag{3.13}$$

is a fundamental postulate in irreversible thermodynamics (deGroot and Mazur 1962). This expression is supported by the Boltzmann equation (deGroot and Mazur 1962) and the revised Enskog theory (Mareschal et.al. 1984). Though a local H-theorem for the square-well KVT III has been demonstrated formally (Blawzdziewicz and

Stell 1989), (3.13) has not been demonstrated explicitly for the KVT III. Based upon formal arguments (Ernst 1966), the relation $e^{(1)} = 0$, which ensures that $e^{(0)}$ has the same form as the equilibrium e, seems sufficient to ensure support of (3.13).

Table 2. Shear Viscosity and Thermal Conductivity for Argon. The units: η in Pa-s $\times 10^{-4}$, λ in W/K/m $\times 10^{-1}$.

State	Transport Coefficient	Theory	E/S
$n^* = 0.844$	η (KRT I)	3.42	2.62(S)
$T^* = 0.722$	λ (KVT I)	1.72	1.24(S)
$n^* = 0.844$	η (KRT III)	2.72	2.76(E)
$T^* = 0.73$	λ (KVT III)	1.46	1.32(E)
$n^* = 0.853$	η (KRT I)	3.85	2.97(S)
$T^* = 0.70$			
$n^* = 0.818$	η (KRT III)	2.19	2.35(E)
$T^* = 0.761$			
$n^* = 0.76$	η (KRT I)	1.765	1.734(S)
$T^* = 0.872$			
$n^* = 0.715$	η (KRT III)	1.26	1.27(E)
$T^* = 0.94$	λ (KVT III)	0.97	0.91(E)

Fluctuations and Short-Time Kinetic Theory

An important class of phenomena to which hydrodynamics does not apply concerns rapid variations in the temporal and spatial behavior of a fluid, for example, decay of density fluctuations. Experimentally, such phenomena are studied by neutron scattering, the cross section for which is directly related to the dynamical structure factor $S(\vec{k},\omega)$. The intermediate scattering function

$F(\vec{k},t)$, the Fourier transform of S, is readily accessible from computer simulations. Kinetic theory is applied to such phenomena as an initial value problem, involving a slight disturbance of the fluid from equilibrium. The equations are expanded about absolute equilibrium to linear order in deviations:

$$f_1(\vec{r},\vec{v},t) = \ell_1(v)[1 + h(\vec{r},\vec{v},t)]$$

$$e(\vec{r},t) = e + \delta e(\vec{r},t)$$

$$\beta(\vec{r},t) = \beta^0 + \delta\beta(\vec{r},t)$$

The quantities h, δe, and $\delta\beta$ represent small departures from equilibrium. Derivatives of F are sum rules which give important information regarding the short time behavior of these decay processes. The linearized KVT III equations yield exact results through third derivative of F (Karkheck 1986). However, it fails to describe time correlation functions at longer times (Wilbertz 1988).

The linearized equations also describe kinetic modes, more rapid processes which underlie hydrodynamic relaxation. For the square-well potential, it is found (Leegwater et.al. 1989) that the exchange between kinetic and potential energy is rather slow, yet nonhydrodynamic. For small well depths this may affect the definition of temperature and may explain the failure of the bulk viscosity (3.9) to reduce to the Enskog result in the limit of vanishing well depth (Joslin et.al. 1990).

BEYOND ENSKOG THEORY

The numerical deficiencies of the bare transport coefficients (3.9) and (3.10) demonstrate the inadequacy of the ensemble (2.13) to approximate the temporal evolution of the fluid over a long time interval. Computer simulations show that velocity correlations build up in time to significant levels in the dense fluid. These affect the values of f_1 and e_p which feed into (2.13), but the ensemble form fails to exhibit them explicitly. Unlike the

Boltzmann theory wherein the constraining function f_1 encompasses both the hydrodynamic variables and the transport coefficients, such correlations are relevant to the transport processes of the dense fluid. Study of (2.3c), (3.5), (3.6) shows that e_p and the fluxes elicit different information from f_2, thus, relying on f_1 and e_p to drive the ensemble leaves out explicit information in f_2 that is needed to accurately describe the transport coefficients.

This situation does not denigrate the maximum entropy formalism. The ensemble (2.13) follows as the most stable inference (Tikochinsky et.al. 1984) drawn from f_1 and e_p that is least sensitive to fluctuations about that information. That ensemble accurately portrays the early stages of relaxation, during which velocity correlations are only starting to build. The difficulty is that the fluctuations that are ignored contain important information about the transport processes, which themselves are fluctuation phenomena. It comes as no surprise that the KVT III produces only bare transport coefficients which provide the minimal structure to ensure a stable hydrodynamics.

By replacing the constraint e_p with f_2 in the maximum entropy procedure, the new constraint subsumes the former and also all the transport coefficient information. The ensemble is obtained (Karkheck and Stell 1982)

$$W_N^m(x^N,t) = \frac{1}{Z} \prod_{i=1}^{N} \exp\{-\lambda_1(x_i,t)\} \prod_{i<j}^{N} \exp\{-\lambda_2(x_1,x_2,t)\}. \tag{4.1}$$

This ensemble bears a form similar to (2.13), but now the Lagrange multiplier λ_2 is velocity dependent. The form (4.1) imposes no condition on f_2, but yields for f_3

$$f_3(x^3,t) = \frac{f_2(x_1,x_2,t)\,f_2(x_1,x_3,t)\,f_2(x_2,x_3,t)}{f_1(x_1,t)\,f_1(x_2,t)\,f_1(x_3,t)}\,Y(x^3,t) \tag{4.2}$$

where Y has the formal cluster expansion (Stell 1963)

$$Y(x^3,t) = 1 + \int dx_4 f_1(x_4,t)h(x_1,x_4,t)h(x_2,x_4,t)h(x_3,x_4,t) + ..$$

and

$$h(x_1,x_2,t) = \frac{f_2(x^2,t)}{f_1(x_1,t)f_1(x_2,t)} - 1.$$

The h does not a priori possess the same cluster property as its equilibrium counterpart. Thus, the nonequilibrium h holds the possibility of long-range velocity correlations. When applied to the hard-sphere fluid (Karkheck and Stell 1982), (4.1) yields a closed kinetic theory for the one and two particle distribution functions. Those equations have not been solved for the transport coefficients. However, the theory is more general than that of Livingston and Curtiss (1961) who in essence set Y = 1. The formal structure of (4.2) suggests the possibilty that this theory can describe mode coupling and more complex dynamic events.

DEDICATION

This article is dedicated to the memory of Jennifer Mills who shared her love for all things, and much of her short life, with many of the faculty and students at GMI.

ACKNOWLEDGMENT

During the last decade, I have had the pleasure of working with George Stell, Henk van Beijeren, Ignatz de Schepper, Jan Sengers, Jiasai Xu, Jan Michels, Chris Joslin, Chris Gray, Rolando Castillo, Esteban Martina, Mariano Lopez de Haro, and of stimulating conversations with Eddie Cohen, John Kincaid, Jim Dufty, Bob Dorfman and Ted Kirkpatrick. NATO Research Grant No. 419/82 has made much of this interaction possible.

REFERENCES

Alder, B.J., Gass, D.M., and Wainwright, T.E. 1970. Studies in molecular dynamics. VIII. The transport coefficients for a hard-sphere fluid. J. Chem. Phys. 53: 3813-26.

Barajas, L., Garcia-Colin, L.S., and Pina, E. 1973. On the Enskog-Thorne theory for a mixture of dissimilar rigid spheres. J. Stat. Phys. 7: 161-83.

Barker, J.A. and Henderson, D. 1967. Perturbation theory and equation of state for fluids. II. A successful theory of liquids. J. Chem. Phys. 47: 4714-21.

Blawzdziewicz, J. and Stell, G. 1989. Local H-theorem for a kinetic variational theory. J. Stat. Phys. 56: 821-40.

Boon, J.P. and Yip, S. 1980. Molecular Hydrodynamics, McGraw-Hill, New York.

Castillo, R.C., Martina, E., Lopez de Haro, M., Karkheck, J., and Stell, G. 1989. Linearized kinetic-variational theory and short-time kinetic theory. Phys. Rev. A 39: 3106-11.

Cercignani, C. and Lampis, M. 1988. On the kinetic theory of a dense gas of rough spheres. J. Stat. Phys. 53: 655-71.

Chapman, S. and Cowling, T.G. 1970. The Mathematical Theory of Nonuniform Gases, Cambridge U., Cambridge, England.

Dahler, J.S. and Theodosopulu, M. 1975. The kinetic theory of dense polyatomic fluids. Adv. Chem. Phys. 31: 155-229.

Davis, H.T., Rice, S.A., and Sengers, J.V. 1961. On the kinetic theory of dense fluids. IX. The fluid of rigid spheres with a square-well attraction. J. Chem. Phys. 35: 2210-33.

deGroot, S.R. and Mazur, P. 1962. Nonequilibrium Thermodynamics North-Holland, Amsterdam.

Dipippo, R., Dorfman, J.R., Kestin, J., Khalifa, H.E., and Mason, E.A. 1977. Composition dependence of the viscosity of dense gas mixtures. Physica 86A: 205-23.

Dymond, J.H. 1976. Modified hard-sphere theory for transport properties of fluids over the whole density range. 2. Viscosity coefficients of diatomic molecules F_2 and O_2. Physica 85A: 175-85.

Enskog, D. 1922. Kinetic theory of thermal conduction, viscosity, and self-diffusion in certain dense gases and liquids. K. Sven. Vetenskapsakad. Handl. 63: No.4.

Ernst, M.H. 1966. Transport coefficients and temperature definition. Physica 32: 252-72.

Ernst, M.H., Dorfman, J.R., Hoegy, W., and van Leeuwen, J.M.J. 1969. Hard-sphere dynamics and binary-collision operators. Physica 45: 127-46.

Grad, H. 1965. On Boltzmann's H-theorem.
J. Soc. Ind. Appl. Math. 13: 259-77.

Hoheisel, C. and Vogelsang, R. 1988. Thermal transport coefficients for one- and two-component liquids from time correlation functions computed by molecular dynamics.
Comp. Phys. Rept. 8: 1-69.

Jaynes, E. T. 1971. Violation of Boltzmann's H theorem in real gases. Phys. Rev. A 4: 747-9.

Joslin, C.G., Gray, C.G., Michels, J.P.J., and Karkheck, J. 1990. The bulk viscosity of a square-well fluid.
Mol. Phys. 69: 535-47.

Karkheck, J. and Stell, G. 1981. Kinetic mean-field theories.
J. Chem. Phys. 75: 1475-87.

Karkheck, J.and Stell, G. 1982. Maximization of entropy, kinetic equations, and irreversible thermodynamics.
Phys. Rev. A 25: 3302-27.

Karkheck, J., Martina, E., and Stell, G. 1982. Kinetic mean-field theory for mixtures: Kac tail limit. Phys. Rev. A 25: 3328-34.

Karkheck, J., van Beijeren, H., de Schepper, I.M., and Stell, G. 1985. Kinetic theory and H-theorem for the dense square-well fluid. Phys. Rev. A 32: 2517-20.

Karkheck, J. 1986. Kinetic theory and ensembles of maximum entropy.
Kinam 7A: 191-208.

Karkheck, J., Stell, G., and Xu, J. 1988. Transport theory for the Lennard-Jones dense fluid. J. Chem. Phys. 89: 5829-33.

Karkheck, J. 1989. Kinetic theory and ensembles of maximum entropy. (different than 1986 paper).In Maximum Entropy and Bayesian Methods, ed. J. Skilling, 491-6. Kluwer Academic Publishers.

Karkheck, J., Stell, G., and Xu, J. 1989. Transport in simple dense fluids. Int. J. Thermophys. 10: 113-23.

Kirkpatrick, T.R. 1988. Microscopic theory of dynamics in an orientationally ordered fluid. J. Chem. Phys. 89: 5020-32.

Konijnendijk, H.H.U. and van Leeuwen, J.M.J. 1973. Kinetic equation for hard-sphere correlation functions. Physica 64A: 342-62.

Lebowitz, J.L., Percus, J.K., Sykes, J. 1969. Kinetic equation approach to time-dependent correlation functions.
Phys. Rev. 188: 487-504.

Leegwater, J.A., van Beijeren, H., and Michels, J.P.J. 1989. Linear kinetic theory of the square-well fluid.
J. Phys. Cond. M. 1: 237-51.

Levine, R. D. and Tribus, M. 1978. The Maximum Entropy Formalism MIT, Cambridge, MA.

Lewis, R.M. 1961. Solution of the equations of statistical mechanics. J. Math. Phys. 2: 222-31.

Lewis, R.M. 1967. A unifying principle in statistical mechanics. J. Math. Phys. 8: 1448-59.

Livingston, P.M. and Curtiss, C.F. 1961. Kinetic theory of moderately dense, rigid sphere gases. Phys. Fluids 4: 816-33.

Mansoori, G.A. and Canfield, F.B. 1969. Variational approach to the equilibrium thermodynamic properties of simple liquids. J. Chem. Phys. 51: 4958-67.

Mareschal, M., Blawzdziewicz, J., and Piasecki, J. 1984. Local entropy production from the revised Enskog equation: general formulation for inhomogeneous fluids. Phys. Rev. Lett. 52: 1169-72.

Michels, J.P.J. and Trappeniers, N.J. 1980. Molecular dynamical calculations on the transport properties of a square-well fluid II. The viscosity above the critical density. Physica 104A: 243-54.

Michels, J.P.J. and Trappeniers, N.J. 1981. Molecular dynamical calculations on the transport properties of a square-well fluid III. The thermal conductivity. Physica 107A: 158-65.

Rasaiah, J. and Stell, G. 1970. Upper bounds on free energies in terms of hard-sphere results. Mol. Phys. 18: 249-60.

Shalitin, D. 1986. Comment on the most-stable-inference approach to maximum entropy. Phys. Rev. A 33: 3575-77.

Stell, G. 1963. The Percus-Yevick equation for the radial distribution function of a fluid. Physica 29: 517-34.

Stell, G., Karkheck, J., and van Beijeren, H. 1983. Kinetic mean field theories: results of energy constraint in maximizing entropy. J. Chem. Phys. 79: 3166-7.

Sung, W. and Dahler, J. 1984. Theory of transport processes in dense fluids. J. Chem. Phys. 80: 3025-37.

Tikochinsky, Y., Tishby, N.Z., and Levine, R.D. 1984. Alternative approach to maximum-entropy inference. Phys. Rev. A 30: 2638-44.

Uhlenbeck, G.E. and Ford, G.W. 1963. Lectures in Statistical Mechanics. Am. Math. Soc., Providence, RI.

van Beijeren, H. and Ernst, M.H. 1973a. The modified Enskog equation. Physica 68: 437-56.

van Beijeren, H. and Ernst, M.H. 1973b. The modified Enskog equation for mixtures. Physica 70: 225-42.

van Beijeren, H., Karkheck, J., and Sengers, J.V. 1988. Nonequilibrium temperature and the bulk viscosity of a dense fluid of square-well molecules. Phys. Rev. A 37: 2247-50.

Weeks, J.D., Chandler, D., and Andersen, H.C. 1971. Role of repulsive forces in determining the equilibrium structure of simple liquids. J. Chem. Phys. 54: 5237-47.

Wilbertz. H., Michels, J., van Beijeren, H., and Leegwater, J.A. 1988. Self diffusion of particles interacting through a square-well or square-shoulder potential. J. Stat. Phys. 53: 1155-76.

Statistical Mechanical Theory of Diffusion and Heat Conduction in Multicomponent Mixtures

Andrzej R. Altenberger and John S. Dahler
Department of Chemical Engineering and Materials Science
University of Minnesota
Minneapolis, MN 55455, USA

ABSTRACT

It often is difficult to establish with certainty molecular definitions of pertinent Onsager coefficients that appear in phenomenological formulations of the linear thermodynamics of non-equilibrium processes. In particular, the question of their symmetry (reciprocal relations) is hard to resolve and has produced considerable discussion in the literature. Recently we proposed a new statistical mechanical method for deriving the linear constitutive relations that does not require the introduction of an entropy source and that allows one to answer unequivocally the question of symmetry of the kinetic coefficients for different frames of reference and for various sets of thermodynamic forces. Our formulation also goes beyond the standard range of applicability of the linear non-equilibrium thermodynamics in the sense that it allows for the spatial and temporal dispersion of transport coefficients.

1. INTRODUCTION

Issues concerning particle and heat transport in multicomponent fluids are still receiving substantial attention in spite of the over fifty-year history of research in the field of non-equilibrium thermodynamics. Admittedly, most current research on these topics is aimed at extending the classical Onsager thermodynamics beyond the conditions to which the traditional constitutive relations apply (Müller and Ruggeri 1987). However, there are important unresolved questions pertaining to the very foundations of the linear non-equilibrium thermodynamics (Truesdell 1969, Miller et al. 1986). These questions often arise in connection with problems of transport and relaxation that fall near the ill-defined boundary separating the domains of applicability of phenomenological thermodynamics and statistical mechanics. Included in this category are such basic theoretical problems as the establishment of correct molecular scale interpretation for macroscopic observables, the symmetries of kinetic coefficients (so-called reciprocal relations) and permissible choices for the thermodynamic forces involved in constitutive relations. These are old questions that often have been answered incorrectly and which still retain the capability of kindling heated discussions. In addition to these, there are newer questions pertaining to the temporal and spatial dispersion effects connected with "local" formulations of traditional non-equilibrium thermodynamics

(Keizer 1987). Interest in this has been aroused by the recent development of spectroscopic techniques for measuring transport coefficients.

Although some specialists prefer to treat phenomenological non-equilibrium thermodynamics as an entirely self-contained discipline, there are some problems such as the symmetry of the kinetic coefficients that require a molecular scale analysis and simply cannot be resolved within the framework of the phenomenological theory. There are other situations where the microscopic theory can be of great value in determining possible ways of generalizing the phenomenological theory.

In the present article we report results obtained recently (Altenberger et al. 1987a,b, Altenberger and Tirrell 1984, 1986) in the field of molecular non-equilibrium thermodynamics of multicomponent mixtures. Our formulation of the non-local thermodynamics of transport processes depends strongly on our modification of the Schofield fluctuation theory (Schofield 1966, 1968) which is summarized in the following section. Sections 3 and 4 are devoted to the outline of the linear theory of fluctuation dynamics and to the problem of various representations of the theory that are possible in different local frames of reference. Sections 5 and 6 provide examples of applications of the theory and in the last section a summary and conclusions are presented.

LOCAL FORMULATION OF THERMODYNAMICS AND SCHOFIELD FLUCTUATION THEORY

Conventional presentations of the thermodynamics of an equilibrium, multicomponent system are based on what may be called a global balance approach. Equilibrium states of the system are described in terms of extensive thermodynamic potentials which are themselves functions of other extensive thermodynamic variables. For example, let us consider an $(s+1)$ component system with the composition $\{N_\alpha; \alpha = 0, 1, \ldots, s\}$ and confined to a volume Ω. Here N_α denotes the number of particles of component α. The internal energy U is treated as a function of the set of variables $\{N_\alpha\}$, volume and of the entropy, S, which also is an extensive variable. It is postulated (Callen 1985) that the internal energy be a first order homogeneous function of these extensive variables. This postulate reflects the requirement of additivity of the extensive thermodynamic quantities with respect to a virtual division of the total system into several subsystems. In agreement with the Euler formula, the internal energy then can be written as

$$U = TS + \sum_{\alpha=0}^{s} \mu_\alpha N_\alpha - p\Omega \tag{2.1}$$

with T, $\{\mu_\alpha\}$ and $-p$, respectively, denoting the temperature, the set of chemical potentials of all components and the negative of the thermodynamic pressure. These quantities are defined as partial derivatives of the internal energy with respect to the corresponding extensive variables $S, \{N_\alpha\}$ and Ω and are intensive quantities, namely, homogeneous functions of zeroth order.

The "global" description is not particularly convenient for describing fluctuations in a small element of volume or "cell" of the equilibrium system. In the first place, for a small open subsystem, the fluctuations caused by the random transport of particles to and from the cell are relatively large. Furthermore, the subsystem may be nonlocal in the sense that its state depends not only on the "local" thermodynamic variables but also on the states of neighboring cells. It is often convenient to use in place of the extensive thermodynamic quantities appearing in (2.1) the corresponding "stoichiometric" densities referred to unit volume. These quantities are defined by the relations

$$
\begin{aligned}
u &= U/\Omega \\
s &= S/\Omega \\
c_\alpha &= N_\alpha/\Omega
\end{aligned}
\tag{2.2}
$$

The total differential of the energy density, treated as a function of the entropy density and of the particle concentrations $\{c_\alpha\}$ is then

$$
du = T\ ds + \sum_{\alpha=0}^{s} \mu_\alpha dc_\alpha \tag{2.3}
$$

and the differentials of the pressure, temperature and the chemical potentials are related to one another by the Gibbs-Duhem formula

$$
s\ dT - dp + \sum_{\alpha=0}^{s} c_\alpha d\mu_\alpha = 0 \tag{2.4}
$$

The differential

$$
dq = T\ ds \tag{2.5}
$$

represents the heat density exchanged during a quasistatic process in the course of which the stoichiometric densities are changed by amounts du, ds and $\{dc_\alpha\}$ from their previous equilibrium values.

In order to extend this formulation to fluctuations that occur due to random transport of particles between "cells" it is useful to define thermodynamic fields $u(\boldsymbol{r})$, $s(\boldsymbol{r})$ and $\{c_\alpha(\boldsymbol{r})\}$ that are related to the total extensive quantities by the formulas

$$
\begin{aligned}
U &= \int_\Omega d\boldsymbol{r}\ u(\boldsymbol{r}) \\
S &= \int_\Omega d\boldsymbol{r}\ s(\boldsymbol{r}) \\
N_\alpha &= \int_\Omega d\boldsymbol{r}\ c_\alpha(\boldsymbol{r})
\end{aligned}
\tag{2.6}
$$

We also assume that the local internal energy density can be treated as a functional of the $s + 2$ fields $s(\boldsymbol{r})$ and $\{c_\alpha(\boldsymbol{r})\}$. It is then possible to write a "local" version of the relation (2.1) of the form

$$
\begin{aligned}
u(\boldsymbol{r}) &= \int d\boldsymbol{r}' \left(\frac{\delta u(\boldsymbol{r})}{\delta s(\boldsymbol{r}')}\right)_{\{c_\alpha(\boldsymbol{r})\}} s(\boldsymbol{r}') + \sum_{\alpha=0}^{s} \int d\boldsymbol{r}' \left(\frac{\delta u(\boldsymbol{r})}{\delta c_\alpha(\boldsymbol{r}')}\right)_{s(\boldsymbol{r})} c_\alpha(\boldsymbol{r}') - p(\boldsymbol{r}) \\
&= \int d\boldsymbol{r}'\ T(\boldsymbol{r} - \boldsymbol{r}')s(\boldsymbol{r}') + \sum_{\alpha=0}^{s} \int d\boldsymbol{r}' \mu_\alpha(\boldsymbol{r} - \boldsymbol{r}')c_\alpha(\boldsymbol{r}') - p(\boldsymbol{r})
\end{aligned}
\tag{2.7}
$$

The variation of this formula leads to the following expression for the functional differential

$$\delta u(\boldsymbol{r}) = \int d\boldsymbol{r}'\ T(\boldsymbol{r}-\boldsymbol{r}')\delta s(\boldsymbol{r}') + \sum_{\alpha=0}^{s} \int d\boldsymbol{r}'\ \mu_\alpha(\boldsymbol{r}-\boldsymbol{r}')\delta c_\alpha(\boldsymbol{r}') \tag{2.8}$$

This corresponds to (2.3) provided that one adopts as a "local" version of the Gibbs-Duhem equation the relationship

$$\int d\boldsymbol{r}'\ s(\boldsymbol{r}-\boldsymbol{r}')\delta T(\boldsymbol{r}') + \sum_{\alpha=0}^{s} \int d\boldsymbol{r}'\ c_\alpha(\boldsymbol{r}-\boldsymbol{r}')\delta \mu_\alpha(\boldsymbol{r}') - \delta p(\boldsymbol{r}) = 0 \tag{2.9}$$

It also is convenient to define a local heat density according to the prescription

$$\delta q(\boldsymbol{r}) = \int d\boldsymbol{r}'\ T(\boldsymbol{r}-\boldsymbol{r}')\delta s(\boldsymbol{r}') \tag{2.10}$$

The convolution form of the relations (2.7)–(2.10) above reflects the requirement that the kernels should be invariant with respect to the translation of the coordinate system during which $\boldsymbol{r} \to \boldsymbol{r}' = \boldsymbol{r} + \boldsymbol{a}$ with $\boldsymbol{a}$ an arbitrary vector.

In many cases it proves to be more convenient to use the temperature rather than entropy as a variable. With the entropy treated as a function of temperature and the stoichiometric concentrations, (2.3) can be rewritten in the form

$$du = C_v\ dT + \sum_{\alpha=0}^{s} \left[\mu_\alpha + T\left(\frac{\partial s}{\partial c_\alpha}\right)_T\right] dc_\alpha \tag{2.11}$$

The functional analogue of this is

$$\delta u(\boldsymbol{r}) = \int d\boldsymbol{r}'\ C_v(\boldsymbol{r}-\boldsymbol{r}')\delta T(\boldsymbol{r}') + \sum_{\alpha=0}^{s} \int d\boldsymbol{r}' \left(\frac{\delta u(\boldsymbol{r})}{\delta c_\alpha(\boldsymbol{r}')}\right)_T \delta c_\alpha(\boldsymbol{r}') \tag{2.12}$$

Here C_v is the specific heat at constant volume and $C_v(\boldsymbol{r})$ its generalization to the functional space. From these relationships it can be seen that the introduction of temperature as a variable permits the local heat density defined by (2.10) to be partitioned into contributions that are dependent and independent of the particle number variations, viz.,

$$\delta q(\boldsymbol{r}) = \int d\boldsymbol{r}'\ C_v(\boldsymbol{r}-\boldsymbol{r}')\delta T(\boldsymbol{r}') + \sum_{\alpha=0}^{s} \int d\boldsymbol{r}' \left(\frac{\delta q(\boldsymbol{r})}{\delta c_\alpha(\boldsymbol{r}')}\right)_T \delta c_\alpha(\boldsymbol{r}') \tag{2.13}$$

In the expressions above the following quantities appear:

$$C_v(\boldsymbol{r}-\boldsymbol{r}') = \int d\boldsymbol{r}''\ T(\boldsymbol{r}-\boldsymbol{r}'')\left(\frac{\delta s(\boldsymbol{r}'')}{\delta T(\boldsymbol{r}')}\right)_{\{c_\alpha(\boldsymbol{r}')\}} \tag{2.14}$$

and

$$\left(\frac{\delta q(\boldsymbol{r})}{\delta c_\alpha(\boldsymbol{r}')}\right)_T = \int d\boldsymbol{r}''\ T(\boldsymbol{r}-\boldsymbol{r}'')\left(\frac{\delta s(\boldsymbol{r}'')}{\delta c_\alpha(\boldsymbol{r}')}\right)_T \tag{2.15}$$

$$\left(\frac{\delta u(\boldsymbol{r})}{\delta c_\alpha(\boldsymbol{r}')}\right)_T = \mu_\alpha(\boldsymbol{r}-\boldsymbol{r}') + \left(\frac{\delta q(\boldsymbol{r})}{\delta c_\alpha(\boldsymbol{r}')}\right)_T \tag{2.16}$$

In a state of equilibrium the temperature can be treated as a strictly local field in the sense that

$$T(\boldsymbol{r}-\boldsymbol{r}') = T_{EQ}\delta(\boldsymbol{r}-\boldsymbol{r}') \quad . \tag{2.17}$$

This significantly simplifies the relations (2.14) and (2.15).

The expressions presented above are generalizations of the familiar formulas of phenomenological thermodynamics. However, from the point of view of statistical mechanics the internal energy and particle density fields are the average values of the corresponding microscopic fields

$$\hat{n}_\alpha(\boldsymbol{r}) = \sum_{i=1}^{N_\alpha} \delta\big[\boldsymbol{r}-\boldsymbol{r}_{i\alpha}\big] \tag{2.18}$$

and

$$\hat{e}(\boldsymbol{r}) = \sum_{\alpha=0}^{s} \hat{h}_\alpha(\boldsymbol{r}) \tag{2.19}$$

with $\hat{h}_\alpha(\boldsymbol{r})$ denoting the microscopic Hamiltonian density specific to the α-th component. The total Hamiltonian of the system is

$$\widehat{H} = \sum_{\alpha=0}^{s} \int d\boldsymbol{r}\ \hat{h}_\alpha(\boldsymbol{r}) \tag{2.20}$$

The symbol $\boldsymbol{r}_{i\alpha}$ appearing here is the location of particle i of component α. Quantities topped with carets are dependent on the set of phase-space variable Γ_0 of all particles in the system at time $t=0$: this dependence is not shown explicitly. Γ_0 can be treated as a set of random variables with a probability distribution function that is known. In particular, the equilibrium value of an arbitrary microscopic field is obtained by averaging over the Gibbs ensemble appropriate to an equilibrium state with a uniform temperature T_{EQ} but for which the composition may be nonuniform due to long-ranged fields of force. Consequently, we write

$$\phi_{EQ}(\boldsymbol{r}) = \langle\hat{\phi}(\boldsymbol{r})\rangle_{EQ} \tag{2.21}$$

and in particular,

$$c_\alpha(\boldsymbol{r}) = \langle\hat{n}_\alpha(\boldsymbol{r})\rangle_{EQ} \tag{2.22}$$

and

$$u(\boldsymbol{r}) = \langle\hat{e}(\boldsymbol{r})\rangle_{EQ} \tag{2.23}$$

The subscript EQ henceforth is omitted.

It is convenient to define the fluctuation fields by

$$\delta\hat{\phi}(\boldsymbol{r}) = \hat{\phi}(\boldsymbol{r}) - \langle\phi(\boldsymbol{r})\rangle \tag{2.24}$$

with

$$\begin{aligned}\delta\hat{n}_\alpha(\boldsymbol{r}) &= \hat{n}_\alpha(\boldsymbol{r}) - c_\alpha(\boldsymbol{r})\\ \delta\hat{e}(\boldsymbol{r}) &= \hat{e}(\boldsymbol{r}) - u(\boldsymbol{r})\end{aligned} \tag{2.25}$$

as particular examples.

The variations of the fields appearing, for example, in the expression (2.12) can be interpreted as deviations of the instantaneous fields from their equilibrium values or, alternatively, as non-equilibrium averages of the fluctuating fields with the functional differentials calculated at the state of equilibrium. Thus, we can rewrite (2.12) as

$$\delta\hat{e}(\boldsymbol{r}) = \int d\boldsymbol{r}' \; C_v(\boldsymbol{r}-\boldsymbol{r}')\delta\widehat{T}(\boldsymbol{r}') + \sum_{\alpha=0}^{s} \int d\boldsymbol{r}' \left(\frac{\delta u(\boldsymbol{r})}{\delta c_\alpha(\boldsymbol{r}')}\right)_T \delta\hat{n}_\alpha(\boldsymbol{r}') \tag{2.26}$$

providing that a molecular definition of the temperature fluctuation can be found.

The local thermodynamic fields can be treated as elements of a linear multidimensional functional space, the dimension of which, at given field location $\boldsymbol{r}$, is determined by the number of linearly independent fields. In the problem considered here these consist of the entropy or temperature and $s+1$ particle density fields. Consequently, the total dimensionality of the functional space is $d = s+2$.

It is useful at this point to introduce a complete set of basic functions $\{\delta\hat{\psi}_i(\boldsymbol{r});\ i = 0,\ldots,d\}$ in terms of which every fluctuation field $\delta\hat{\phi}(\boldsymbol{r})$ can be expressed as a linear combination:

$$\delta\hat{\phi}(\boldsymbol{r}) = \sum_{i,j=0}^{d} \iint d\boldsymbol{r}'d\boldsymbol{r}''\delta\hat{\psi}_i(\boldsymbol{r}')\Gamma_{ij}^{-1}(\boldsymbol{r}',\boldsymbol{r}'')\langle\delta\hat{\psi}_j(\boldsymbol{r}'')\delta\hat{\phi}(\boldsymbol{r})\rangle \tag{2.27}$$

The functions $\Gamma_{ij}^{-1}(\boldsymbol{r}',\boldsymbol{r}'')$ appearing in (2.27) are components of the matrix inverse, defined by

$$\sum_{j=0}^{d} \int d\boldsymbol{r}''\Gamma_{ij}^{-1}(\boldsymbol{r},\boldsymbol{r}'')\Gamma_{jk}(\boldsymbol{r}'',\boldsymbol{r}') = \delta_{ik}\delta(\boldsymbol{r}-\boldsymbol{r}') \quad , \tag{2.28}$$

of the correlation function

$$\Gamma_{ij}(\boldsymbol{r},\boldsymbol{r}') = \langle\delta\hat{\psi}_i(\boldsymbol{r})\delta\hat{\psi}_j(\boldsymbol{r}')\rangle \quad . \tag{2.29}$$

The inverse matrix $\Gamma_{ij}^{-1}(\boldsymbol{r},\boldsymbol{r}')$ plays a role of a metric tensor in the curvilinear set of coordinates $\{\delta\hat{\psi}_i(\boldsymbol{r})\}$ defined in the functional Hilbert space spanned by the basic functions. The scalar product of two fields, $\delta\hat{\phi}_1$ and $\delta\hat{\phi}_2$, in this space is given by

$$(\delta\hat{\phi}_1(\boldsymbol{r}),\delta\hat{\phi}_2(\boldsymbol{r}''')) = \sum_{i,j=0}^{d} \iint d\boldsymbol{r}'d\boldsymbol{r}''\langle\delta\hat{\phi}_1(\boldsymbol{r})\delta\hat{\psi}_i(\boldsymbol{r}')\rangle\Gamma_{ij}^{-1}(\boldsymbol{r}',\boldsymbol{r}'')\langle\delta\hat{\phi}_2(\boldsymbol{r}''')\delta\hat{\psi}_j(\boldsymbol{r}'')\rangle \tag{2.30}$$

The microscopic density fluctuations can be used as a basis for a subspace of $d-1$ dimensions. The projector on this subspace can be written in the form

$$P\delta\hat{\phi}(\boldsymbol{r}) = (k_BT)^{-1} \sum_{\alpha,\beta=0}^{s} \iint d\boldsymbol{r}'d\boldsymbol{r}''\delta\hat{n}_\alpha(\boldsymbol{r}')\left(\frac{\delta\mu_\alpha(\boldsymbol{r}'')}{\delta c_\beta(\boldsymbol{r}')}\right)_T\langle\delta\hat{n}_\beta(\boldsymbol{r}'')\delta\hat{\phi}(\boldsymbol{r})\rangle \tag{2.31}$$

where the functional derivative

$$\left(\frac{\delta\mu_\alpha(\boldsymbol{r})}{\delta c_\beta(\boldsymbol{r}')}\right)_T = k_BT\left[\frac{\delta_{\alpha\beta}\delta(\boldsymbol{r}-\boldsymbol{r}')}{c_\alpha(\boldsymbol{r})} - C_{\alpha\beta}(\boldsymbol{r},\boldsymbol{r}')\right] \tag{2.32}$$

is expressed in terms of the equilibrium density profile $c_\alpha(\boldsymbol{r})$ and the direct correlation function $C_{\alpha\beta}(\boldsymbol{r},\boldsymbol{r}')$ specific to the two species α and β (Evans 1979). The direct correlation function is related to the irreducible pair correlation function $h_{\alpha\beta}(\boldsymbol{r},\boldsymbol{r}')$ through the Ornstein-Zernicke relation which can be written in the form of (2.28), namely

$$\sum_{\gamma=0}^{s}\int d\boldsymbol{r}''\left(\frac{\delta\mu_\alpha(\boldsymbol{r})}{\delta c_\gamma(\boldsymbol{r}'')}\right)_T\left(\frac{\delta c_\gamma(\boldsymbol{r}'')}{\delta\mu_\beta(\boldsymbol{r}')}\right)_T=\delta_{\alpha\beta}\delta(\boldsymbol{r}-\boldsymbol{r}') \tag{2.33}$$

with

$$\begin{aligned} k_BT\left(\frac{\delta c_\alpha(\boldsymbol{r})}{\delta\mu_\beta(\boldsymbol{r}')}\right)_T &= \langle\delta\hat{n}_\alpha(\boldsymbol{r})\delta\hat{n}_\beta(\boldsymbol{r}')\rangle \\ &= c_\alpha(\boldsymbol{r})\left[\delta_{\alpha\beta}\delta(\boldsymbol{r}-\boldsymbol{r}')+c_\beta(\boldsymbol{r}')h_{\alpha\beta}(\boldsymbol{r},\boldsymbol{r}')\right] \end{aligned} \tag{2.34}$$

We now define the variation of the chemical potential at constant temperature by the formula

$$\delta_T\hat{\mu}_\alpha(\boldsymbol{r})=\sum_{\beta=0}^{s}\int d\boldsymbol{r}'\left(\frac{\delta\mu_\alpha(\boldsymbol{r})}{\delta c_\beta(\boldsymbol{r}')}\right)_T\delta\hat{n}_\beta(\boldsymbol{r}') \tag{2.35}$$

The projector (2.31) then can be written in the more compact form

$$P\delta\hat{\phi}(\boldsymbol{r})=(k_BT)^{-1}\sum_{\alpha=0}^{2}\int d\boldsymbol{r}'\langle\delta\hat{n}_\alpha(\boldsymbol{r}')\delta\hat{\phi}(\boldsymbol{r})\rangle\delta_T\hat{\mu}_\alpha(\boldsymbol{r}') \tag{2.36}$$

Although the energy density could be selected as the remaining basis function, temperature is the variable that fits more naturally into the traditional description of continuum mechanics. An important distinction between energy density and temperature is that the latter is independent of composition. Therefore, the corresponding basis function should be orthogonal to the subspace spanned by compositional fluctuations. This suggests that we use the Schmidt orthogonalization procedure (Morse and Feshbach 1953) to define an energy related microscopic field

$$\delta\hat{\sigma}(\boldsymbol{r})=(1-P)\delta\hat{e}(\boldsymbol{r}) \tag{2.37}$$

that is orthogonal to the fluctuations of composition. In terms of this field we then can construct a projection operator P', orthogonal to P and with the action

$$P'\delta\hat{\phi}(\boldsymbol{r})=\iint d\boldsymbol{r}'\,d\boldsymbol{r}''\,\delta\hat{\sigma}(\boldsymbol{r}')\Gamma_{dd}^{-1}(\boldsymbol{r}',\boldsymbol{r}'')\langle\delta\hat{\sigma}(\boldsymbol{r}'')\delta\hat{\phi}(\boldsymbol{r})\rangle \tag{2.38}$$

In order that this projection satisfy the condition

$$P'\delta\hat{e}(\boldsymbol{r})=\delta\hat{\sigma}(\boldsymbol{r}) \tag{2.39}$$

it is necessary for $\Gamma_{dd}^{-1}(\boldsymbol{r}',\boldsymbol{r}'')$ to be so defined that

$$\int d\boldsymbol{r}''\,\Gamma_{dd}^{-1}(\boldsymbol{r}',\boldsymbol{r}'')\langle\delta\hat{\sigma}(\boldsymbol{r}'')\delta\hat{e}(\boldsymbol{r})\rangle=\delta(\boldsymbol{r}-\boldsymbol{r}') \tag{2.40}$$

This is consistent with the identification of Γ_{dd}^{-1} as the inverse of the correlation function

$$\Gamma_{dd}(\boldsymbol{r},\boldsymbol{r}') = \langle \delta\hat{\sigma}(\boldsymbol{r})\delta\hat{\sigma}(\boldsymbol{r}')\rangle = \langle \delta\hat{\sigma}(\boldsymbol{r})\delta\hat{e}(\boldsymbol{r}')\rangle \tag{2.41}$$

We now define the microscopic temperature fluctuation field $\delta\widehat{T}(\boldsymbol{r})$ by the formula

$$\delta\hat{\sigma}(\boldsymbol{r}) = \int d\boldsymbol{r}'\ C_v(\boldsymbol{r}-\boldsymbol{r}')\delta\widehat{T}(\boldsymbol{r}') \tag{2.42}$$

and identify $C_v(\boldsymbol{r}-\boldsymbol{r}')$ with

$$C_v(\boldsymbol{r}-\boldsymbol{r}') = (k_B T^2)^{-1}\langle \delta\hat{\sigma}(\boldsymbol{r})\delta\hat{\sigma}(\boldsymbol{r}')\rangle \tag{2.43}$$

The action of the projection operator P' of (2.38) then can be written in the form

$$\begin{aligned} P'\delta\hat{\phi}(\boldsymbol{r}) &= (k_B T^2)^{-1} \iint d\boldsymbol{r}' d\boldsymbol{r}''\, \delta\widehat{T}(\boldsymbol{r}') C_v(\boldsymbol{r}'-\boldsymbol{r}'')\langle \delta\widehat{T}(\boldsymbol{r}'')\delta\hat{\phi}(\boldsymbol{r})\rangle \\ &= (k_B T^2)^{-1} \int d\boldsymbol{r}'\ \delta\widehat{T}(\boldsymbol{r}')\langle \delta\hat{\sigma}(\boldsymbol{r}')\delta\hat{\phi}(\boldsymbol{r})\rangle \end{aligned} \tag{2.44}$$

To insure the property

$$P'\delta\widehat{T}(\boldsymbol{r}) = \delta\widehat{T}(\boldsymbol{r}) \tag{2.45}$$

it is necessary and sufficient that

$$\int d\boldsymbol{r}''\ C_v(\boldsymbol{r}'-\boldsymbol{r}'')\langle \delta\widehat{T}(\boldsymbol{r}'')\delta\widehat{T}(\boldsymbol{r})\rangle = k_B T^2\ \delta(\boldsymbol{r}-\boldsymbol{r}') \tag{2.46}$$

It should be noted that, unlike the correlations of fluctuations of the particle density fields, the correlations between fluctuations of temperature and energy are local, viz.,

$$\langle \delta\widehat{T}(\boldsymbol{r})\delta\hat{e}(\boldsymbol{r}')\rangle = k_B T^2\ \delta(\boldsymbol{r}-\boldsymbol{r}') \tag{2.47}$$

The two projection operators P and P' span a complete state space for the thermodynamic fields. Thus, every thermodynamic field can be represented as a projection on the set of basic functions (2.27) of the form

$$\delta\hat{\phi}(\boldsymbol{r}) = \sum_{\alpha=0}^{2}\int d\boldsymbol{r}'\left(\frac{\delta\phi(\boldsymbol{r})}{\delta c_\alpha(\boldsymbol{r}')}\right)_T \delta\hat{n}_\alpha(\boldsymbol{r}') + \int d\boldsymbol{r}'\left(\frac{\delta\phi(\boldsymbol{r})}{\delta T(\boldsymbol{r}')}\right)_{\{c_\alpha\}} \delta\widehat{T}(\boldsymbol{r}') \tag{2.48}$$

with

$$\begin{aligned} \left(\frac{\delta\phi(\boldsymbol{r})}{\delta c_\alpha(\boldsymbol{r}')}\right)_T &= (k_B T)^{-1}\sum_{\beta=0}^{s}\int d\boldsymbol{r}''\langle \delta\hat{n}_\beta(\boldsymbol{r}'')\delta\hat{\phi}(\boldsymbol{r})\rangle\left(\frac{\delta\mu_\alpha(\boldsymbol{r}'')}{\delta c_\beta(\boldsymbol{r}')}\right)_T \\ &= (k_B T)^{-1}\langle \delta_T\hat{\mu}_\alpha(\boldsymbol{r}')\delta\hat{\phi}(\boldsymbol{r})\rangle \end{aligned} \tag{2.49}$$

and

$$\left(\frac{\delta\phi(\boldsymbol{r})}{\delta T(\boldsymbol{r}')}\right)_{\{c_\alpha\}} = (k_B T^2)^{-1} \int d\boldsymbol{r}'' \langle \delta\widehat{T}(\boldsymbol{r}'')\delta\hat{\phi}(\boldsymbol{r})\rangle C_v(\boldsymbol{r}' - \boldsymbol{r}'')$$

$$= (k_B T^2)^{-1} \langle \delta\hat{\sigma}(\boldsymbol{r}')\delta\hat{\phi}(\boldsymbol{r})\rangle \tag{2.50}$$

For a mechanical system whose Hamiltonian is translationally and rotationally invariant the average value of a field is constant and the equilibrium correlation functions of two fields depend only upon the relative distance between the two field-points. For systems of this sort it is convenient to use Fourier representations of the fields and functional derivatives. The Fourier transform of the field $\phi(\boldsymbol{r})$ is defined by

$$\phi(\boldsymbol{k}) = \Omega^{-1/2} \int d\boldsymbol{r} e^{-i\boldsymbol{k}\cdot\boldsymbol{r}} \phi(\boldsymbol{r})$$

$$\phi(\boldsymbol{r}) = \Omega^{-1/2} \int d\boldsymbol{k} e^{i\boldsymbol{k}\cdot\boldsymbol{r}} \phi(\boldsymbol{k}) \tag{2.51}$$

and, for example, the Fourier transform of the functional derivative appearing in (2.35) can be written as

$$\left(\frac{\delta c_\alpha(\boldsymbol{k})}{\delta\mu_\beta(\boldsymbol{k})}\right)_T = \Omega^{-1} \iint d\boldsymbol{r}\, d\boldsymbol{r}' e^{-i\boldsymbol{k}\cdot(\boldsymbol{r}-\boldsymbol{r}')} \left(\frac{\delta c_\alpha(\boldsymbol{r})}{\delta\mu_\beta(\boldsymbol{r}')}\right)_T$$

$$= (k_B T)^{-1} \left[c_\alpha \delta_{\alpha\beta} + c_\alpha c_\beta h_{\alpha\beta}(\boldsymbol{k})\right] \tag{2.52}$$

As previously stated, the functional formulation of equilibrium thermodynamics is a generalization of the standard "global" phenomenological and statistical thermodynamic descriptions of a multicomponent system. Consistency demands that the results of the functional theory reduce to those of classical thermostatics in some well-defined limit. The Fourier representation is particularly convenient for this purpose since one expects to recover the classical results in the long-wave limit, where gradients in the system are small. For example, (2.52) reduces to the formula

$$\lim_{k\to 0}\left(\frac{\delta c_\alpha(\boldsymbol{k})}{\delta\mu_\beta(\boldsymbol{k})}\right)_T = \left(\frac{\partial c_\alpha}{\partial\mu_\beta}\right)_T = \frac{c_\alpha}{k_B T}\left[\delta_{\alpha\beta} + c_\beta \int d\boldsymbol{r}\, h_{\alpha\beta}(\boldsymbol{r})\right] \tag{2.53}$$

which is a well known result of equilibrium statistical thermodynamics. Other examples can be found in the seminal papers of Schofield (Schofield 1966, 1968).

3. THE DYNAMICS OF RELAXATION PROCESSES

According to the Onsager regression hypothesis, spontaneous fluctuations of the microscopic thermodynamic fields obey the same equations of evolution as do macroscopic perturbations of the average fields, with the exception of additional "random" terms that may contribute to the microscopic fluctuation dynamics but which vanish when averaging is performed. The basic fluctuations $\delta\hat{\psi}_i(\boldsymbol{r})$ $(i = 0, \ldots, s+2)$ of the previous section depend not only on the field location variable $\boldsymbol{r}$ but also on the time, by virtue of the time dependence of the particles phase space variables. The microscopic particle densities and the microscopic heat density defined by (2.42) satisfy the following equations of evolution

$$\partial_t \delta\hat{n}_\alpha(\boldsymbol{r},t) + \nabla_{\boldsymbol{r}} \cdot \hat{\boldsymbol{\jmath}}_\alpha(\boldsymbol{r},t) = 0 \tag{3.1}$$

$$\partial_t \delta\hat{\sigma}(\boldsymbol{r},t) + \nabla_{\boldsymbol{r}} \cdot \hat{\boldsymbol{\jmath}}_\sigma(\boldsymbol{r},t) = 0 \tag{3.2}$$

Here $\hat{\jmath}_\alpha(\boldsymbol{r},t)$ and $\hat{\jmath}_\sigma(\boldsymbol{r},t)$, respectively, are the microscopic particle current of the α-th component and the microscopic heat current

$$\hat{\jmath}_\sigma(\boldsymbol{r},t) = \hat{\jmath}_e(\boldsymbol{r},t) - \sum_{\alpha=0}^{s} \int d\boldsymbol{r}' \hat{\jmath}_\alpha(\boldsymbol{r}',t) \left(\frac{\delta u(\boldsymbol{r})}{\delta c_\alpha(\boldsymbol{r}')} \right)_T \tag{3.3}$$

The quantity $\hat{\jmath}_e(\boldsymbol{r},t)$ is the microscopic energy flux occurring in the conservation equation

$$\partial_t \hat{e}(\boldsymbol{r},t) + \nabla_{\boldsymbol{r}} \cdot \hat{\jmath}_e(\boldsymbol{r},t) = 0 \tag{3.4}$$

and the functional derivative appearing in (3.3) is defined by the expression (2.16). Explicit, microscopic expressions for the particle and energy currents will not be needed here (see e.g. Schofield 1966, 1968).

Time-dependent fields such as $\delta\hat{n}_\alpha(\boldsymbol{r},t)$ and $\delta\hat{\sigma}(\boldsymbol{r},t)$ and the corresponding currents can be treated formally as products of the evolution operator of the system acting on the initial values of the fields. In the absence of a time-dependent external forces this evolution operator may be expressed as follows:

$$G(t) = \exp(tK) \tag{3.5}$$

with K denoting the (anti-self-adjoint) Liouville operator. The time-dependent field $\hat{\phi}(\boldsymbol{r},t)$ is then given by

$$\hat{\phi}(\boldsymbol{r},t) = G(t)\hat{\phi}(\boldsymbol{r}) \tag{3.6}$$

Here it must be remembered that the evolution operator acts on the set of the phase-space variables Γ_0, upon which the field $\hat{\phi}(\boldsymbol{r})$ depends implicitly, transforming them from $\Gamma_0 = \Gamma(t=0)$ to $\Gamma(t)$. The field coordinates $\boldsymbol{r}$ are not affected by this transformation.

Let us now consider a system which at the initial instant of time is in a non-equilibrium state. Thus, at $t = 0$ the average values of the fluctuations

$$\delta\psi_i(\boldsymbol{r}) = \langle \psi_i(\boldsymbol{r}) \rangle - \langle \psi_i(\boldsymbol{r}) \rangle_{EQ} \tag{3.7}$$

of the basic state function [$\delta\psi_i(\boldsymbol{r})$ means either $\delta\hat{n}_\alpha(\boldsymbol{r})$ for $\alpha = 0,\ldots,s$ or $\delta\widehat{T}(\boldsymbol{r})$ or $\delta\hat{\sigma}(\boldsymbol{r})$] are different from zero. For times $t > 0$ we expect relaxation to occur with the currents being functionals of the fluctuating fields:

$$\hat{\jmath}_i(\boldsymbol{r},t) = \hat{\jmath}_i^R(\boldsymbol{r},t) + \sum_{k=0}^{s+2} \int_0^t dt' \int d\boldsymbol{r}' \frac{\delta j_i(\boldsymbol{r},t)}{\delta\psi_k(\boldsymbol{r}',t')} \delta\hat{\psi}_k(\boldsymbol{r}',t')$$

$$+ \frac{1}{2} \sum_{k,\ell=0}^{s+2} \int_0^t dt' \int_0^{t'} dt'' \int d\boldsymbol{r}' d\boldsymbol{r}'' \frac{\delta^2 j_i(\boldsymbol{r},t)}{\delta\psi_k(\boldsymbol{r}',t')\delta\psi_\ell(\boldsymbol{r}'',t'')} \delta\hat{\psi}_k(\boldsymbol{r}',t')\delta\hat{\psi}_\ell(\boldsymbol{r}'',t'') + \ldots \tag{3.8}$$

where the kernels may be generalized functions. In this and later expressions index values $i = 0,1,\ldots,s$ denote the various species and the value $i = s+1$ corresponds to the heat transport index $i = \sigma$. The quantity $\hat{\jmath}_i^R(\boldsymbol{r},t)$, often imprecisely identified as a random current, is that portion of the current which is orthogonal to the thermodynamic state space spanned by the functions $\delta\hat{\psi}_i(\boldsymbol{r})$. It consists of contributions to the transport process due to causes other than fluctuations of the basic fields. The functional derivatives of the currents, which at present are unknown, are to be calculated for the equilibrium values of the fields

$\psi_i(\boldsymbol{r})$. Once the currents (3.8) are expressed as functionals of the fluctuating fields, the continuity equations (3.1) and (3.2) become equations of evolution for the fluctuating fields $\delta\hat{\psi}_i(\boldsymbol{r},t)$.

One of many possible ways of determining the kernels appearing in the constitutive relations (3.8) is to use the operator identity (Mori 1965)

$$G(t) = G_R(t)(1-\Pi) + G(t)\Pi + \int_0^t dt' G(t-t')\Pi K(1-\Pi)G_R(t')(1-\Pi) \tag{3.9}$$

Here Π is the projection operator on the thermodynamic state space

$$\Pi = P + P' \tag{3.10}$$

and $G_R(t)$ is the modified (projected) evolution operator defined by

$$G_R(t) = \exp[t(1-\Pi)K] \tag{3.11}$$

By acting with the operator $G(t)$ of (3.10) on the initial values of the currents we obtain the particle current constitutive equations ($\alpha = 0, \ldots, s$)

$$\hat{\jmath}_\alpha(\boldsymbol{r},t) = \hat{\jmath}_\alpha^R(\boldsymbol{r},t) + \sum_{\beta=0}^{s} \int_0^t dt' \int d\boldsymbol{r}'\, \boldsymbol{L}_{\alpha\beta}(\boldsymbol{r},t'|\boldsymbol{r}',0) \cdot \widehat{\boldsymbol{X}}_\beta^T(\boldsymbol{r},t-t')$$
$$+ \int_0^t dt' \int d\boldsymbol{r}'\, \boldsymbol{L}_{\alpha\sigma}(\boldsymbol{r},t'|\boldsymbol{r}',0) \cdot \widehat{\boldsymbol{X}}_\sigma(\boldsymbol{r}',t-t') \tag{3.12}$$

and the heat flux

$$\hat{\jmath}_\sigma(\boldsymbol{r},t) = \hat{\jmath}_\sigma^R(\boldsymbol{r},t) + \sum_{\beta=0}^{s} \int_0^t dt' \int d\boldsymbol{r}'\, \boldsymbol{L}_{\sigma\beta}(\boldsymbol{r},t'|\boldsymbol{r}',0) \cdot \widehat{\boldsymbol{X}}_\beta^T(\boldsymbol{r},t-t')$$
$$+ \int_0^t dt' \int d\boldsymbol{r}'\, \boldsymbol{L}_{\sigma\sigma}(\boldsymbol{r},t'|\boldsymbol{r}',0) \cdot \widehat{\boldsymbol{X}}_\sigma(\boldsymbol{r}',t-t') \tag{3.13}$$

The quantities

$$\hat{\jmath}_\alpha^R(\boldsymbol{r},t) = G_R(t)\hat{\jmath}_\alpha(\boldsymbol{r},0) \tag{3.14}$$

and

$$\hat{\jmath}_\sigma^R(\boldsymbol{r},t) = G_R(t)\hat{\jmath}_\sigma(\boldsymbol{r},0) \tag{3.15}$$

are the so-called "random" parts of the particle and heat fluxes, and the second rank, tensor valued functions

$$\boldsymbol{L}_{ij}(\boldsymbol{r},t|\boldsymbol{r}',0) = (k_B T)^{-1}\langle \hat{\jmath}_i^R(\boldsymbol{r},t)\hat{\jmath}_j^R(\boldsymbol{r}',0)\rangle_{EQ} \tag{3.16}$$

with $i,j = 0,1,\ldots,s+1$, are nonlocal, Onsager kinetic coefficients. The $s+2$ independent thermodynamic forces $\widehat{\boldsymbol{X}}_\beta^T$ ($\beta = 0,\ldots,s$) and $\widehat{\boldsymbol{X}}_\sigma$ are related by the expressions

$$\widehat{\boldsymbol{X}}_\beta^T(\boldsymbol{r},t) = -\nabla_{\boldsymbol{r}}\delta_T\hat{\mu}_\beta(\boldsymbol{r},t) \tag{3.17}$$

$$\widehat{\boldsymbol{X}}_\sigma(\boldsymbol{r},t) = -T_{EQ}^{-1}\nabla_{\boldsymbol{r}}\delta\widehat{T}(\boldsymbol{r},t) \tag{3.18}$$

to the time-dependent fluctuations of the chemical potentials (at constant temperature) and temperature defined by

$$\delta_T\hat{\mu}_\alpha(\boldsymbol{r},t) = G(t)\delta_T\hat{\mu}_\alpha(\boldsymbol{r})$$
$$\delta\widehat{T}(\boldsymbol{r},t) = G(t)\delta\widehat{T}(\boldsymbol{r}) \tag{3.19}$$

Finally, the initial values of these functions are defined by (2.35) and (2.42) of the preceding section.

The so-called "absolute" fluxes, given by the formulas (3.12) and (3.13), are measured with respect to the laboratory frame of reference. These are formally exact expressions belonging to the general class of functional relationships (3.8). They closely resemble the functional formulation of thermodynamics due to Vojta (1967) who, however, derived equations of evolution using the more traditional, variational principle approach to thermodynamics (Gyarmati 1970). The apparent linearity of the constitutive relations (3.12) and (3.13) is misleading: the kinetic coefficients $\boldsymbol{L}_{ij}$ of (3.16) are, in fact, nonlinear functionals of the thermodynamic forces. This point has been examined in more detail by Altenberger et al. (1987a,b). To obtain truly linear constitutive relations one must discard the projection operators Π from the evolution operators $G_R(t)$ that appear in the definitions (3.16) of the random current time-correlation functions. Consequently, the constitutive relations of the linearized theory are expressed in terms of the familiar, linear response (L.R.) coefficients

$$\boldsymbol{L}_{ij}^{L.R.}(\boldsymbol{r},t|\boldsymbol{r}',0) = (k_BT)^{-1}\langle \boldsymbol{\jmath}_i(\boldsymbol{r},t)\boldsymbol{\jmath}_j(\boldsymbol{r}',0)\rangle \tag{3.20}$$

Significant simplifications are to be found when the reference equilibrium system is spatially homogeneous. Thus, due to the translational and rotational invariance, the current time-correlation functions and equilibrium correlation functions become dependent upon the distance between two field points and the tensors $\boldsymbol{L}_{ij}$ become isotropic. In these circumstances it is convenient to use the Fourier-Laplace representation of the absolute currents defined by

$$\boldsymbol{\jmath}_i(\boldsymbol{k},\omega) = \Omega^{-1/2}\int d\boldsymbol{r}\; e^{-i\boldsymbol{k}\cdot\boldsymbol{r}}\int_0^\infty dt\; e^{-i\omega t}\boldsymbol{\jmath}_i(\boldsymbol{r},t) \tag{3.21}$$

The constitutive relations then can be expressed in much simpler form

$$\boldsymbol{\jmath}_i(\boldsymbol{k},\omega) = \boldsymbol{\jmath}_i^R(\boldsymbol{k},\omega) + \sum_{i=0}^{s+1} L_{ij}(\boldsymbol{k},\omega)\widehat{\boldsymbol{X}}_j(\boldsymbol{k},\omega) \tag{3.22}$$

with the kinetic coefficients

$$L_{ij}(\boldsymbol{k},\omega) = (k_BT)^{-1}\hat{\boldsymbol{k}}\cdot\langle \boldsymbol{\jmath}_i(\boldsymbol{k},\omega)\boldsymbol{\jmath}_j(-\boldsymbol{k})\rangle\cdot\hat{\boldsymbol{k}} \tag{3.23}$$

In the expression (3.22) we define the thermodynamic forces $\widehat{\boldsymbol{X}}_j(\boldsymbol{k},\omega)$ using the convention that the index $i = s+1$ applies to the heat transfer related quantities, i.e., for $i = 0,\ldots,s$

$$\widehat{\boldsymbol{X}}_i(\boldsymbol{k},\omega) = -i\boldsymbol{k}\;\delta_T\hat{\mu}_i(\boldsymbol{k},\omega) \tag{3.24}$$

and for $i = s+1$

$$\widehat{\boldsymbol{X}}_{s+1}(\boldsymbol{k},\omega) = -i\boldsymbol{k}\;\delta\widehat{T}(\boldsymbol{k},\omega)/T_{EQ} \tag{3.25}$$

The Onsager kinetic coefficients are symmetric in the laboratory frame, i.e. $L_{ij}(\boldsymbol{k},\omega) = L_{ji}(\boldsymbol{k},\omega)$.

The transport coefficients of traditional, linear Onsager thermodynamics are obtained from those of the present theory by passing to the low frequency, long-wavelength limit, viz.,

$$L_{ij}^{CL} = \lim_{\omega\to 0} \lim_{k\to 0} L_{ij}(\boldsymbol{k},\omega) \tag{3.26}$$

In this same limit the equilibrium functional derivatives that contribute to the thermodynamic forces become ordinary partial derivatives as illustrated, for example, by (2.53) of the preceding section.

The thermodynamic forces used in the constitutive relations of this section are linearly independent and all components of the solution are treated equally. However, there are situations for which alternative sets of thermodynamic forces are more convenient. For example, the experimental conditions may be such that some of the forces vanish identically. Several of the most frequently used sets of forces are considered in detail by Altenberger et al. (1987a,b). One of these is obtained by using the Gibbs-Duhem relation to eliminate the fluctuation of the chemical potential of the "solvent" in favor of the fluctuation of thermodynamic pressure. The chemical potentials of "solute" components are then treated as functionals of the pressure field and of the solute particle density fields. The new set of thermodynamic forces then consists of the negative gradients of the "solute" chemical potentials and of the pressure and temperature. This set is particularly convenient for the study of diffusion processes under isothermal and isobaric conditions. The corresponding kinetic coefficients are linear combinations of those found previously, namely the L_{ij}'s of (3.20) or (3.23). These coefficients do not satisfy the Onsager reciprocal relations; they are unsymmetric. That such unsymmetric coefficients can indeed appear seems to support Truesdell's (1969) critical remarks about the validity of reciprocal relations under arbitrary conditions.

4. FLOW REFERENCE FRAMES

The fluxes of the previous section are measured with respect to the stationary, laboratory frame of reference. However, it sometimes is convenient to refer the equations of evolution for the $s+2$ basic thermodynamic fields to a local frame of reference connected with the motion of some characteristic property of the fluid or with the motion of one of its components. Well known examples include the local center of mass or barycentric velocity and the velocity of one specific component of the mixture. The velocity field of the local reference frame will be denoted by $\hat{\boldsymbol{u}}^g(\boldsymbol{r},t)$ with the superscript g labelling a particular choice of this frame. The frame velocity is, in general, a linear functional of the absolute particle fluxes and depends on the choice of the equilibrium reference state. From here on, we assume that the equilibrium state is spatially homogeneous. The frame velocity then is related to the absolute fluxes as follows;

$$\hat{\boldsymbol{u}}^g(\boldsymbol{r},t) = \sum_{\alpha=0}^{s} z_\alpha^g \hat{\boldsymbol{\jmath}}_\alpha(\boldsymbol{r},t) \ . \tag{4.1}$$

The weight factors z_α^g depend on the particular choice of local frame and satisfy the condition

$$\sum_{\alpha=0}^{s} c_\alpha z_\alpha^g = 1 \ . \tag{4.2}$$

For example, in the case of the solvent-fixed frame

$$z_\alpha^s = c_0^{-1}\delta_{\alpha 0} \tag{4.3}$$

with the index $\alpha = 0$ indicating the solvent component. In the case of the barycentric frame

$$z_\alpha^b = \frac{m_\alpha}{\sum_{\beta=0}^{s} m_\beta c_\beta} \tag{4.4}$$

with m_α denoting the mass of an α-component particle.

The absolute fluxes are related to the reference frame velocity and the relative fluxes $\hat{\boldsymbol{\jmath}}_\alpha^g(\boldsymbol{r},t)$ and $\hat{\boldsymbol{\jmath}}_\sigma^g(\boldsymbol{r},t)$ by the connections

$$\hat{\boldsymbol{\jmath}}_\alpha(\boldsymbol{r},t) = c_\alpha \hat{\boldsymbol{u}}^g(\boldsymbol{r},t) + \hat{\boldsymbol{\jmath}}_\alpha^g(\boldsymbol{r},t) \tag{4.5}$$

and

$$\hat{\boldsymbol{\jmath}}_\sigma(\boldsymbol{r},t) = \sigma \hat{\boldsymbol{u}}^g(\boldsymbol{r},t) + \hat{\boldsymbol{\jmath}}_\sigma^g(\boldsymbol{r},t) \tag{4.6}$$

with σ denoting the equilibrium heat density defined by the formula

$$\sigma = u - \sum_{\alpha=0}^{s} c_\alpha \left(\frac{\partial u}{\partial c_\alpha}\right)_T \tag{4.7}$$

The relative currents then can be expressed in terms of the kinetic coefficients and thermodynamic forces of the laboratory frame by substitution into the equations

$$\hat{\boldsymbol{\jmath}}_\alpha^g(\boldsymbol{r},t) = \hat{\boldsymbol{\jmath}}_\alpha(\boldsymbol{r},t) - c_\alpha \sum_{\beta=0}^{s} z_\beta^g \hat{\boldsymbol{\jmath}}_\beta(\boldsymbol{r},t)$$

$$\hat{\boldsymbol{\jmath}}_\sigma^g(\boldsymbol{r},t) = \hat{\boldsymbol{\jmath}}_\sigma(\boldsymbol{r},t) - \sigma \sum_{\beta=0}^{s} z_\beta^g \hat{\boldsymbol{\jmath}}_\beta(\boldsymbol{r},t) \tag{4.8}$$

Starting with a given set of laboratory frame thermodynamic forces [e.g., the set used explicitly in Section 3 or the alternative set mentioned near the end of Section 3], one obtains new kinetic coefficients that are linear combinations of the lab frame coefficients. These new coefficients generally do not satisfy the Onsager reciprocal relations. Examples of this are provided by Altenberger et al. (1987a). However, symmetry can be restored by introducing the new set of thermodynamic forces consisting of:

$$\begin{aligned}\widehat{\boldsymbol{X}}_0^g(\boldsymbol{r},t) &= \sum_{\beta=0}^{s} c_\beta \widehat{\boldsymbol{X}}_\beta^T(\boldsymbol{r},t) + \sigma \widehat{\boldsymbol{X}}_\sigma(\boldsymbol{r},t) \\ &= -\nabla_{\boldsymbol{r}} \delta_T \hat{p}(\boldsymbol{r},t) - \frac{\sigma}{T_{EQ}} \nabla_{\boldsymbol{r}}\, \delta\widehat{T}(\boldsymbol{r},t)\end{aligned} \tag{4.9}$$

$$\begin{aligned}\widehat{\boldsymbol{X}}_\beta^{Tg}(\boldsymbol{r},t) &= \widehat{\boldsymbol{X}}_\beta^T(\boldsymbol{r},t) - (z_\beta^g/z_0^g)\widehat{\boldsymbol{X}}_0^T(\boldsymbol{r},t) \\ &= -\nabla_{\boldsymbol{r}} \delta_T \left[\hat{\mu}_\beta(\boldsymbol{r},t) - (z_\beta^g/z_0^g)\hat{\mu}_0(\boldsymbol{r},t)\right]\end{aligned} \tag{4.10}$$

for $1 \le \beta \le s$ and

$$\widehat{\boldsymbol{X}}^g_\sigma(\boldsymbol{r},t) = \widehat{\boldsymbol{X}}_\sigma(\boldsymbol{r},t) = -\frac{1}{T_{EQ}}\ \nabla_{\boldsymbol{r}}\delta\widehat{T}(\boldsymbol{r},t) \tag{4.11}$$

The number of independent thermodynamic forces is the same as before, equal to the dimension of the thermodynamic functional space. The quantity $\delta_T\hat{p}(\boldsymbol{r},t)$ appearing in (4.9) is the variation, at constant temperature, of the thermodynamic pressure.

The kinetic coefficients associated with the forces (4.9)–(4.11) are the following: for $i,j = \alpha$ or σ with $1 \le \alpha \le s$

$$\overline{\boldsymbol{\Omega}}^g_{i,j}(\boldsymbol{r},t|\boldsymbol{r}',0) = (k_BT)^{-1}\langle\hat{\boldsymbol{\jmath}}^{R,g}_i(\boldsymbol{r},t)\hat{\boldsymbol{\jmath}}^{R,g}_j(\boldsymbol{r}',0)\rangle \tag{4.12}$$

and for $i = \alpha$ or σ

$$\overline{\boldsymbol{\Omega}}^g_{i,0}(\boldsymbol{r},t|\boldsymbol{r}',0) = (k_BT)^{-1}\langle\hat{\boldsymbol{\jmath}}^{R,g}_i(\boldsymbol{r},t)\hat{\boldsymbol{u}}^g(\boldsymbol{r}',0)\rangle \tag{4.13}$$

Here the quantities $\hat{\boldsymbol{\jmath}}^{R,g}_i(\boldsymbol{r},t)$ are the random parts of the relative currents, obtained by collecting together the random flux contributions from the absolute currents according to (4.5) and (4.6).

The constitutive relations for the relative currents are ($i = \alpha$ or σ; $\alpha = 1,2,\ldots,s$)

$$\begin{aligned}\hat{\boldsymbol{\jmath}}^g_i(\boldsymbol{r},t) &= \hat{\boldsymbol{\jmath}}^{R,g}_i(\boldsymbol{r},t) + \int_0^t dt' \int d\boldsymbol{r}'\overline{\boldsymbol{\Omega}}^g_{i0}(\boldsymbol{r},t'|\boldsymbol{r}',0)\cdot\widehat{\boldsymbol{X}}^g_0(\boldsymbol{r}',t-t')\\ &+\sum_{\beta=1}^{s}\int_0^t dt' \int d\boldsymbol{r}'\overline{\boldsymbol{\Omega}}^g_{i\beta}(\boldsymbol{r},t'|\boldsymbol{r}',0)\cdot\widehat{\boldsymbol{X}}^{T,g}_\beta(\boldsymbol{r}',t-t')\\ &+\int_0^t dt' \int d\boldsymbol{r}'\overline{\boldsymbol{\Omega}}^g_{i\sigma}(\boldsymbol{r},t'|\boldsymbol{r}',0)\cdot\widehat{\boldsymbol{X}}^g_\sigma(\boldsymbol{r}',t-t')\end{aligned} \tag{4.14}$$

It should be noted that the reference frame velocity field $\hat{\boldsymbol{u}}^g(\boldsymbol{r},t)$ must be determined by some additional equation of motion or from the known absolute currents. As before, the constitutive relations simplify when the system is translationally invariant, allowing us to replace the tensorial kinetic coefficients by scalar multiples of the unit tensor. The number of independent forces remains equal to the dimension of the state space. However, we now can identify a special force, $\widehat{\boldsymbol{X}}^g_0$, which describes the influence of the motion of the reference frame on the particle and heat fluxes.

5. TRANSPORT IN THE PRESENCE OF A WEAK GRAVIATIONAL FIELD

A system of particles acted upon by an external gravitational field develops an inhomogeneous particle distribution. There are two possible choices for the equilibrium reference state of such a system and, correspondingly, two choices for the Onsager kinetic coefficients and related transport quantities. One choice corresponds to the uniform distribution of particles that existed in the homogeneous equilibrium system prior to the establishment of the field. The other corresponds to the fully developed inhomogeneous equilibrium state caused by the external potential. The form of the equations of evolution depends on which is selected. We choose the former. The derivation of the constitutive equations for the laboratory frame of reference then proceeds analogously to that presented by Altenberger et al. (1987b). The gravitational potential is indicated by the symbol $V(\boldsymbol{r})$ and the associated force on a mass m_α is written as $m_\alpha \boldsymbol{E}(\boldsymbol{r})$ with $\boldsymbol{E}(\boldsymbol{r}) = -\nabla_{\boldsymbol{r}} V(\boldsymbol{r})$ denoting the field intensity.

The particle and heat fluxes can be written as follows (with $i = \alpha$ or σ and $\alpha \in [0, s]$):

$$\hat{\boldsymbol{\jmath}}_i(\boldsymbol{r},t) = \hat{\boldsymbol{\jmath}}_i^R(\boldsymbol{r},t) + \sum_{\gamma=0}^{s} \int_0^t dt' \int d\boldsymbol{r}' \; \boldsymbol{L}_{i\gamma}(\boldsymbol{r},t'|\boldsymbol{r}',0) \cdot \widehat{\boldsymbol{X}}_\gamma^T(\boldsymbol{r}',t-t')$$

$$+ \int_0^t dt' \int d\boldsymbol{r}' \; \boldsymbol{L}_{i\sigma}(\boldsymbol{r},t'|\boldsymbol{r}',0) \cdot \widehat{\boldsymbol{X}}_\sigma(\boldsymbol{r}',t-t')$$

$$+ (k_B T)^{-1} \sum_{\gamma,\nu=0}^{s} \int_0^t dt' \int d\boldsymbol{r}' \int d\boldsymbol{r}'' \; \boldsymbol{M}_{i\gamma\nu}(\boldsymbol{r},t'|\boldsymbol{r}',\boldsymbol{r}'',0) \cdot \boldsymbol{E}(\boldsymbol{r}'') m_\nu \delta_T \hat{\mu}_\gamma(\boldsymbol{r}',t-t')$$

$$+ (k_B T)^{-1} \sum_{\nu=0}^{s} \int_0^t dt' \int d\boldsymbol{r}' \int d\boldsymbol{r}'' \; \boldsymbol{M}_{i\sigma\nu}(\boldsymbol{r},t'|\boldsymbol{r}',\boldsymbol{r}'',0) \cdot \boldsymbol{E}(\boldsymbol{r}'') m_\nu \delta \widehat{T}(\boldsymbol{r}',t-t') \qquad (5.1)$$

Due to the presence of the field, the thermodynamic forces of (3.17) are replaced with the modified forces

$$\widehat{\boldsymbol{X}}_\alpha^T(\boldsymbol{r},t) = -\left[\nabla_{\boldsymbol{r}} - \frac{m_\alpha}{k_B T} \boldsymbol{E}(\boldsymbol{r})\right] \delta_T \hat{\mu}_\alpha(\boldsymbol{r},t) \;\; ; \;\; \alpha \in [0,s] \qquad (5.2)$$

The thermal force (3.18) remains unchanged. The Onsager kinetic coefficients $\boldsymbol{L}_{ij}$ are defined by the formulas (3.16) but now depend additionally on the external field. A new feature, which seems to have gone unrecognized in most of the non-equilibrium thermodynamics literature, is the appearance of the two additional, explicitly field-dependent terms involving the three-index coefficients

$$\boldsymbol{M}_{i\beta\gamma}(\boldsymbol{r},t|\boldsymbol{r}',\boldsymbol{r}'',0) = (k_B T)^{-1} \langle \hat{\boldsymbol{\jmath}}_i^R(\boldsymbol{r},t) \hat{n}_{\beta\gamma}^{\#}(\boldsymbol{r}',\boldsymbol{r}'') \hat{\boldsymbol{\jmath}}_\gamma(\boldsymbol{r}'') \rangle \qquad (5.3)$$

and

$$\boldsymbol{M}_{i\sigma\gamma}(\boldsymbol{r},t|\boldsymbol{r}',\boldsymbol{r}'',0) = (k_B T)^{-1} \langle \hat{\boldsymbol{\jmath}}_i^R(\boldsymbol{r},t) \hat{\sigma}(\boldsymbol{r}') \hat{\boldsymbol{\jmath}}_\alpha(\boldsymbol{r}'') \rangle \qquad (5.4)$$

The quantities $\hat{n}_{\beta\gamma}^{\#}(\boldsymbol{r}',\boldsymbol{r}')$ and $\hat{\sigma}(\boldsymbol{r}')$ appearing in these two formulas are defined by

$$\hat{n}_{\beta\gamma}^{\#}(\boldsymbol{r}',\boldsymbol{r}'') = \hat{n}_\beta(\boldsymbol{r}') - \delta_{\beta\gamma}\delta(\boldsymbol{r}'-\boldsymbol{r}'') \qquad (5.5)$$

and

$$\hat{\sigma}(\boldsymbol{r}) = \hat{e}(\boldsymbol{r}) - \sum_{\alpha=0}^{s} \int d\boldsymbol{r}' \; \hat{n}_\alpha(\boldsymbol{r}') \left(\frac{\delta u(\boldsymbol{r})}{\delta c_\alpha(\boldsymbol{r}')}\right)_T , \qquad (5.6)$$

respectively.

The particle and heat fluxes of the phenomenological thermodynamics of irreversible processes correspond to the ensemble average values of the microscopic expressions given above. The initial, non-equilibrium distribution of the phase-space variables can be written in the form

$$\rho(\Gamma) = \rho_{EQ}(\Gamma) + \delta\rho(\Gamma) \qquad (5.7)$$

with $\rho_{EQ}(\Gamma)$ denoting the equilibrium canonical distribution of the spatially homogeneous reference state and $\delta\rho(\Gamma)$ a function that satisfies the condition

$$\int d\Gamma\ \delta\rho(\Gamma) = 0 \ . \tag{5.8}$$

This function $\delta\rho(\Gamma)$ represents the perturbation of the reference state that exists at the initial instant of time. The distribution function of (5.6) is independent of the external field which is switched on adiabatically at times $t > 0$.

In the linear, thermodynamic limit the influence of the thermodynamic forces on the kinetic coefficients is neglected (Altenberger et al. 1987a,b). The Onsager coefficients are then expressed in terms of equilibrium time-correlation functions of appropriate microscopic currents. By averaging the currents (5.1) using the distribution function (5.7) we obtain the formulas

$$\begin{aligned}\boldsymbol{j}_i(\boldsymbol{r},t) &= \boldsymbol{j}_i^R(\boldsymbol{r},t) + \sum_{\gamma=0}^{s}\int_0^t dt' \int d\boldsymbol{r}'\ \boldsymbol{L}^0_{i\gamma}(\boldsymbol{r},t'|\boldsymbol{r}',0)\cdot \boldsymbol{X}^T_\gamma(\boldsymbol{r},t-t') \\ &+ \int_0^t dt \int d\boldsymbol{r}'\ \boldsymbol{L}^0_{i\sigma}(\boldsymbol{r},t'|\boldsymbol{r}',0)\cdot \boldsymbol{X}_\sigma(\boldsymbol{r}',t-t') \\ &+ (k_BT)^{-1}\sum_{\gamma,\nu=0}^{s}\int_0^t dt' \int d\boldsymbol{r}' d\boldsymbol{r}''\, \boldsymbol{M}^0_{i\gamma\nu}(\boldsymbol{r},t'|\boldsymbol{r}',\boldsymbol{r}'',0)\cdot \boldsymbol{E}(\boldsymbol{r}'')m_\nu \delta_T\mu_\gamma(\boldsymbol{r}',t-t') \\ &+ (k_BT)^{-1}\sum_{\nu=0}^{s}\int_0^t dt' \int d\boldsymbol{r}' d\boldsymbol{r}''\, \boldsymbol{M}^0_{i\sigma\nu}(\boldsymbol{r},t'|\boldsymbol{r}',\boldsymbol{r}'',0)\cdot \boldsymbol{E}(\boldsymbol{r}'')m_\nu \delta T(\boldsymbol{r}',t-t')\end{aligned} \tag{5.9}$$

The superscript zeros appearing here indicate the linear-response limit. The Onsager coefficients still depend on the external field, and since this field may be inhomogeneous, the coefficients are not translationally invariant. The "random" part of the flux, $\boldsymbol{j}_i^R(\boldsymbol{r},t)$, can be written as the sum of one contribution,

$$\delta \boldsymbol{j}_i^R(\boldsymbol{r},t) = \int d\Gamma\ \delta\rho(\Gamma)\hat{\boldsymbol{j}}_i^R(\boldsymbol{r},t) \tag{5.10}$$

due to the initial deviation from equilibrium and another,

$$\boldsymbol{j}_i^{R,EXT}(\boldsymbol{r},t) = \sum_{\gamma=0}^{s}\int_0^t dt' \int d\boldsymbol{r}'\ \boldsymbol{L}^0_{i\gamma}(\boldsymbol{r},t'|\boldsymbol{r}',0)\cdot \boldsymbol{E}(\boldsymbol{r}')m_\gamma \tag{5.11}$$

caused by the external field.

Further simplifications of the constitutive relations are possible if we assume that at the initial time the system was in a homogeneous equilibrium state (i.e., $\delta\rho(\Gamma) = 0$) and that the external field is so weak that only contributions linear with respect to the field intensity need be retained. Under these circumstances the terms involving the three-index transport coefficients (5.3) and (5.4) can be neglected as being at least second order with respect to the external field. All the remaining kinetic coefficients become field independent and translationally invariant so that ($i, j = 0, 1, \ldots, s$ or σ)

$$\boldsymbol{L}_{ij}^{0}(\boldsymbol{r}, t'|\boldsymbol{r}', 0) = \boldsymbol{I}\Big(\frac{1}{3}\ Tr\ \boldsymbol{L}_{ij}^{0}\Big) = \boldsymbol{I}L_{ij}^{0}(\boldsymbol{r}-\boldsymbol{r}', t') \tag{5.12}$$

with $L_{ij}^{0}(\boldsymbol{r}-\boldsymbol{r}', t')$ a scalar valued function.

The constitutive relations in the laboratory frame now assume the forms

$$\begin{aligned}\boldsymbol{j}_i(\boldsymbol{r}, t) = &\sum_{\gamma=0}^{s}\int_0^t dt' \int d\boldsymbol{r}'\ L_{i\gamma}^{0}(\boldsymbol{r}-\boldsymbol{r}', t')\boldsymbol{E}(\boldsymbol{r}')m_\gamma \\ &+\sum_{\gamma=0}^{s}\int_0^t dt' \int d\boldsymbol{r}'\ L_{i\gamma}^{0}(\boldsymbol{r}-\boldsymbol{r}', t')\boldsymbol{X}_\gamma^T(\boldsymbol{r}', t-t') \\ &+\int_0^t dt' \int d\boldsymbol{r}'\ L_{i\sigma}^{0}(\boldsymbol{r}-\boldsymbol{r}', t')\boldsymbol{X}_\sigma(\boldsymbol{r}', t-t')\end{aligned} \tag{5.13}$$

where X_γ^T and X_σ denote the average values of the thermodynamic forces defined by (3.17) and (3.18), respectively.

If sedimentation is taking place in an incompressible mixture, the absolute particle currents will be subject to the additional constraint

$$\sum_{\alpha=0}^{s}\overline{\omega}_\alpha \boldsymbol{j}_\alpha(\boldsymbol{r}, t) = 0 \tag{5.14}$$

Here $\overline{\omega}_\alpha$, the partial volume per particle of component α, can be expressed in terms of the direct correlation functions as follows:

$$\overline{\omega}_\alpha = \frac{\sum_{\beta=0}^{s}\left[\delta_{\alpha\beta} - c_\beta \int d\boldsymbol{r}\ C_{\alpha\beta}(\boldsymbol{r})\right]}{\sum_{\alpha,\beta=0}^{s}\left[c_\alpha\delta_{\alpha\beta} - c_\alpha c_\beta \int d\boldsymbol{r}\ C_{\alpha\beta}(\boldsymbol{r})\right]} \tag{5.15}$$

The ensemble average velocity of the reference system fixed on the local center of volume is given by the expression

$$\boldsymbol{u}^v(\boldsymbol{r}, t) = \sum_{\alpha=0}^{s}\overline{\omega}_\alpha \boldsymbol{j}_\alpha(\boldsymbol{r}, t) \tag{5.16}$$

Consequently, the meaning of (5.14) is that for an incompressible mixture there is no difference between the laboratory frame of reference and the local frame of reference moving with the velocity $\boldsymbol{u}^v(\boldsymbol{r}, t)$. This together with the reciprocal relations implies the following relationship between the kinetic coefficients ($\beta \in [1, s]$ or $\beta = \sigma$):

$$\sum_{\alpha=0}^{s}\overline{\omega}_\alpha L_{\alpha\beta}^{0}(\boldsymbol{r}, t) = \sum_{\alpha=0}^{s} L_{\beta\alpha}^{0}(\boldsymbol{r}, t)\overline{\omega}_\alpha = 0 \tag{5.17}$$

These permit us to rewrite the constitutive equations (5.13) in the alternative forms

$$j_i(\boldsymbol{r},t) = \sum_{\gamma=1}^{s} \int_0^t dt' \int d\boldsymbol{r}' \; L^0_{i\gamma}(\boldsymbol{r}-\boldsymbol{r}',t')\boldsymbol{E}(\boldsymbol{r}')m_\gamma \left[1 - \frac{\overline{\omega}_\gamma m_0}{\overline{\omega}_0 m_\gamma}\right]$$

$$+ \sum_{\gamma=0}^{s} \int_0^t dt' \int d\boldsymbol{r}' \; L^0_{i\gamma}(\boldsymbol{r}-\boldsymbol{r}',t')\boldsymbol{X}^T_\gamma(\boldsymbol{r}',t-t')$$

$$+ \int_0^t dt' \int d\boldsymbol{r}' \; L^0_{i\sigma}(\boldsymbol{r}-\boldsymbol{r}',t')\boldsymbol{X}_\sigma(\boldsymbol{r}',t-t') \tag{5.18}$$

which incorporate the buoyancy factors $(1-\overline{\omega}_\gamma m_0/\overline{\omega}_0 m_\gamma)$. From these equations we conclude that the microscopic definition of the sedimentation coefficient should be $(\alpha = 1,\ldots,s)$

$$s_\alpha = \lim_{k\to 0} \lim_{\omega\to 0} \sum_{\gamma=1}^{s} L^0_{\alpha\gamma}(\boldsymbol{k},\omega)m_\gamma \left[1 - \frac{\overline{\omega}_\gamma m_0}{\overline{\omega}_0 m_\gamma}\right] \Big/ c_\alpha \;, \tag{5.19}$$

in agreement with the result reported by Altenberger and Tirrell (1986). It should be noted that $L^0_{\alpha\gamma}(k,\omega)$ is the double Fourier-Laplace transform of the appropriate kinetic coefficient and that the sedimentation coefficient refers to the particle (not mass) flux. Modification to the mass flux is straightforward. We also can define the "thermal" sedimentation coefficient

$$s_\sigma = \lim_{k\to 0} \lim_{\omega\to 0} \sum_{\gamma=1}^{s} L^0_{\sigma\gamma}(\boldsymbol{k},\omega)m_\gamma \left[1 - \frac{\overline{\omega}_\gamma m_0}{\overline{\omega}_0 m_\gamma}\right] \Big/ \overline{e} \tag{5.20}$$

which reflects the influence of sedimentation on the flux of heat.

An alternative to the constitutive relation (5.13) can be obtained by using the relative current representation. This leads to the following expression for particle and heat currents measured relative to the volume fixed frame of reference:

$$j^v_i(\boldsymbol{r},t) = \sum_{\gamma=1}^{s} \int_0^t dt' \int d\boldsymbol{r}' \; \overline{\Omega}^v_{i\gamma}(\boldsymbol{r}-\boldsymbol{r}',t')m_\gamma \boldsymbol{E} \left[1 - \frac{\overline{\omega}_\gamma m_0}{\overline{\omega}_0 m_\gamma}\right]$$

$$+ \sum_{\gamma=1}^{s} \int_0^t dt' \int d\boldsymbol{r}' \; \overline{\Omega}^v_{i\gamma}(\boldsymbol{r}-\boldsymbol{r}',t')\boldsymbol{X}^{Tv}_\gamma(\boldsymbol{r}',t-t')$$

$$+ \int_0^t dt' \int d\boldsymbol{r}' \; \overline{\Omega}^v_{i\sigma}(\boldsymbol{r}-\boldsymbol{r}',t')\boldsymbol{X}^v_\sigma(\boldsymbol{r}',t-t')$$

$$+ \int_0^t dt' \int d\boldsymbol{r}' \; \overline{\Omega}^v_{i0}(\boldsymbol{r}-\boldsymbol{r}',t')\boldsymbol{X}^v_0(\boldsymbol{r}',t-t') \tag{5.21}$$

The kinetic coefficients $\overline{\Omega}^v_{i\gamma}$ appearing here are defined by

$$\overline{\Omega}^v_{i\gamma}(\boldsymbol{r}-\boldsymbol{r}',t) = (k_B T)^{-1} \, \frac{1}{3} \, \langle \hat{\jmath}^v_i(\boldsymbol{r},t) \cdot \hat{\jmath}^v_\gamma(\boldsymbol{r}') \rangle \tag{5.22}$$

where $\hat{\jmath}_i^v$ denotes the microscopic current relative to the microscopic, volume fixed frame of reference defined according to (3.6). The coefficient $\overline{\Omega}_{i0}^v$ is defined analogously as

$$\overline{\Omega}_{i0}^v(\boldsymbol{r}-\boldsymbol{r}',t) = (k_B T)^{-1}\ \frac{1}{3}\ \langle \hat{\jmath}_i^v(\boldsymbol{r},t)\cdot \hat{\boldsymbol{u}}^v(\boldsymbol{r}')\rangle \tag{5.23}$$

Finally, the thermodynamic forces appearing in (5.21) are given by:

$$\boldsymbol{X}_0^v(\boldsymbol{r},t) = -\nabla_{\boldsymbol{r}}\delta_T p(\boldsymbol{r},t) - \frac{\sigma}{T}\ \nabla_{\boldsymbol{r}}\delta T(\boldsymbol{r},t) + \boldsymbol{E}(r)\sum_{\gamma=0}^{s} m_\gamma c_\gamma \tag{5.24}$$

$$\boldsymbol{X}_\beta^v(\boldsymbol{r},t) = -\nabla_{\boldsymbol{r}}\delta_T\left[\mu_\beta(\boldsymbol{r},t) - \frac{\overline{\omega}_\beta}{\overline{\omega}_0}\ \mu_0(\boldsymbol{r},t)\right] \quad ; \quad \beta\in[1,s] \tag{5.25}$$

and

$$\boldsymbol{X}_\sigma^v(\boldsymbol{r},t) = -\frac{1}{T}\ \nabla_{\boldsymbol{r}}\delta T(\boldsymbol{r},t) \tag{5.26}$$

The thermodynamic force $\boldsymbol{X}_0^v$ of (5.24) should vanish when mechanical equilibrium is established in the system. When this occurs and the condition (5.14) is satisfied one obtains the following expression for the sedimentation coefficient:

$$s_\alpha = \lim_{k\to 0}\ \lim_{\omega\to 0}\sum_{\gamma=1}^{s}\overline{\Omega}_{\alpha\gamma}^v(\boldsymbol{k},\omega)m_\gamma\left[1-\frac{\overline{\omega}_\gamma m_0}{\overline{\omega}_0 m_\gamma}\right]\Big/ c_\alpha\ , \tag{5.26}$$

This does not agree with the result recently reported by Raineri and Timmermann (1989), but agrees with the formulas obtained by Kops-Werkhoven et al. (1983).

6. FLUCTUATION DYNAMICS IN A SINGLE COMPONENT SYSTEM

During recent years a great deal of effort has been devoted to modifying the Onsager thermodynamics in order to account for the finite rates of propagation of thermodynamic disturbances. The "wave equation" approach to this problem was considered earlier by Sandler and Dahler (1969). More recent developments can be found in the review paper by Gyarmati (1977) and in other, related references (see e.g., Sieniutycz 1979). Here we simply wish to point out that the "wave equation" formulation follows naturally from the constitutive equations of the form (3.12)-(3.13). As an example, let us consider the relaxation of the particle density and temperature fluctuations in a single component system. We assume that the equilibrium reference state is translationally and rotationally invariant so that the two constitutive relations in the laboratory frame of reference are:

$$\begin{aligned}\hat{\jmath}_0(\boldsymbol{r},t) = \hat{\jmath}_0^R(\boldsymbol{r},t) + \int_0^t dt'\int d\boldsymbol{r}'\ \Big[&L_{00}(\boldsymbol{r}-\boldsymbol{r}',t-t')\widehat{\boldsymbol{X}}_0^T(\boldsymbol{r}',t') \\ &+ L_{0\sigma}(\boldsymbol{r}-\boldsymbol{r}',t-t')\widehat{\boldsymbol{X}}_\sigma(\boldsymbol{r}',t')\Big] \\ \hat{\jmath}_\sigma(\boldsymbol{r},t) = \hat{\jmath}_\sigma^R(\boldsymbol{r},t) + \int_0^t dt'\int d\boldsymbol{r}'\ \Big[&L_{\sigma 0}(\boldsymbol{r}-\boldsymbol{r}',t-t')\widehat{\boldsymbol{X}}_0^T(\boldsymbol{r}',t') \\ &+ L_{\sigma\sigma}(\boldsymbol{r}-\boldsymbol{r}',t-t')\widehat{\boldsymbol{X}}_\sigma(\boldsymbol{r}',t')\Big]\end{aligned} \tag{6.1}$$

The thermodynamic forces involved here are defined by the eqs. (3.17) and (3.18).

The kinetic coefficients $L_{00}(\boldsymbol{r},t)$ and $L_{\sigma\sigma}(\boldsymbol{r},t)$ correspond to particle transport caused by the gradient of chemical potential and to heat transport caused by the gradient of temperature. The cross-kinetic coefficients $L_{0\sigma}(\boldsymbol{r},t)$ and $L_{\sigma 0}(\boldsymbol{r},t)$, are connected respectively with thermal convection, namely particle transport due to the temperature gradient, and the Dufour effect, heat transfer caused by the gradient of the chemical potential. It should be noted that the relations (6.1) are not the standard expressions of the classical thermodynamics of nonequilibrium processes, which usually are restricted to the relative fluxes. In our opinion, it is often more convenient to formulate the constitutive relations in the laboratory frame of reference. Also, most observations of fluctuation-relaxation processes, for example light or neutron scattering experiments, are made in the laboratory frame of reference.

The equations of evolution for the fluctuating fields are most conveniently expressed in the Fourier-Laplace transform representation:

$$\begin{aligned}\left[i\omega + \boldsymbol{k}^2 D_{00}(\boldsymbol{k},\omega)\right]\delta\hat{n}_0(\boldsymbol{k},\omega) + \boldsymbol{k}^2 L_{0\sigma}(\boldsymbol{k},\omega)/T\ \delta\widehat{T}(\boldsymbol{k},\omega) &= \delta\hat{n}_0(\boldsymbol{k},t=0)\\ &\quad - i\boldsymbol{k}\cdot\hat{\jmath}_0^R(\boldsymbol{k},\omega)\end{aligned}\tag{6.2}$$

and

$$\begin{aligned}\left[i\omega C_v(\boldsymbol{k}) + \boldsymbol{k}^2 L_{\sigma\sigma}(\boldsymbol{k},\omega)/T\right]\delta\widehat{T}(\boldsymbol{k},\omega) + \boldsymbol{k}^2 D_{\sigma 0}(\boldsymbol{k},\omega)\delta\hat{n}_0(\boldsymbol{k},\omega) &= C_v(\boldsymbol{k})\delta\widehat{T}(\boldsymbol{k},0)\\ &\quad - i\boldsymbol{k}\cdot\hat{\jmath}_\sigma^R(\boldsymbol{k},\omega)\end{aligned}\tag{6.3}$$

The laboratory frame "diffusivities," $D_{00}(\boldsymbol{k},\omega)$ and $D_{\sigma 0}(\boldsymbol{k},\omega)$, are defined by the formulas

$$D_{00}(\boldsymbol{k},\omega) = L_{00}(\boldsymbol{k},\omega)\left(\frac{\delta\mu_0(\boldsymbol{k})}{\delta n_0(\boldsymbol{k})}\right)_T \tag{6.4}$$

and

$$D_{\sigma 0}(\boldsymbol{k},\omega) = L_{\sigma 0}(\boldsymbol{k},\omega)\left(\frac{\delta\mu_0(\boldsymbol{k})}{\delta n_0(\boldsymbol{k})}\right)_T \tag{6.5}$$

The thermodynamic factor [defined by (2.32)]

$$\left(\frac{\delta\mu_0(\boldsymbol{k})}{\delta n_0(\boldsymbol{k})}\right)_T = k_B T\left[c_0^{-1} - C_{00}(\boldsymbol{k})\right] = \frac{1}{c_0}\left(\frac{\delta p(\boldsymbol{k})}{\delta n_0(\boldsymbol{k})}\right)_T = B_T(\boldsymbol{k})/c_0^2 \tag{6.6}$$

is directly related to the isothermal compressibility, B_T.

Since the present formulation allows us to take into account the temporal and spatial dispersion of the transport coefficients, it is interesting to examine the asymptotic forms of the equations of evolution (6.2) and (6.3) appropriate to the short-time (or high frequency) and long-time (or small frequency) limits.

In the high frequency limit the kinetic coefficients can be approximated by the expressions

$$L_{00}(\boldsymbol{k},\omega\to\infty) = \frac{1}{i\omega}\,(3k_BT)^{-1}\langle\hat{\jmath}_0(\boldsymbol{k})\cdot\hat{\jmath}_0(-\boldsymbol{k})\rangle \equiv \frac{1}{i\omega}\,\ell_{00}(k) \tag{6.7}$$

$$\begin{aligned}L_{0\sigma}(\boldsymbol{k},\omega\to\infty) = L_{\sigma 0}(k,\omega\to\infty) &= \frac{1}{i\omega}\,(3k_BT)^{-1}\langle\hat{\jmath}_0(\boldsymbol{k})\cdot\hat{\jmath}_\sigma(-\boldsymbol{k})\rangle\\ &\equiv \frac{1}{i\omega}\,T\ell_{0\sigma}(\boldsymbol{k})\end{aligned}\tag{6.8}$$

$$L_{\sigma\sigma}(\boldsymbol{k},\omega\to\infty) = \frac{1}{i\omega}\,(3k_BT)^{-1}\langle\hat{\jmath}_\sigma(\boldsymbol{k})\cdot\hat{\jmath}_\sigma(-\boldsymbol{k})\rangle \equiv \frac{1}{i\omega}\,T\,C_v(\boldsymbol{k})\ell_{\sigma\sigma}(k) \tag{6.9}$$

and the equations of evolution become the pair of coupled wave equations:

$$[k^2 w_n^2(k) - \omega^2]\delta\hat{n}_0(k,\omega) + k^2 \ell_{0\sigma}(k)\delta\widehat{T}(k,\omega) = i\omega\ \delta\hat{n}_0(k,t=0) - ik\cdot\hat{\jmath}_0(k,t=0) \tag{6.10}$$

and

$$[k^2 w_\sigma^2(k) - \omega^2]\delta\widehat{T}(k,\omega) + k^2 w_n^2 \frac{T}{C_v(k)} \frac{\ell_{0\sigma}(k)}{\ell_{00}(k)} \delta\hat{n}_0(k,\omega) = i\omega\ \delta\widehat{T}(k,t=0) - ik\cdot\hat{\jmath}_0(k,t=0)/C_v(k) \tag{6.11}$$

The two quantities

$$w_n^2(k) = \ell_{00}(k)\left(\frac{\delta\mu_0(k)}{\delta n_0(k)}\right)_T = \frac{B_T(k)}{m_0 c_0} \tag{6.12}$$

and

$$w_\sigma^2(k) = \ell_{\sigma\sigma}(k) \tag{6.13}$$

appearing in these equations are, respectively, the squares of the propagation velocities for the particle and thermal waves. The former obviously is related to the appearance of the Brillouin doublet characteristic of the high frequency portion of the light scattering spectrum, although our results indicate that temperature fluctuations also should contribute to this part of the spectrum.

The asymptotic behavior of the fluctuations in the low-frequency (quasi-elastic) part of the spectrum is described by the two first order (in the frequency) equations

$$[i\omega + k^2 D(k,\omega=0)]\delta\hat{n}_0(k,\omega) + k^2 L_{0\sigma}(k,\omega=0)/T\ \delta\widehat{T}(k,\omega) = \delta\hat{n}_0(k,t=0) - ik\cdot\hat{\jmath}_0^R(k,\omega) \tag{6.14}$$

$$[i\omega C_v(k) + k^2 L_{\sigma\sigma}(k,\omega=0)/T]\delta\widehat{T}(k,\omega) + k^2 D_{\sigma 0}(k,\omega=0)\delta\hat{n}_0(k,\omega) = C_v(k)\delta\widehat{T}(k,t=0) - ik\cdot\hat{\jmath}_\sigma^R(k,\omega) \tag{6.15}$$

These describe a superposition of two Lorentzian lines. Spatial dispersion (the wave-vector dependence) may still appear for particles that are relatively large.

Since the convolution type constitutive equations are sufficiently general to account both for the finite rate of propagation of fluctuations and their long-time decay, there appears to be no need for introducing generalized Cattaneo-type constitutive relations into the formalism of the thermodynamics of irreversible processes.

7. CONCLUSIONS

The linear functional generalization of Onsager thermodynamics presented here provides a convenient and logically consistent description of transport processes with temporal and spatial dispersions. The theory is intimately related to Schofield's ingenious formulation of thermodynamic fluctuation theory and to the statistical mechanics of inhomogeneous fluids. The theory also can be extended to systems of particles with intrinsic degrees of freedom (Altenberger and Dahler 1987). However, only moderate progress has been made in this direction, a direction which includes liquid crystalline fluids, dipolar particles and polarizable molecules.

In its present formulation the theory is linear. Consequently, the non-equilibrium states whose dynamics one wants to describe must be close to the reference equilibrium state for which the transport coefficients are calculated. There is, however, the distinct possibility of extending this linear description to situations far from equilibrium. If the probability density of the phase-space variables can be found for stationary, non-equilibrium states, then these can be used as reference states for the calculation of transport coefficients. Deviations from the stationary states are then expected to have similar, linear relaxational dynamics providing of course that the stationary non-equilibrium states are stable. The transport coefficients then would be dependent on the parameters characterizing the non-equilibrium stationary states. Similar problems, encountered in connection with the non-Newtonian hydrodynamics of ferromagnetic suspensions recently has been considered by the present authors (Altenberger and Dahler, 1989).

ACKNOWLEDGEMENT

The research reported here was supported by a grant from the National Science Foundation.

REFERENCES

Altenberger, A.R. and Tirrell, M. 1984. Friction coefficients in self-diffusion, velocity sedimentation and mutual diffusion. J. Polym. Sci. (Phys. Ed.) 22: 909–910.

Altenberger, A.R. and Tirrell, M. 1986. Comments on remarks on the mutual diffusion of Brownian particles by N. Yoshide. J. Chem. Phys. 84: 6527–6528.

Altenberger, A.R., Dahler, J.S. and Tirrell, M. 1987a. On the molecular theory of diffusion and heat conduction in multicomponent solutions. J. Chem. Phys. 86: 2909–2921.

Altenberger, A.R., Dahler, J.S. and Tirrell, M. 1987b. A statistical mechanical theory of transport processes in charged particle solutions and electrophoretic fluctuation dynamics. J. Chem. Phys. 86: 4541–4547.

Altenberger, A.R. and Dahler, J.S. 1987. Kinetic theory of a concentrated suspension of axisymmetric solute particles. Molec. Phys. 60: 1015–1036.

Altenberger, A.R. and Dahler, J.S. 1989. Theoretical rheology of suspensions of ferromagnetic rod-like particles. Int. J. of Thermophys. 10: 183–197.

Callen, H.B. 1985. Thermodynamics. New York: Wiley.

Evans, R. 1979. The nature of the liquid-vapour interface and other topics in the statistical mechanics of non-uniform classical liquids. Adv. in Phys. 28: 143–200.

Gyarmati, I. 1970. Non-Equilibrium Thermodynamics. Berlin: Springer.

Gyarmati, I. 1977. On the wave approach of thermodynamics and some problems of nonlinear theories. J. Non-Equilib. Thermod. 2: 233–260.

Keizer, J. 1987. Statistical Thermodynamics of Non-Equilibrium Processes. New York: Springer.

Kops-Werkhoven, M.M., Vrij, A. and Lekkerkerker, H.N.W. 1983. On the relation between diffusion, sedimentation and friction. J. Chem. Phys. 78: 2760–2763.

Miller, D.G., Vitagliano, V. and Sartorio, R. 1986. Some comments on multicomponent diffusion: negative main term diffusion coefficients, second law constraints, solvent choices, and reference frame transformations. J. Phys. Chem. 90: 1509–1519.

Mori, H. 1965. Transport, collective motion and Brownian motion. Progr. Theor. Phys. 33: 423–455.

Morse, P.M. and Feshbach, H. 1953. Methods of Theoretical Physics. Vol. 1, Chap. 8. New York: McGraw-Hill.

Müller, I. and Ruggeri, T. (eds.) 1987. Kinetic Theory and Extended Thermodynamics. I.S.I.M.M. Symposium. Bologna: Pitagora.

Raineri, F.O. and Timmermann, E.O. 1989. A Green-Kubo formula for the sedimentation coefficients. J. Chem. Phys. 91: 3685–3688.
Sandler, S.I. and Dahler, J.S. 1964. Nonstationary diffusion. Phys. Fluids 7: 1743–1746.
Schofield, P. 1966. Wavelength-dependent fluctuations in classical fluids. Proc. Phys. Soc. 88: 149–170.
Schofield, P. 1968. Experimental knowledge of correlation functions in simple liquids. In Physics of Simple Liquids, Chap. 13, eds., H.N.V. Temperley, J.S. Rowlinson and G.S. Rushbrooke. Amsterdam: North Holland.
Sieniutycz, S. 1979. Wave equations for simultaneous heat and mass transfer. Int. J. Heat Mass Transfer 22: 585–599.
Truesdell, C. 1969. Rational Thermodynamics, Chap. 7. New York: McGraw-Hill.
Vojta, G. 1967. Hamiltonian formalisms in the thermodynamic theory of irreversible processes in continuous systems. Acta Chim. Acad. Sci. Hung. 54: 55–64.

Application of Nonequilibrium Thermodynamics to Heat and Mass Transport Properties: Measurement and Prediction in Nonelectrolyte Liquid Mixtures

Richard L. Rowley
Department of Chemical Engineering
Brigham Young University, Provo, UT 84601, USA

ABSTRACT
This review focuses on one of many possible approaches toward understanding, measuring and predicting heat-mass transport coefficients in nonelectrolyte liquid mixtures. The approach is a combined experimental and theoretical examination of diffusion, thermal conductivity, and heat of transport. Nonequilibrium thermodynamics principles have been used in both the development of the theories and the design of the experiments. Mutual diffusion coefficients were measured in a variety of liquid mixtures chosen to elucidate the effects of molecular structural. Molecular dynamics simulations of Lennard-Jones fluids have been found to predict these structural effects, and new simulations on more realistic models are expected to yield more accurate and general results. Experiments near the liquid-liquid critical point have also revealed new information about the critical exponents of diffusion coefficients, heats of transport, and Onsager coefficients. The diffusion thermoeffect, previously thought to be small in liquids, has been successfully measured and developed into an accurate experimental technique for measurement of heats of transport. Furthermore, the method has also been extended to ternary mixtures for which there is currently no other way to measure heats of transport. A new thermal conductivity predictive method which requires no adjustable parameters has been developed and tested against multicomponent data measured in this laboratory. The results of the predictions generally are within the experimental uncertainty. While individual calculation methods have been developed for each of the heat-mass transport properties discussed in this review, they are yet far from the goal of an accurate prediction method for transport properties based on a single set of universal, fundamental constants. However, current development of nonequilibrium molecular dynamics simulations for these properties may produce just such a unified method.

1. INTRODUCTION

To the chemical engineer involved in process development and plant design, accurate and conveniently available thermophysical properties are essential for efficient and cost-effective process designs. Today there is a greater demand for experimental transport data and prediction methods for inclusion in design simulators. Not only have rising energy costs necessitated the need for tighter designs and the acquiescence to nonequilibrium design, but the emergence of new materials, such as biochemical solutions, composites, semiconductors and superconductors, which often require small, batch processes has also increased the demand for accurate transport coefficients. As an illustrative example, consider semiconductor material processing. The size and quality of single-crystal semiconductor material grown from the melt require control of the crystal-melt interface geometry. Hanks (1988) showed that the shape of the growing interface is strongly dependent upon the fluid viscosity and the relative thermal conductivities of the solid and melt. After the crystal is grown, n-p junctions are made by diffusional drives of dopant material. The worth of the device rests upon the quality of the formed junctions and the later etches, processes requiring mutual diffusion coefficient data.

The author's research over the past fifteen years has focused on measurement and prediction of transport properties in liquid mixtures of nonelectrolytes. The experimental approach has focused on diffusion, thermal conductivity and heat-of-transport measurements, either at conditions designed to improve understanding of the molecular transport process or on systems designed to test new predictive equations for transport coefficients. Theoretical efforts have focused primarily on predictive treatments which require only fundamental constants and other, more plentiful, thermophysical properties. Currently molecular dynamics simulations are being developed as a more universal predictive method. As more realistic intermolecular potential models are used in the simulations, accurate prediction of all the transport properties may be obtainable from a single set of pure-component interaction constants.

This review focuses solely on the author's contributions to the measurement and prediction of diffusion coefficients, thermal conductivities and heats of transport in nonelectrolyte liquid mixtures. The underlying principles of nonequilibrium thermodynamics have been useful in developing the experimental techniques used and providing insights into development of predictive methods. Because much of the design for the diffusion and thermal conductivity experiments was developed by other workers and is adequately reviewed elsewhere, only the results of these studies will be reviewed. On the other hand, diffusion thermoeffect experiments used to measure heats of transport were developed by the author and are reviewed here along with the obtained results.

2. HEAT AND MASS TRANSPORT COEFFICIENTS

Unlike empirical formulations of flux-force relations, nonequilibrium thermodynamics explicitly identifies the driving forces that can couple to produce various fluxes. For simplicity, the equations presented here are for binary mixtures, but they shall be extended, as needed, in later sections to multicomponent systems. Symbols used throughout are defined in the nomenclature section at the end of this review. From nonequilibrium thermodynamics, vector heat and mass fluxes can be produced in binary, isotropic, nonelectrolyte, field-free, liquid mixtures by both gradients of temperature and chemical potential. For example, Rowley and Horne (1980) write

$$-\boldsymbol{q} = L_{00}\nabla \ln T + L_{01}\nabla_T(\mu_1 - \mu_2) \tag{2.1}$$

$$-j_1 = L_{10}\nabla \ln T + L_{11}\nabla_T(\mu_1-\mu_2) \ . \tag{2.2}$$

Four heat-mass transport phenomena are apparent from these equations. Isothermal mutual diffusion is the induction of a mass flux of component 1 relative to the center of mass due to an isothermal chemical potential gradient. It is characterized by the Onsager coefficient L_{11}. Thermal conduction is the heat flux produced by a temperature gradient, characterized by L_{00}. The cross transport phenomena are less well known. Thermal diffusion (Soret effect) is a mass flux produced in mixtures by a temperature gradient, characterized by L_{10}, and the diffusion thermoeffect (Dufour effect) is the conjugate phenomenon of a heat flux produced by an isothermal chemical potential gradient, characterized by L_{01}. Only three of these coefficients are independent, however, since Onsager reciprocity requires that the matrix of Onsager coefficients be symmetric when defined in terms of independent fluxes and forces. In this binary example,

$$L_{01} = L_{10} \tag{2.3}$$

The above flux equations can also be written in terms of the transport coefficients more commonly used by experimentalists. While several definitions of the cross transport coefficients exist in the literature, we use here those of Rowley and Horne (1980),

$$\begin{aligned} L_{00} &= kT & L_{01} &= \rho D Q_1^* w_1/\mu_{11} \\ L_{10} &= -\rho D_T & L_{11} &= \rho w_2 D/\mu_{11} \end{aligned} \tag{2.4}$$

where

$$\mu_{11} \equiv \left(\frac{\partial \mu_1}{\partial w_1}\right)_{T,P} \ . \tag{2.5}$$

Elimination of the Onsager coefficients in Equations (2.1) and (2.2) in favor of the transport coefficients yields,

$$-q = k\nabla T + \rho D Q_1^* \nabla w_1 \tag{2.6}$$

$$-j_1 = \rho D \nabla w_1 + \rho D_T \nabla \ln T \ . \tag{2.7}$$

Experiments to measure the four transport coefficients could be designed based solely on Equations (2.6) and (2.7). In such steady-state experiments, one would impose the appropriate constant gradient, measure the steady-state flux and calculate the corresponding transport coefficient. Although some steady state methods are occasionally used, they are usually quite difficult to implement because of the time required to establish the steady state and the propensity for continual decay of the driving force. With the advent of high-speed affordable computers for rapid data acquisition, transient experiments have developed into the more accurate and generally preferred techniques.

Transient measurement methods are based on substitution of Equations (2.6) and (2.7) into the hydrodynamic conservation equations which relate gradients in the local variables of $\boldsymbol{v}$, w_1 and T to their time rate of change. Again following the notation of Rowley and Horne (1980),

$$\left(\frac{\partial \rho}{\partial t}\right) = -\nabla\cdot(\rho \boldsymbol{v}) \tag{2.8}$$

$$\rho\left(\frac{\partial w_1}{\partial t}\right) = -\nabla\cdot \boldsymbol{j}_1 + \rho \boldsymbol{v}\cdot\nabla w_1 \tag{2.9}$$

$$\rho C_P\left(\frac{\partial T}{\partial t}\right) = -\rho C_P \boldsymbol{v}\cdot\nabla T + \Phi_1 - \nabla\cdot \boldsymbol{q} - \boldsymbol{j}_1\cdot\nabla(\bar{H}_1 - \bar{H}_2)\ . \tag{2.10}$$

When the flux equations are substituted into these conservation equations, partial differential equations are obtained which are explicit in the measurable variables ρ, w_1 and T. These equations form the basis of the transient experimental methods employed in the studies reviewed below.

3. MUTUAL DIFFUSION AND THE L_{11} ONSAGER COEFFICIENT

Structural and Compositional Effects

Experimental Study. There is currently no accurate method to predict the composition dependence of the mutual diffusion coefficient in binary liquid mixtures. Empirical correlations such as those by Darken (1948) and Vignes (1966) generally correlate the kinetic portion of the mutual diffusion coefficient in terms of intradiffusion coefficients. For example, Darken proposed,

$$D = (x_1 D_2 + x_2 D_1)\, Q\ . \tag{3.1}$$

The relationship between the various diffusion coefficients is shown in Figure 1. Note that the mutual diffusion coefficient at infinite dilution is equal to the intradiffusion coefficient at infinite dilution. However, the self diffusion coefficient is the pure component limit of the intradiffusion coefficient and has no known relationship to the mutual diffusion coefficient. While correlations based on D_i provide reasonably good estimates for D in ideal mixtures, they often fail in nonideal mixtures.

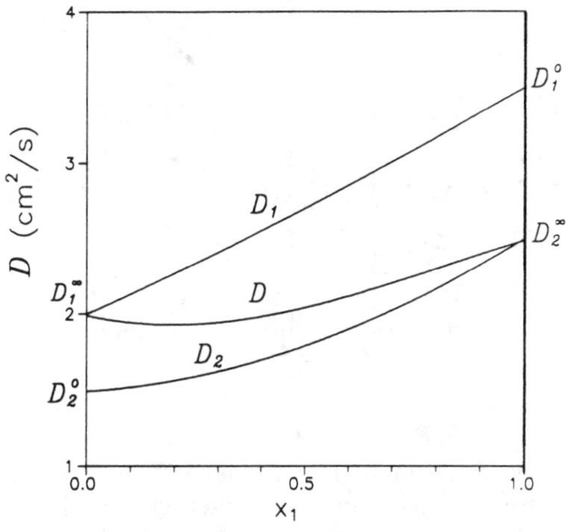

Figure 1. Schematic showing relationship of intra-, self and mutual diffusion coefficients.

In order to better understand the compositional dependence of D, a systematic study of both compositional and structural effects was undertaken by

Rowley *et al.* (1987 and 1988a). First, measurements were made of D over the entire composition range in the fairly ideal mixtures of *n*-hexane, *n*-heptane, *n*-octane, 2-methylpentane, 2,3-dimethylpentane and 2,2,4-trimethylpentane in carbon tetrachloride. Similar measurements were then made in substantially nonideal mixtures containing the same alkanes in chloroform.

The diffusion apparatus used for these measurements is a version of the Taylor (1953) dispersion method. The theory and use of this method, design constraints, and correction terms have been well established by Aris (1956), Wakeham (1981) and Matos Lopes and Nieto de Castro (1986). The apparatus itself is adequately described by Rowley *et al.* (1987 and 1988a). Because the concentration differences used in this apparatus are so small, the flowing pulse reference frame used in the experiment is equivalent to the center of mass reference frame in Equations (2.6) and (2.7).

Figure 2 shows the results obtained for the study of alkanes in carbon tetrachloride; similar results for the chloroform mixtures are shown in Figure 3. The mutual diffusion coefficient decreases with increasing chain length for straight-chain molecules. Likewise, the addition of more methyl groups onto the pentane chain decreases the mutual diffusivity. A comparison between pairs of alkanes having the same molecular weight shows that the more branched alkane has a smaller value of D in chloroform than the straight chain. It appears from these results that the infinite dilution value of the intradiffusion coefficient of chloroform in the alkane solvent differs with molecular weight, but is apparently the same for both branched- and straight-chain molecules. On the other hand, the intradiffusion coefficient of alkane in nearly pure chloroform decreases regularly with the solute's molecular weight and its branched character.

Molecular-Dynamics (MD) Simulations. In an effort to develop a predictive technique for mutual diffusion coefficients which can correctly predict the compositional and structural

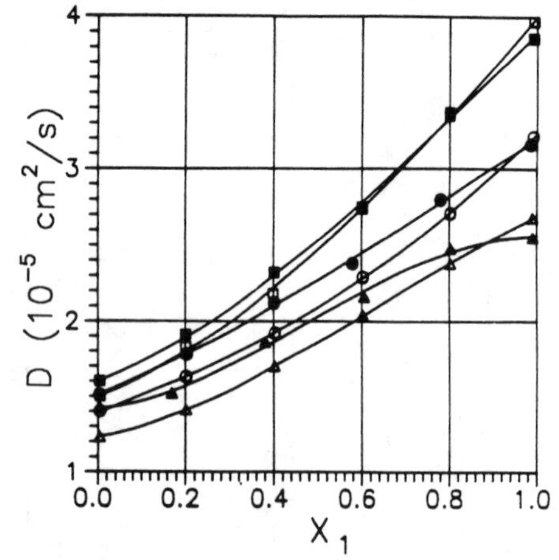

Figure 2. Experimental points and smoothed curves for mutual diffusion coefficients as a function of alkane mole fraction in CCl_4 at 30°C: *n*-hexane (■), *n*-heptane (•), *n*-octane (▲), 3-methylpentane (□), 2,3-dimethylpentane (○), and 2,2,4-trimethylpentane (△).

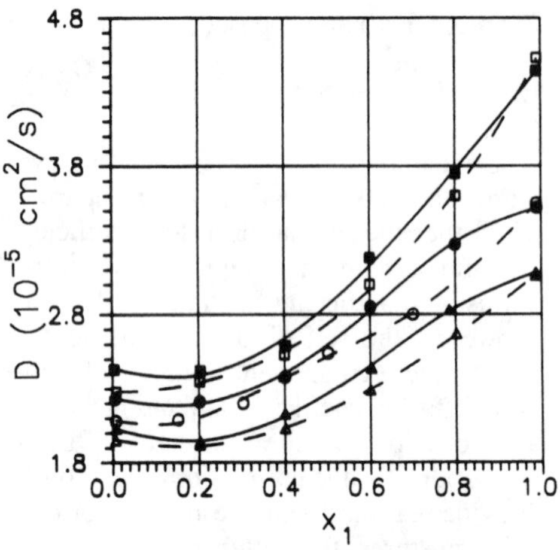

Figure 3. Experimental points and smoothed curves for mutual diffusion coefficients as a function of alkane mole fraction in $CHCl_3$ at 30°C: *n*-hexane (■), *n*-heptane (•), *n*-octane (▲), 3-methylpentane (□), 2,3-dimethylpentane (○), and 2,2,4-trimethylpentane (△).

effects in both ideal and nonideal mixtures, Stoker and Rowley (1989) and Rowley *et al.* (1990) modeled the above mixtures as Lennard-Jones (LJ) fluids and obtained values for D from MD simulations. The pure-component LJ values for ϵ were obtained from the literature. Diffusion coefficients were found to be very sensitive to the value of σ used in the simulation so σ values were regressed by comparing simulated values of self diffusion coefficients to experimental. Self diffusion and intradiffusion coefficients were calculated from the single-particle velocity correlation function

$$D_i = \frac{1}{3}\int_{t_o}^{\infty}\langle \mathbf{v}_i(t)\cdot \mathbf{v}_i(t_o+t)\rangle\, dt\ . \tag{3.2}$$

Mutual diffusion coefficients were obtained from integration of the collective time correlation function,

$$D_{ij} = \frac{Q}{3Nc_ic_j}\int_{t_o}^{\infty}\langle \mathbf{J}_{ij}(t_o)\cdot \mathbf{J}_{ij}(t_o+t)\rangle\, dt \tag{3.3}$$

where

$$\mathbf{J}_{ij}(t) = c_j\sum_{k=1}^{N_i}\mathbf{v}_k(t) - c_i\sum_{m=1}^{N_j}\mathbf{v}_m(t) \tag{3.4}$$

It should be noted that care must be used in calculating mutual diffusion coefficients from MD simulations because of the long-time tail of the correlation function which can contribute to the above integral even after it appears that all correlations have decayed. Additionally, the collective nature of mutual diffusion expressed in Equation (3.3) permits only one contribution to the correlation average per time step, while a contribution to the correlation average can be made for each particle at each time step using Equation (3.2) for the intradiffusion coefficients. Thus, simulations of mutual diffusion coefficients must include at least a factor of N additional independent time origins in evaluating the correlation function to achieve accuracy comparable to that of intradiffusion simulations.

Stoker and Rowley found that the Lorentz-Berthelot (LB) combining rule for σ was inadequate to reproduce the observed composition dependence of the mutual diffusion coefficient. Nevertheless, they found that a fixed, weighted combining rule could be used for all of the alkane + CCl_4 mixtures that they studied. A comparison of their predictions for the straight-chain alkanes in CCl_4 is shown in Figure 4. Their MD simulations indicate that mutual diffusion coefficients can be predicted reasonably well from LJ fluid simulations if the appropriate combining rule is used. While other types of mixtures will require a different cross combining rule, Rowley *et al.* (1990) did find that the same combining rule could be used for the much more nonideal mixtures of alkanes in chloroform as shown in Figure 5. Thus, the MD simulation method can be used to predict D for mixtures in a particular class or to correlate the composition dependence of D from a single experimental value. Failure of the LB combining rule is attributed to the fact that alkane molecules are hardly spherical. New MD simulations using LJ potentials at each methyl or methylene site for the alkanes are in progress. It is hoped that the LB combining rules can be used for such site-site models to make the method totally predictive.

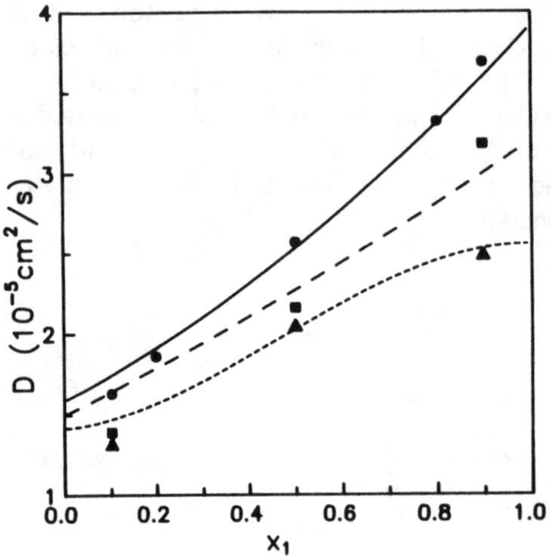

Figure 4. Comparison of MD simulated (points) and measured (lines) diffusion coefficients for n-alkane + CCl_4 mixtures at 30°C: n-hexane (•), n-heptane (■) and n-octane (▲).

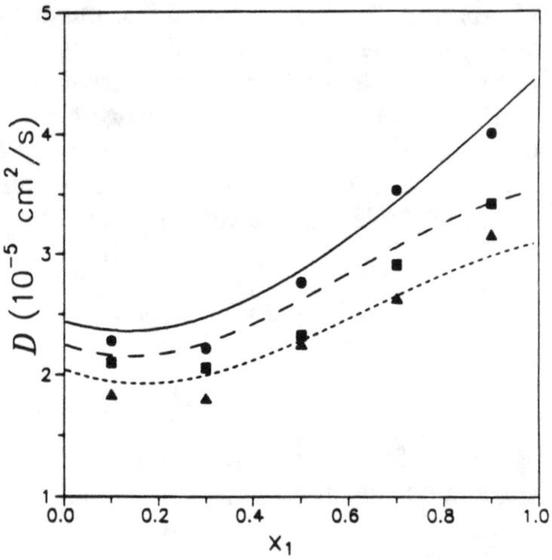

Figure 5. Comparison of MD simulated (points) and measured (lines) diffusion coefficients for n-alkane + $CHCl_3$ mixtures at 30°C: n-hexane (•), n-heptane (■) and n-octane (▲).

Rowley *et al.* (1990) used their simulations to test the validity of equations such as (3.1). If Equation (3.3) is expanded and written as

$$D = \left[x_1 D_2 + x_2 D_1 + kT \left(\frac{x_2}{m_1} + \frac{x_1}{m_2} \right) \int_0^\infty \Upsilon(t)\mathrm{d}t \right] Q \ , \tag{3.5}$$

where $\Upsilon(t)$ is a sum of all cross-correlations of the type $<v_i(t) \cdot v_j(t)>$ for $i \neq j$, then it becomes apparent that Equation (3.1) is valid only when cross correlations are small. Rowley *et al.* found that the cross correlation terms are substantial, particularly in the nonideal mixtures of alkanes in chloroform and that Equation (3.1) is in general not realistic. In fact, the kinetic portion of the mutual diffusion coefficient (sans Q) obtained from the simulations was often outside the range of values bracketed by the simulated intradiffusion coefficients, as illustrated in Figure 6 for mixtures of n-octane and chloroform.

Diffusion Near Mixture Liquid-Liquid Critical Points

Partially miscible liquids may exhibit a region in the composition-temperature domain in which two liquid phases are in equilibrium. The location of the binodal (or coexistence) curve, the locus of coexisting compositions at each temperature, is governed by the phase equilibrium requirement $\mu_1' = \mu_1''$. The binodal curve for methanol + n-hexane measured by Clark and Rowley (1986) is shown in Figure 7. For this system, the binodal curve terminates at an upper critical solution temperature or consolute point. The spinodal curve, or mathematical dividing line between stable and unstable states, is located at $\mu_{11} = 0$. The spinodal curve lies everywhere within the binodal curve except at the critical point where it is coincident.

Diffusion near the upper critical solution temperature is of particular interest because of the dependence of D on both μ_{11} and L_{11} shown in Equation (2.4). It is well known that μ_{11} vanishes with a critical exponent of about $+4/3$ as the critical temperature is approached from above along the critical composition line. Thus, the diffusion coefficient should be identically zero at all points on the spinodal curve including the critical point. Indeed the spinodal line can be thought of as the demarkation between positive diffusion coefficients (propensity to eliminate concentration differences) and negative diffusion coefficients (since the solution demixes in the unstable region). Interestingly, the diffusion coefficient has only a $+2/3$ critical exponent. This implies that the Onsager coefficient L_{11} must diverge in the critical region with a $-2/3$ exponent.

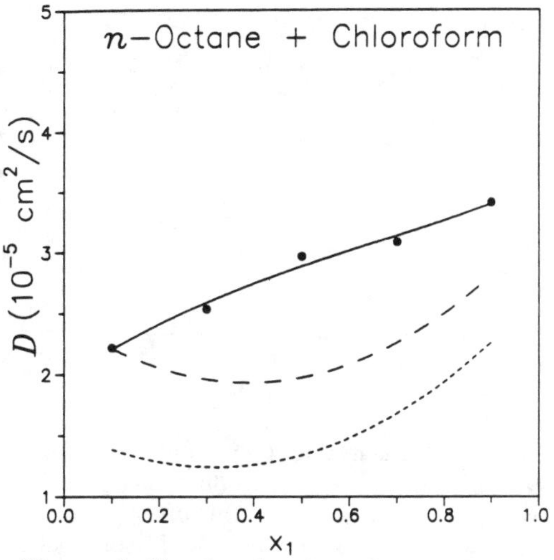

Figure 6. Simulated values for the kinetic portion of D (•,——), D_1 (- - -) and D_2 (—— ——) for *n*-octane in chloroform at 30°C.

Binary Measurements. Clark and Rowley (1986) devised a clever method for studying diffusion coefficients in the vicinity of a liquid-liquid critical point. Their initial sharp concentration gradient was created by the phase split of a partially miscible mixture prepared at the critical composition and equilibrated at a temperature below T_c. The initial compositions were determined from the known initial temperature and the coexistence curve which had been previously measured as a function of temperature. The diffusion process was then initiated by rapidly (relative to the diffusion process) jumping the temperature of the mixture above T_c. Clark and Rowley used a Gouy interferometer to monitor the transient compositions following the temperature jump and then regressed the diffusion coefficient from the timed interferograms of the decaying composition gradient. A particular advantage of the temperature-jump method is the ability to repeat measurements on the same critical mixture by simply re-equilibrating the fluid at $T < T_c$. Clark and Rowley used this method to measure D as a function of distance from the critical temperature, $(T - T_c)/T_c$, along the line of constant (critical) composition for the methanol + *n*-hexane system. Contrary to a previously reported value by Skripov *et al.* (1980), the critical exponent of D for this system was found to be 0.685; i.e.,

$$D \sim \left(\frac{T-T_c}{T_c}\right)^{0.685}, \tag{3.6}$$

which is in excellent agreement with theory (Sengers, 1973) and experimental results on other systems.

Clark and Rowley also measured the composition dependence of D near T_c in methanol + *n*-hexane mixtures. For these measurements, a boundary sharpening or Tiselius-type diffusion cell replaced the temperature-jump apparatus. From their measurements of D as a function of both composition and temperature in the critical region, Clark and Rowley

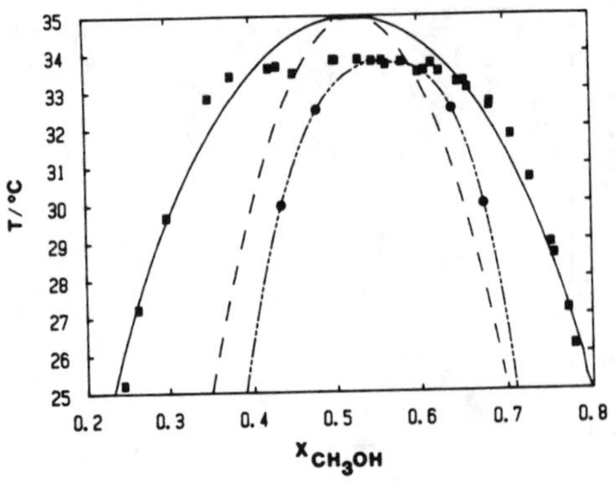

Figure 7. Measured binodal points (■), their NRTL fit (———), spinodal points extrapolated from the diffusion study (•, — - - —), and the NRTL spinodal prediction (— — —) for methanol + *n*-hexane mixtures.

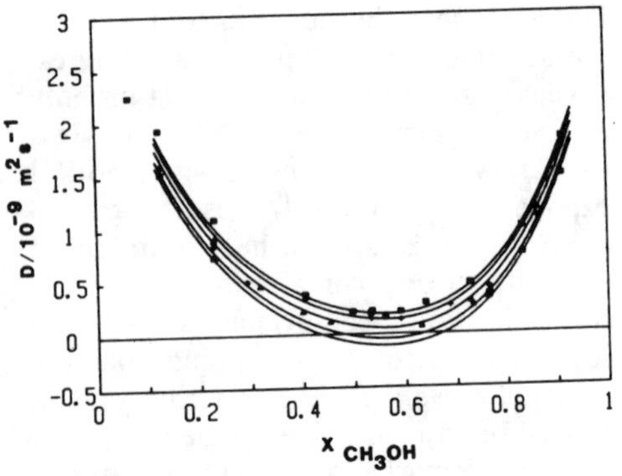

Figure 8. Determination of spinodal curve from the smoothed composition dependence of diffusion coefficients measured at 40°C (□), 37.5°C (+), 34.6°C (▲), 32.5°C (♦) and 30°C (■).

were able to find smooth polynomial descriptions for the composition dependence of D along each isotherm. Measurements were made in the one-phase region above T_c and to a limited extent for $T < T_c$. These latter measurements were particularly interesting because extrapolation of D as a function of composition to the points where $D = 0$ provided direct experimental determination of the spinodal line as illustrated in Figure 8. Spinodal loci cannot be measured by equilibrium measurements because of experimental difficulties in the metastable region which separates the binodal and spinodal curves. In this region, one can never be sure that density, composition and temperature fluctuations have been damped enough to approach the spinodal adequately closely. On the other hand, the method developed by Clark and Rowley uses transport measurements to establish the location of the instability line.

By careful analysis of the composition and temperature dependence of D in the homogeneous one-phase region, Clark and Rowley were able to extrapolate, with reasonable accuracy, to the stability limit where $D = 0$. The spinodal points located with this procedure are shown in Figure 7 as solid circles; the dotted-dashed line represents the spinodal curve connecting these points with the critical point. Prior to this study, the consistency of binodal and spinodal loci predicted by popular activity coefficient models (see Reid *et al.*, 1987) could not be tested. Models such as UNIQUAC and NRTL contain at least two adjustable parameters. To model liquid-liquid equilibria, the parameters are generally regressed from experimental data on the equilibrium compositions. One was forced to simply assume for lack of additional data that the composition dependence was correct and that activity coefficients at other compositions were then accurately predicted. If the binodal data of Figure 7 is used to obtain NRTL parameters, the resultant spinodal curve predicted by the model is shown by the dashed line. The disagreement observed by Clark and Rowley between the measured spinodal curve and the values calculated from the NRTL model indicates that the composition dependence incorporated in such empirical models may be incapable of correctly correlating both the spinodal and the binodal curves of liquid mixtures with the same set of parameters.

Ternary Measurements. In a separate study, Clark and Rowley (1985) were able to generalize their temperature-jump technique to measurement of diffusion coefficients near plait points in ternary mixtures. Extension of Equation (2.2) to ternary mixtures yields four independent mutual diffusion coefficients. In terms of the derivatives of chemical potential with respect to composition, the conditions for stability of a single phase are in ternary mixtures are

$$\mu_{11}\mu_{22} - \mu_{12}\mu_{21} \geq 0; \qquad \mu_{11} > 0; \qquad \mu_{22} > 0 \tag{3.7}$$

where the equality defines the spinodal curve. Kirkaldy and Purdy (1969) were able to show from Equation (3.7) that the corresponding stability restrictions on the four independent diffusion coefficients are

$$D_{11} + D_{22} > 0; \; D_{11}D_{22} - D_{12}D_{21} > 0; \; (D_{11} + D_{22})^2 > 4(D_{11}D_{22} - D_{12}D_{21}) \; . \tag{3.8}$$

Note that there is no restriction on the sign of any individual diffusion coefficient.

Clark and Rowley (1985) examined the critical exponents and interrelationship of the four mutual diffusion coefficients near the 30°C plait point of water + 2-propanol + cyclohexane mixtures. The four independent diffusion coefficients were all found to rapidly decrease as the consolute or plait temperature was approached from above along the constant plait composition line. Effective critical exponents were determined and found to be +0.55 for the individual diffusion coefficients and about twice that, +1.31, for the determinant of the diffusion coefficient matrix. Although all diffusion coefficients apparently go to zero at the critical point, the coupling between the cross diffusion coefficients is very large in the near-critical region. Clark and Rowley found that the cross diffusion coefficients were nearly as large as the main diffusion terms. This coupling produces interesting effects. For example, even though all of the stability requirements of Equation (3.8) were satisfied by the values of D_{ij} measured by Clark and Rowley, a negative value was obtained for the water *main* diffusion coefficient. This indicates that the coupling between the driving forces of the various components is so large that water actually diffuses against its *composition* (not chemical potential) gradient. This large coupling occurs because of the large dependence of the chemical potential and its gradients on the ternary composition near the plait point. The large nonidealities of the mixture that cause phase separation also produce the significant coupling between the driving forces.

4. THERMAL CONDUCTIVITY AND THE L_{00} ONSAGER COEFFICIENT

Experimental

Thermal conductivity measurements in liquid mixtures have primarily focused on two-component systems. In fact, thermal conductivity correlations have often been of a closed form, valid only for binary mixtures. Moreover, lack of multicomponent experimental data has inhibited testing of those few correlations with forms amenable to multicomponent mixtures. As part of an overall effort to develop a multicomponent prediction method which requires no adjustable parameters to calculate liquid mixture thermal conductivity, systematic measurements of thermal conductivity in ternary liquid mixtures and their constituent binaries were initiated. Two new thermal conductivity apparatuses, based on the well-established principles of the transient hot-wire method, were designed for this purpose. The equipment has been adequately described by Rowley and White (1987), Rowley *et al.* (1987, 1988a and 1988b) and Rowley and Gubler (1988). The measurements by Rowley and White (1987) on six ternary mixtures at 25°C and by Rowley *et al.* (1988b) on eighteen ternary mixtures at 25°C covered the entire ternary composition range, including the constituent binaries. The purpose of these studies was to test the composition dependence of a new thermodynamic model for k and the appropriateness of the model's assumption that only pair interactions are needed for ternary predictions. Later, Rowley and Gubler (1988) measured the thermal conductivity of seven ternary mixtures at two different temperatures to test the temperature dependence of the new model.

A Local Composition Model for Multicomponent Mixtures

Most mixture thermal conductivity estimation methods are correlations of pure component thermal conductivities (Reid *et al.* 1987). Deviation from an "ideal" composition average constitutes an excess thermal conductivity. Most correlations attempt to correlate this excess in terms of the pure component k values and fixed parameters. Rowley (1982) suggested that fixed relationships involving only pure k values do not contain the flexibility needed to model the wide variety of mixture interactions observed in nature. A fixed form of the correlation is destined to only model an average type of mixture. Authors of these correlations suggest that a constant in the model be used as an adjustable parameter to gain mixture specificity, but this requires experimental data and the model is no longer predictive. Rowley reasoned that mixture specificity could be incorporated by inclusion of information available from activity coefficient models about the mixture's nonidealities. He attributed deviations from an "ideal" mass fraction average value to nonrandom mixing; i.e., deviations of local from bulk compositions. Local compositions can be obtained from standard local composition models such as NRTL (Reid *et al.*, 1987). Parameters for these models have been obtained for a wide variety of fluids from vapor-liquid equilibrium data and are readily available in the literature. Thus, the concept is to use equilibrium information to calculate the local compositions. These can in turn be used to obtain the excess thermal conductivity. In this way, the method adjusts itself for each system through the binary interaction information included in the local composition parameters, but the method is still predictive since mixture thermal conductivity data are not used.

Using these ideas, Rowley (1982) showed that

$$k = \sum_{i=1}^{n} w_i k_i + \sum_{i=1}^{n}\sum_{j=1}^{n} w_i w_j G_{ji}(k_{ji}-k_i) \Bigg/ \left(\sum_{l=1}^{n} w_l G_{li}\right), \tag{4.1}$$

where

$$k_{ji} = \frac{w_i^* w_{ii}^* k_i + w_j^* w_{jj}^* k_j}{w_i^* w_{ii}^* + w_j^* w_{jj}^*}, \tag{4.2}$$

$$w_{ii}^* = \frac{w_i^*}{\sum_{k=1}^{n} w_k^* G_{ki}}, \tag{4.3}$$

$$w_i^* = \frac{M_i\sqrt{G_{ji}}}{M_i\sqrt{G_{ji}} + M_j\sqrt{G_{ij}}}, \tag{4.4}$$

and G_{ij} are nonrandom mixing parameters obtained from equilibrium information. While the method can be used for other models, the above equations were derived from the NRTL model. Thus, G_{ij} are obtained from tabulated values (the DECHEMA VLE and LLE Data Collection volumes are an excellent source) of the NRTL constants α, A_{ij} and A_{ji} using

$$G_{ij} = \exp\left(-\frac{\alpha A_{ij}}{RT}\right) . \tag{4.5}$$

To use the method, NRTL constants are obtained from a literature source, such as DECHEMA Data Collection, or regressed from vapor-liquid equilibrium data. Equation (4.5) is used to obtain G_{ij} values. These are then used with Equations (4.2) through (4.4) to obtain thermal conductivity cross interactions, k_{ij}. The k_{ij} and G_{ij} values determine the deviation from the ideal mass fraction average value as shown in Equation (4.1).

Rowley (1982) compared experimental data to the predictions of the local composition method for eighteen different binary mixtures at various temperatures and over the entire composition range. He found an average deviation from experiment of about 1.0%. Similar tests by Rowley and White (1987), Rowley *et al.* (1988b) and Rowley and Gubler (1988) have shown the model to be as efficacious for ternary mixtures. Generally the percent deviation between predicted and experimental values is within the experimental uncertainty.

Wei and Rowley (1985) extended this local composition model to liquid mixture shear viscosity. This permits prediction of both thermal conductivity and shear viscosity in liquid mixtures from the same NRTL constants. The shear viscosity model uses volume fractions instead of weight fractions, but the final equations are similar in form to the equations for k. The viscosity equation corresponding to Equation (4.1) is

$$\xi = \sum_{i=1}^{n} \phi_i \xi_i + \sum_{i=1}^{n} \left[\frac{\sum_{j=1}^{n} \phi_j G_{ji} (\xi_{ji} - \xi_i)}{\sum_{l=1}^{n} \phi_l G_{li}} \right] + \sigma \frac{H^E}{RT} , \tag{4.6}$$

where

$$\xi \equiv \ln(\eta V) , \tag{4.7}$$

H^E is the heat of mixing at the desired composition, and σ is a transition effectiveness parameter usually taken to be equal to 0.25. As in the thermal conductivity model,

$$\xi_{21} = \frac{\phi_1^* \phi_{11}^* \xi_1 + \phi_2^* \phi_{22}^* \xi_2}{\phi_1^* \phi_{11}^* + \phi_2^* \phi_{22}^*} , \tag{4.8}$$

$$\phi_{ii}^* = \frac{\phi_i^*}{\sum_{k=1}^{n} \phi_i^* G_{ki}} , \tag{4.9}$$

and

$$\phi_i^* = \frac{V_1\sqrt{G_{21}}\ \exp(\xi_2/2)}{V_1\sqrt{G_{21}}\ \exp(\xi_2/2) + V_2\sqrt{G_{12}}\ \exp(\xi_1/2)} . \tag{4.10}$$

Comparisons of predicted and experimental viscosity data by Wei and Rowley (1983a, 1983b and 1985) indicate that the model performs quite well for a wide range of mixtures covering a diverse range of mixture viscosity behavior. An example of the method's predictive capability is shown in Figure 9.

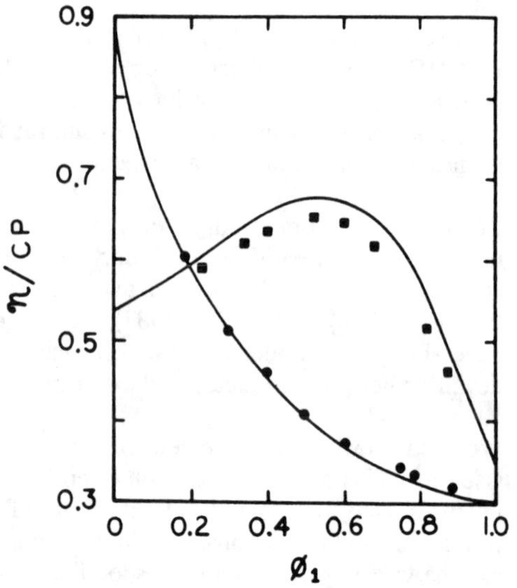

Figure 9. Predicted (———) and experimental viscosity for acetone + cyclohexane and triethylamine + chloroform mixtures at 25°C.

5. DIFFUSION THERMOEFFECT, L_{01}, AND THERMAL DIFFUSION, L_{10}, COEFFICIENTS

Measurement Methods

The diffusion thermoeffect or Dufour effect is the heat flux produced by an isothermal chemical potential gradient, the second term in Equations (2.1) and (2.6). A quantitative measure of the heat flux produced under isothermal conditions is given by the heat of transport formally defined as

$$Q_i^* \equiv \left(\frac{q}{j_1}\right)_{\Delta T=0} \tag{5.1}$$

The inverse effect, thermal diffusion, is a diffusional flux induced by a temperature gradient in a system of initially uniform chemical potential. Because Onsager reciprocity guarantees that $L_{10} = L_{01}$ in an isotropic system, only one of these coefficients needs to be measured, unless redundancy is desired as a consistency check. The focus here is on the diffusion thermoeffect.

Prior to the late 1970's, the diffusion thermoeffect had been measured only in gases. The effect in liquids was long thought to be too small to measure quantitatively because of the relatively larger thermal conductivity and heat of mixing in liquid mixtures. Rastogi and Madan (1965) did observe nonuniform temperatures in a diffusion mixture, but because their cell design did not conform to a geometry tractable to a solution of the energy equation, proper analysis of the experiment could not be made. Ingle and Horne (1973) solved the coupled, nonlinear, nonhomogeneous partial differential equations which describe the diffusion thermoeffect in a one-dimensional cell with adiabatic ends and found that the induced temperature differences in liquids, although small, should indeed be measurable. Their analysis indicated that the largest transient temperature responses would be generated with an initially sharp interface between two phases. Additionally, the initial composition of the two phases should be as disparate as possible in order to maximize the incipient temperature response.

Based on those results, Rowley and Horne (1980) designed the diffusion thermoeffect cell shown in Figure 10. An initially sharp interface between two mixtures of different composition is created by withdrawing an immiscible fluid of intermediate density from between the two layers. Rowley and Horne made measurements on carbon tetrachloride and cyclohexane mixtures using water as the withdrawable liquid gate due to its insolubility in either component and its intermediate density. Initially the cell was thermostatted at the desired temperature and the measurement cell (C) was filled with the more dense, carbon tetrachloride-rich mixture up to the bottom of reservoir (A). Reservoir (A) was then filled with water, and the glass syringe (J) was filled with the cyclohexane-rich, upper phase. Water was then slowly siphoned from the cell through the withdrawal ports (F) in order to gradually lower the upper phase into reservoir (A) and then into the measurement cell (C). When contact between the upper and lower phases occurred, any residual water around the edge of the cell (due to the meniscus) was pulled into the equatorial ring (E) by the wetting action of the water on the glass. This capturing ring prevented residual water from restricting the diffusional cross-sectional area. Once contact between the two phases occurred, the external jacket around cell (B) was evacuated via vacuum connection (H) in order to impose adiabatic wall conditions. The transient temperature response was measured with a bank of thermocouples positioned above and below the interface. Thermocouple leads (K) passed through a ground-glass fitting and were connected to a measuring potentiometer and reference junctions in an ice-point bath.

The heat of transport was determined from measurements of the temperature difference, ΔT, between upper and lower thermocouples located equidistant from the initial interface. The analysis by Ingle and Horne showed that heat of mixing effects are primarily symmetric about the interface while those due to the heat of transport are mainly antisymetric. Therefore, measurement of ΔT maximizes the response due to the diffusion thermoeffect relative to that of the heat of mixing. The

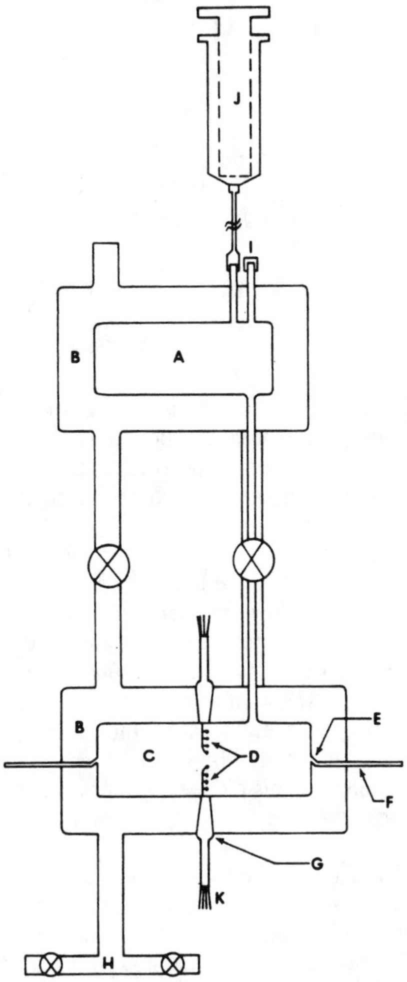

Figure 10. Withdrawable liquid gate diffusion thermoeffect cell.

heat of transport was obtained by adjusting Q_1^* until the best agreement, in the least squares sense, was found between calculated and measured $\Delta T(t)$ profiles.

The boundary value problem obtained from the equations of change consistent with this experimental apparatus can be found by substituting the flux expressions of Equations (2.6) and (2.7) into the component continuity and energy conservation equations, Equations (2.9) and (2.10), and simplifying for one-dimensional geometry. Rowley and Horne (1980) were further able to simplify the resulting equations by showing that the explicit center of mass velocity terms are small except at very short times, that bulk flow terms are small, and that the thermal diffusion term can be neglected due to the small temperature differences produced. The final equations are coupled partial differential equations that describe the temporal and spatial profiles of the mixture's composition and temperature:

$$\rho\left(\frac{\partial w_1}{\partial t}\right) = \left\{\frac{\partial[\rho D(\partial w_1/\partial z)]}{\partial z}\right\} \tag{5.2}$$

and

$$\begin{aligned}\rho C_p\left(\frac{\partial T}{\partial t}\right) = \left\{\frac{\partial[k(\partial T/\partial z)]}{\partial z}\right\} + \left\{\frac{\partial[\rho D Q_1^*(\partial w_1/\partial z)]}{\partial z}\right\} \\ + \frac{\rho D M^3}{(M_1 M_2)^2}\left(\frac{\partial^2 H^E}{\partial x_1^2}\right)\left(\frac{\partial w_1}{\partial z}\right)^2\end{aligned} \tag{5.3}$$

Rowley and Horne used a numerical technique to solve these equations subject to appropriate initial and boundary conditions. Numerical solutions showing the temporal and spatial features of the composition and temperature surfaces shown in Figures 11 and 12 illustrate the diffusion thermoeffect response observed in experiments.

The measurements performed by Rowley and Horne on carbon tetrachloride - cyclohexane mixtures constituted the first accurate measurements of the diffusion thermoeffect in liquid mixtures. They obtained Q_i^* by adjusting its value in the numerical solution until agreement was obtained between calculated and measured $\Delta T(t)$ profiles. The resultant values of Q_i^* also provided the first verification of Onsager reciprocity for L_{01} and L_{10} (Rowley and Horne, 1978). They converted measured Q_i^* values to thermal diffusion coefficients, D_T, and compared them to literature values measured via thermal diffusion techniques. The agreement found between diffusion thermoeffect and thermal diffusion measurements was excellent. This pioneering work established the diffusion thermoeffect as a reliable means for measuring Q_i^* or D_T values in liquid mixtures.

Unfortunately, only a few mixtures can be studied with the original diffusion thermoeffect cell design. The cell is very restrictive because a fluid gate must be used which is both insoluble in both components and intermediate in density between the two starting layers. To study additional systems, Platt *et al.* (1982b) and Rowley and Hall (1986) designed a new cell which created the initial boundary using a boundary-sharpening technique similar to that employed by Tiselius diffusion cells. The cell, shown schematically in Figure 13, was thermostatted in an air bath. The cell body (C) was machined from a cylindrical rod of

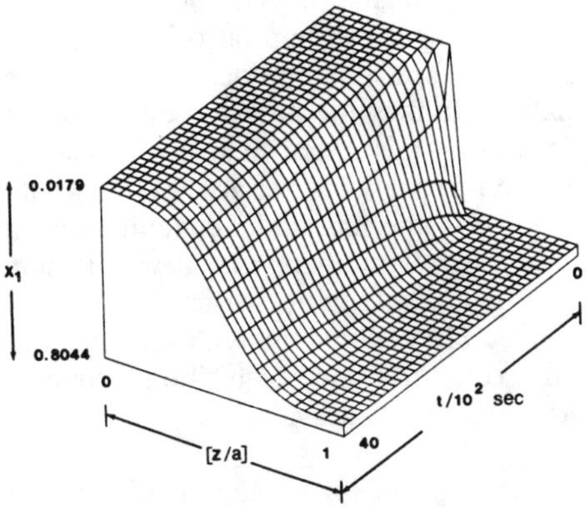

Figure 11. Composition surface for CCl_4 + cyclohexane at 23°C with intially x_1 = 0.0179 and x_1 = 0.8044 in the upper and lower phases, respectively.

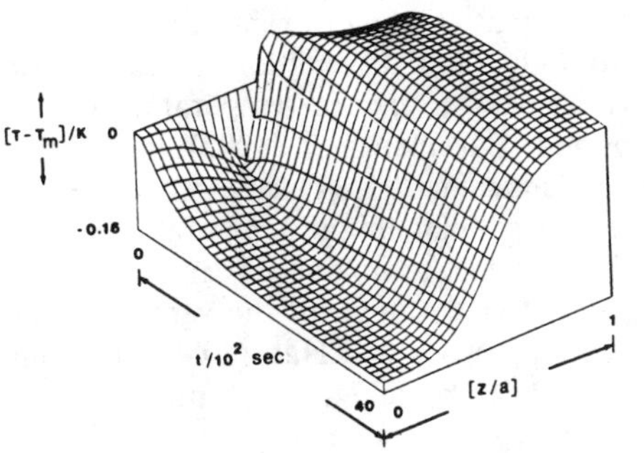

Figure 12. Temperature surface for CCl_4 + cyclohexane at 23°C with intially x_1 = 0.0179 and x_1 = 0.8044 in the upper and lower phases, respectively.

teflon. End caps (D) sealed the cell by compression of O-rings (F) when brass securing rings (H) were tightened. In lieu of the vacuum jacket in the original cell, adiabatic boundary conditions were naturally maintained by low heat transfer rates through the end caps due both to the thickness and low thermal conductivity of the teflon and to the very small temperature differences between the test fluid and the outside air bath. Thermocouple leads were threaded through a thin stainless steel tube imbedded in a groove in the inner wall of the cell. The thermocouple sheaths were bent at right angles in order to extend the thermocouple measurement junctions laterally to the cell center. The other ends of the thermocouple sheaths extended through septum (G) and were attached to micrometer heads which could be used to accurately position the thermocouple leads at desired distances from the interface.

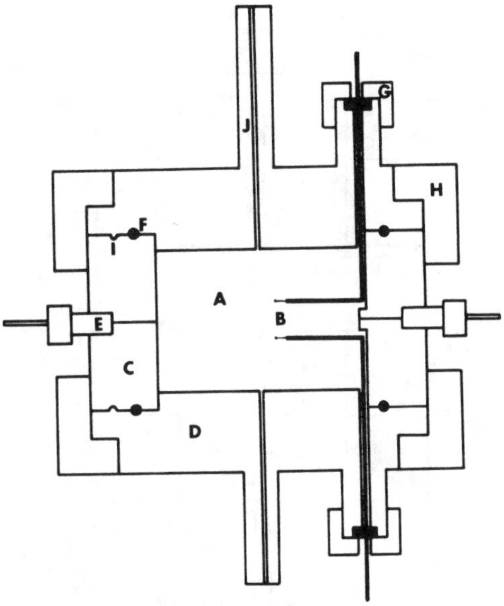

Figure 13. Schematic diagram of boundary-sharpening diffusion thermoeffect cell.

In operation, the cell was first filled from the bottom with the lower, more dense fluid past a three-way stopcock located above the filling port (J). The filled cell and the upper layer contained in a separate reservoir were then thermostatted at the desired experimental temperature. The stopcock was turned to form an initial, rough interface which was gradually lowered by siphoning fluid through equatorial ports (E). Siphoning at rates faster than the dispersion process kept the interface sharp until the diffusion thermoeffect experiment was begun by shutting off the pumps. Heat transfer is more rapid than mass transfer and so this technique does not guarantee isothermal initial conditions during interface formation. In fact, the diffusion thermoeffect is expected to induce temperature nonuniformities during this initial interface formation period. But, Platt *et al.* (1982) showed that measured $\Delta T(t)$ profiles after about 800 s would be identical to those calculated mathematically based on isothermal initial conditions, even for substantial initial temperature nonuniformities. In effect then, this new diffusion thermoeffect technique trades the capability of making measurements within the first 800 s of the experiment for the versatility and convenience. In conjunction with this new cell design, Rowley and Hall (1986) also implemented a computer data acquisition system to automatically make temperature measurements at preset time intervals.

Using this new apparatus, Rowley and Hall measured Q_1^* in binary liquid mixtures of toluene, chlorobenzene and bromobenzene, and calculated and L_{01} using thermodynamic factors obtained from the literature. These mixtures were nearly ideal in the thermodynamic sense, and the thermodynamic factor was very nearly unity. Yi and Rowley (1987a) studied Q_1^* in nonideal binary liquid mixtures. Measurements were made on binary mixtures of carbon tetrachloride with benzene, toluene, 2-propanone, *n*-hexane and *n*-octane, from which both Q_1^* and L_{01} were obtained. Rowley *et al.* (1987 and 1988a) also used diffusion thermoeffect measurements to investigate the effect of structure on the heat of transport. They studied the systems *n*-hexane, *n*-heptane and *n*-octane in both carbon tetrachloride and chloroform to determine the effects of chain-length and molecular weight on Q_1^* within a homologous series. They also studied the effect of chain-branching by performing similar measurements on mixtures of 3-methylpentane, 2,3-dimethylpentane and 2,2,4-trimethylpentane, again in both carbon tetrachloride and chloroform. Their results are shown in Figures 14 and 15.

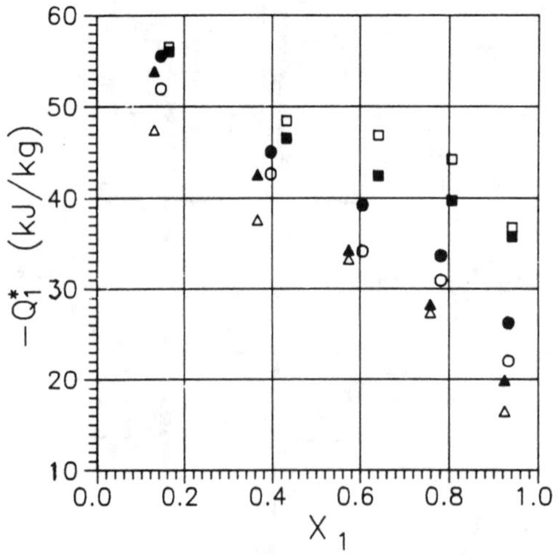

Figure 14. Measured values of Q_1^* at 30°C in mixtures of *n*-hexane (■), *n*-heptane (•), *n*-octane (▲), 3-methylpentane (□), 2,3-dimethylpentane (○) and 2,2,4-trimethylpentane (△) in carbon tetrachloride.

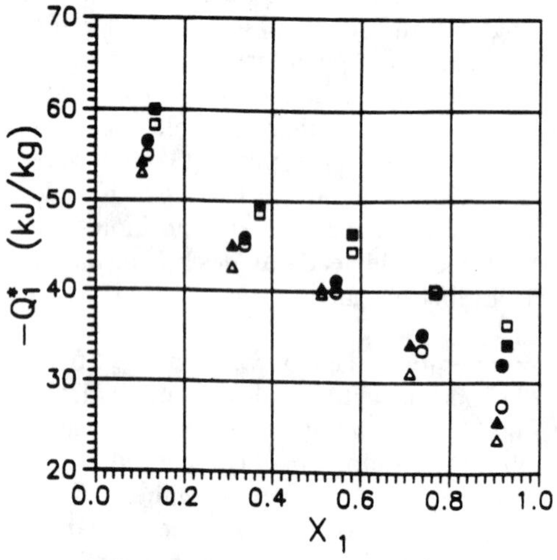

Figure 15. Measured values of Q_1^* at 30°C in mixtures of *n*-hexane (■), *n*-heptane (•), *n*-octane (▲), 3-methylpentane (□), 2,3-dimethylpentane (○) and 2,2,4-trimethylpentane (△) in chloroform.

Multicomponent Heats of Transport

One of the particular advantages of diffusion thermoeffect experiments over thermal diffusion is that temperature rather than composition measurements are made. Platt *et al.* (1982a) realized that this is a particular advantage in studying multicomponent systems where transient compositions of the independent components are impractical to measure.

The extension of Equations (2.1) - (2.10) to ternary mixtures was made by Platt *et al.* (1982a). In general for n components, there are n independent fluxes and n independent driving forces defined by

$$-q = L_{00}\nabla \ln T + \sum_{j=1}^{n-1} L_{0j}\nabla_T(\mu_j-\mu_n) \tag{5.4}$$

$$-j_i = L_{i0}\nabla \ln T + \sum_{j=1}^{n-1} L_{ij}\nabla_T(\mu_j-\mu_n) \ ; \quad (i=1,2,...,n-1) \ . \tag{5.5}$$

These equations can be reformulated into a single matrix equation by using the constant temperature and pressure Gibbs-Duhem equation and the chain rule. Thus,

$$-\boldsymbol{J} = \boldsymbol{A}\boldsymbol{B}\boldsymbol{X} \ , \tag{5.6}$$

where

$$\boldsymbol{J} = \begin{bmatrix} q \\ j_1 \\ \vdots \\ j_{n-1} \end{bmatrix} ; \quad \boldsymbol{A} = \begin{bmatrix} L_{00} & L_{01} & \cdots & L_{0,n-1} \\ L_{10} & L_{11} & \cdots & L_{1,n-1} \\ \vdots & \vdots & & \vdots \\ L_{n-1,0} & L_{n-1,1} & \cdots & L_{n-1,n-1} \end{bmatrix} \tag{5.7}$$

$$\boldsymbol{B} = \begin{bmatrix} 1 & 0 & \cdots & 0 \\ 0 & b_{11} & \cdots & b_{1,n-1} \\ \vdots & \vdots & & \vdots \\ 0 & b_{n-1,1} & \cdots & b_{n-1,n-1} \end{bmatrix} ; \quad \boldsymbol{X} = \begin{bmatrix} \nabla \ln T \\ \nabla w_1 \\ \vdots \\ \nabla w_{n-1} \end{bmatrix} \tag{5.8}$$

and

$$b_{ij} = \sum_{k=1}^{n-1} (\delta_{ik}+w_k/w_n)(\partial\mu_k/\partial w_j)_{T,P} \ ; \quad (i,j=1,2,...,n-1) \ . \tag{5.9}$$

Equation (5.6) expresses the flux matrix in terms of the product of Onsager, thermodynamic and force matrices.

Heats of transport in multicomponent systems were defined by Platt *et al.* (1982a and 1983) analogously to the definition in Equation (5.1) for binary mixtures. Thus,

$$Q_i^* = \left(\frac{q}{j_i}\right)_{\Delta T=0,\, j_{k\neq i}=0} \qquad (5.10)$$

In terms of Onsager coefficients,

$$Q_i^* = \frac{|C_i|}{|A'|} \qquad (5.11)$$

where **A'** is the principal submatrix of **A** formed by elimination of the first row and column, and $\mathbf{C}_i$ are matrices formed by replacement of the ith row of **A'** with the elements L_{0j} ($j=1,2,...,n$-1). Consistent with Fick's law, the diffusion coefficient matrix is defined by

$$\rho D = A' B' \qquad (5.12)$$

where **B'** is the principal submatrix of **B** obtained upon elimination of the first row and column. The definitions for k and $D_{T,i}$ are analogous to those in Equation (2.4).

Equations (5.4) and (5.5) can be reformulated in terms of the transport coefficients by using these inter-relationships between transport and Onsager coefficients to yield,

$$-q = k\nabla T + \sum_{k=1}^{n-1} \sum_{l=1}^{n-1} \rho Q_k^* D_{kl} \nabla w_l \qquad (5.13)$$

$$-j_i = \rho D_{T,i} \nabla \ln T + \sum_{l=1}^{n-1} \rho D_{il} \nabla w_l \; ; \qquad i=1,2,...,n-1 \; . \qquad (5.14)$$

Platt *et al.* (1982b) measured the diffusion thermoeffect in ternary mixtures of toluene, chlorobenzene and bromobenzene. These measurements are the only measurements in ternary mixtures known to the author. These particular components were chosen because their ternary mixtures are reasonably ideal, the mutual diffusion coefficients for this system had already been measured by Burchard and Toor (1962) and because heats of transport for the binary mixtures of these three components were measured by Hall and Rowley. This permitted Hall and Rowley to develop a correlation between the heats of transport in the ternary mixture and values extrapolated to infinite dilution from the binary data.

Analysis of the transient temperature profile in terms of the heat of transport coefficient is considerably more difficult than in the binary case because two independent heats of transport must be regressed from the $\Delta T(t)$ measurements. The boundary value problem for the experiment is similar to that previously described for the binary case, but using Equations (5.13) and (5.14) in the conservation Equations (2.9) and (2.10) yields

$$\rho\left(\frac{\partial w_i}{\partial t}\right) = \frac{\partial}{\partial z}\left[\sum_{j=1}^{2} \rho D_{ij}\left(\frac{\partial w_j}{\partial z}\right)\right] ; \quad i=1,2 \qquad (5.15)$$

and

$$\rho C_p\left(\frac{\partial T}{\partial t}\right) = \frac{\partial}{\partial z}\left[k\left(\frac{\partial T}{\partial z}\right) + \rho\sum_{i=1}^{2}\sum_{j=1}^{2} Q_j^* D_{ji}\left(\frac{\partial w_i}{\partial z}\right)\right]$$
$$+ \rho\sum_{i=1}^{2}\sum_{j=1}^{2} H_{jj}^{E} D_{ji}\left(\frac{\partial w_i}{\partial z}\right)\left(\frac{\partial w_j}{\partial z}\right) . \tag{5.16}$$

To obtain heats of transport from diffusion thermoeffect temperature measurements, Equations (5.15) and (5.16) must be solved subject to appropriate initial and boundary conditions. The solution of the temperature equation can then be fitted to measured temperature profiles using Q_1^* and Q_2^* as adjustable parameters. Platt *et al.* (1982b) were able to use an analytical perturbation solution of this problem to obtain Q_1^* and Q_2^* from their experimental measurements. The analytical solution can realistically only be carried out to first order in the perturbation scheme; this permits only a linear composition dependence for the thermophysical properties. Yi and Rowley (1989) generalized the analysis method by using a numerical method which allows the full composition and temperature dependence of the thermophysical properties to be included. This is the currently recommended procedure.

Platt *et al.* (1982a) showed that Q_1^* and Q_2^* are generally quite tightly coupled in the normal analysis of ternary diffusion thermoeffect experiments. In order to decouple them, at least two experiments using disparate starting composition differences between the two phases must be performed at each desired overall composition. By doing a parametric analysis of the ternary diffusion thermoeffect equations, Platt *et al.* were able to show the effect of the heats of transport on the measured temperature response. They were also able to find optimum thermocouple locations for highest resolution and decoupling of Q_1^* from Q_2^*. Implementation of these techniques produced values of Q_i^* for toluene - chlorobenzene - bromobenzene mixtures with an estimated uncertainty of about 10%. An example of the agreement between measured and experimental $\Delta T(t)$ profiles when Q_1^* and Q_2^* are simultaneously obtained from two experiments with different initial composition differences is shown in Figure 16.

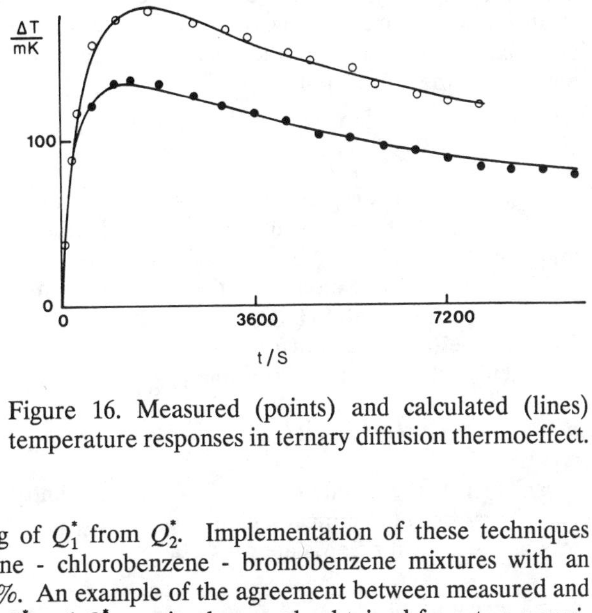

Figure 16. Measured (points) and calculated (lines) temperature responses in ternary diffusion thermoeffect.

Equation (5.10) can be used to relate the cross Onsager coefficients to the measured heats of transport. Platt *et al.* (1983) showed that

$$L_{0j} = L_{1j}Q_j^* + L_{2j}Q_j^* \tag{5.17}$$

or, upon inverting Equation (5.11), that

$$L_{ij} = \frac{\rho(D_{ij}b_{kk} - D_{ik}b_{kj})}{|B'|} \quad ; \quad (i,j,k = 1,2 \text{ but } j \neq k) \ . \tag{5.18}$$

Using these equations, Platt *et al.* were able to obtain all of the independent heat-mass Onsager coefficients for mixtures of the aforementioned three components from diffusion thermoeffect measurements and literature values for the diffusion coefficients. Their results were consistent with the constraints

$$L_{11}L_{22} \geq L_{12}L_{21} \qquad L_{00}L_{ii} \geq L_{0i}^2 \tag{5.19}$$

imposed by the positive semidefinite requirement of entropy production.

Heat of Transport Near the Liquid-Liquid Critical Point
Rowley and Horne (1979) used a temperature-jump technique, similar in principle to that described above for Clark and Rowley's diffusion experiments, to initiate diffusion thermoeffect experiments near the liquid-liquid critical point of water-isobutyric acid mixtures. Because thermal relaxation times are an order of magnitude shorter than diffusional, the temperature jump in diffusion thermoeffect experiments must be very quick and spatially uniform. Rowley and Horne used microwave absorption for the temperature jump, again measuring temperatures equidistant from the initial interface as a function of time following the jump. However, unlike previous analyses in binary systems, Q_1^* was not treated as a constant. The overall temperature in the experiment was allowed to gradually decay back to T_c. This permitted the constants A and λ in the simple power law expression for Q_1^* near the critical point

$$Q_1^* = A\epsilon^\lambda \tag{5.20}$$

to be simultaneously regressed from the resultant temperature profile. From an average of seven independent experiments, Rowley and Horne determined a critical exponent of 0.65 for Q_1^*, or approximately 2/3. As previously mentioned, it is well established that D vanishes with a +2/3 critical exponent in the critical region. Thus, $\mu_{11} \sim \epsilon^{4/3}$, $L_{11} \sim \epsilon^{-2/3}$, and, in accordance with Equation (2.4), $L_{01} \sim \epsilon^{o}$. It therefore appears that L_{11} is the only heat-mass Onsager coefficient for binary liquid mixtures with a nonzero critical exponent. This is consistent with the general criterion

$$L_{00}L_{11} - L_{10}L_{01} \geq 0 \tag{5.21}$$

derivable from the semidefinite positive nature of entropy production. Since this expression holds away from the critical region and only L_{11} diverges in the critical region, it remains true as the critical point is approached.

Development of Theory
While there have been several kinetic models formulated for the heat of transport (Wirtz 1939; Wirtz and Hiby 1943; Denbigh 1952, Prigogine *et al.* 1950, Dougherty and Drickamer 1955, Mortimer and Eyring 1980, and Guy 1986) and some statistical mechanical formulations of the heat of transport for simple mixtures (Bearman *et al.*, 1958), none of these models are quantitatively correct for nonideal mixtures. For the mixtures tested by Yi and Rowley (1987b), none of the methods produced even qualitatively correct results. Yi and Rowley were able to obtain qualitative predictions of Q_1^* without adjustable

parameters using a modified square-well model in conjunction with the revised Enskog theory. Agreement was generally improved if the potential well width was treated as an adjustable parameter. Efforts to develop a quantitative model continue.

Rowley and Horne (1979) and Platt *et al.* (1982a) proposed a mechanism for the diffusion thermoeffect. They suggested that the heat of transport can not be calculated directly from equilibrium properties, but is intimately connected to the diffusion mechanism itself. In an initially isothermal liquid mixture, the velocities of each molecule vary but obey a distribution law as shown in Figure 17. Hence, there are molecules which at any one time have velocities larger than the average velocity characterized by the system temperature (shown as a dotted line). For convenience of discussion, these molecules have an "excess energy" over and above that of their average counterparts. Now when two isothermal phases at different compositions are brought into contact such that mutual diffusion begins, the more energetic molecules will diffuse more rapidly than the others. Although collisions continuously randomize the energy and velocity of the molecules, the net effect is the diffusion of molecules carrying the excess energy. This same phenomenon occurs in the opposite direction, but the excess energy carried from the two separate original phases is not equivalent because the breadths of the distributions are not the same. Collisional re-equilibration shifts the average energy of each phase in opposite directions as shown. The connection between the diffusion process and the diffusion thermoeffect results from the fact that the more energetic molecules are those which both diffuse farther and carry excess energy. The difference between the excess energy carried in the two directions is directly related to the difference in the relative widths of the energy distributions in the two phases and corresponds to the temperature rise and decrease in the two halves of the cell during a diffusion thermoeffect experiment. This is but a model to explain the phenomenon, but nonequilibrium MD simulations are currently being made to investigate its viability.

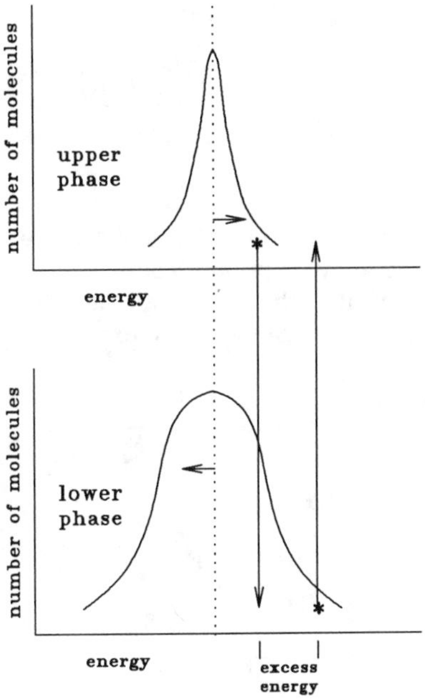

Figure 7. Schematic of connection between diffusion mechanism and the heat of transport.

6. MOLECULAR DYNAMICS AS A UNIFIED PREDICTIVE APPROACH

Current research efforts are aimed at development of MD simulations as a unified approach to prediction of transport properties from a single set of molecular interaction parameters. At the present time, nonequilibrium MD simulations of structured molecules are being used with site-site LJ potentials. The LJ parameters are being obtained exclusively from pure-component properties. Unlike molecular interaction models, the LB combining rules have generally provided good results when applied to site-site models, presumably because the interaction fields around sites are more spherical in geometry and similar in size than effective fields around molecules. This permits LB combining rules to be used to obtain the cross interaction parameters. The structure of *n*-alkane molecules is particularly amenable to these kinds of studies using the methodology of Edberg *et al.* (1986) and the group LJ interaction constants determined by Jorgensen *et al.* (1984).

The algorithms being used in the author's current MD studies were developed by Evans and co-workers (Evans *et al.*, 1983 and Evans and Morriss, 1984). They are particularly convenient because Gauss's principle is used for all the constraints: to impose the constraints due to bond lengths and angles, to constrain the system to isothermal mechanics, and to impose a synthetic field which acts to produce the desired flux. If the Green-Kubo relation for a transport coefficient L_{ij} is $\int <J_i(t) \cdot J_j(0)> dt$, then a fictitious field, F_e, that couples to the system with the dissipative flux $J = J_j$ is applied. The steady state response flux $<J_i(t)>_{ss} = \lim_{t \to \infty} <J_i(t)>$ is obtained from the simulation, the transport coefficient calculated from

$$L_{ij} = \lim_{F_e \to 0} \left[\frac{<J_1(t)>_{ss}}{F_e} \right] . \tag{6.1}$$

Current MD efforts are similar to the previously described experimental work - all four heat-mass transport coefficients are determined. This is done in the simulations by calculating the steady state fluxes induced in each of two different simulations corresponding to the application of the external mass, F_1, and heat, F_q, driving forces, respectively. Thus,

$$L_{qq} = \lim_{F_q \to 0} \left[\frac{<J_q(t)>_{ss}}{F_q} \right] \qquad L_{1q} = \lim_{F_q \to 0} \left[\frac{<J_1(t)>_{ss}}{F_q} \right]$$

$$L_{q1} = \lim_{F_1 \to 0} \left[\frac{<J_q(t)>_{ss}}{F_1} \right] \qquad L_{11} = \lim_{F_1 \to 0} \left[\frac{<J_1(t)>_{ss}}{F_1} \right] \tag{6.2}$$

Multiple simulations must be made at each state point using different values of F_1 and F_q in order to extrapolate the response functions to zero field. Equations for F_1, F_q and the flux algorithms are available in MacGowan and Evans (1986).

7. CONCLUSIONS

The increasing demand for transport coefficients in chemical process design requires better experimental data and methods for prediction of transport properties from a few fundamental constants. The author's research has focused on measurement of heat-mass transport coefficients in nonelectrolyte liquid mixtures. The principles of nonequilibrium thermodynamics have been valuable in designing experiments to measure complete sets of heat-mass Onsager and transport coefficients and in developing predictive equations.

Experimental work has focused on elucidation of molecular effects upon the transport coefficients that can then be used to test and discriminate between predictive techniques. For example, molecular structural effects upon the transport coefficients were studied in mixtures of *n*-hexane, *n*-heptane, *n*-octane, 3-methylpentane, 2,3-dimethylpentane and 2,2,4-trimethylpentane with both carbon tetrachloride and chloroform. It was found that the effects of chain length and branching upon the mutual diffusion coefficient were appropriately produced by MD simulations using molecular LJ potentials. However, these simulations are not entirely predictive in nature since the cross interaction combining rule was obtained from a single experimental value of the mutual diffusion coefficient in the *n*-hexane + carbon tetrachloride system. Current research focuses on using MD simulations with the site-site LJ model which should enable prediction of all of the heat-mass transport coefficients from a single set of parameters. With this model, the cross interactions should not require an experimental datum, but should be obtainable from the LB combining rules.

Other experimental efforts have produced the first successful measurements of the diffusion thermoeffect in liquid mixtures, establishing this method as an accurate way to measure cross heat-mass transport coefficients. Onsager reciprocity has been experimentally verified using this experimental technique, and extension of the measurements to ternary mixtures provides the only known method for obtaining heats of transport or thermal diffusion coefficients in multicomponent systems. Efforts to calculate these cross transport coefficients with a modified Enskog model yields values in qualitative agreement with experiment. This is an improvement over existing models which are often in considerable error, even wrong in sign. However, quantitative calculations of heats of transport without adjustable parameters cannot yet be made. Currently MD simulations are being studied as a method to provide such quantitative calculations and as a test of a new molecular transport model which explains the diffusion thermoeffect.

A local composition theory has also been developed which permits accurate prediction of liquid mixture thermal conductivity (and viscosity) from pure component data and activity coefficient information in the form of NRTL parameters. No adjustable parameters are contained in this model. The method has been verified by comparison with experimental data measured in this laboratory on a wide variety of binary and ternary liquid mixtures. It appears that the method has applicability to any number of components.

Experimental measurements of heat-mass transport coefficients near the liquid-liquid critical point in binary mixtures have revealed that only L_{11} has a nonzero critical exponent. Measurements of L_{01} near the critical point help confirm current theoretically calculated values of critical exponents. These exponents provide a universal description of transport coefficients in the near critical region. Interestingly, diffusion coefficients measured in the critical region were able to supply equilibrium thermodynamic information not measurable with equilibrium experiments. Spinodal loci determined from diffusion meausrements provide important information for testing and improving liquid mixture free energy models.

Finally, it appears from current work that nonequilibrium MD simulations of site-site LJ fluids may provide a more general and universal method for predicting transport coefficients from a single set of constants. Efforts are currently in progress to test values of the three independent heat-mass transport coefficients in binary mixtures obtained from MD simulations against the experimental data described in this review.

8. REFERENCES

Aris, R., 1956. On the dispersion of a solute in a fluid flowing through a tube. *Proc. R. Soc. A* **235**: 67-77.

Bearman, R.J., J.G. Kirkwood and M. Fixman, Statistical-mechanical theory of transport processes. X. The heat of transport in binary liquid solutions. *Advances in Chemical Physics* (Interscience, New York, 1958), Vol.1, pp. 1-13.

Burchard, J.K., and H.L. Toor 1962. Diffusion in an ideal mixture of three completely miscible non-electrolytic liquids - toluene, chlorobenzene, bromobenzene. *J. Chem. Phys.* **66**: 2015-2022.

Clark, W.M., and R.L. Rowley 1985. Ternary liquid diffusion coefficients near plait points. *Int. J. Thermophys.* **6**: 631-642.

Clark, W.M., and R.L. Rowley 1986. The mutual diffusion coefficient of methanol-*n*-hexane near the consolute point. *AIChE J.* **32**: 1125-1131.

Darken, L.S., 1948. *Trans. Am. Inst. Mining Metall. Eng.* **175**: 184.

Denbigh, K.G., 1952. *Trans. Faraday Soc.* **48**: 1.

Dougherty, E.L., and H.G. Drickamer 1955. A theory of thermal diffusion in liquids. *J. Chem. Phys.* **23**: 295-309.

Edberg, R.E., G.P. Morriss and D.J. Evans 1986. Constrained molecular dynamics: simulations of liquid alkanes with a new algorithm. *J. Chem. Phys.* **84**: 6933-6939.

Evans, D.J., and G.P. Morriss 1984. Non-Newtonian molecular dynamics. *Computer Physics Reports* **1**: 297-343.

Evans, D.J., W.G. Hoover, B.H. Failor, B. Moran, A.J.C. Ladd 1983. Nonequilibrium molecular dynamics via Gauss's principle of least constraint. *Phys. Rev. A* 28: 1016-1021.

Guy, A.G., 1986. Prediction of thermal diffusion in binary mixtures of nonelectrolyte liquids by the use of nonequilibrium thermodynamics. *Int. J. Thermophys.* **7**: 563-572.

Hanks, M. 1988. Mathematical model for the shape of the melt/solid interface during horizontal bridgman growth of compound semiconductor crystals. M.S. thesis, Brigham Young University, Provo.

Ingle, S.E. and F.H. Horne 1973. The Dufour effect. *J. Chem. Phys.* **59**: 5882-5894.

Jorgensen, W.L., J.D. Madura and C.J. Swenson 1984. Optimized Intermolecular Potential Functions for Liquid Hydrocarbons. *J. Am. Chem. Soc.* **106**:6638-6646.

Kirkaldy, J.S., and G.R. Purdy 1969. Diffusion in multicomponent metallic systems. X. Diffusion at and near ternary critical states. *Can. J. Phys.* **47**: 865-871.

MacGowan, D., and D.J. Evans 1986. Heat and matter transport in binary liquid mixtures. *Phys. Rev. A* **34**: 2133-2141.

Matos Lopes, M.L., and C.A. Nieto de Castro 1986. Liquid mutual diffusivities of the H_2O/D_2O system. *Int. J. Thermophys.* **7**: 699-708.

Mortimer, R.G., and H. Eyring 1980. Elementary transition-state theory of the Soret and Dufour effects. *Proc. Natl. Acad. Sci. U.S.A.* **77**: 1728-1731.

Platt, G., T. Vongvanich, and R.L. Rowley 1982a. The diffusion thermoeffect in ternary, nonelectrolyte liquid mixtures. *J. Chem. Phys.* **77**: 2113-2120.

Platt, G., G. Fowler, T. Vongvanich, and R.L. Rowley 1982b. Diffusion thermoeffect measurements of heats of transport in ternary liquid toluene-chlorobenzene--bromobenzene mixtures at 25°C and 35°C. *J. Chem. Phys.* **77**: 2121-2129.

Platt, G., G. Fowler, T. Vongvanich, and R.L. Rowley 1983. Heat-mass Onsager coefficients in ternary nonelectrolyte liquid mixtures. *J. Noneq. Thermo.* **8**, 1-18.

Prigogine, I., L. deBrouckere and R. Armand 1950. Recherches sur la thermodiffusion en phase liquide. *Physica* **16**: 577-598.

Rastogi, R.P., and G.L. Madan 1965. Dufour effect in liquids. *J. Chem. Phys.* **43**: 4179-4180.

Reid, R.C., J.M. Prausnitz, and B.E. Poling, 1987. *The Properties of Gases and Liquids, Fourth Edition*, McGraw-Hill.

Rowley, R.L., 1982. A local composition model for multicomponent liquid mixture thermal conductivities. *Chem. Eng. Sci.* **37**: 897-904.

Rowley, R.L., and M.D. Hall 1986. Heats of transport from the diffusion thermoeffect in binary liquid mixtures of toluene, chlorobenzene, and bromobenzene. *J. Chem. Phys.* **85**: 3550-3555.

Rowley, R.L., and F.H. Horne 1978. The Dufour effect. II. Experimental confirmation of the Onsager heat-mass reciprocal relation for a binary liquid mixture. *J. Chem. Phys.* **68**: 325-326.

Rowley, R.L., and F.H. Horne 1979. The behavior of the heat of transport near the critical solution temperature of isobutyric acid-water mixtures. *J. Chem. Phys.* **71**: 3841-3850.

Rowley, R.L., and F.H. Horne 1980. The Dufour effect. III. Direct experimental determination of heats of transport for the carbon tetrachloride-cyclohexane system. *J. Chem. Phys.* **72**: 131-139.

Rowley, R.L., and V. Gubler 1988. Thermal conductivities in seven ternary liquid mixtures at 40°C and 1 atm. *J. Chem. Eng. Data* **33**: 5-8.

Rowley, R.L., and G.L. White 1987. Thermal conductivities of ternary liquid mixtures. *J. Chem. Eng. Data* **32**: 63-69.

Rowley, R.L., S.C. Yi, V. Gubler, and J.M. Stoker 1987. Mutual diffusivity, thermal conductivity, and heat of transport in binary liquid mixtures of alkanes in carbon tetrachloride. *Fluid Phase Equilib.* **36**: 219-233.

Rowley, R.L., S.C. Yi, D.V. Gubler, and J.M. Stoker 1988a. Mutual diffusivity, thermal conductivity, and heat of transport in binary liquid mixtures of alkanes in chloroform. *J. Chem. Eng. Data* **33**: 362-366.

Rowley, R.L., G.L. White, and M. Chiu 1988b. Ternary liquid mixture thermal conductivities. *Chem. Eng. Sci.* **43**: 361-371.

Rowley, R.L., J.M. Stoker, and N.F. Giles 1990. Molecular dynamics simulations of mutual diffusion in nonideal mixtures. *J. Chem. Phys.* (in press).

Sengers, J.V. 1973. Transport properties of gases and binary liquids near the critical state. In *Transport Phenomena - AIP Conference Proceedings No. 11*, AIP, New York.

Skripov, V.P., V.S. Vitkalov and Y.D. Kolpakov 1980. Kinetics of the approach to equilibrium in systems in the vicinity of the critical point according to light scattering. *Russ. J. Phys. Chem.* **54**: 997-999.

Stoker, J.M., and R.L. Rowley 1989. Molecular dynamics simulation of real-fluid mutual diffusion coefficients with the Lennard-Jones potential model. *J. Chem. Phys.* **91**: 3670-3676.

Taylor, G., 1953. Dispersion of soluble matter in solvent flowing slowly through a tube. *Proc. R. Soc. A* **219**: 186-203.

Vignes, A., 1966. Diffusion in binary solutions - variation of diffusion coefficient with composition. *Ind. Eng. Chem. Fundam.* **5**: 189.

Wakeham, W.A., 1981. Diffusion coefficient measurements by the chromatographic method. *Faraday Symp. Chem. Soc.* **15**: 145-154.

Wei, I.C., and R.L. Rowley 1983a. Ternary liquid mixture viscosities and densities. *J. Chem. Eng. Data* **29**: 336-340.

Wei, I.C., and R.L. Rowley 1983b. Binary liquid mixture viscosities and densities. *J. Chem. Eng. Data* **29**: 332-335.

Wei, I.C., and R.L. Rowley 1985. A local composition model for multicomponent liquid mixture shear viscosity. *Chem. Eng. Sci.* **40**: 401-408.

Wirtz, K. and J.W. Hiby, 1943. *Phys. Z. Leipzig* **44**: 369.

Wirtz, K. 1939. *Ann. Phys. Leipzig* **36**: 295.

Yi, S.C., and R.L. Rowley 1987a. Heats of transport from diffusion thermoeffect measurements on binary liquid mixtures of carbon tetrachloride with benzene, toluene, 2-propoanone, *n*-hexane, and *n*-octane. *J. Chem. Phys.* **87**: 7208-7213.

Yi, S.C., and R.L. Rowley 1987b. On the use of a modified square-well model for prediction and correlation of thermal diffusion factors in binary liquid mixtures. *J. Chem. Phys.* **87**: 7214-7220.

Yi, S.C., and R.L. Rowley 1989. Measurement of heats of transport in ternary liquid mixtures via the diffusion thermoeffect. *J. Nonequilib. Thermodyn.* **14**, 293-297.

9. NOMENCLATURE

α	NRTL constant
γ_i	activity coefficient of component i
∇_T	isothermal gradient; $= \nabla - (\partial/\partial T)$
ϵ	Lennard-Jones energy parameter
λ	critical exponent for Q_1^*
μ_{11}	$= (\partial\mu_1/\partial w_1)_{T,P}$; defined in Equation (2.5)
μ_i	specific chemical potential of component i
ξ	viscosity variable defined in Equation (4.7)
ρ	mass density
σ	Lennard-Jones size parameter
σ	transition effectiveness parameter, usually taken to be 0.25
Φ_1	entropy source strength
ϕ_i	volume fraction of component i
ϕ_{ij}	local volume fraction of molecule type i around j
$\boldsymbol{A}$	Onsager coefficient matrix
A	multiplicative factor in Equation (5.20)
$\boldsymbol{A}'$	principal submatrix of $\boldsymbol{A}$
A_{ij}, A_{ji}	NRTL parameters
$\boldsymbol{B}$	thermodynamic matrix defined in Equations (5.8) and (5.9)
$\boldsymbol{B}'$	principal submatrix of $\boldsymbol{B}$
c_i	number density of component i
C_p	constant pressure heat capacity
D	binary mutual diffusion coefficient
D_i^o	self diffusion coefficient of component i
D_i	intradiffusion coefficient of component i
D_{ij}	mutual diffusion coefficient for flux i and driving force j in multicomponent systems
D_T	thermal diffusion coefficient
$\boldsymbol{F}_1$	driving force for molecular mass flux
$\boldsymbol{F}_e$	general external driving force in nonequilibrium MD simulation
$\boldsymbol{F}_q$	driving force for molecular heat flux
G_{ij}	NRTL nonrandomness factor defined by Equation (4.5)
$\bar{H}_i$	partial specific enthalpy of component i
H^E	excess enthalpy
$\boldsymbol{J}$	flux matrix defined in Equation (5.6)
$\boldsymbol{J}_1$	molecular mass flux
j_1	mass flux of component 1 relative to center of mass velocity
$\boldsymbol{J}_{ij}$	collective particle flux defined by Equation (3.4)
$\boldsymbol{J}_q$	molecular heat flux

k	thermal conductivity
k_{ij}	thermal conductivity cross interaction defined by Equation (4.2)
L_{0i}	Onsager heat flux coefficients
L_{i0}	Onsager mass flux coefficients
M	mean molecular mass of the mixture
M_i	molecular weight of component i
N	total number of molecules
Q	thermodynamic factor = $1 + (\partial \ln\gamma_1/\partial \ln x_1)_{T,P}$
$\boldsymbol{q}$	heat flux
Q_1^*	heat of transport of component 1
R	ideal gas constant
t	time
T	temperature
T_c	upper critical solution temperature
V	mean molar volume
V_i	molar volume of component i
$\boldsymbol{v}$	center of mass velocity
$\boldsymbol{v}_i$	velocity vector of molecule i
w_i	mass fraction of component i
w_i	mass fraction of component i
w_{ij}	local mass fraction of component i around j
$\boldsymbol{X}$	driving force matrix
x_i	mole fraction of component i

Dissipative Thermohydrodynamic Oscillators

Manuel G. Velarde and Xiao-Lin Chu
U.N.E.D., Facultad de Ciencias
Apartado 60 141, Madrid 28 080, Spain

ABSTRACT
The *harmonic* oscillator is the building block in the study of *oscillatory* phenomena and *standing* simple waves in Physics. Based upon this a unifying picture of different albeit not unrelated thermohydrodynamic oscillatory convective instabilities can be given. We describe how dissipation / damping competes with restoring forces thus leading to oscillations. We also illustrate how *nonlinearity* helps sustaining these oscillations in the form of *limit cycle* motions. Finally we show how the Korteweg-deVries *soliton* description is to some *traveling* nonlinear wave phenomena another building block and again we illustrate the role played by dissipation in triggering and eventually sustaining nonlinear motions.

CONTENTS

INTRODUCTION

Oscillatory convection is generally the consequence of instability when two or more competing causes produce overshooting at an instability threshold (overstability). When, say, buoyancy or some other body force, heat and viscous dissipation compete, either oscillatory or steady cellular convection may appear (Velarde & Normand 1980; Legros, Sanfeld & Velarde 1987; Velarde 1988). This is the typical situation in Rayleigh-Bénard convection where buoyancy dominates dissipation by heating a liquid layer from below. If the liquid layer is, however, heated from above buoyancy rather favours stability of an initially motionless state. Yet this is not sufficient to ensure static stability. It has indeed been shown that for a two-component Bénard layer, where buoyancy can be induced by temperature and solute gradients instability is possible even when the density stratification clearly favours stability. On the other hand, if a liquid layer is rotated with, say, uniform speed, it has also been shown that overstability is possible for appropriate values of the Prandtl and Taylor (rotation) numbers. For viscoelastic liquids when heating a layer from below or above overstability has also been predicted. For unstable interfaces between two liquids or liquid-air competing mechanisms include gravity, non-uniform surface tension (Marangoni stresses), electric fields, etc. In all mentioned cases there is, generally, a synchronous coupling of the various agents.

In the present chapter, we propose to look at some of these overstabilities from an *intuitively appealing* perspective by considering that, as a consequence of the above mentioned competition among forces and dissipation, a fluid property like temperature, velocity, interfacial deformation or the concentration of a surfactant can be made to oscillate with damping that may or not vanish. Suppose that we are able to write down the time evolution of one of these quantities in the form of a *dissipative* harmonic oscillator equation. Then we expect that at vanishing damping there is the overstability threshold which is in fact the condition of *free* oscillation in a *dissipative system far away from equilibrium* (Chu & Velarde 1989a, 1989b, 1989c, 1989d; Jimenez-Fernandez & Velarde 1989; Legros, Sanfeld & Velarde 1987; Velarde & Normand 1980; Velarde 1988; Velarde & Chu 1988, 1989a, 1989b, 1990).

Nonlinear corrections would eventually sustain the oscillation in the form of a *limit cycle* with possible space modulation. The same approach is also taken for *soliton* waves obeying a Korteweg-de Vries description (Drazin & Johnson 1989; Lamb Jr. 1980; Lamb 1932; Whitham 1974).

For simplicity, with the exception of a rather short excursion to the Maxwell's viscoelastic model, we shall restrict consideration to the Boussinesquian limit of a Newtonian liquid layer described by the Navier-Stokes and Fourier or Fick equations. Moreover, we just refer to the standard Bénard model problem, first in the Raylegh approach for buoyancy-driven convection and subsequently in the case of a layer open to air where Marangoni surface tension stresses are the key ingredient in the physics of the problem, as in the original Bénard experiments.

1. OSCILLATORY RAYLEIGH-BENARD BUOYANCY-DRIVEN CONVECTION

1.1. NEWTONIAN LIQUIDS

Let us consider a horizontal liquid layer, thickness h, enclosed between the heights $z = 0$ and $z = h$. In this section we assume the temperature uniformly established in both upper and lower boundaries that are taken undeformable and, for simplicity, stress-free. When the layer is heated from below it tis well known that, for a suitable value of the temperature gradient, β, the system exhibits steady convection in the form of rolls. For a single-component liquid layer we take (Landau & Lifshitz 1959; Legros, Sanfeld & Velarde 1987; Levich 1962)

$$\nabla * \mathbf{v} = 0 \tag{1.1}$$

$$\rho \frac{\partial \mathbf{v}}{\partial t} = -\nabla p + \eta \nabla^2 \mathbf{v} \tag{1.2}$$

and

$$\frac{\partial \theta}{\partial t} = \beta w + \kappa \nabla^2 \theta \tag{1.3}$$

where $\mathbf{v} = (u,w)$, p and θ denote velocity, pressure and temperature, respectively. ρ is the density of the liquid, $\eta = \rho\nu$ the dynamic viscosity and ν the kinematic viscosity, κ is the heat diffusivity (thermometric conductivity) and β is the temperature gradient taken positive when heating from below. Note that for a first approximation to the problem suffices a linear analysis. Then a Fourier normal mode decomposition is the adequate approach:

$$w(\bar{r}, z; t) = W(t) \exp[i\bar{r}a] \sin(\pi z/h) \quad (1.4)$$

$$\theta(\bar{r}, z; t) = \theta(t) \exp[i\bar{r}a] \sin(\pi zh) \quad (1.5)$$

where r =(x,y) is the two-dimensional horizontal vector and a= (a_x, a_y) is the wave vector in the Fourier reciprocal lattice, W(t) and θ(t) are the time-dependent amplitudes of the expected convective mode. For infinitesimal disturbances these amplitudes obey the following evolution equations:

$$\frac{dW}{dt} = -\nu\gamma W + \frac{\alpha g a^2}{\gamma}\theta \quad (1.6)$$

$$\frac{d\theta}{dt} = \beta W - \kappa\gamma\theta \quad (1.7)$$

$$a^2 = a_x^2 + a_y^2 \quad (1.8)$$

$$\gamma = a^2 + \pi^2/h^2 \quad (1.9)$$

The term $\alpha g a^2 \theta/\gamma = F$ is the buoyancy / body force induced by the heating. g is the gravitational acceleration. α is the thermal expansion coefficient. Then system (1.6)- (1.7) can be rewritten as

$$\frac{dW}{dt} = -\nu\gamma W + F \quad (1.10a)$$

$$\frac{dF}{dt} = \frac{\alpha\beta g a^2}{\gamma} W - \kappa\gamma F \quad (1.10b)$$

Equation (1.10a) describes the dynamic balance between inertia, viscous damping and buoyancy. Note that the coefficient of W in Eq. (1.10b) has spring constant dimension (s^{-2}), whereas κγ has dimension of s^{-1}. The latter factor is the buoyancy force relaxation time scale.

Introducing now $F_v = -\nu\gamma W$ (viscous damping force) and $f = F + F_v$, the system becomes:

$$\frac{dW}{dt} = f \qquad (1.11a)$$

$$\frac{df}{dt} = KW - bf \qquad (1.11b)$$

with $K = \alpha\beta a^2/\gamma - \gamma^2\nu\kappa$ and $b = \gamma(\nu + \kappa)$. The system (1.11) can be reduced to a single second-order equation:

$$\frac{d^2W}{dt^2} + b\frac{dW}{dt} - cW = 0 \qquad (1.12)$$

which is the *dissipative* harmonic-oscillator equation describing the evolution of vertical velocity disturbances in the liquid layer (Jimenez-Fernandez & Velarde 1989).

The following consequences can be extracted:

i) When buoyancy is equilibrated by the viscous damping, f vanishes and the only solution is steady convection (W = const.) at a = 0. Then

$$\frac{\alpha\beta g}{\nu\kappa} = \frac{\gamma^3}{a^2} \qquad (1.13)$$

defines the *critical* Rayleigh number, *i.e.*, the *threshold* for the onset of convection which is already well known.

ii) When β is negative, *i.e.*, we heat the layer from *above* Eq. (1.12) defines a *damped* harmonic oscillator (b is always positive) with an initial frequency

$$-\gamma^2\nu\kappa + \frac{\alpha\beta g a^2}{\gamma} \qquad (1.14)$$

iii) When the heating is from *below* (β is positive), we have "exchange of instabilities" leading to steady cellular convection and no oscillation is expected unless we proceed to a second instability threshold.

If now we consider a *two-component* liquid layer subjected to temperature and solutal / surfactant concentration gradients, the only difference with the description given earlier comes from the addition of N (mass-fraction disturbance of component "one", say) and its corresponding Fick's evolution equation. We add

$$N(r\ z;t) = \Gamma(t)\exp[ira]\sin(\pi z/h) \qquad (1.15)$$

and

$$\frac{d\Gamma}{dt} = \frac{\alpha_s \beta_s a^2}{\gamma} W - D\gamma F \tag{1.16}$$

with $\Gamma(t)$ the amplitude and D the mass diffusivity, α_s and β_s are the solutal expansion coefficient and concentration gradient, respectively. Note that we could have considered rather than two thermal gradients only the temperature gradient with the cross-thermal-diffusion Soret effect creating the concentration gradient.

The buoyancy force now is

$$F = \alpha g \frac{a^2 \theta}{\gamma} + \frac{\alpha_s g a^2 \Gamma}{\gamma} \tag{1.17}$$

The previous argument applies here *verbatim* and Eq. (1.12) is still formally valid. We obtain *two* Rayleigh numbers that are to be algebraically added thus allowing for competition of *two* body forces. To the *thermal* Rayleigh number, the *solutal* Rayleigh number $R_s = \alpha_s \beta_s g h^4 / \nu D$ is added. Then what really matters is the buoyancy ratio $\alpha\beta / \alpha_s\beta_s$ and particularly its sign.

Thus for a two-component liquid layer the conclusions become:
ii) * When β is negative, *i.e.*, we heat the layer from *above*, oscillations are possible provided β_s is suitably chosen.

iii) ** When the heating is from *below* (β positive), oscillations are possible provided β_s is suitably chosen. Only under very limited conditions there is "exchange of stabilities".

1.2. VISCOELASTIC LIQUIDS

For simplicity let us consider the simplest Maxwell's model of constitutive equation

$$\lambda \frac{d\tau}{dt} + \tau = 2\eta D \tag{1.18}$$

where λ is the characteristic relaxation time, τ is the standard stress tensor and D is the rate of strain tensor. As before η is the limiting Newtonian dynamic viscosity. For infinitesimal disturbances in a liquid layer heated from below or above the problem is to solve an extension of the system (1.11)

$$\frac{dW}{dt} = F_\lambda + \frac{\alpha g a^2}{\gamma}\theta \qquad (1.19a)$$

$$\frac{dF_\lambda}{dt} = \beta W - \kappa\gamma\theta \qquad (1.19b)$$

$$\frac{dF_\lambda}{dt} = -\frac{\nu\gamma}{\lambda}W - \frac{1}{\lambda}F_\lambda \qquad (1.19c)$$

with $F_\lambda=(1/\rho)\mathrm{div}\,\tau$ the z-component of the divergence of the stress tensor. Eq. (1.19c) shows quite clearly the role played by the relaxation time constant λ and once more indicates together with Eq. (1.19a) the merely additive role of the viscoelasticity. F_λ is a force analogous to buoyancy. As in the preceding section a bit of algebra permits reducing Eqs (1.19) to a single *dissipative* harmonic oscillator equation. Note, however, that directly from Eqs. (1.19) a conclusion can be drawn. As the coefficient of W in Eq. (1.19c) is always negative, it follows that *if there is no buoyancy* when a Maxwellian liquid layer is excited it can only decay to the initial motionless state with a damped oscillatory motion of initial frequency $(\nu\gamma/\lambda)^{1/2}$ and damping time scale $1/\lambda$ (Jimenez-Fernandez & Velarde 1989).

2. INTERFACIAL OSCILLATORS

2.1. INTRODUCTION

In this section we leave buoyancy-driven flows and turn to interfacial phenomena. Capillary waves -or ripples, gravity waves and solitons have been well studied since Laplace, Kelvin, Stokes and Korteweg and de Vries (Drazin & Johnson 1989; Lamb Jr 1980; Whitham 1974). While these waves are rather *transverse* motions due to the deformation of the surface there is yet another type of wave. It refers to mostly *longitudinal* motion along the surface, in the limit along a flat surface. Their existence is not surprising considering that a strong analogy is expected between, say, a monolayer-covered surface and a stretched elastic membrane. The coverage with a surfactant monolayer -either by adsorption from solution or by spreading- gives indeed elastic properties to a surface so that it tends to resist the periodic surface *expansion and compression* which appears as *wave* motion. Normally, any surface wave motion has both transverse and longitudinal components, and they are not separable. Only in the case of *small* amplitude excitations (corresponding to *linear* theory) *transverse* and *longitudinal* waves can be considered as two genuinely different modes of oscillation.

Longitudinal waves are, to a major extent, related to the boundary condition for *tangential* stress with a frequency that depends on the viscosity and surface "elasticity" (Legros, Sanfeld & Velarde 1987; Levich 1962). Gravity-capillary waves, however, are related to the b.c. for *normal* stress to the surface and have a frequency that depends on gravity and on surface tension (Laplace overpressure) but not on viscosity. The latter only appears in the damping factor and frequency-deviation in the dispersion relation.

Generally these waves, these oscillatory motions, are damped albeit differently by viscosity. However, if a non equilibrium distribution of surfactant / temperature is imposed in the liquid, with mass / energy transfer across the surface, the *Marangoni effect* may transform the "thermal" or "chemical" energy into convective motion, overtaking the viscous dissipation and thus sustaining the wave motions.

2.2. THE CASE OF A LIQUID OPEN TO AIR AND THE MARANGONI EFFECT

2.2.1. DISTURBANCE EQUATIONS

Let us, for simplicity, consider a liquid layer of "infinite" depth. The undisturbed surface is located at z=0. As in section 1 the linearized equations that disturbances upon the quiescent state obey are:

$$\nabla * \mathbf{v} = 0 \tag{2.1}$$

$$\rho \frac{\partial \mathbf{v}}{\partial t} = -\nabla p + \eta \nabla^2 \mathbf{v} \tag{2.2}$$

$$\frac{\partial c}{\partial t} - \beta^c w = D \nabla^2 c \tag{2.3}$$

where besides the quantities earlier introduced here we have D, the mass diffusivity and $\beta^c = (\partial c/\partial z)_0$, the gradient of surfactant concentration at the quiescent state. Here rather than heat we have taken Fickian mass transport for we shall illustrate the role of surfactants. Note that we have neglected the body / buoyancy force in eq. (2.2). For a linearized analysis suffices to consider linearized boundary conditions (Chu & Velarde 1988; Levich 1962) :

$$\frac{\partial \zeta}{\partial t} = w \tag{2.4}$$

$$-T_0\nabla_\Sigma^2\zeta + g\rho\zeta - p - 2\eta\frac{\partial w}{\partial z} = 0 \qquad (2.5)$$

$$\left(\frac{\partial T}{\partial \Gamma}\right)_0 \nabla_\Sigma \gamma - \eta\left(\nabla_\Sigma w + \frac{\partial u}{\partial z}\right) = 0 \qquad (2.6)$$

$$\frac{\partial \gamma}{\partial t} + \Gamma_0 \nabla_\Sigma * u_\Sigma - D_\Sigma \nabla_\Sigma^2 \gamma + D\frac{\partial c}{\partial z} = 0 \qquad (2.7)$$

$$\gamma = k^1 (c - \beta^c \zeta)_\Sigma \qquad (2.8)$$

where ζ is the surface deviation from the z=0 level and Γ is the excess surfactant concentration at the surface. The subscript "0" indicates a value in a reference state, here γ is the disturbance upon Γ_0 and T is the surface tension. Subscript "Σ" accounts for either a value taken on the surface or a derivative along the surface. Note that to avoid a trivial long-wavelength (a=0) divergence or instability we have retained "g" in the b.c. (2.5). Note also that b.c. (2.6) accounts for the *Marangoni effect*, i.e., convective motions triggered by the nonuniform surface tension distribution along the surface.

For universality in the presentation, we introduce new units to rescale the quantities in the equations. The capillary length:

$$l = \sqrt{\frac{T_0}{g\rho}} \qquad (2.9)$$

is chosen as "unit"; ν/l, l^2/ν, $\nu^2\rho/l^2$, $\beta^c l$ and Γ_0 are used as "units" for velocity, time, pressure, surfactant concentration and excess surface concentration, respectively. Thus Eqs. (2.1)-(2.8) become:

$$\nabla * \mathbf{v} = 0 \qquad (2.10)$$

$$\frac{\partial \mathbf{v}}{\partial t} = -\nabla p + \nabla^2 \mathbf{v} \qquad (2.11)$$

$$\frac{\partial c}{\partial t} - w = S^{-1}\nabla^2 c \tag{2.12}$$

with boundary conditions at $z = \zeta$

$$\frac{\partial \zeta}{\partial t} = w \tag{2.13}$$

$$-\frac{1}{SC}\nabla^2_{\Sigma}\zeta + \frac{Bo}{SC}\zeta - p - 2\frac{\partial w}{\partial z} = 0 \tag{2.14}$$

$$\frac{HE}{SH_z}\nabla_{\Sigma}\gamma + \left(\nabla_{\Sigma}v + \frac{\partial u}{\partial z}\right) = 0 \tag{2.15}$$

$$HS\left(\frac{\partial \gamma}{\partial t} + \nabla_{\Sigma} * u_{\Sigma} - S_{\Sigma}^{-1}\nabla_{\Sigma}^{-2}\gamma\right) + \frac{\partial c}{\partial z} = 0 \tag{2.16}$$

$$\gamma = \frac{H_z}{H}(c - \zeta)_{\Sigma} \tag{2.17}$$

Note that although we have used the same notation for **v**, p, c, γ and ζ in these equations as in the dimensional case, all the units have changed and they are dimensionless now. The following dimensionless parameters have also been used : S= v/D,Schmidt number; $S_{\Sigma} = v/D_{\Sigma}$, surface Schmidt number; $C = \eta D/lT_0$, capillary number; $B0 = \rho g l^2/T_0$, Bond number (for simplicity is taken equal to unity); $E = -(\partial c/\partial \Gamma)_0\, k^l \beta l^2/(\eta D)$, surfactant (elasticity) Marangoni number; $H = \Gamma/(\beta^c l^2)$, surface excess surfactant number; and $H_z = k^l/l$, Langmuir adsorption number. As in the buoyancy-driven problems discussed in section 1 we choose Fourier modes

$$w = (A\,e^{az} + B\,e^{mz})\,e^{iax+\lambda t} \tag{2.18}$$

$$u = (iA\,e^{az} + i\,mBe^{mz})e^{iax+\lambda t} \tag{2.19}$$

$$p = -\left(\frac{\lambda A e^{az}}{a}\right) e^{iax + \lambda t} \qquad (2.20)$$

$$c = \left(\frac{A}{\lambda} e^{az} + \frac{SB}{\lambda (S-1)} e^{mz} + F e^{qz}\right) e^{iax + \lambda t} \qquad (2.21)$$

$$\text{where } m = \sqrt{a^2 + \lambda} \text{ and } q = \sqrt{a^2 + S\lambda}$$

"a" is the Fourier wavenumber and λ denotes the time constant. The real part of λ determines the stability of the system, whereas the imaginary part determines the convective mode. An *oscillatory* motion requires the imaginary part of λ, $\omega = \mathrm{Im}(\lambda)$, not being zero. If, however, $\omega=0$, the transition is expected from a motionless steady state to a steady cellular convective pattern. A, B and F are arbitrary constants to be obtained using the boundary conditions.

2.2.2. TRANSVERSE GRAVITY-CAPILLARY OSCILLATIONS

Assuming for simplicity that both surface adsorption and accumulation have negligible effect for the *transverse* surface waves Eqs. (2.15)-(2.17) reduce to

$$\frac{Ea^2}{S}(c - \zeta) + \left(\frac{\partial^2}{\partial z^2} + a^2\right) w = 0 \qquad (2.23)$$

$$\frac{\partial c}{\partial z} = 0 \qquad (2.24)$$

Equation (2.11) gives the dynamic evolution equation of the liquid layer. It is valid in the volume as well as at the surface - a part of the liquid. On the other hand, we have a kinematic relation (2.13) on the surface. Substituting w in the left side of Eq. (2.11) by (2.13) a relation on the surface $z=\zeta$ is written

$$\frac{\partial^2 \zeta}{\partial t^2} = -\frac{\partial p}{\partial z} + \nabla^2 w \qquad (2.25)$$

From Eq. (2.20) one has : $\partial p/\partial z = ap$. Using Eqs. (2.14) and (2.23), Eq. (2.25) becomes

$$\frac{\partial^2 \zeta}{\partial t^2} + \frac{Bo + a^2}{SC} a\zeta = -2a \frac{\partial w}{\partial z} - 2a^2 w - \frac{Ea^2}{S}(c - \zeta) \quad (2.26)$$

To proceed further we estimate the coefficients A, B and F. From Eq. (2.14) one has

$$B = \frac{\lambda f(B_0) + 2a}{\lambda f(B_0) - \lambda/a - 2p} A \quad (2.27)$$

where

$$f(B_0) = \frac{B_0 + a^2}{SC\lambda^2} + \frac{1}{a} \quad (2.28)$$

Now for transverse waves the Laplace law f(Bo) = 0 holds and as the interfacial disturbance penetrates little in the liquid ; $\omega \gg a^2$. Then from Eq. (2.27) follows

$$B \approx -\frac{2a^2}{\lambda} A \ll A \quad (2.29)$$

Thus we take in fact the *potential* part as the major ingredient in the velocity field which is consistent with the fact that transverse waves exist in ideal non-viscous fluids. Taking advantage of this simplification Eq. (2.26) becomes

$$\frac{d^2 \zeta}{dt^2} + \frac{B_0 + a^2}{SC} a\zeta = -4a^2 \frac{d\zeta}{dt} - \frac{Ea^2}{S}(c - \zeta) \quad (2.30)$$

The coefficient F can be obtained from Eq. (2.24)

$$F = -\frac{1}{q}\left(\frac{a}{\lambda} A + \frac{mS}{\lambda (S-1)} B\right) \approx -\frac{a}{q\lambda} A \quad (2.31)$$

that leads to

$$c - \zeta = \left(F + \frac{B}{\lambda (S-1)} \right) e^{iax + \lambda t} \approx -\frac{aA}{q\lambda} e^{iax + \lambda t} \approx a \frac{d\zeta/dt - \omega\zeta}{\omega\sqrt{2S\omega}} \quad (2.32)$$

Thus Eq. (2.30) becomes

$$\frac{d^2\zeta}{dt^2}+\left(\frac{Bo+a^2}{SC}a-\frac{Ea^3}{S\sqrt{2s\omega}}\right)\zeta=-\left(4a^2+\frac{Ea^3}{S\omega\sqrt{2S\omega}}\right)\frac{d\zeta}{dt} \tag{2.33}$$

Assuming now that the surfactant Marangoni number E is not exceedingly high, albeit of order $O(C^{-1})$, it follows that

$$(Bo+a^2)/(SC)a \gg (Ea^3)/(S\sqrt{2s\omega})$$

Therefore Eq. (2.33) can be reduced to

$$\frac{d^2\zeta}{dt^2}+\frac{Bo+a^2}{SC}a\zeta=-\left(4a^2+\frac{Ea^3}{S\omega\sqrt{2S\omega}}\right)\frac{d\zeta}{dt} \tag{2.34}$$

Equation (2.34) describes the oscillatory convective motion of the surface and takes the form of a simple oscillator, Eq. (2.22). One can directly judge that the l.h.s. terms of Eq. (2.34), that is the potential part, is the leading part of the equation. In the absence of dissipation and Marangoni effect, the equation appears as an *ideal/free* oscillator governed by the Laplace law. Thus potential motion is the main character of a transverse wave. The r.h.s. of Eq. (2.34) reflects the competition between viscous dissipation and Marangoni stresses. According to our earlier discussion when

$$4a^2+\frac{Ea^3}{s\omega\sqrt{2S\omega}}>0\left(\text{or } -E<\frac{4S\omega\sqrt{2S\omega}}{a}\right) \tag{2.35}$$

the oscillatory convection will be damped out. In contrast, when

$$4a^2+\frac{Ea^3}{s\omega\sqrt{2S\omega}}<0\left(\text{or } -E>\frac{4S\omega\sqrt{2S\omega}}{a}\right) \tag{2.36}$$

the oscillation grows exponentially. At the *neutral* state

$$-E=\frac{4S\omega\sqrt{2S\omega}}{a} \tag{2.37}$$

The kinetic energy dissipated by viscosity and that produced by surface tension work just compensate each other, giving a *free* harmonic oscillation. The oscillation frequency is given by the dispersion relation

$$\frac{B_0 + a^2}{SC} a - \omega^2 = 0 \tag{2.38}$$

TABLE 1 Interfacial oscillations : Typical threshold values for an open liquid layer heated from the air side. β is negative for liquids with no minimum (nor maximum) in their surface tension. Note the relevance of the results to spacebound low / micro-gravity experiments.

	Water		Mercury		Tin	
β (K/cm)	$1.9\cdot10^3$	5.85	441	1.39	55	1.8
Period (sec)	0.14	143	0.12	119.50	0.17	146
Penetration depth (cm)	$1.5\cdot10^{-2}$	0.48	$0,5\cdot10^{-2}$	0.15	$6\cdot10^{-3}$	0.18
Gravitational acceleration	g	10^{-4}g	g	10^{-4}g	g	10^{-4}g

By taking dE(a,w(a))/da = 0, the necessary condition for minimum yields the neutral curve, *i.e.*, the critical values for sustained transverse waves :

$$E_c^T = 4\sqrt{10}\left(\frac{6S}{5\sqrt{5}C}\right)^{3/4} \approx -7.931\left(\frac{S}{C}\right)^{3/4} \tag{2.39}$$

$$\omega_c^T = \sqrt{\frac{6}{5^{3/2}}}\frac{1}{\sqrt{SC}} \approx \frac{0.7326}{\sqrt{SC}} \tag{2.40}$$

$$a_c^T = \frac{1}{\sqrt{5}} \approx 0.4472 \tag{2.41}$$

2.2.3. LONGITUDINAL WAVES

For simplicity let us consider that the deformability of the surface has negligible influence on the *longitudinal* wave motion. Thus we set the capillary number, C, to zero. Also, it is known that surfactant accumulation on the surface affects mainly high fequency oscillatory convection and the frequency of longitudinal waves normally is small, so that the surfactant accumulation number H can be neglected. With these assumptions Eqs. (2.13)-(2.17) reduce to

$$w = 0 \tag{2.42}$$

$$\frac{Ea^2}{S} c + \frac{\partial^2 w}{\partial z^2} = 0 \tag{2.43}$$

$$SH_z \frac{\partial c}{\partial t} = -\frac{\partial c}{\partial z} \tag{2.44}$$

The surfactant concentration on the surface is chosen as the relevant variable in the harmonic oscillator. Differentiating Eq. (2.12) with respect to z and using Eq. (2.44) we have

$$-SH_z \frac{\partial^2 c}{\partial t^2} = \frac{\partial w}{\partial z} + S^{-1} \nabla^2 \frac{\partial c}{\partial z} \tag{2.45}$$

Now we use the boundary conditions (2.42)-(2.44). From (2.42) it follows

$$B = -A \tag{2.46}$$

Note that here the *rotational* part is of the same order as the *potential* one near the surface. Thus we ignore neither rotational nor potential terms in the longitudinal wave case. Using Eq. (2.46) and the long-wavelength assumption $a^2 \ll \omega \ll 1$, we obtain

$$\frac{\partial w}{\partial z} \approx \frac{1}{m}\left(1 - \frac{a}{\sqrt{\lambda}}\right)\frac{\partial^2 w}{\partial z^2}$$

Replacing $\partial^2 w/\partial z^2$ by Eq. (2.43)

$$\frac{\partial w}{\partial z} \approx Ea^2 q\left(\frac{\partial c/\partial t}{S^{3/2}\omega^2} - \frac{ac}{Sqm\sqrt{\lambda}}\right) \tag{2.47}$$

It also follows

$$S^{-1}\frac{\partial}{\partial z}\nabla^2 c \approx q\frac{\partial c}{\partial t} \tag{2.48}$$

Substitution of (2.47) and (2.48) in (2.45) yields

$$-\frac{SH_z}{q}\frac{\partial^2 c}{\partial t^2} - \frac{Ea^3}{Sqm\sqrt{\lambda}}c = \left(1 + \frac{Ea^2}{S^{3/2}\omega^2}\right)\frac{\partial c}{\partial t} \tag{2.49}$$

Here we have

$$-\frac{SH_z}{q}\frac{\partial^2 c}{\partial t^2} \approx \frac{H_z\sqrt{S}}{\sqrt{2\omega}}\left(\frac{\partial^2 c}{\partial t^2} + \omega\frac{\partial c}{\partial t}\right) \tag{2.50}$$

$$-\frac{Ea^3}{Sqm\sqrt{\lambda}}c \approx \frac{Ea^3}{S\omega\sqrt{2S\omega}}\left(c + \frac{1}{\omega}\frac{\partial c}{\partial t}\right) \tag{2.51}$$

Using Eqs. (2.50) and (2.51) in Eq. (2.49), and neglecting high-order small quantities, we obtain

$$\frac{1}{\sqrt{2S\omega}}\left(SH_z\frac{\partial^2 c}{\partial t^2} - \frac{Ea^3}{S\omega}c\right) = -\left(1 + \frac{Ea^2}{S^{3/2}\omega^2}\right)\frac{\partial c}{\partial t} \tag{2.52}$$

This is the equation we expected. In contrast with transverse waves, the leading part in Eq. (2.52) corresponds to the terms with first order derivative. Indeed

$$\left|\frac{1}{\sqrt{2S\omega}}\left(SH_z\,\omega^2 + \frac{Ea^3}{S\omega}\right)\right| \ll \left|\omega\left(1 + \frac{Ea^2}{S^{3/2}\omega^2}\right)\right| \tag{2.53}$$

This relation reveals the intrinsic character of the longitudinal wave: it is necessarily related to viscous convection. When energy dissipation and Marangoni work compensate each other

$$1+\frac{Ea^2}{S^{3/2}\omega^2}=0 \tag{2.54}$$

there appears the threshold for longitudinal oscillations at the surface. The stability analysis for the oscillator gives

$$-E\begin{cases} > \dfrac{S^{3/2}\omega^2}{a^2} & \text{explosion} \\ < \dfrac{S^{3/2}\omega^2}{a^2} & \text{damped motion} \end{cases}$$

In the *neutral* case we also have

$$SH_z\,\omega^2+\frac{Ea^3}{S\omega}=0 \tag{2.55}$$

Thus Eqs. (2.54) and (2.55) characterize the longitudinal oscillation of the air-liquid interface.

2.2.4. TRANSVERSE OSCILLATIONS NOT OBEYING LAPLACE LAW

Numerical exploration of the problem posed in section 2.1 shows that there exists a mode of *transverse* oscillation that does not obey the Laplace law. From Eq. (2.26) we get

$$\frac{\partial^2\zeta}{\partial t^2}+\frac{Bo+a^2}{SC}a\zeta=-2a^2\frac{\partial z}{\partial t}-2a\frac{\partial w}{\partial z}-\frac{Ea^2}{S}(c-\zeta) \tag{2.56}$$

whereas from Eqs. (2.13)-(2.21)

$$\frac{\partial w}{\partial z}=\frac{f(Bo)(a-m)-1}{2m-2a-\lambda/a}\lambda\frac{\partial\zeta}{\partial t}=\frac{a}{\lambda}[\lambda+2ma+2a^2+f(Bo)(m+a)\lambda]\frac{\partial\zeta}{\partial t} \tag{2.57}$$

$$c - \zeta = -\frac{1}{q\lambda}\,\frac{\lambda f(Bo)\left(a + \dfrac{q - mS}{S - 1}\right) - \lambda + 2a\dfrac{q - m}{S - 1}}{2m - 2a - \lambda/a}\,\frac{\partial\zeta}{\partial t}$$

$$= \frac{a}{q\lambda^3}(m + a)\left[\lambda f(Bo)\left(a + \frac{q - mS}{S - 1}\right) - \lambda + 2a\frac{q - m}{S-1}\right]\frac{\partial\zeta}{\partial t} \qquad (2.58)$$

Using these relations, Eq. (2.56) becomes

$$\frac{\partial^2\zeta}{\partial t^2} + \frac{Bo + a^2}{SC}a\zeta = -4a^2\left(1 + \frac{a(a + m)}{\lambda}\right)\frac{\partial\zeta}{\partial t}$$

$$-2a^2 f(Bo)(m + a)\frac{\partial\zeta}{\partial t} - \frac{Ea^3}{Sq\lambda^3}(m + a^2)$$

$$\times\left[\lambda f(Bo)\left(a + \frac{q - mS}{S - 1}\right) - \lambda + 2a\frac{q-m}{S-1}\right]\frac{\partial\zeta}{\partial t} \qquad (2.59)$$

Assume also here the high frequency approximation $\omega \gg a^2 = 0(1)$. This permits a simplified exploration of the problem and is still reasonable as we search for *transverse* oscillatory modes only. Then Eq. (2.59) becomes

$$\frac{\partial^2\zeta}{\partial t^2} + \frac{Bo + a^2}{SC}a\zeta - \sqrt{2}\,\omega^{3/2}a^2 f(Bo)\zeta$$

$$= -4a^2\frac{\partial\zeta}{\partial t} - \sqrt{2\omega}\;a^2 f(Bo)\frac{\partial z}{\partial t}$$

$$+\frac{Ea^3}{Sq\lambda}\left[1 - f(Bo)\frac{q - a S\sqrt{\lambda}}{S - 1}\right]\frac{\partial\zeta}{\partial t} \qquad (2.60)$$

As a particular case of this equation, one can easily obtain the harmonic equation for Laplace transverse waves, by taking the Laplace approximation f(Bo)=0. Now let us explore the case $f(Bo)\neq 0$. For simplicity assume the case of low Schmidt numbers, $S \ll a^2/\omega$. Then we obtain

$$\frac{\partial^2 \zeta}{\partial t^2} + \left[\frac{Bo + a^2}{SC} a - \frac{Ea^2}{S}\right]\zeta$$

$$= \left\{ -4a^2 + \left(\frac{Ea}{2} - a^2\sqrt{2\omega}\right) f(Bo) \right\} \frac{\partial \zeta}{\partial t} \tag{2.61}$$

This equation describes once more a *dissipative* harmonic oscillator. When the damping factor, *i.e.*, the coefficient of the first derivative with respect to time vanishes we obtain

$$\frac{Bo + a^2}{SC} a - \omega^2 = \frac{Ea^2}{S} \tag{2.62}$$

$$\left[\frac{Ea}{2} - \sqrt{2\omega}\, a^2\right] \left[\frac{1}{a} - \frac{Bo + a^2}{SC\,\omega^2}\right] - 4a^2 = 0 \tag{2.63}$$

It follows

$$E^2 - 2a\sqrt{2\omega}\, E + 8S\omega^2 = 0 \tag{2.64}$$

There are two roots (both of them are positive)

$$E = a\sqrt{2\omega}\ (1 \pm \sqrt{1 - 4S\omega/a^2}\,) \approx a\sqrt{2\omega}\left(1 \pm 1 \pm \frac{2S\omega}{a^2}\right) \tag{2.65}$$

It appears that the branch $E \approx 2a(2\omega)^{1/2}$ has no minimum when the wavelength varies in the capillary length range. Therefore it has no physical meaning and is a spurious consequence of our simplifications. For the other root, $E= 2(2)^{1/2}S\omega^{3/2}/a$, the dispersion relations are

$$\frac{Bo + a^2}{SC} a - \omega^2 = \frac{Ea^2}{S} \tag{2.66}$$

$$E = \frac{2\sqrt{2}\, S\omega^{3/2}}{a} \tag{2.67}$$

These two equations determine the *neutral* state. By taking

$$\frac{dE(a, \omega(a))}{da} = 0 \tag{2.68}$$

we obtain the critical Marangoni number

$$E_c = \frac{5a^2 - B_0}{2aC} \tag{2.69}$$

for the threshold of *transverse* oscillations at the air-liquid surface. However, these oscillations are not controlled by the Laplace law. Interesting enough it has been shown (Chu & Velarde 1989c) that the threshold for this mode of instability is lower than the threshold for steady cellular Bénard convection provided the capillary number is large enough and the Schmidt number (or the Prandtl number in the equivalent heat transfer problem) is small enough, as in the case of a Helium-4 layer slightly above the *lambda* line.

3. AN INTERFACIAL EHD OSCILLATOR

In section 2 we have considered liquid layers under Marangoni stresess. The possibility exists, however, of sustaining transverse waves, *i.e.*, transverse oscillatory motions at the interface of a dielectric liquid, say, subjected to an electric field. In this case we shall see that the dispersion relation incorporates the capillary-gravity terms together with an *electrocapillary* component (Chu, Velarde & Castellanos 1989).

With electric fields the disturbance equations also include Maxwell's equations in the EHD approximation. The magnetic field effects and the current are assumed weak enough to be neglected; in practice this is achieved when the current is of nano- or micro-ampere order. Boundary conditions are also needed for the electric field and the pressure balance incorporates the electrical stresses. It is from these electrical stresses and thus from the applied electric field that like in the case of Marangoni stresses we can obtain enough energy to overcome the viscous damping of the expected oscillatory motion. Indeed for a suitable value of the strength of the applied electric field, the force acting on the interface due to the surface charge density compensates the viscous damping and gravity-electro-capillary waves can be sustained.

Disturbances upon the velocity, pressure and density fields act upon the electrical field and now the disturbance equations in the absence of other constraints than an applied field are

$$\nabla * \mathbf{v} = 0 \tag{3.1}$$

$$\rho \frac{\partial \mathbf{v}}{\partial \tau} = -\nabla p + \eta \nabla^2 \mathbf{v} \tag{3.2}$$

$$\nabla^2 e = 0 \; (\nabla * \mathbf{e} = \nabla \times \mathbf{e} = 0) \tag{3.3}$$

together with the boundary conditions that for the case of *tangential* field to the interface are

$$[e_x] + [E_{0z}] \frac{\partial \zeta}{\partial x} = 0 \tag{3.4}$$

$$[w] = \left[\frac{\partial w}{\partial z}\right] = 0 \tag{3.5}$$

$$\frac{\partial \zeta}{\partial t} = w \tag{3.6}$$

$$-\frac{\partial \zeta}{\partial x} E_{0x} [\sigma + \varepsilon\lambda] + [(\sigma + \varepsilon\lambda)e_z] + [\varepsilon E_{0z}] \frac{\partial u}{\partial x} = 0 \tag{3.7}$$

$$\frac{\partial \zeta}{\partial x} [\varepsilon(E_{0z}^2 - E_{0x}^2)] + [\varepsilon(e_z E_{0x} + e_x E_{0z})] + \left[\eta\left(\frac{\partial u}{\partial z} + \frac{\partial w}{\partial x}\right)\right] = 0 \tag{3.8}$$

$$[p] = 2[\varepsilon e_z E_{0z}] - [\varepsilon e E_0] + 2\left[\eta \frac{\partial w}{\partial z}\right] + T\frac{\partial^2 \zeta}{\partial x^2} + [\rho] g\zeta \tag{3.9}$$

where we have chosen E_0 as the strength of the applied field and **e** its disturbance, ε and σ are the dielectric constant and the electric conductivity, respectively. T is once more the surface tension. The bracket denotes the jump at the interface

$$[f] = f_2 - f_1$$

Subscripts 1 and 2 indicate the liquids below and above the interface. The remaining symbols have the meaning earlier introduced.

When an electric field is applied *tangential* to the interface ($E_{0x} = E_0$ and $E_{0z} = 0$), a straightforward, albeit lengthy analysis following the pattern discussed in Section 2 yields the *dissipative* harmonic oscillator equation for the interfacial deformation, ζ (Velarde & Chu 1989a; Chu, Velarde & Castellanos 1989)

$$(\Sigma\rho)\frac{\partial^2\zeta}{\partial t^2} + \left\{ Ta^3 - g[\rho]a + 2a^2E_0^2\frac{([\sigma^2] + \omega^2[\varepsilon^2])[\varepsilon\rho\sqrt{v}\,]}{(\Sigma\sigma)^2 + \omega^2(\Sigma\varepsilon)^2(\Sigma\,\rho)\sqrt{v}}\right.$$

$$\left. - a^2E_0^2\frac{[\varepsilon][\rho\sqrt{v}\,]}{(\Sigma\rho)\sqrt{v}}\right\}\zeta = \frac{1}{(\Sigma\rho)\sqrt{v}}\left\{ - 2a\rho_1\rho_2\sqrt{2\omega v_1 v_2}\right.$$

$$\left. - 4a^2E_0^2\frac{[\varepsilon\rho\sqrt{v}\,](\varepsilon_2\sigma_1 - \varepsilon_s\sigma_1)}{(\Sigma\sigma)^2 + \omega^2(\Sigma\varepsilon)^2}\right\}\frac{\partial\zeta}{\partial t} \qquad (3.10)$$

with $\Sigma\rho = \rho_1 + \rho_2$, $\Sigma\sigma = \sigma_1 + \sigma_2$, $\Sigma\varepsilon = \varepsilon_1 + \varepsilon_2$. Note that in this section ω and "a" have units of s^{-1} and cm^{-1}.

For simplicity, as in earlier cases, we have used the high frequency approximation $\omega/v \gg a^2$ in deducing Eq. (3.10). Here are some specific cases

(i) $\sigma_1 = \sigma_2 = 0$

$$(\Sigma\rho)\frac{\partial^2\zeta}{\partial t^2} + \left\{ Ta^3 - g[\rho]a + \frac{a^2E_0^2[\varepsilon]}{(\Sigma\rho)\sqrt{v}}\left(2\frac{[\varepsilon\rho\sqrt{v}\,]}{(\Sigma\varepsilon)} - [\rho\sqrt{v}\,]\right)\right\}\zeta$$

$$= \frac{1}{(\Sigma\rho)\sqrt{v}}\{- 2a\rho_1\rho_2\sqrt{2\omega v_1 v_2}\}\frac{\partial\zeta}{\partial t} \qquad (3.11)$$

(ii) $\varepsilon_1 = \varepsilon_2 = \varepsilon$, $\sigma_1 = \sigma$, $\sigma_2 = 0$

$$(\Sigma\rho)\frac{\partial^2\zeta}{\partial t^2}+\left\{Ta^3-g[\rho]a-2a^2E_0^{\,2}\frac{\sigma^2\varepsilon[\rho\sqrt{v}\,]}{(\sigma^2+4\omega^2\varepsilon^2)(\Sigma\rho)\sqrt{v}}\right\}\zeta$$

$$=\frac{1}{(\Sigma\rho)\sqrt{v}}\left\{-2a\rho_1\rho_2\sqrt{2\omega v_1 v_2}-4a^2E_0^2\frac{\varepsilon^2[\rho\sqrt{v}\,]}{\sigma^2+4(\omega\varepsilon)^2}\right\}\frac{\partial\zeta}{\partial t} \qquad (3.12)$$

(iii) $\rho_2 = 0, \sigma_2 = 0, \varepsilon_2 = \varepsilon_0, \sigma_1 = \sigma, \rho_1 = \rho, v_1 = v, \varepsilon_1 = \varepsilon_0\ \varepsilon_r = \varepsilon$

$$\rho\frac{\partial^2\zeta}{\partial t^2}+\left\{Ta^3+g\rho a-2a^2E_0^2\varepsilon\frac{\sigma^2-\omega^2(\varepsilon_0^2-\varepsilon^2)}{\sigma^2+\omega^2(\varepsilon+\varepsilon_0)^2}+a^2E_0^2(\varepsilon_0-\varepsilon)\right\}\zeta$$

$$=\left\{-4a^2\eta+4a^2E_0^2\frac{\varepsilon_0\varepsilon\sigma}{\sigma^2+\omega^2(\varepsilon+\varepsilon_0)^2}\right\}\frac{\partial\zeta}{\partial t} \qquad (3.13)$$

It appears that in cases (ii) and (iii) sustained transverse oscillations are expected when the damping factor in their respective r.h.s. vanishes. In Eq. (3.11), however, the damping factor is always negative and the oscillations appear only in the form of damped motions.

In general terms, we can conclude from Eq. (3.10) that the necessary conditions in order to sustain transverse oscillatory motions are

$$\frac{\sigma_2}{\sigma_1}>\frac{\varepsilon_2}{\varepsilon_1}>\frac{\rho_1\sqrt{v_1}}{\rho_2\sqrt{v_2}}\quad\text{or}\quad\frac{\sigma_2}{\sigma_1}<\frac{\varepsilon_2}{\varepsilon_1}<\frac{\rho_1\sqrt{v_1}}{\rho_2\sqrt{v_2}} \qquad (3.14)$$

that for the case $\rho_1 = \rho_2$ and $v_1 = v_2$ reduce to

$$\frac{\sigma_2}{\sigma_1}>\frac{\varepsilon_2}{\varepsilon_2}>1\quad\text{or}\quad\frac{\sigma_2}{\sigma_1}<\frac{\varepsilon_2}{\varepsilon_1}<1 \qquad (3.15)$$

In the simplest case $\rho_2 = 0, \sigma_2 = \sigma_1 = 0, \varepsilon_2 = \varepsilon_0$ and $\rho_1 = \rho$, the oscillatory frequency is given by

$$\omega^2 = ag + a^2 E_0^2 \frac{(\varepsilon - \varepsilon_0)^2}{\rho(\varepsilon + \varepsilon_0)} + a^3 \frac{T}{\rho} \qquad (3.16)$$

The *electrowave* scales with a^2, an intermediate wavenumber range between the ranges of gravity and capillary waves. The result (3.16) is to be expected based on dimensional analysis. Taken separately $\omega^2 \approx ag$ and $\omega^2 \approx Ta^3/\rho$ are the heuristic estimates for gravity and capillary waves, respectively. With electrical forces and zero conductivity, dimensional analysis gives $\omega^2 \approx EDa^2 / \rho$ ($D = \varepsilon E$). For the case of a field *orthogonal, i.e., transverse* to the interface a similar study has been developped. Once more we have been able to obtain the *dissipative* harmonic oscillator description of the interfatial motions (Chu, Velarde & Castellanos 1989).

4. LIMIT CYCLE OSCILLATIONS.

If we now consider the nonlinear extension of the problem posed in section 2.1 thus limiting ourselves to transverse oscillations we have in dimensionless form

$$\nabla * \underline{v} = 0 \qquad (4.1)$$

$$\frac{\partial w}{\partial t} + u\frac{\partial w}{\partial x} + w\frac{\partial w}{\partial z} = -\frac{\partial p}{\partial z} + \nabla^2 w \qquad (4.2a)$$

$$\frac{\partial u}{\partial t} + u\frac{\partial u}{\partial x} + w\frac{\partial u}{\partial z} = -\frac{\partial p}{\partial x} + \nabla^2 u \qquad (4.2b)$$

and

$$\frac{\partial \theta}{\partial t} + u\frac{\partial \theta}{\partial x} + w\frac{\partial \theta}{\partial z} = w + P^{-1}\nabla^2\theta \qquad (4.3)$$

where we have explicitely indicated, yet in a two-dimensional problem, the full *nonlinear* disturbance system. Besides, in order to rely our analysis to the standard Bénard problem we take here ß as temperature gradient rather than surfactant gradient. Please recall that ß is *positive* when heating the layer from *below*.

The b.c (2.4)-(2.8) at the open surface need to be extended to the nonlinear case. Thus with M, the Marangoni number, $M = -(d\sigma/dT)_0 \, \beta \, l^2/\eta\kappa$, and C the capillary number based upon κ, we now have

$$\frac{\partial \zeta}{\partial t} = w - u\frac{\partial \zeta}{\partial x} \tag{4.4}$$

$$p - \frac{Bo}{CP}\zeta + \frac{1}{N^3}\left[\frac{1}{CP} - \frac{M}{P}(\theta - \zeta)\right]\frac{\partial^2 \zeta}{\partial x^2}$$

$$= \frac{2}{N^2}\left[\frac{\partial w}{\partial z} - \left(\frac{\partial u}{\partial z} + \frac{\partial w}{\partial x}\right)\frac{\partial \zeta}{\partial x} + \frac{\partial u}{\partial x}\left(\frac{\partial \zeta}{\partial x}\right)^2\right] \tag{4.5}$$

$$-\frac{M}{P}\left[\frac{\partial(\theta - \zeta)}{\partial x} + \frac{\partial \theta}{\partial z}\frac{\partial \zeta}{\partial x}\right]$$

$$= \frac{1}{N}\left\{\left(\frac{\partial u}{\partial z} + \frac{\partial w}{\partial x}\right)\left[1 - \left(\frac{\partial \zeta}{\partial x}\right)^2\right] + 4\frac{\partial w}{\partial z}\frac{\partial \zeta}{\partial x}\right\} \tag{4.6}$$

and

$$\frac{\partial \theta}{dz} = 0 \tag{4.7}$$

For later convenience we have denoted the surface deformation with ζ rather than ξ. We see nonlinear contributions like the second term in the r.h.s. of the kinematic b.c. (4.4). Eq.(4.7) prescribes the heat flux at the open surface. $N= (1 + |\partial\zeta/\partial x|^2)^{1/2}$. As in earlier parts of this chapter the air is assumed to be passive and weigthless with respect to the liquid.

The simplest approach to the above posed nonlinear problem is the *single-mode* solution which is expected to be a useful description in a small enough neighborhood of the onset of overstability. Note that if we move into *low* -gravity the capillary length becomes larger and larger thus providing greater relevance to this *single-mode* approximation. On the other hand, as said many times, transverse interfacial disturbances are expected to penetrate little in the liquid; the penetration depth depends indeed on the wavelength and frequency excited and on the viscosity of the liquid. Thus once more the latter assumption gives relevance to the "potential" flow approximation in time-dependent convection or in other terms it justifies the relevance of a study limited to high-frequency motions only. Thus for an arbitrary disturbance f(x,z,t) we set f(x,z,t) ≈ f(z,t) exp(iax) and $\omega_0{}^2 = (Bo + a^2)a/CP$ (Laplace's law for potential flow). Using them, Eq. (4.4) becomes

$$\frac{\partial\zeta}{\partial t} = w + \zeta\frac{\partial w}{\partial z} \tag{4.8}$$

On the other hand Eq. (4.1) at z=ξ is

$$\frac{\partial w}{\partial t} = -\frac{Bo + a^2/N^3}{PC}a\zeta + \frac{M}{PN^3}(\theta - \zeta)a^3\zeta$$

$$-2a\frac{1 + a^2\zeta^2}{N^2}\frac{\partial w}{\partial z} - a^2\left(1 + \frac{2a\zeta}{N^2}\right)w$$

$$+\left(1 - \frac{2a\zeta}{N^2}\right)\frac{\partial^2 w}{\partial z^2} \tag{4.9}$$

Note that neglecting the nonlinear terms and using the high frequency approximation, $\omega_0 >> a^2$, Eq. (4.9) yields, as expected, the (Laplace) *ideal / dissipation-free* harmonic oscillator equation

$$\frac{\partial^2\zeta_0}{\partial t^2} + \frac{Bo + a^2}{CP}a\zeta_0 = 0 \tag{4.10}$$

The zeroth-order (linear) disturbances are ξ_0,

$$w_0 = \frac{\partial\zeta_0}{\partial t}e^{az} \tag{4.11}$$

and

$$\theta_0 \approx \zeta_0 e^{az} + \frac{a}{\sqrt{2P\omega_0}}\left(\frac{1}{\omega_0}\frac{\partial\zeta_0}{\partial t} - \zeta_0\right)\exp\left(\sqrt{2P\omega_0}\, z\right) \tag{4.12}$$

where ω_0 denotes the harmonic frequency in (4.10).

Consideration now of the nonlinear contributions in Eqs. (4.8) and (4.9) up to *cubic* terms leads to

$$\frac{\partial \zeta}{\partial t} = w\,(1 + a\zeta) \tag{4.13}$$

and

$$\frac{dw}{dt} = -\frac{B_0 + a^2}{CP} a\zeta$$

$$+ \frac{Ma^3}{P\omega_0 \sqrt{2Pg\omega_0}} \left(\frac{d\zeta}{dt} - \omega_0 \zeta \right) \left(-1 + 3a\zeta + \frac{5}{2} a^2 \zeta^2 \right)$$

$$- 4a^2 [\, 1 + 2a\zeta(1 - 2a\zeta)] w + \frac{3a^5}{2CP} \zeta^3 \tag{4.14}$$

which after reduction to a single differential equation become

$$\frac{d^2\zeta}{dt^2} + \delta \frac{d\zeta}{dt} + [\omega_0^2 - \omega_0(\delta - 4a^2)]\zeta$$

$$= -\omega_0^2 a\zeta^2 + a\left(\frac{d\zeta}{dt} \right)^2$$

$$+ (\delta - 4a^2)\left(\frac{d\zeta}{dt} - \omega_0 \zeta \right) 2a\zeta - 8a^3 \zeta \frac{d\zeta}{dt}$$

$$- a^2 \zeta \left(\frac{d\zeta}{dt} \right)^2 + \frac{11}{2} a^2 \zeta^2 (\delta - 4a^2) \left(\frac{d\zeta}{dt} - \omega_0 \zeta \right) + 16a^4 \zeta^2 \frac{d\zeta}{dt} + \frac{3a^5}{2CP} \zeta^3 \tag{4.15}$$

where

$$\delta = \frac{Ma^3}{\sqrt{2}\,(P\omega_0)^{3/2}} + 4a^2 \tag{4.16}$$

At $\delta=0$ we have *overstability* from the linear analysis. Positive (respectively, negative) values of δ account for subcritical (respectively, supercritical) motions.

It is useful to rescale both space and time. Thus, with $\xi = a\,\zeta$ and $\tau = \omega_0 t$, and using the high frequency limit (here $\omega_0 \gg a^2 \approx 1$) with $\omega_0^2 \approx 1/CP$, Eq.(4.15) reduces to

$$\frac{d^2\xi}{d\tau^2} + \Delta\frac{d\xi}{d\tau} + \xi$$

$$= -\,\xi^2 + \alpha\xi^3 + \left(\frac{d\xi}{d\tau}\right)^2 - \beta\zeta\frac{d\xi}{d\tau} - \xi\left(\frac{d\xi}{d\tau}\right)^2 - \gamma\xi^2\frac{d\xi}{d\tau} \tag{4.17}$$

where $\Delta=\delta/\omega_0$, $\alpha=3a^3/2CP\omega_0^2$, $\beta=16a^2/\omega_0$ and $\gamma=6a^2/\omega_0$. Thus Eq. (4.17) is the simplest *nonlinear* equation for a *dissipative* oscillator describing the *limit cycle* oscillations of the air-liquid interface. The first two terms in the r.h.s. account for the Duffing aharmonic potential, as expected. We have checked that indeed Eq. (4.15) as well as Eq. (4.17) possess *limit cycle* solution. This has been done both with the computer and using the time-derivative expansion procedure (Chu & Velarde 1989d). Note that the next step is the extension of Eq.(4.17) to the space-modulated case thus providing a set of Boussinesq-like equations for the nonlinear evolution of the open surface (Velarde & Chu 1990).

The time-derivative expansion procedure permits one to obtain in a perturbative scheme both the amplitude and the period of the oscillation in terms of the initial condition and thus to assess the stability of the limit cycle (Chu & Velarde 1989d). On the one hand it can be shown that the limit cycle bifurcates supercritically for Δ (or δ) negative, i.e., we have a supercritical Hopf bifurcation, and on the other hand one obtains the amplitude,

$$\zeta^2_{max} = -\,\frac{2\delta}{3a^2} \tag{4.18}$$

5. SOLITONS EXCITED BY THE MARANGONI EFFECT. AN EXTENSION OF THE KdV EQUATION

5.1. INTRODUCTION

In the preceding sections we have discussed the onset and eventual nonlinear sustainment of some buoyancy-driven and interfacial oscillations. In the latter case these oscillations were either transverse or longitudinal waves. In a closed container these excitations are expected to develop as *standing* periodic motions at the open surface of the liquid when the Marangoni effect is operating. The small viscous penetration depth or the high frequency limit approximations used tacitely restricted consideration to an *infinte* layer depth. If, however, we turn to *shallow* layers with quite a similar viewpoint we may consider as the *ideal / dissipation-free* building block not a harmonic oscillator but rather the nonlinear and dispersive Korteweg-de Vries *soliton* equation (Drazin & Johnson 1989; Lamb Jr 1980; Whitham 1974) to which we can again add dissipation (viscosity, heat diffusivity, ...) and the Marangoni effect. Then the latter effect here triggers and eventually sustains a *traveling* interfacial soliton along the open surface of a liquid layer heated from above (or below, according to the liquid used).

5.2. DISSIPATIVE KdV EQUATION

We consider a *shallow* horizontal liquid layer of "relatively small" thickness "h" as Bénard did. Disturbances upon the quiescent state obey the same continuity and Navier-Stokes equations earlier written to which we add either Fourier's heat equation or Fick's mass diffusion equation. These equations are now suplemented with the corresponding *nonlinear* b.c. at the bottom and at the open surface. As before, for simplicity we shall restrict consideration to the x and z geometry. Here "l" is the "wavelength" or maximum horizontal extent of the expected interfacial deformation. For self-consistency in this section we recall the equations to be used. They are

$$\nabla * \mathbf{v} = 0 \tag{5.1}$$

$$\frac{\partial}{\partial t}\mathbf{v} + \mathbf{v} * \nabla \mathbf{v} = -\frac{1}{\rho} \nabla p + \nu \nabla^2 \mathbf{v} \tag{5.2}$$

$$\frac{\partial}{\partial t}\theta + \mathbf{v} * \nabla \theta = \beta w + \kappa \nabla^2 \theta \tag{5.3}$$

with b.c. at the rigid insulating bottom z=0

$$w = u = 0 \tag{5.4}$$

and

$$\frac{\partial}{\partial z}\theta = 0 \tag{5.5}$$

and b.c. at the surface $z=h+\xi$ (again for convenience we use now ξ and later on ζ)

$$\frac{\partial}{\partial t}\xi = w - u\frac{\partial}{\partial x}\xi \tag{5.6}$$

$$p = \rho g\xi + 2\mu\frac{\partial}{\partial z}w - \sigma_0 \nabla^2_\Sigma \xi \tag{5.7}$$

$$\left[\frac{\partial}{\partial T}\sigma\right] \nabla_\Sigma(\theta - \beta\xi) - \eta\left[\frac{\partial}{\partial z}u + \frac{\partial}{\partial x}w\right] = 0 \tag{5.8}$$

and

$$\frac{\partial}{\partial z}\theta = 0 \tag{5.9}$$

For a first-order analysis we shall disregard the *nonlinear* convective term in the heat equation (5.3) and thus we consider *linear* thermal disturbances upon the viscous and *nonlinear* Navier-Stokes problem. The velocity **v** can be decomposed into its *potential* (φ) and *rotational* (ψ) parts

$$\mathbf{v} = \nabla\varphi + \nabla \times \underline{\psi} \tag{5.10}$$

For the two dimensional problem

$$\underline{\psi} = \psi\underline{i} \tag{5.11}$$

Thus

$$u = \partial\varphi/\partial x - \partial\psi/\partial z \tag{5.12}$$

$$w = \partial\varphi/\partial z + \partial\psi/\partial x \tag{5.13}$$

Noting that $\nabla^2\varphi=0$ we set

$$\varphi = \sum_{j=0}^{\infty} \frac{(-)^j}{(2j)!} z^{2j} \varphi_0^{(2j)} \tag{5.14}$$

On the other hand

$$\psi = \sum_{i=0}^{\infty} z^i \psi_i \tag{5.15}$$

The simplest choices to fit the b.c. are

$$\psi_0 = 0$$

$$\psi_1 = \varphi_{0,x}$$

and $\psi 2 = \psi 3 = \ldots = 0$ or

$$\psi = z\varphi_{0,x} \tag{5.16}$$

with the condition

$$\varphi_{0,xxxt} = \nu\varphi_{0,xxxx} \tag{5.17}$$

The subscripts x, z and t denote derivative with respect to x, z and t , respectively. Thus

$$u = -\frac{z^2}{2}\varphi_0^{(3)} + \frac{z^4}{24}\varphi_0^{(5)} - \ldots \tag{5.18}$$

$$w = \frac{z^3}{6}\varphi_0^{(4)} - \frac{z^5}{120}\varphi_0^{(6)} + \ldots \tag{5.19}$$

Assume now

$$\theta = \theta_0 + z\theta_1 + z^2\theta_2 + \ldots \tag{5.20}$$

Then (5.6) and (5.3) yield

$$\theta_0 = \theta_2 = 0 \tag{5.21}$$

Besides, in (5.3) w has *odd* powers of z only. Thus

$$\theta = z\theta_1 + z^3\theta_3 + \ldots \tag{5.22}$$

or else, using (5.10)

$$\theta = z(3h^2 - z^2)\theta_3 + \dots \qquad (5.23)$$

Differentiating now (5.3) with respect to z gives $\kappa\theta_{zzz}=\beta u_x$ at z=h, that after using (5.18) and (5.23) yields

$$3\kappa\theta_3 = \beta h^2 \varphi_0^{(4)}/2 \qquad (5.24)$$

Then (5.19) becomes

$$\theta = z^5 (\frac{7}{5}h^2 - z^2)\frac{\beta}{252\, h^2 \kappa}\varphi_0^{(4)} + \dots \qquad (5.22)$$

Using the nonlinear kinematic b.c. (5.6)

$$\xi_t = \frac{1}{6}h^3 \varphi_0^{(4)} + \frac{1}{2}h^2 \xi\varphi_0^{(4)} - \frac{1}{120}h^5\varphi_0^{(6)} + \frac{1}{2}h^2 \varphi_0^{(3)}\xi x \qquad (5.23)$$

Using now (5.17) and (5.18) together with (5.8) in (5.2) taken at the open surface and defining $f = \varphi_0^{(3)}$ we have

$$\xi_t = \frac{1}{6}h^3 f_x + \frac{1}{2}h^2\xi f_x - \frac{1}{120}h^5 f_{xxx} + \frac{1}{2}h^2 f\xi_x \qquad (5.24)$$

and

$$f_t - \frac{2g}{h^2}\xi_x + 2\frac{\sigma}{\rho h^2}\xi_{xxx} + \frac{2}{h}\xi f_t - \frac{1}{12}h^2 f_{txx} - \frac{1}{6}h^2 ff_x$$

$$= 2\nu (f_{xx} + f/h^2) \qquad (5.25)$$

Using now the b.c. (5.8), Eq. (5.25) becomes

$$f_t - \frac{2g}{h^2}\xi_x + 2\frac{\sigma}{\rho h^2}\xi_{xxx} + \frac{2}{h}\xi f_t - \frac{1}{12}h^2 f_{xx} - \frac{1}{6}h^2 f_{xx}$$

$$= \frac{2\nu}{h^2}\left[\frac{4}{3}h^2 f_{xx} - \left(\frac{\partial\sigma}{\partial T}\right)\frac{1}{\rho\nu h}(\theta_x - \beta\xi_x)\right] \tag{5.26}$$

For convenience we rescale the unknowns. With "a" (not to be confused with the earlier notation for Fourier modes) the maximum value of ξ we set

$$\theta = \beta h \Theta,\ \xi = a\zeta\ \xi t = \tau/C_0,\ x = ly \text{ and } f = \frac{aC_0}{h^3}v,\quad \text{wih } C_0^2 = 2\,gh$$

Then eqs. (5.28) and (5.30) become

$$\zeta_\tau = \frac{1}{6}v_y + \frac{1}{2}\varepsilon\zeta v_y - \frac{1}{120}\delta^2 v_{yyy} + \frac{1}{2}\varepsilon v\zeta_y \tag{5.27}$$

$$v_\tau - \frac{2gh}{C_0^2}\zeta_y + \frac{2\sigma h}{\rho l^2 C_0^2}\zeta_{yyy} + 2\varepsilon\varphi v_\tau - \frac{\delta^2}{12}v_{\tau yy} - \frac{\varepsilon}{6}v v_y$$

$$= \frac{2\nu}{C_0 l}\left[\frac{4}{3}v_{yy} - \left(\frac{\partial\sigma}{\partial T}\right)\frac{\beta l h}{a\rho\nu C_0}\theta'_y\right] \tag{5.28}$$

with

$$\varepsilon = a/h,\ \delta = h/l \text{ and } \theta'_y = \frac{aC_0 h}{3kl}v_{yy}$$

Thus recalling the definitions of the Bond and Marangoni numbers earlier given and denoting Re= $C_0 l/\nu$ we obtain

$$\zeta_\tau - \frac{1}{6}v_y - \frac{1}{2}\varepsilon\zeta v_y - \frac{1}{2}\varepsilon v\zeta_y + \frac{1}{120}\delta^2 v_{yyy} = 0 \tag{5.29}$$

and

$$v_\tau - \zeta_y + \frac{\delta^2}{Bo}\zeta_{yyy} + 2\varepsilon\zeta v_\tau - \frac{\varepsilon}{6} vv_y - \frac{\delta^2}{12} v_{\tau yy}$$

$$= \frac{2}{630\, Re}[840 + M]\, v_{yy} \tag{5.30}$$

Now we rescale again τ and v. We set $\tau=(6)^{1/2}t$ and $v = -(6)^{1/2}V$. Then Eqs. (5.29) and (5.30) become

$$\zeta_t + V_y + 3\varepsilon\zeta V_y + 3\varepsilon V\zeta_y - \frac{1}{20}\delta^2 V_{yyy} = 0 \tag{5.31}$$

and

$$V_t + \zeta_y + 2\varepsilon\zeta V_t + \varepsilon VV_y - \frac{\delta^2}{12} V_{tyy} - \frac{\delta^2}{Bo}\zeta_{yyy}$$

$$= \frac{\sqrt{6}}{315\, Re}[840 + M]\, V_{yy} \tag{5.32}$$

At zeroth-order we have

$$V^{(0)} = \zeta \quad (V_t + V_y = 0)$$

so that the general solution can be assumed in the form

$$V = V^{(0)} + \varepsilon V^{(1)} + \delta^2 V^{(2)} + \gamma V^{(3)} \tag{5.33}$$

with $\varepsilon = a/h$ and

$$\gamma = \frac{\sqrt{6}}{315\, Re}[\, 840 + M\,]$$

Introducing (5.37) in both equations (5.35) and (5.36) and substracting them the result is

$$\zeta_t + \xi_y + \frac{5}{2}\varepsilon\zeta\zeta_y - \frac{30 - Bo}{60Bo}\delta^2\zeta_{yyy} = \frac{\gamma}{2}\zeta_{yy} \tag{5.34}$$

Thus we see that setting $\gamma=0$, eq.(5.34) provides the Korteweg-de Vries equation

(5-8). $\gamma=0$ demands either $\nu=0$, i.e., no viscosity, which is an irrelevant case in our analysis, or M = -840. The latter result indicates that for such negative value of the Marangoni number the open surface of the liquid layer heated from the air side is excitable in the form of a KdV soliton (Chu & Velarde, 1990; Velarde, Chu & Garazo, 1990). Whether or not the soliton is stable can only be decided by studying the role of the *nonlinear* part in Eq. (5.3) left out here. However, what we can safely say is that, due to the Marangoni effect, M= -840 defines the onset of a soliton excitation in a quiescent shallow liquid layer subjected to a transverse thermal gradient (Weidman, Linde & Velarde, 1990; Garazo & Velarde, 1990). If, however, M is positive we are in the case of Bénard convection and we know that at M= 80 there is the onset of steady cellular Bénard convection.

ACKNOWLEDGMENTS

This chapter is based upon research sponsored by CICYT (Spain) Grant PB 86-651 and by an EEC Grant. Part 1 has been carried out in cooperation with Prof. J. Jiménez-Fernández and part 3 with Prof. A. Castellanos, with whom we also had numerous fruitful discussions. Both authors also acknowledge fruitful discussions and correspondence with Professors J. P. Chabrerie, Ph. Drazin, J. K. Koster, J. C. Legros, H. Linde, A. Sanfeld, R. Sani and P. Weidman and Dr. M. Hennenberg and Dr. A. N. Garazo.

REFERENCES

Chu, X.-L. & Velarde, M.G. 1988, Physicochem. Hydrodyn. **10**, 727

Chu, X.-L. & Velarde, M.G. 1989a, J. Colloid Interface Sci. **131**, 471

Chu, X.-L. & Velarde, M.G. 1989b, Il Nuovo Cimento **D 11**, 1615

Chu, X.-L. & Velarde, M.G. 1989c, Il Nuovo Cimento **D 11**, 1631

Chu, X.-L. & Velarde, M.G. 1989d, Phys. Lett. **A136**, 126

Chu, X.-L., Velarde, M.G. & Castellanos, A. 1989, Il Nuovo Cimento **D 11**, 726

Chu, X.-L. & Velarde, M.G. 1990, Phys. Rev. A (to appear)

Drazin, P.G. & Johnson, R.S. 1989, "Solitons : An Introduction", Cambridge Univ. Press, Cambridge

Garazo, A.N. & Velarde, M.G. 1990, J. Fluid Mech. (submitted)

Jimenez-Fernandez, J. & Velarde, M.G. 1989, Il Nuovo Cimento **D 11**, 717

Lamb Jr, G.L. 1980, "Elements of Soliton Theory", John Wiley, New York

Lamb, H. 1932, "Hydrodynamics", Dover, New York

Landau, L.D. & Lifshitz, E.M. 1959, "Fluid Mechanics", Pergamon, Oxford

Legros, J.C., Sanfeld, A., & Velarde, M.G. 1987, in "Fluid Sciences and Materials Science in Space" (H.U. Walter, Ed.) pp. 83-140, Springer-Verlag, New York and references therein

Levich, B.G., 1962, "Physicochemical Hydrodynamics", Prentice-Hall, Inc., Englewood Cliffs, N.J.
Velarde, M.G. & Normand,C. 1980, Sci. American **243**, 78
Velarde, M.G.(Ed) 1988, "Physicochemical Hydrodynamics. Interfacial Phenomena", Plenum Press, New York.
Velarde, M.G. & Chu, X.-L. 1988, Phys. Lett. A **131**, 403
Velarde, M.G. & Chu, X.-L. 1989a, Phys. Scripta **T25**, 231
Velarde, M.G. & Chu, X.-L. 1989b, Il Nuovo Cimento **D 11**, 707
Velarde, M.G., Chu, X.-L. & Garazo, A.N. 1990, Physica Scripta (to appear)
Velarde, M.G. & Chu., X.-L. 1991, "Interfacial Instabilities", World Scientific, London & Singapore (in preparation)
Weidman, P., Linde, H. & Velarde, M.G. 1990, Phys. Fluids (submitted)
Whitham,G.B. 1974, "Linear and Non-linear Waves", John Wiley, New York

Wave Equations of Heat and Mass Transfer

Stanislaw Sieniutycz
Institute of Chemical Engineering
Warsaw Technical University
00-645 Warsaw, 1 Warynskiego Street, Poland

ABSTRACT
It is shown that under highly nonstationary conditions, where relaxation of diffusive fluxes is essential, an approach based on irreversible thermodynamics leads to a conclusion about the existence of extra terms of inertial type in phenomenological extended model of a viscoelastic body which takes into account the relaxation of diffusional fluxes of mass, heat, and momentum. As a result, the equations of change assume a hyperbolic form, leading to an overall transport model which is non-Fourier and non-Fick in character. Various forms of hyperbolic equations of change are presented. Stability of dissipative wave systems and variational principles of classical (least action) type are discussed. A transfer of harmonic thermal disturbances in solids and the qualitative role of the relaxation effects are briefly characterized.

1. INTRODUCTION
When describing heat, mass and momentum transfer it is customary to typically use one of Fourier's, Fick's or Newton's, constitutive equations. These equations relate irreversible (diffusional) fluxes with respective gradients. In this case the application of conservation laws leads to parabolic equations of change. However, all such standard equations of change (with parabolic terms) have an absurd physical property: a disturbance (thermal, concentrational etc.) at any point in the medium is felt instantly at every other point; that is the velocity of propagation of disturbances is infinite. This paradox is clearly seen certain routine solutions of parabolic equations; for instance, in the case of heat conduction in a semi-infinite solid on the surface of which the temperature may suddenly increase from, for example, $T=0$ to $T= T_{wall}$. The solution, which is based on the error integral, provides $T=0$ for the time $t=0$, but for the

nonvanishing T(x, t) in the whole space implying infinitely fast propagation of the disturbance.

2. THE RELAXATION THEORY OF HEAT FLUX

The above mentioned nonphysical behavior has been pointed out by many authors (Cattaneo 1958, Vernotte 1958, Luikov 1966 and 1969, Chester 1963, Kaliski 1965, Baumeister and Hamil 1968, Bubnov 1976, Lebon 1978, Sieniutycz 1977 and 1979), and others, and the dilemma was resolved by the acceptance of the hypothesis of heat flux relaxation. The link between the hypothesis and certain results of the nonequlibrium statistical mechanics, such as Grad's solution of the Boltzmann kinetic equation (Grad 1958) was found, either in the context of phenomenological equations (Lebon 1978, Jou, Casas-Vazquez and Lebon 1988), or conservation laws (Sieniutycz and Berry 1989 and 1990).

The hypothesis is based on the position that Fourier's law (as well as Fick's and Newton's) is an approximation to a more exact equation, called Maxwell-Cattaneo equation,

$$\mathbf{J}_h = -\lambda \nabla T - \tau_h \frac{\partial \mathbf{J}_h}{\partial t} \qquad (v=0), \tag{1}$$

where τ_h is the relaxation time of heat flux.

The analogous equations for the irreversible fluxes of mass and momentum have also been found. The important notions are:
a) In the case of mass diffusion an equation identical with equation (1) results from the nonstationary version of the Maxwell- Stefan equation under certain broad conditions (Sandler and Dahler 1964).
b) In the case of a momentum diffusion equation of the type similar to equation (1), one finds simply Maxwell's equation for viscoelastic fluids, describing shear stress versus velocity gradient.
c) Experiments in acoustic dispersion and absorption (Carrassi and Morro 1972 and 1973) show the superiority of the generalized (wave) Navier-Stokes equation with respect to the classical Navier-Stokes equation in the high frequency regime.

Eq. (1) when combined with the simplest conservation law for the classical thermal energy (the case of a rigid solid) leads to an equation of change

$$\rho C_p \frac{\partial T}{\partial t} = \lambda\left(\nabla^2 T - \frac{\partial^2 T}{c_0^2 \partial t^2}\right) \quad , \tag{2}$$

where

$$c_0 = \sqrt{a_h/\tau_h} = \sqrt{\lambda/(\rho C_p \tau_h)} \tag{3}$$

is called the propagation speed of the thermal wave. When c_0 approaches infinity, eq. (2) simplifies to the well-known parabolic equation of heat. On the other hand eq. (2) is of hyperbolic type and its solution (Baumeister and Hamil 1968, Luikov 1969) for the above-mentioned case of a semi-infinite solid has the following property: temperature T(x,t) has a jump at the distance $x= c_0t$ from the source of heat placed at x=0 (heating of solid by an ideally mixed fluid). Therefore two regions exist in a solid: the first in which the heat transfer takes place (disturbed region), and the second where the disturbance is not present (undisturbed region). In contrast, as mentioned above, Fourier theory predicts the appearance of the disturbances everywhere, even for distances x_c greater than ct (c is the light speed) which is of course unphysical behavior.

For the hyperbolic case considered, an interesting effect appears for the wall heat flux (x=0) when the "driving force is being turned on". Namely, the wall heat flux, $\mathbf{J}_h(0)$, does not start instantaneously, but rather grows gradually (Baumeister and Hamil 1968) with the rate which depends on the relaxation time τ_h. This is just relaxation in the current and not in the state, the latter being known better than the former one. The acoustic and chemical reaction phenomena may serve as examples of state relaxation whereas the heat and viscous stress relaxation and also current relaxation in electric circuits, (associated with a change in the magnetic energy) are the examples of current relaxation. After some time the wall heat flux arrives at a maximum and then it decreases in time, similarly as in Fourier case. This decrease is a classical effect and it occurs since the temperature gradient at the wall decreases in the course of heating of the solid. Consequently the Fourier and Fick theories are inappropriate for description of the short-time effects, and although relaxation times are typically very small such effects can still have theoretical importance.

Examples of relaxation time data are (τ_h, τ_d, and τ_m, mean the relaxation time for heat, mass diffusion and momentum diffusion, respectively):
Metals $\tau_h=10^{-12}$ s
Gases at normal conditions $\tau_h=10^{-9}$ s
Typical liquids $\tau_h = 10^{-11}\text{-}10^{-13}$ s.

Relaxation times can be much greater in rarefied gases, viscoelastic liquids, capillary porous bodies, dispersed systems, Brownian systems, helium II: for capillary porous bodies, for example, Luikov evaluated effective $\tau_d=10^{-4}$s (Luikov 1966). Calculations exploiting some drying experiments (Mitura 1981) seem to confirm the order of magnitude of these evaluations, (Suminska 1983, Sieniutycz 1989).

For other special systems one has the following values of τ:
Brownian diffusion in the diameter range $10^{-7}\text{-}10^{-3}$m:
$\tau_d= 3\times10^{-8}\text{-}3$ s

(Sieniutycz 1984). These evaluations were based on use of some experimental data (Davies 1966) and the experimentally confirmed Stokes term of the equation of motion.
Liquid helium:
$\tau_h = 4.7 \times 10^{-3}$ s
(Peshkov 1960, Keesom 1949, Bubnov 1982)
Turbulent flows:
$\tau = 10^{-3} - 10^{3}$ s
(Luikov 1966, Monin and Yaglom 1967, Builties 1977).
Kinetic theory (Natanson 1986, Uhlenbeck 1949, Wang Cheng and Uhlenbeck 1951, Sieniutycz 1977) predicts the following upper limit velocities for the propagation of disturbances (propagation speeds):

$$c_h \equiv \sqrt{\frac{D_h}{\tau_h}} = c_d \equiv \sqrt{\frac{D_d}{\tau_d}} = c_m \equiv \sqrt{\frac{\nu}{\tau_m}} = \sqrt{\frac{P}{\rho}} \tag{4}$$

where ν is the kinematic viscosity or the diffusivity of the momentum. Hence one can compute the relaxation times as

$$\tau_h = \frac{\rho D_h}{P} \qquad \tau_d = \frac{\rho D_d}{P} \qquad \tau_m = \frac{\rho \nu}{P} = \frac{\eta}{P} \tag{5}$$

where η is the dynamic viscosity. For liquids the shear modulus G appears instead of pressure in the formula for the propagation speed

$$c_0 = \sqrt{\frac{G}{\rho}} \tag{6}$$

At present no exact general criterion is available when it is necessary to include relaxation terms in equations of change. Usually one assumes that these terms are essential when the frequency of the fast variable transients is comparable (or greater) to the reciprocal of a longest relaxation time. The above-mentioned cases are usually cited when these terms should be practically significant for: viscoelastic fluids, capillary porous bodies, dispersed systems, rarefied gases, and helium II; for high rate, unsteady state processes. Experiments confirming the wave nature of heat are available (Ascroft 1976). Among the solid state physicists a frequent opinion prevails, that underestimation of the so called "balistic" regime of the phonon transfer, which is in fact our wave regime, lead researches to the unusually large number of invalid statements about the heat transfer in solids.

3. EXTENDED THERMODYNAMICS OF COUPLED HEAT AND MASS TRANSFER

As pointed out earlier, the nonstationary equation of diffusive motion, i.e. Maxwell-Stefan equation of diffusion, leads to the presence of extra (relaxation terms in the phenomenological equations. Below we will show that the relaxation terms are also justified by irreversible thermodynamics. However, the classical expression for entropy and entropy source do not apply in the present (relaxation) case as every relaxation phenomenon is a consequence of the local nonequilibrium in the macroscopic medium. Both phenomenological (Sieniutycz 1981) and statistical approaches (Lebon 1978, Jou, Lebon and Casas-Vazquez 1988, Sieniutycz and Berry 1989), the latter based on Enskog and Grad's iteration methods, lead to the conclusion that the entropy of a locally nonequilibrium medium differs from the static (i.e. equilibrium) entropy, and the difference depends on all diffusiving fluxes $\mathbf{J}_1, \mathbf{J}_2, ...\mathbf{J}_{n-1}, \mathbf{J}_q$. This is the objective of the so-called extended thermodynamics, Nettleton 1960, 1969, 1986, 1987, Mc Lennan 1974, Gyarmati 1977, Nonnenmacher 1980, Lebon, Jou and Casas-Vazquez 1980, Jou, Lebon and Casas-Vazquez 1988, Garcia -Colin at al 1987, Eu 1982, 1986, and 1987, Sieniutycz 1979 and 1981, Sieniutycz and Berry 1989 and 1990), and many others. Despite various differences in diverse approaches the basic procedure can be characterized by the representative scheme outlined below. For brevity the momentum diffusion is neglected.

The difference between the true local entropy of a nonequilibrium state, s', and the termostatic entropy, s, (evaluated for the same internal energy e), called relaxation entropy, is associated with the tendency of every element of a continuum to recover thermodynamic equilibrium during the vanishing of diffusive fluxes of heat and mass, or the relaxation of the viscous stresses (maximum s' equals to s for $\mathbf{J}=0$ and for $\Pi=0$). Since the relaxation is an irreversible process the relaxation entropy, s'-s, corresponding with a stable equilibrium, is always the negative quantity, the consequence of the concavity of the entropy around the macroscopically stable equilibrium. The presence of the relaxation entropy in the total nonequilibrium entropy expression can be described by the following formula (Sieniutycz 1981)

$$ds' = T^{-1}de + pT^{-1}d\rho^{-1} + \sum_{1}^{n-1} (\mu_n - \mu_i)T^{-1}dy_i + \sum_{1}^{n}\sum_{1}^{n} g_{ik}\mathbf{J}_i.d\mathbf{J}_k \tag{7}$$

This is a simplest possible generalization of the classical Gibbs equation which in the linear case can be transformed to the concise integrated matrix form

$$s'(\mathbf{z},\mathbf{J}) = s(\mathbf{z}) + \frac{1}{2\rho G}\mathbf{J}^T\mathbf{C}^{-1}\mathbf{J} \tag{8}$$

In the above equations:

$$\mathbf{z} = \text{col}\,(y_1, y_2, ...y_{n-1}, e) \qquad \text{- state matrix} \tag{9}$$

$$\mathbf{J} = \mathrm{col}\,(\mathbf{J}_1, \mathbf{J}_2, \ldots \mathbf{J}_{n-1}, \mathbf{J}_e) \qquad \text{- flux matrix} \tag{10}$$

$$\mathbf{u} = \mathrm{col}\,\left(\frac{\mu_n - \mu_1}{T}, \frac{\mu_n - \mu_2}{T}, \ldots \frac{\mu_n - \mu_{n-1}}{T}, \frac{1}{T} \right) \qquad \text{- transfer potential matrix} \tag{11}$$

$$\mathbf{C} = \frac{\partial \mathbf{z}}{\partial \mathbf{u}} = \left(\frac{\partial^2 s}{\partial z_i \partial z_j}\right)^{-1} \leq 0 \qquad \text{- entropy capacity matrix} \tag{12}$$

$$g_{ik} = \mathbf{C}/G = \mathbf{C}/(\rho c_0^2) \qquad \text{- inertial matrix} \tag{13}$$

One may ask about the entropy source form which corresponds to the Gibbs equation (7) or with its integrated counterpart, eq.(8). The answer is found analogously, as in classical thermodynamics, by combining the Gibbs equation with the equations describing the conservation laws for mass and energy:

$$\rho\frac{dy_i}{dt} = -\nabla.\mathbf{J}_i \tag{14}$$

$$\rho\frac{de}{dt} = -\nabla.\mathbf{J}_q \tag{15}$$

As a result, the following entropy balance is obtained

$$\rho\frac{ds'}{dt} = -T^{-1}\nabla.\mathbf{J}_q - \sum_1^{n-1}\left(\frac{\mu_n - \mu_k}{T}\right)\nabla.\mathbf{J}_k + G^{-1}\mathbf{J}^T.\mathbf{C}^{-1}.\frac{d\mathbf{J}}{dt} \tag{16}$$

This equation can be split into the sum of divergence and source terms

$$\rho\frac{ds'}{dt} = -\nabla.[T^{-1}(\mathbf{J}_q - \sum_1^{n} \mu_k \mathbf{J}_k)] - \mathbf{J}_q.\nabla T^{-1} + \sum_1^{n-1} \mathbf{J}_k.\nabla\left(\frac{\mu_n - \mu_k}{T}\right) + G^{-1}\mathbf{J}^T.\mathbf{C}^{-1}.\frac{d\mathbf{J}}{dt} \tag{17}$$

or more concisely

$$\rho \frac{ds'}{dt} = - \nabla . \mathbf{J}_s - \mathbf{J}^T . [\nabla \mathbf{u} + (GC)^{-1} . \frac{d\mathbf{J}}{dt}] \tag{18}$$

where the diffusive entropy flux $\mathbf{J}_S$ has been defined as

$$\mathbf{J}_s = T^{-1} (\mathbf{J}_q - \sum_1^n \mu_k \mathbf{J}_k) \tag{19}$$

The condition of nonnegativeness of the entropy source in eq. (18) leads to the following matrix phenomenological equation

$$\mathbf{J} = \mathbf{L} . [\nabla \mathbf{u} + (GC)^{-1} . \frac{d\mathbf{J}}{dt}] \tag{20}$$

which simplifies to the well-known Onsager's relationship

$$\mathbf{J} = \mathbf{L} . \nabla \mathbf{u} \tag{20a}$$

when G and c_0 tend to infinity. The result obtained, eq.(20), can be written in the form of the equations

$$\mathbf{J} + \tau . \frac{d\mathbf{J}}{dt} = \mathbf{L} . \nabla \mathbf{u} \tag{21}$$

$$\tau = - \mathbf{L} . \mathbf{C}^{-1} / G \qquad \text{and} \qquad G = \rho c_0^2 \qquad (G^{\text{ideal gas}} = \rho RT/M) \tag{22}$$

which constitute the phenomenological equation and the relaxation time definition, respectively. Thus we have found the matrix generalization of the Maxwell-Cattaneo equation, eq. (20), describing coupled heat and mass diffusion with the finite wave speed, as well as the formula, eq. (22), which is suitable to compute the elements of the relaxation matrix. One may see that the matrix τ can be computed on the basis of the two important thermodynamic matrices, the capacity matrix, **C**, (the reverse of entropy hessian in its natural frame), and Onsager's matrix **L**, eq. (20a). The latter is usually found from an experiment. Since the product $-\mathbf{L}\mathbf{C}^{-1}\rho^{-1}$ represents the general matrix of diffusion defined in a classical manner (including thermal diffusion terms), equation (22) can be written in the simple alternative form as

$$\tau = -\mathbf{D}/c_0^2 \tag{23}$$

The special case of this result, for the pure heat transfer, was previously known (Luikov 1969, Chester 1963).

4. VARIOUS FORMS OF WAVE EQUATIONS FOR COUPLED HEAT AND MASS TRANSFER

The conservation equations

$$\rho\frac{d\mathbf{z}}{dt} = -\nabla.\mathbf{J} \tag{24}$$

constitute the matrix form of the conservation laws for mass and energy, eqs.(14) and (15). Taking the divergence of eq. (24) and using phenomenological relationships, eqs.(21) through (23), the matrix system of the wave equations is obtained,

$$\frac{d\mathbf{z}}{dt} = \mathbf{D}.(\nabla^2\mathbf{z} - \frac{d^2\mathbf{z}}{c_0^2 dt^2}) \tag{25}$$

If, instead of the state variables, z, eq.(19), the transport potentials, eq.(11), are used an altermative form is found,

$$\rho C.\frac{d\mathbf{u}}{dt} = -\mathbf{L}.(\nabla^2\mathbf{u} - \frac{d^2\mathbf{u}}{c_0^2 dt^2}) \tag{26}$$

In the resting systems the substantial derivatives simplify to the partial derivatives and eqs. (25) and (26) contain on their right sides d'Alembertians instead of the usual Laplacians. The form of the wave equations representing resting media is useful for finding relativistic generalization of the wave equations (25) and (26). To obtain this generalization it is sufficient (Sieniutycz 1979) to perform the Lorentz transformation in every equation describing resting system. Of course when Gallilean transformations are used the set, eqs.(25) and (26) is recovered. The relativistic counterparts of the eqations (25) and (26) are,

$$\gamma.\frac{d\mathbf{z}}{dt} = \mathbf{D}.(\nabla^2\mathbf{z} - \frac{\partial^2\mathbf{z}}{c_0^2\partial t^2}) - (\frac{1}{c_0^2} - \frac{1}{c^2})\gamma^2\mathbf{D}.\frac{d^2\mathbf{u}}{dt^2} \tag{27}$$

$$\rho\gamma \mathbf{C}.\frac{d\mathbf{u}}{dt} = - \mathbf{L}.(\nabla^2\mathbf{u} - \frac{\partial^2\mathbf{u}}{c_0^2\partial t^2}) + (\frac{1}{c_0^2} - \frac{1}{c^2})\gamma^2\frac{d^2\mathbf{u}}{dt^2} \tag{28}$$

where,

$$\gamma = 1/(\sqrt{1 - v^2/c^2})$$

and c is the light speed. See references (Sieniutycz 1979, Pavon, Jou and Casas-Vazquez 1980, Kranys 1966) for more information.

The physical structure of eq.(25) is obscured by its matrix form. We operate here with the not so common quantities, **z** and **u**. Therefore another approach was made (Sieniutycz 1981) where the fluxes, forces and diffusivities were transformed to the quantities related to the pure heat flux rather than to the energy flux (as in eqs.(21) and (25)). As a result of such transformations the following wave system is obtained,

$$\frac{d\mathbf{y}}{dt} = \mathbf{D}_m.(\nabla^2\mathbf{y} - \frac{d^2\mathbf{y}}{c_0^2 dt^2}) + \frac{\mathbf{D}_T}{T}(\nabla^2 T - \frac{d^2T}{c_0^2 dt^2}) \tag{29}$$

$$\rho C_p\frac{dT}{dt} = -\rho T\mathbf{D}_T^T\mathbf{C}_m^{-1}.(\nabla^2\mathbf{y} - \frac{d^2\mathbf{y}}{c_0^2 dt^2}) + \lambda(\nabla^2 T - \frac{d^2T}{c_0^2 dt^2}) \tag{30}$$

with $\mathbf{C} = \text{diag}\,(\mathbf{C}_m^{-1}, -C_p^{-1}T^2)$ and $\mathbf{D}_T = (\rho T)^{-1}\mathbf{L}_T$

where $\mathbf{L}_T$ is the part of Onsager's matrix related to the thermal diffusion. This set operates with the most common variables, temperature T and concentrations y_i and simplifies to the classical set for c_0 approaching infinity. It is expected that equations (29) and (30) will describe heat and mass transport better than the classical ones especially in the case of highly nonstationary cases, (e.g. during the travel of sound, electromagnetic waves through the medium when the thermal diffusion is being intensified, or when the ultrasonic or dielectric drying of solutions of solids takes place.

Analogously the inclusion of relaxation terms into the momentum transport equation can be performed. Clearly the Maxwell's equation of viscoelastic body is then obtained as the phenomenological equation. It is seen that the relaxation theory of heat and mass transfer exhibits a certain connection to the rheological concepts. An extension of these concepts was also observed in the so-called power models of diffusion

$$\sigma_{ik}^{0} = -\tilde{a}_m(\nabla^0 \mathbf{v})^p \tag{31}$$

$$\mathbf{J}_h = -\tilde{a}_h(\nabla^0 T)^p \tag{32}$$

$$\mathbf{J}_d = -\tilde{a}_d\,(\nabla y\,)^m \tag{33}$$

These equations may also lead to the finite propagation of disturbances as the simplest equations of the wave theory,

$$\sigma_{ik}^{0} = -2\eta\nabla^0\mathbf{v} - \tau_m\frac{d\sigma_{ik}^{0}}{dt} \tag{34}$$

$$\mathbf{J}_h = -\lambda\nabla T - \tau_h\frac{d\mathbf{J}_h}{dt} \tag{35}$$

$$\mathbf{J}_d = -\rho D\nabla y - \tau_d\frac{d\mathbf{J}_d}{dt} \tag{36}$$

which can be derived from the memory concepts (Berkovsky and Bashtovoi 1972, Luikov and Berkovsky 1974, Nunziato 1971).

Since $\tau=D/c_0^2$, the knowledge of the propagation velocity c_0 is essential. The phonon gas theory has been applied (Chester 1963, Guyer and Krumhansl1964, Prohofsky and Krumhansl 1964) to evaluate c_0. The random walk model (Goldstein 1953) also leads to values of c_0. The use of sound dispersion and absorption experiments for the computation of c_0 have also been proposed (Bubnov 1976).

STABILITY OF DISSIPATIVE WAVE SYSTEMS

Although the classical (parabolic) and wave equations have the same steady state solutions the stability depends on the transient behavior which is different in both cases. It is well known (Glansdorff and Prigogine 1971) that the stability of the local equilibrium transients with respect to small perturbations described by parabolic equations can be proved by considering the second differential of entropy. An analogous method was used (Lebon and Casas-Vazquez 1976, Lebon and Boukary 1982, Sieniutycz 1981, Sieniutycz and Berry 1990) for the wave equations of heat (equations of hydrodynamics were included in the last three references cited, but we will neglect them here). Taking into account that in the relaxation (wave) case the entropy expression contains the flux term one has

$$\delta^2 S = \int_V \rho(\delta T^{-1}\delta e + \sum_1^{n-1} \delta\frac{\mu_i}{T}\delta y_i + \sum_1^{n-1} \delta\zeta_i . \delta \mathbf{J}_i)dV \tag{37}$$

where ξ_i is the adjoint variable of $\mathbf{J}_i$ in the expression describing perfect differential of entropy. Using this expression as the Liapounov functional for the disturbances of e, y_i, and J_i related by the conservation and phenomenological equations (14) (15) and (20) it was found (Sieniutycz 1981) that if

$$C \leq 0 \tag{38}$$

the following holds

$$\delta^2 S \leq 0 \quad \text{and} \quad \frac{\partial \delta^2 S}{\partial t} \equiv \int_V (\delta \mathbf{J}.\mathbf{L}^{-1}.\delta \mathbf{J})dV \geq 0 \tag{39}$$

for the constant potentials at the boundaries of the system. This means that the negativeness of the entropy capacity **C** (well known in classical thermodynamics) is the sufficient stability condition.

Alternatively an excess entropy production functional was tested as a Liapunov functional (Sieniutycz 1981a)

$$\widetilde{V} = \frac{1}{2}\int_V \{\sum_i \sum_k L_{ik} \nabla\delta u_i \nabla \delta u_k + \sum_i \sum_k \frac{L_{ik}}{c_0^2} \frac{\partial \delta u_i}{\partial t}\frac{\partial \delta u_k}{\partial t} \}dV \geq 0 \tag{40}$$

For the transients of the perturbations governed by eq. (26) it was obtained that

$$\frac{\partial \widetilde{V}}{\partial t} = \int_V \{\sum_i \sum_k \rho C_{ik} \frac{\partial \delta u_i}{\partial t}\frac{\partial \delta u_k}{\partial t} \}dV \leq 0 \tag{41}$$

which again proves that the negativeness of **C** is the sufficient condition of stability.

Thus similarly as in the classical situation of infinite c_0 it has been concluded that the trajectories of simultaneous heat and mass transfer processes are always asymptotically stable if they are described by our wave model. Broadly speaking this means that the two arbitrary solutions $\mathbf{u}(\mathbf{x},t)$ and $\mathbf{u}_0(\mathbf{x}, t)$ will approach one another despite the essential inertial terms existing for wave equations. This also means that one should not expect principally new results concerning stability of transients around equilibrium (where the inequality $\mathbf{C}< 0$ always holds) when the wave equations are used.

6. VARIATIONAL PRINCIPLES

An intryguing fact is (Sieniutycz 1977, 1978, 1979, 1980, 1981c, 1982, 1983, 1985, 1987,1988,1989, Sieniutycz and Berry 1989, 1990) that the set of the wave equations of heat and mass transfer can be derived from a certain modification of Hamilton's principle of least action taken in the form containing the relaxation time τ. When, for example, the pure heat transfer is considered an action functional of the resting solid has the form

$$\tilde{S} = \int_{V \times t} \varepsilon^{-1}\left(\frac{1}{2}\mathbf{J}_h^2 c_0^{-2} - \frac{1}{2}\rho_h^2 + \mathbf{J}_h . \frac{\partial \mathbf{H}_h}{c_0^2 \partial t} + \rho_h \nabla . \mathbf{H}_h\right) \exp(t/\tau) dV dt \tag{42}$$

where ε is the equilibrium energy of an undisturbed state, $\mathbf{J}_h$ is pure heat flux, $\mathbf{H}_h$ is the heat (Biot's) vector, $\rho_h = \rho C_p(T - T_{eq})$ is the "volumetric charge of the nonequilibrium enthalpy" and ρ is the mass density. The stationarity conditions for the action (42) are

$$\mathbf{J}_h = \frac{\partial \mathbf{H}}{\partial t} \tag{43}$$

$$\rho_h = -\nabla . \mathbf{H} \tag{44}$$

$$\frac{\partial \rho_h}{\partial t} + \nabla . \mathbf{J}_h = 0 \tag{45}$$

$$\frac{\partial \mathbf{J}_h}{\partial t} + \frac{\mathbf{J}_h}{\tau} = -\nabla \rho_h = -\rho C_p \nabla T \tag{46}$$

the last result being an equation of heat conduction equivalent to the Maxwell-Cattaneo equation. From eqs(45) and (46) the wave equation (2) is obtained. This means that a complete description of the thermal field is obtained from a modified principle of least action including the relaxation exponential term. In a similar manner many new variational formulations leading to wave equations of heat, mass and electric charge transport have been found (Vujanovic 1971, Vujanovic and Djukic 1972, Vujanovic and Jones 1988, Lebon 1976, Bhattacharya 1982, and Sieniutycz 1978, 1979, 1980, 1982, 1983, 1984, 1985, 1987, 1988). For example, the functional which leads directly to the wave equations (25) or (26) is of the following form

$$\tilde{S} = \int_{V \times t} \frac{1}{2} [\exp(\mathbf{C}\mathbf{L}^{-1} \rho c_0^2 t).\mathbf{L}]:[\nabla\mathbf{u}\nabla\mathbf{u} - c_0^{-2}\dot{\mathbf{u}}\dot{\mathbf{u}}] dV dt \tag{47}$$

which holds for a resting medium, **v**=0. The Gallilean and relativistic generalizations of this functional for the moving systems were found (Sieniutycz 1979). A certain relation to the minimum entropy production theorem can be noted. Namely, for **C** tending to zero i.e. when the process becomes purely dissipative and for the stationary state the principle of minimum entropy production

$$\delta S = \delta \int_V (\mathbf{L}:\nabla\mathbf{u}\nabla\mathbf{u}\, dV) = 0 \tag{48}$$

is recovered from eq. (47). Thus the variational approach based on the modified principle of least action leads to a counterpart of the minimum entropy production theorem for the highly nonstationary processes in which both dissipative and dynamical terms are essential. The Lagrangian representation of diffusion (replacing Euler representation used here) has also been used (Sieniutycz 1983) in the least action type formulations.

Broad applications of the variational methods are only simply and briefly characterized here. For example, one can study the drying process and a moisture content distribution Y(x, t) in a resting infinite plate the sides of which are maintained at a constant concentration (i.e. $Y=Y_{eq}$ and Biot's number is tending to infinity) and the initial moisture content is parabolic (Sieniutycz 1985). Thus, the initial and boundary conditions pertain to the onset of the so-called second drying period i.e.

$$Y = Y_e + (Y_{max} - Y_e)[1 - (x/X)^2] \quad \text{at } t = 0 \quad \text{for } -X<x<X \tag{49}$$

$$Y = Y_e \quad \text{at } x = +X \text{ and } x = -X \text{ for } t>0 \text{ if } Bi^{-1} = 0 \tag{50}$$

On has to determine approximate solution of the equation

$$\frac{\partial Y}{\partial t} = D(\nabla^2 Y - c_0^{-2}\frac{\partial^2 Y}{\partial t^2}) \tag{51}$$

by minimizing the functional

$$\tilde{S} = \frac{1}{2}\rho D \int_0^{t_f} \int_{-X}^{X} [c_0^{-2}(\frac{\partial Y}{\partial t})^2 - (\frac{\partial Y}{\partial x})^2] \exp(t/\tau) dx dt \tag{52}$$

the stationarity condition of which is the wave equation (51). Applying the Kantorovich method, (Kantorovich and Krylow 1958), one may assume the following form for the moisture content field

$$Y(x, t) = Y_e + (Y_{max} - Y_e)[1 - (x/X)^2]h(t) \qquad h(0) = 1 \tag{53}$$

where h(t) is an unknown function of time which must extremize the functional (52). Applying eq.(53) into eq.(52) one finds

$$\tilde{S} = \frac{8\rho D(Y_{max} - Y_e)^2 X}{15} \int_0^{t_f} [c_0^2 \dot{h}(t)^2 - 2.5X^{-2}h^2(t)] \exp(t/\tau) dt \tag{54}$$

The best approximating function h(t) must obey the Euler equation for the integral (54) which is

$$\tau \ddot{h} + \dot{h} = -2.5\, h\, D\, X^{-2} \tag{55}$$

Hence, our moisture content field is determined by eqs. (53) and (55). An averaging procedure applied for eqs. (53) and (55) gives

$$h(t) = \frac{\overline{Y}(t) - Y_e}{\overline{Y}(0) - Y_e} \qquad \text{with } \overline{Y} \equiv \frac{1}{2X}\int_{-X}^{X} Y(x, t) dx \tag{56}$$

and hence eq. (55) assumes the form containing the average moisture content

$$\tau\frac{d^2\overline{Y}}{dt^2} + \frac{d\overline{Y}}{dt} = -K(\overline{Y} - Y_e) \qquad (57)$$

This is none other than the generalized drying equation for the second period of drying taking into account relaxation (τ term), with the well-known drying coefficient $K=2.5DX^2$. Obviously for $\tau=0$ the conventional drying equation is recovered. Eq. (57) was used for experimental evaluation of the drying relaxation times from the drying kinetics curves (Mitura 1981, Suminska, 1983). Effective values of τ are of the order of magnitude 10^{-4} seconds in agreement with the Luikov's evaluations (Luikov 1966). A comprehensive investigation of the frequency dependent properties of eq.(57) was made (Sieniutycz 1989). Many other applications of the variational principles with exponential relaxation terms, (for example, in the context of the fluid mechanics and boundary layer theory), are presented in the book (Vujanovic and Jones 1988).

With a limited success the variational techniques of this kind can be applied to chemically reacting systems where a coupled set of equations containing a non-Fick's diffusion equation and non-Gulberg and Waage kinetics is obtained for a resting fluid (Sieniutycz 1987a). However an extension of the variational method using the relaxation time τ to the case of moving fluids with an arbitrary variable velocity $\mathbf{v}(\mathbf{x},t)$ is probably impossible. In fact, considerable difficulties have been observed to make this approach consistent with the exact nontruncated form of the conservation (or balance) laws for the energy, momentum, mass, and entropy. Therefore, it seems that the use of the relaxation time terms in the functionals like eq.(42) or (47) is more mathematical than physical concept, and the physical variational structure taking into account irreversibility has yet to come. Recent results in variational treatment of conservation laws (Sieniutycz 1988, 1990, Sieniutycz and Berry 1989, 1990) indicate the power of the entirely new approach in the variational description of nonequilibrium processes. Its physical origin is based on the independent transfer of the entropy and mass. This approach treats the matter and the entropy as the two fundamental entities, distinguishes the material and termal degrees of freedom, indicates the appropriateness of the momentum carried by either of these degrees of freedom, and allows to construct a new general structure of nonequilibrium dynamics using the material and thermal momenta as well as corresponding quantum phases as the natural variables of the Gibbs fundamental equation out of equilibrium. It is presented as the separate paper in this series (Sieniutycz 1990).

7. OTHER APPLICATIONS

Various other applications are:

Analytical description of thermal laser shocks (Domanski 1978)
Impulsive radiative heating of solids (Grigoriew 1979)
Short time heat conduction in thin surface layers (Kao 1977)
Short time contacts and collisions (Kazimi and Erdman 1973)
Fast phase changes; melting (Sadd and Didlake 1977)
Short time heat conduction transients (Wiggert 1977)

Experiments in solid helium and ionic crystals, confirming wave nature of heat (Ascroft 1976)
Interference and diffraction of heat waves (Pellam 1961)
Wave behavior of atmospheric isotherms; Shuleikin works (Bubnov 1976)
Brownian systems (Davies 1966, Sieniutycz 1984)
Acoustic dispersion in high frequency regime (Carrasi and Morro 1972, 1973)
Chemical transients and extended affinities (Sieniutycz 1987a)
"Hyperbolic reactors"(Holderith and Reti 1979, Reti 1980)
Multiphase chemical reactions (Reti and Holderith 1980)

8. HIGH FREQUENCY BEHAVIOR OF THERMODYNAMIC SYSTEMS

One may note that the drying equation (57) has the form typical of a damped oscillator. The notion that the (dissipative) drying system can be described by an equation of this kind is rather surprising because such an equation may often imply damped or undamped oscillations. However the comprehensive analysis of the material constant magnitude (Sieniutycz 1989) indicates that the physical nature of the drying is such that only forced oscillations are possible, no free oscillations can occur, which is, of course, in agreement with our experience. Clearly our wave drying equation does not imply the free oscillations for the real material constants. Consequently, although our thermodynamic system is described in the same manner as the mechanical or electrical system, the qualitative behavior of this system is quite different; it is overdamped. In general the overdamping effect prevail, thus, as an analysis of the steady forced oscillation shows, the maximum of the rate amplitude occurs for the frequency $\omega_0=(K/\tau)^{1/2}$ in the form of plateau which extends in the vast range of frequencies ω between K and τ (Sieniutycz 1989). When $\tau=0$ (conventional theory) the plateau extends to infinity. Therefore the discrepancies between the two theories, wave and conventional, start at the frequencies being of the order of τ^{-1}. For lower ω an excellent agreement between the two theories is obtained i.e. each theory is acceptable. It can be shown that the efficiency of energy transmission due to fast disturbances is practically equal to zero in the classical regime $K < \omega < \tau^{-1}$. In the nonclassical regime, described by the relaxation equation, this efficiency can approach even unity in the high frequency range due to the predominant role of the inertial terms. This fact serves as a justification of a growing interest in the investigation of high frequency phenomena with a finite wave speed.

The thermodynamics of these phenomena can be investigated in the context of the kinetic theory (Grad 1958, Eu 1986, 1987, Jou, Casas-Vazquez and Lebon 1988) and hydrodynamic principle of least action by taking into account an additional kinetic term in the kinetic potential, (Sieniutycz 1988, 1990). The origin of this term can also be found in the kinetic theory (Sieniutycz and Berry 1989, 1990). In the least action approach, which is believed to be a physical approach, the exponential relaxation term, introduced here in Section 6, is abandoned, and the primary role is played by a generalized kinetic energy of diffusion involving heat terms and quantum phases. The partial derivative of the kinetic potential density, L, with respect to the velocity of entropy transfer, having the significance of a thermal momentum, is an important physical variable of the extended Gibb's equation involving heat terms (Sieniutycz and Berry 1989, 1990, Sieniutycz 1990).

The idea of thermal momentum is the physical idea emerging from the fact that any heat flux, an ordered quantity, causes a decrease of the system entropy [(eq. 8), for instance] from its isergetic equilibrium value, or the corresponding increase of the internal energy taken at given entropy (an increase of the availability of the system). Since the nonequilibrium increase of the internal energy Δe is well defined in terms of Grad's solution of Boltzmann kinetic equation (Sieniutycz and Berry 1989) one has to expect to find the corresponding momenta as every Lagrangian approach requires. In the frame moving with the fluid the thermal momenta are the internal momenta similar to the diffusional momenta of mass. All internal momenta compensate in the fluid frame giving rise to the vanishing hydrodynamic velocity. They are also essential in evolution equations, preserving the finite propagation of heat and diffusion. An apparent redefinition of the hydrodynamic velocity is required when taking into account the contribution of the thermal momenta (Sieniutycz 1990). Neglect the thermal momentum leads inevitably to the local equilibrium description, excluding any deviation of the internal energy or entropy from the equilibrium Gibbs surface. However such exclusion is in disagreement with the results of the kinetic theory. Consequently, the meaning of the thermal momentum is an an implicit consequence of the kinetic theory, but it may also be viewed independently as a basic starting point of the Lagrangian approach based on the kinetic potential involving the material and thermal coordinates,velocities, momenta and quantum phases. Due to the independence property of the thermal terms from those involving masses, all dissipative phenomena, heat, mass diffusion and viscous motion, emerge naturally from the motion of the entropy independent of mass, and any dissipation can be simply regarded as an effect of the motion of various degrees of freedom with respect to the center of mass of the system. The theory changes essentially the structure of the energy-momentum tensor and conservation laws in the system far from equilibrium preserving at the same time its correspondence with the well known structure of these laws for close-to-equilibrium and nondissipative equilibrium fluids (perfect fluids).

As predicted by the theory, the invariant nonequilibrium temperature of the system, defined in terms of the kinetic potential density or energy density is

$$T = -\left(\frac{\partial L}{\partial S}\right)_{v_s} = \left(\frac{\partial E}{\partial S}\right)_{i_s} \tag{58}$$

where $\mathbf{v}_S$ is the entropy transfer velocity and $\mathbf{i}_S$ is the corresponding thermal momentum. T obeys the requirements of the canonical formalism and it is, at the same time, the appropriate coefficient linking the heat flux and diffusional entropy flux (Sieniutycz 1988, 1990, Sieniutycz and Berry 1989, 1990). This fact allows one to construct a consistent theory of the nonequlibrium potentials remaining in agreement with the modern concepts of some "microscopic" thermodynamics (de Broglie 1964).

Some of these ideas are presented in another paper in this volume (Sieniutycz 1990). The recently discovered basic role played by the quantum phases in the fundamental Gibbs equation describing the physical systems out of equilibrium (Sieniutycz 1990), and the direct link between the invariance of the action, second law inequality, and the chemical stoichiometry, indicate that the thermal inertia effects may have a quantum origin.

ACKNOWLEDGEMENT

Helpful discussion with Prof. Christopher Essex, University of Western Ontario, is gratefuly acknowledged. The author acknowledge a partial support of the U.S. Department of Energy, under contract No. DOE-FG02-86ER13488, as well as the fruitful research atmosphere and Prof. Peter Salamon hospitality in Telluride's Summer Research Center and San Diego State University, Department of Mathematics, where a part of this work was done.

REFERENCES

Ascroft, N. W.1976. Solid State Physics. New York: J.Wiley.

Baumeister, K. J., and T. D. Hamil. 1968. Hyperbolic heat conduction equation - a solution for the semiinfinite body problem. J. Heat Transfer Trans. ASME. 91: 543-548.

Berkovsky, B. M., and V. G. Bashtovoi. 1972. The finite velocity of heat propagation from the view point of the kinetic theory. Intern. Jl. Heat Mass Transfer 20: 621-627.

Bhattacharya, D. K. 1982. A variational principle for thermodynamic waves. Annal. Phys. 7 (39): 252-332.

Bubnov, V. A. 1976. On the nature of heat transfer in acoustic wave. Inzh. Fiz. Zh. 31: 531-536.

Bubnov, V. A. 1976a. Wave concepts in theory of heat. Intern. Jl Heat Mass Transfer 19: 175-195.

Bubnov V. A. 1982. Toward the theory of thermal waves. Inzh. Fiz. Zh. 43: 431-438.

Builtjes, P. J. H. 1977. Memory Effects in Turbulent Flows. W. T. H. D. 97: 1- 145.

Cattaneo, C. 1958. Sur une forme d l'equation eliminant le paradoxe d'une propagation instantance. C. R. Hebd. Seanc. Acad. Sci. 247: 431-433.

Carrasi, M., and A. Morro. 1972. A modified Navier-Stokes equation and its consequence on sound dispersion. Nuovo Cim. 9B: 321-343.

Carrasi, M. and A. Morro. 1973. Some remarks about dispersion and adsorption of sound in monoatomic rarefied gases. Nuovo Cim. 13B: 281-249.

Chester, M. 1963. Second sound in solids. Phys.Rev. 131: 2013-2115.

Davies, P. 1966. Deposition from moving areosols. In Areosol Science, ed. P. Davies. London: Academic Press.

De Broglie, L. V. 1964. La thermodynamique "cache" des particules. Ann. Inst. Henri Poincare 1: 1-19.

Domanski, R. 1978. 6-th Intern. Heat Transfer Conf. Paper CO-10: 275.

Domanski, R. 1978a. Analytical description of temperature field cased by heat pulses. Arch. Termod. i Spalania 9: 401-413.

Eu, B. Ch. 1982. Irreversible thermodynamics of fluids, Annals Phys. 140: 341-371.

Eu, B. Ch. 1986. On the modified moment method and irreversible thermodynamics. Jl Phys Chem. 85: 1592-1602.

Eu, B. Ch. 1987. Kinetic theory and irreversible thermodynamics of dense fluids subject to an external field. Jl Phys. Chem. 87: 1220-1237.

Garcia-Collin, L.S. , M. Lopez de Haro, R. F. Rodriguez, J. Casas-Vazquez and D. Jou. 1984. On the foundations of extended thermodynamics. Jl Statist. Phys. 37:465-484.

Glansdorff, P., and I. Prigogine.1971. Thermodynamic theory of structure, stability and fluctuations. New York: J.Wiley.

Guyer, R. A., and J. A. Krumhansl.1964. Dispersion relation for second sound in solids. Phys. Rev. A 133: 1141-1415.

Grad, H. 1958. Principles of the theory of gases. In Handbook der Physik vol 12, ed. S. Flugge. Berlin: Springer.

Grigoriew, B. A. 1979. Simplification of one-dimensional heat conduction problem at radiation impulses. Heat Phys. High Temp. 11: 133-137.

Gyarmati, I. 1977. On the wave approach of thermodynamics and some problems of nonlinear theories. Jl Nonequilib. Thermodyn. 2: 233-243.

Goldstein, S. 1953. On diffusion by discontinuous movements and on the telegraph equation. Q. Mech. Appl. Math. 6:290-312.

Holderith, J., and P. Reti. 1979. On the boundary conditions for hyperbolic reactors. Annal. Univers. Scient. Budapest. XV: 39-43.

Jou, D., J. Casas-Vazquez, and G. Lebon. 1988. Extended thermodynamics. Rep. Progr. Phys. 51: 1105-1080.

Kaliski, S. 1965. Wave equation of heat conduction. Bull. Acad. Pol. Sci. Ser. Sci. Tech.13: 211-219.

Kantorovich, L.V., and V. I. Krylow. 1958. Approximate Methods of Higher Analysis. Groningen: Nordhoff.

Kao, T. 1977. Non-Fourier Heat Conduction in Thin Surface Layers. Jl of Heat Transfer Trans ASME 99: 343-345.

Kazimi, M. S., and C. A. Erdman. 1973. On the interface temperature of two suddenly contacting materials. Jl of Heat Transfer Trans ASME 97: 617-619.

Keesom, W. H., and B. F. Sarris.1940. Further measurements of the heat conductivity of liquid helium II. Physica7: 241-252.

Keesom, W. H. 1949. Helium. Moscow: IL.

Kranys, M. 1966. Relativistic hydrodynamics without paradox of infinite velocity of heat conduction. Nuovo Cim. 42B: 1125-1194.

Lebon, G., and J. Casas-Vazquez. 1976. On the stability condition for heat conduction with finite wave speed. Phys. Lett. 55A: 93-94.

Lebon, G. 1976. A new variational principle for the nonlinear unsteady heat conduction problem. Q. Jl Mech. Appl. Math. 29: 499-509.

Lebon, G. 1978. Derivation of generalized Fourier and Stokes-Newton equations based on thermodynamics of irreversible processes. Bull. Acad. Soc. Belg. Cl. Sci. LXIV: 456-460.

Lebon, G., D. Jou, J. Casas-Vazquez.1980. An extension of the local equilibrium hypothesis. Jl. Phys. A. 13: 275-280.

Lebon, G. and M. S. Boukary.1982. Lyapounov functions in extended irreversible thermodynamics.Phys. Lett. 88A: 391-393.

Luikov, A. V. 1966. Application of irreversible thermodynamics to investigation of heat and mass transfer. Intern. Jl Heat Mass Transfer 9:138-152.

Luikov, A. V. 1969. Analytical Heat Diffusion Theory. New York: Academic Press.

Luikov, A.V., and B. M. Berkovsky.1974. Convection and thermal waves. Moscow:Energy.

Luikov, A.V., V. A. Bubnov, and I.A. Soloview.1976. On wave solutions of the heat conduction equations. Int.Jl Heat Mass Transfer 19:245-249.

Mitura, E. 1981. Relaxation Effects of Heat and Mass Fluxes as a Basis for Classification of Dried Solids. Ph D Thesis. Lodz Technical Univ.,Poland.

Monin, A. S., and A. M. Yaglom. 1967. The Mathematical Problems of Turbulence. Moscow: Nauka.

Mc Lenan, J.A. 1974. Onsagers theorem and higher order hydrodynamic equations. Phys. Rev. A. 10: 1272-1276.

Natanson, L.1896. Uber die Gezetze nicht umkerbarer Vorgange. Z. Phys. Chem. 21: 193-200.

Nettleton, R. E. 1960. Relaxation theory of heat conduction in liquids. Phys. Fluids 3: 216-226.

Nettleton, R. E. 1969. Thermodynamics of viscoelasticity in liquids. Phys. Fluids 2: 256-263.

Nettleton, R. E. 1986. Lagrangian formulation of non-linear extended irreversible thermodynamics. Jl Phys. A: Math. Gen.19: L 295-L297.

Nettleton, R. E. 1987. Nonlinear heat conduction in dense liquids. Jl Phys. A: Math. Gen.20: 4017-4025.

Nunziato, J.W. 1971. On the heat conduction in materials with memory. Quart. Appl. Math. 29: 187-204.

Nonnenmacher, T. F. 1980. On the derivation of second order hydrodynamic equations. Jl Non-Equilib. Thermodyn. 5: 361-378.

Pavon, D., D. Jou, and J. Casas-Vazquez. 1980. Heat conduction in relativistic extended thermodynamics. Jl Phys. A: Math Gen. 13: L67-L79.

Pellam, R. 1961. Chapter 6 in Modern Physics for Engineer. New York: Mc Graw Hill.

Prohovsky, E.W., and J. A. Krumhansl. 1964. Second sound propagation in dielectric s solids. Phys.Rev. A. 133: 1411-1415.

Putterman, S. J. 1974. Superfluid Hydrodynamics. Amsterdam: North Holland.

Reti, P. 1980. Stability of hyperbolic reactors. React. Kinet. Catalyst. Lett. 15: 215-220.

Reti, P., and J. Holderith. 1980. The modelling of multiphase hyperbolic chemical reactors by averaging methods. Hung. Jl Ind. Chem. 1:337- 382.

Sandler, S. I., and J. S. Dahler. 1964. Nonstationary diffusion. Phys. Fluids 2: 1743-1746.

Sadd, M. H. and J. E. Didlake. 1977. Non-Fourier melting of a semiinfinite solid. Jl of Heat Transfer ASME 99: 25-28.

Sieniutycz, S. 1977. The variational principles of classical type for non-coupled non-stationary irreversible transport processes with convective motion and relaxation. Intern. Jl Heat Mass Transfer 20: 1221-1231.

Sieniutycz, S. 1978. Optimization in Process Engineering. Warsaw:WNT.

Sieniutycz, S. 1979. The wave equations for simultaneous heat and mass transfer in moving media - structure testing, time space transformations and variational approach. Intern. Jl Heat Mass Transfer 22:585-599.

Sieniutycz, S. 1980. The variational principle replacing the principle of minimum entropy production for coupled nonstationary heat nd mass transfer processes with convective motion and relaxation. Intern. Jl Heat Mass Transfer 23: 1183-1193.

Sieniutycz, S. 1981. Thermodynamics of coupled heat, mass and momentum transport with finite wave speed. I. Basic ideas of theory. II.Examples of transformations of fluxes and forces. Intern. Jl Heat Mass Transfer 24: 1723-1732 and 24:1759-1769.

Sieniutycz, S. 1981a. The thermodynamic stability of coupled heat and mass transfer described by linear wave equations. Chem. Eng. Sci. 36:621-624.

Sieniutycz, S. 1981b. On the applicability of classical thermodynamic stability conditions for the coupled heat and mass transfer with finite wave speed. Jl Non-Equilib. Thermodyn. 6:79-84.

Sieniutycz, S. 1981c. Action functionals for linear wave dissipative systems with coupled heat and mass transfer. Physics Lett. 84A: 98-102.

Sieniutycz, S. 1982. Entropy of flux relaxation and variational theory of simultaneous heat and mass transfer governed by non-Onsager phenomenological equations. Appl. Sci. Res. 39: 81-103.

Sieniutycz, S. 1983. The inertial relaxation terms and the variational principles of least action type for nonstationary energy and mass diffusion. Intern. Jl Heat Mass Transfer 26: 55-63.

Sieniutycz, S. 1984. The variational approach to nonstationary Brownian and molecular diffusion described by wave equations. Chem. Eng. Sci. 39:71-80.

Sieniutycz, S. 1985. A synthesis of some variational theorems of extended irreversible thermodynamics of nonstationary heat and mass transfer. Appl. Sci. Res. 44: 211-228.

Sieniutycz, S. 1987. Variational approach to the fundamental equations of heat, mass and momentum transport in strongly unsteady-state processes. I. Intern. Chem. Engng. 27: 545-555.

Sieniutycz, S. 1987a. From a last action principle to mass action law and extended affinity. Chem. Engng. Sci. 42: 2697-2711.

Sieniutycz, S. 1988. Variational approach to the fundamental equations of heat, mass and momentum transport in strongly unsteady-state processes. II. Intern. Chem. Engng. 28: 353-361.
Sieniutycz, S. 1988a. Hamiltonian energy-momentum tensor in extended thermodynamics of one-component fluid. Inz. Chem. Proc. 4: 839-861.
Sieniutycz, S. 1989. Experimental relaxation times, drying- moisturizing cycles and the relaxation drying equation. Chem Eng. Sci. 44: 727-740.
Sieniutycz, S., and R. S. Berry.1989. Conservation laws from Hamilton's principle for nonlocal thermodynamic equilibrium fluids with heat flow. Phys.Rev.A. 40: 348-361.
Sieniutycz, S. 1990/91. Thermal momentum and a macroscopic extension of de Broglie microthermodynamics I. The multicomponent fluids with sourceless continuity constraints. II. The conservation laws for continuity constraints with sources. Adv. in Thermodyn. Vol 3 and 7., eds. S.Sieniutycz and P.Salamon. New York: Taylor and Francis.
Sieniutycz, S., and R. S. Berry. 1991. Field thermodynamics potentials and geometric thermodynamics with heat transfer. Phys. Rev. A.43: 2807-2818.
Suminska, B. 1983. Investigation of relaxation effects in drying. Ms D thesis. Warsaw Technical University. Poland.
Uhlenbeck, G.E. 1949. Transport phenomena in very dilute gases. Univ. Mich. Engng. Research Inst. Rept. CM: 579.
Vernotte, P. Les paradoxes de la theorie continue de l'equation de la chaleur. C. R. Hebd. Seanc. Acad. Sci. 246: 3154-3155.
Vujanovic, B. 1971. An approach to linear and nonlinear heat transfer problem. A. I. A. A. Jl 9: 131-134.
Vujanovic, B., and D. J. Djukic. 1972. On the variational principle of Hamilton's type for nonlinear heat transfer problem. Intern. Jl Heat Mass Transfer 15: 1111- 11115.
Vujanovic, B., and S. E. Jones. 1988. Variational Methods in Nonconservative Phenomena. New York: Academic Press.
Wang-Chang, C.S., and G.E. Uhlenbeck.1951. Transport phenomena in polyatomic gases. Univ. Mich. Engng. Research Inst. Rept. CM:681.
Wiggert, D. C. 1977. Analysis of early time transient heat conduction by method of characteristics. 1977. Jl of Heat Transfer Trans ASME 99: 35-40.

Diffusion in Elastic Media with Stress Fields

B. Baranowski
Institute of Physical Chemistry
Polish Academy of Sciences, 01-224 Warsaw, Poland

Summary:

Influence of stress fields on diffusion is treated in isotropic elastic media, mainly when the stresses are developed by the diffusing component (self- stresses). A simple Gibbs' model is taken for the chemical potential of the mobile component. Detailed equations are derived for one - dimensional plates. An integro - differential equation results for the balance equation. Under suitable boundary and initial conditions the stress induced diffusion flow can be separated for short times. Experimental results of hydrogen diffusion in a PdPt alloy are compared with the theory. The general importance of stress field contributions to diffusion in solids is outlined.

1). Introduction;

Changes of concentration in a continuous system are usually accompanied by volume changes. In most simple cases the volume of the solution is a sum of the volumes of the constituent components (no volume of solution). In fluids the volume changes during the solution process are relaxed by local hydrodynamic flows. A quite different situation arises in solids: The interdiffusion of components causes displacements in the initial intermolecular distances of the amorphous or crystalline solids involved. These displacements are the origin of stresses which - due to the elastic coupling existing between lattice sites - can be propagated to large distances from the original location of displacement. Underway the stresses can be relaxed, at least partially, on dislocations or grain boundaries. Below we shall disregard such reductions, limiting ourselves to ideal elastic media. This implies that the stresses involved don't exceed the elasticity range beyond which plastic (irreversible) deformations take place.

A stress field in solids can arise due to two reasons:
1). It can be imposed from outside for instance by dilatation (stretching) or compression. 2). It can arise during diffusion, due to the volume changes accompanying the mixing process.
The stresses arising here may be termed self - stresses and will form the main object of our treatment. We shall limit ourselves to consideration of isotropic solid bulk phases, thus disregarding the variety of phenomena which may occur on surfaces, mixture of heteregeneous phases, phase transitions or due to anisotropy of crystals. Our main advantage will be the combination of theory with experimental results.

2). Thermodynamics of solids with stress fields:

Since Gibbs (1906) the problem of thermodynamic significance of stress fields in solids is known. Contrary to fluids where a reasonable mobility is mostly guarantied, the formulation of the contribution of stress fields to the free energy of a solid many component system may be a source of serious discussions (Larche, Cahn 1973, 1978, 1985). As at least some components of a solid mixture may be practically immobile, the achievement of thermodynamic equilibrium and the formulation of chemical potentials in stress fields become controversial.

For our purpose let us adopt the following simplified model: The system consists of a solid isotropic body (crystaline or amorphous). The elements of this solid matrix are practically immobile with the exception of one interstitial component which is mobile. Thus the components of the solid matrix can serve as frame of reference for the transport of the interstitial particles. These assumptions allow to take over the thermodynamics from Gibbs' solid (Gibbs 1906) and similar approaches from (Larche and Cahn 1973) and (Li et al. 1969). The basic quantity for our further treatment is the chemical potential μ_i of the mobile interstitial component. In stress free conditions it is expressed by the relation

$$\mu_i = \mu_i^o + RT \ln f_i c_i \qquad (1)$$

where μ_i^o denotes its standart value, R, T have the conventional meaning and f_i, c_i are the activity coefficient and concentration of the interstitial. The chemical potential in the presence of stresses μ_i^σ will be given by

$$\mu_i^\sigma = \mu_i - V_i\sigma = \mu_i + RT \ln f_i c_i - V_i\sigma \qquad (2)$$

where V_i denotes the partial molar volume of the interstitial and σ is the hydrostatic (thermodynamic) part of the stress tensor, being identical with the sum of the diagonal elements

$$\sigma = \sigma_{xx} + \sigma_{yy} + \sigma_{zz} \qquad (3)$$

In fluid systems the correspondance with the hydrostatic pressure exists

$$\sigma = -p \qquad (4)$$

Equation (3) is equivalent with the omission of non - diagonal terms of the total pressure tensor in the chemical potential. The justification of this omission was recently proved experimentally in Pd-H and Pd-Si-H systems (Kirchheim 1986) contrary to previous statements (Hirth 1977). Equation (4) is simultaneously the equivalence between the internal stress in a solid and the hydrostatic pressure imposed from outside (for instance by a pressure transmitting fluid). In fact the mobile component considered does not distinguish between the stress field originated in the lattice and externaly varied hydrostatic pressure. Both factors exhibit identical thermodynamic significance.

As we are interested in diffusion, the phenomenological background has to be taken from linear non - equilibrium thermodynamics (de Groot and Mazur 1962, Fitts 1962, Baranowski 1975). In our case of only one mobile interstitial we can add to its flow the flow of electrons. This is appropriate if the mobile component is partially ionic. As experimentally we shall treat Me - H,D systems such an approach is reasonable. Movement of hydrogen particles dissolved in a metallic matrix in an electrostatic field (electrotransport) is a well known phenomenon (Wipf 1978). It proves that hydrogen particles exhibit a non - vanishing effective electrical charge. The most probable model for it is the screened proton. Thus the mentioned two flows can be written as

$$\underline{J}_i = L_{ie}\,\underline{X}_e + L_{ii}\,\underline{X}_i \tag{5}$$

$$\underline{J}_e = L_{ee}\,\underline{X}_e + L_{ei}\,\underline{X}_i \tag{6}$$

where $\underline{J}_i$, $\underline{J}_e$ are flows of the interstitial component and the electrons; $\underline{X}_i$, $\underline{X}_e$ are the conjugated thermodynamic forces and L_{ij} the corresponding phenomenological coefficients. Limiting ourselves to cases where the electronic conductivity prevails, the L_{ee} coefficient in (6) will be related to the electrical conductivity. In a similar way the L_{ii} coefficient can be expressed in terms of the diffusion coefficient, characterizing the mobility of the interstitial component. The cross coefficients L_{ie} and L_{ei} describe the interaction between both flows: L_{ie} expresses the flow of component "i" which arises at vanishing value of $\underline{X}_i$ due to the action of $\underline{X}_e$. An inverse symmetrical interpretation can be attached to the coefficient L_{ei}. Due to Onsager reciprocal relations

$$L_{ie} = L_{ei} \tag{7}$$

only one of these coefficients has to be determined. The II. low requires

$$L_{ii}\,L_{ee} \geq L_{ei}^2 \tag{8}$$

This inequality is mostly fulfiled in a way that in many cases the influence of cross effects can be neglected. As our interest is limited to the diffusion of the interstitial component, let us apply as sufficient approximation

$$\underline{J}_i \cong L_{ii}\,\underline{X}_i \tag{9}$$

According to non-equilibrium thermodynamics the thermodynamic force $\underline{X}_i$ can be expressed as

$$\underline{X}_i = -\operatorname{grad}\mu_i^{\sigma} + e_i \operatorname{grad}\xi \tag{10}$$

where e_i is the molar charge of component "i" and ξ the electrostatic potential. Being mainly not interested in the presence of electrostatic fields we can assume as sufficient approximation

$$\underline{X}_i \cong -\operatorname{grad}\mu_i^{\sigma} \tag{11}$$

what is equivalent to the absence of external electrical fields and to negligible diffusion potentials. In isothermal conditions (11) can be given in explicite form after introducing (2)

$$\underline{X}_i = -RT\operatorname{grad}\ln f_i\,c_i + V_i \operatorname{grad}\sigma \tag{12}$$

The last equation implies

$$\text{grad } V_i = 0 \tag{13}$$

what means that the partial volume of the interstitial component is space independant.

Combining (12) with (9) and reformulating the first term on the right hand side of (12) we derive

$$\underline{J}_i = - L_{ii} \left\{ \frac{RT}{c_i} \left(1 + \frac{\delta \ln f_i}{\delta \ln c_i}\right) \text{grad } c_i - V_i \text{ grad } \sigma \right\} \tag{14}$$

Let us now express the L_{ii} coefficient by a direct measurable quantity. For this purpose we introduce the definition of Fick's diffusion coefficient D

$$\underline{J}_i = - D_i \text{ grad } c_i \tag{15}$$

Equation (14) reduces to (15) if we consider an ideal solution (f_i=1) under negligible stress gradient (grad σ = 0). This leads to

$$L_{ii} = \frac{D_i c_i}{RT} \tag{16}$$

Thus the flow of the interstitial component (14) will be expressed by

$$\underline{J}_i = - D_i \left\{\left(1 + \frac{\delta \ln f_i}{\delta \ln c_i}\right) \text{grad } c_i - \frac{V_i c_i}{RT} \text{ grad } \sigma\right\} \tag{17}$$

This is the local flow of the component considered in the space dependant fields of concentration and stress. The next step will be the formulation of the balance equation for the mobile interstitial component. Besides the flow, given by (17), we have to clear up the question of creation or destruction of interstitial particles inside the immobile matrix. It is not an unrealistic question as such particles can be as well immobilized in some traps as well becoming free from such traps. This should find evidence in the source term of the balance equation. If we assume that all interstitial particles are participating in diffusion without changing their number (conservation of particle number), the balance equation for the interstitial component will be

$$\frac{\delta c_i}{\delta t} = - \text{div } \underline{J}_i \tag{18}$$

or making use of (17)

$$\frac{\delta c_i}{\delta t} = D_i\left\{\left(1+\frac{\delta \ln f_i}{\delta \ln c_i}\right)\right\} \text{div grad } c_i - \frac{D_i V_i c_i}{RT} \text{ div grad } \sigma - $$

$$- \frac{D_i V_i}{RT} \text{ grad } c_i \cdot \text{grad } \sigma \tag{19}$$

In stress free conditions only the first term on the right hand side of (19) appears, being identical with the classical II. Fick's law. The presence of a stress gradient adds to the traditional diffusion equation a term similar (analytically) to the dependance on concentration (second term on the right hand side of (19)) and a term proportional to the scalar product of gradients of concentration and stress (third term on the right hand side of (19)). Balance equation (19) demonstrates clearly a considerable complication of the diffusion equation if gradients of stress don't vanish. Unfortunatly two variables (besides the time) - concentration and the stress - appear in (19). Therefore one needs an analytical dependance between the concentration and stress fields in order to reduce (19) to a two variable differential equation. Even if disregarding the possibility of imposing a stress field from outside, for instance by bending or streching, there does not exist a general relation between self - stresses and the distribution of concentration in a diffusion process. The distribution of stresses depends strongly on the geometry of the specimen and is furtheron strongly varying if changing the initial and boundary conditions of the diffusion problem considered (Li 1978, Larche and Cahn 1982, Stephenson 1988). Therefore let us not go into more details in generalities, but let us discuss below some concrete results observed recently in metal - hydrogen systems.

3). Diffusion of hydrogen in metals with stress fields:
Hydrogen (deuterium) particles as an interstitial component are very mobile at room temperature in transition metals (Volkl and Alefeld 1978). In most cases the diffusion coefficients at about 300K are found in the range 10^{-5} - 10^{-9} $\frac{cm^2}{sec}$. In vanadium hydrogen particles perform 10^{12} jumps per second at this temperature. This high mobility allows to treat the metallic lattice components as being at rest playing the role of frame of reference for the displacement of hydrogen particles. As metals exhibit in a certain range elastic properties, the metal - hydrogen systems mentioned can serve as models for the influence of stresses on diffusion phenomena. In fact this found application in the past in two main directions: 1). If introducing into a metallic sample with homogeneous distribution of hydrogen an external stress - for example by bending the sample - a redistribution of hydrogen takes place , leading to creation of a concentration gradient. This theoretically forseen phenomenon (Gorsky 1935) is termed Gorsky effect. It is a trivial consequence of equation (12):
If introducing by an external constraint a non vanishing grad σ the equilibrium condition

$$\underline{X}_i = 0 \tag{20}$$

requires

$$\text{grad } \sigma = \frac{RT}{V_i} \text{ grad } \ln f_i c_i \tag{21}$$

This means that the initial homogeneous distribution of particles has to go over to a non-homogeneous as long as the stress gradient is maintained. Equation (21) is in a certain sense equivalent to the distribution of gases in a gravitational field, that is to the non-homogeneity developed in a gradient of hydrostatic pressure which is created by the action of an external (gravitational) field. Gorsky effect found in metal - hydrogen systems a useful aplication in determination of the diffusion coefficients (Volkl and Alefeld 1978, 1979). 2). The introduction of hydrogen into a metallic sample creates self - stresses due to the lattice expansion accompanying the absorption process. If the hydrogen distribution is asymmetrical - for instance only one surface of a flat plate is exposed to the hydrogen source (for instance contact with a gas volume or a cathodic deposition of hydrogen) the sample considered will change its initial shape. It can result in a time depending bending caused by the diffusion kinetics of the ab- or desorbed hydrogen. This "diffusion - elastic" effect can serve as a method for the determination of diffusion coefficients (Cermak et al. 1985, Kufudakis et al. 1989) whereby a higher sensitivity is claimed as compared with the method based on the Gorsky effect.

In global terms the diffusion - elastic effect can be treated as an inverse phenomenon to the Gorsky effect: Namely in the first case we create a gradient of concentration what causes the development of a gradient of stress (see eq. (12)), demonstrating itself - for instance - in the bending of the sample. In the Gorsky effect the externally imposed bending creates a concentration gradient. Thus the cause and its result are interchanged in both effects.

Recently a much more spectacular effect of stresses due to hydrogen diffusion in metals was found (as it often happens it did not result from a systematic investigation but was initially observed as an unexpected side effect (Lewis et al. 1983)): Let us imagine a flat metallic membrane through which hydrogen can diffuse. This is shown schematically on Fig.1.

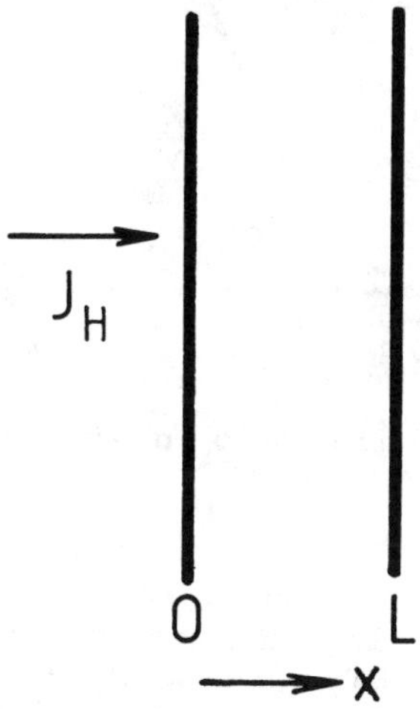

Fig.1. Schema of a one - dimensional hydrogen diffusion through a metallic plate.

If initially a negligible hydrogen concentration ($c_H \cong 0$) characterizes the whole plate ($L \geq x \geq 0$) and if we increase instantenously ($t \geq 0$) the chemical potential of hydrogen at the entrance surface ($x=0$), we force effectively a certain diffusional inflow of hydrogen (J_H) into the membrane. The time course of hydrogen concentration in the membrane will be described by a proper solution of II. Fick's law, taking the corresponding initial and boundary conditions into account (Barrer 1956). It implies that hydrogen penetrates the membrane with a characteristic break through time (τ_B, see Fig.2), dependant on the diffusion coefficient of the hydrogen particles and the thickness (L) of the plate. The simplest realization of the process discussed will be the increase of hydrogen pressure in the volume contacting with the entrance wall ($x=0$) and measuring simultaneously the time course of hydrogen pressure in a volume contacting with the opposite surface of the plate ($x=L$). Taking in experimental realization a metallic tube closed at one rounded end, the outer surface plays the role of the entrance (x=0) and the inner surface exhibits the coordinate x=L of Fig.1. The hydrogen supply from the gaseous phase was exercised more recently (Lewis et al. 1987, 1988, Baranowski 1989, Baranowski et al. 1989). An alternative way of hydrogen supply is the electrolysis (Lewis et al. 1983, Lewis et al. 1988, Baranowski 1989).The process discussed will be characterized by a negligible hydrogen pressure inside the tube, up to the time when the diffusion front of hydrogen reaches its inner surface. From this time a continuous increase of hydrogen pressure follows, reflecting the increase of hydrogen permeation. The behavior described is schematically presented on Fig.2.

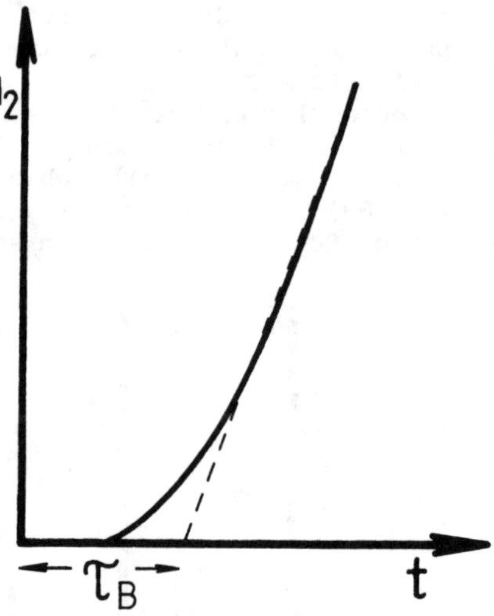

Fig.2. Schematical course of the hydrogen pressure inside the tube, when starting with negligible initial hydrogen content and increasing the hydrogen pressure at the outer wall.

A quite different time course is observed when initially a considerable homogeneous hydrogen content is inside the membrane. This corresponds to keeping both surfaces of the tube in contact with gaseous hydrogen of the same pressure for times longer than the relaxation time of hydrogen diffusion. If we now increase instantaneously the hydrogen pressure in the volume surrounding the outer surface of the tube, the hydrogen pressure inside the tube may follow the behavior presented schematically on Fig.3.

The instantaneous increase of the hydrogen pressure at the outer surface is immediatly reflected by a decrease of the hydrogen pressure in the tube. Thus the hydrogen moves in the opposite direction (into the tube wall!) than that imposed from outside and observed after passing a minimum. The very characteristic of the curve presented on Fig.3 is the reaction of the inner tube wall practically contemporary with the change of the hydrogen pressure at the outer wall of the tube. Anyway it seems obvious that the signal invoking the reduction of hydrogen pressure in the tube is transfered by a higher velocity than the time characteristic for diffusion.

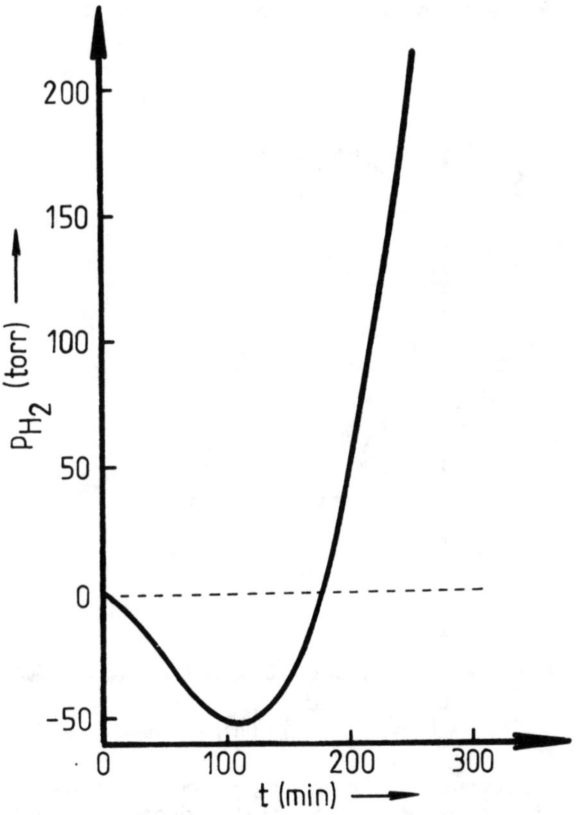

Fig.3. Schematical course of the hydrogen pressure inside the tube, when starting with considerable initial hydrogen content and increasing the hydrogen pressure at the outer wall.

A symmetrical - in a certain sense - curve to Fig.3 may be observed if, starting again from a homogeneous considerable hydrogen content in the tube wall, we decrease instantaneously the chemical potential of hydrogen at the outer wall. This can be easily carried through by a quick pumping off the hydrogen volume surrounding the outer surface of the tube. What will be observed is schematically shown on Fig.4.
Again the response of the inner wall seems strange as the hydrogen pressure inside the tube is first increasing and after passing through a maximum it takes over the expected decrease of the hydrogen activity. The classical course of the internal pressure should be a period of constancy, and after a characteristic break - through time, as shown for the absorption process on Fig.2, a continuous reduction of the hydrogen pressure should follow. The inner wall reacts again immediately without any delay to be expected from pure diffusion kinetics.

How to explain qualitatively the behavior presented on Fig.3 and 4? The entrance of hydrogen into the membrane causes an increase of the solid sample volume and what follows the creation of stresses. Due to the elastic coupling existing between lattice sites of the metallic membrane, these stresses are propagated with sound velocity to regions where so far homogeneous hydrogen particle distribution exists. What follows is the creation of a chemical potential gradient of the mobile hydrogen particles which, in the initial period of the diffusion process, causes a stress induced diffusion (SID) flow, leading to the "anomalies" presented in Fig.3 and 4. Below we shall go over to a quantitative description of these phenomena.

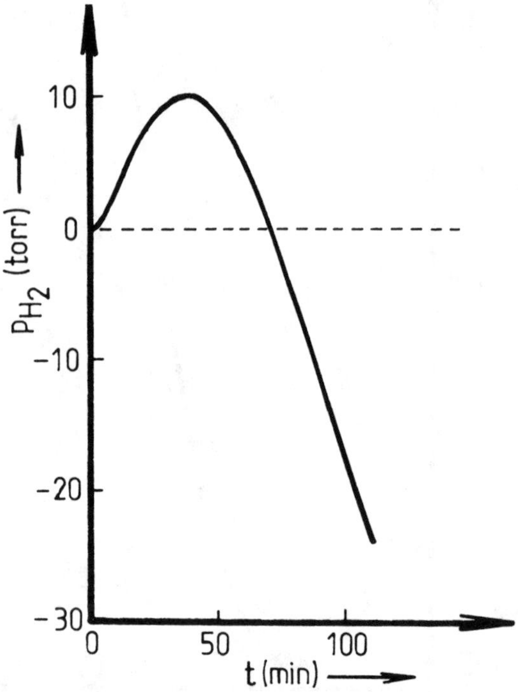

Fig.4. Schematical course of the hydrogen pressure inside the tube, when starting with considerable initial hydrogen content and decreasing the hydrogen pressure at the outer wall.

4). <u>Theory of stress induced diffusion (SID) in flat membranes:</u>
The general balance equation of the mobile interstitial component in a stress field (19) requires an expression for the stresses in terms of the concentration field. As we have already mentioned in the discussion of this equation (see 2) the explicite expression for the stress field is strongly dependant on the geometry of the sample as well as on the initial and boundary conditions imposed. Coming back to the experiments described in 3). we are looking for the distribution of stresses in a one - dimensional flat membrane of thickness L. In fact a tube was used but assuming the wall thickness L as small in comparison to the internal radius of the tube (in our case this ratio equals about 6×10^{-2}, (Baranowski 1989)) we can treat our problem as a one - dimensional. Let us remark that the stresses created by concentration changes of the interstitial component in an elastic matrix can be compared with that arising due to temperature gradients appearing in the same sample. "Thermo - stresses" developed are due to thermal expansion of the elastic matrix; thus the molecular mechanism of stresses in temperature and concentration gradients is the same, namely the change of local distances between the lattice sites. The similarity discussed means that analytical expressions for thermo - stresses can be taken over for the description of self - stresses arising in concentration fields of diffusing components. Some authors are using the term "chemical stresses" for stress fields arising in concentration gradients (Li 1978). Is this direct terminological analogy to thermal stresses justified? A serious difference consists in the possibility of freezing in the chemical stresses by suitable lowering of the temperature - what means the

existance of such stress fields without any continuous dissipation of energy - whereby thermal stresses exist only in experimental conditions in which energy dissipation, due to thermal conductivity, is taking place. But this difference does not eliminate mathematical analogies of which we can take advantage. The situation is similar to the formal analogies existing between Fick's and Fourier's law in diffusion and heat conductance problems in continuous media. Here very often analytical solutions known for one transport phenomena can be taken over to the other by suitable reformulation of the initial and boundary conditions involved. But let us underline a further difference between thermal and "chemical stresses". Usually for elastic solids the characteristic relaxation times for heat conductivity are much higher than that for diffusion. As examples metallic palladium and platinum, whose alloys will be treated in the experimental part of this review, are characterized by heat diffusivities at room temperature of the order $10^{-1}\frac{cm^2}{sec}$ whereby the diffusion coefficient of interstitial hydrogen is - in the same units - six orders of magnitude lower. As consequence of this difference effects similar to that presented on Fig.3 and 4 for "chemical stresses" could not be observed for thermal stresses in the same media when applying the similar geometrical conditions.

The problem of thermal stresses in a thin plate with one - dimensional temperature gradient across the thickness L (see Fig.1) and with infinite extensions in y and z directions was solved long time ago (Timoshenko 1934). The expression derived was - by analogy - taken over for the stresses developed during diffusion in a plate (Prussin 1961). Timoshenko's result was also applied later (Li 1978, Larche and Cahn 1982, Larche 1988). The most exact version for the stresses developed is

$$\sigma_{yy}=\sigma_{zz}=-\frac{V_i\Delta c_i E}{3(1-\nu)}\frac{1}{3L(1-\nu)}\int_o^L \Delta c_i V_i E dx - $$

$$-\frac{4(x-\frac{L}{2})}{L^3(1-\nu)}\int_o^L \Delta c_i V_i E(x-\frac{L}{2})dx \qquad (22)$$

where σ_{yy} and σ_{zz} are the components of the stress tensor in y and z directions ΔC_i is the difference in concentrations

$$\Delta c_i = c_i - c_i^o$$

where c_i^o denotes the concentration in stress - free conditions. We can start, for instance, from a homogeneous (gradientless) distribution of concentration which does not invoke any stress at all and only additional changes of the concentration distribution will give rise to a stress field.

E - in (22) denotes Young's modulus and ν the Poisson's ratio. Equation (22) takes into account the possibility of a coordinate dependant partial molar volume V_i of the mobile component and of the Young's modulus of the elastic matrix. The last dependance may be of importance if the interstitial component is influencing the elastic properties of the immobile matrix in a considerable degree. For our purpose we shall disregard the coordinate dependance of both quantities, giving (22) in a simplified version

$$\sigma_{yy} = \sigma_{zz} = -\frac{V_i Y}{3}\{\Delta c_i - \frac{1}{L}\int_o^L \Delta c_i dx - $$

$$- \frac{12(x-\frac{L}{2})}{L^3}\int_o^L \Delta c_i (x-\frac{L}{2})dx\} \quad (23)$$

whereby Y means

$$Y = \frac{E}{1-\nu} \quad (24)$$

No stresses are assumed in the direction of the concentration gradient

$$\sigma_{xx} = 0 \quad (25)$$

Thus the total hydrostatic part of the stress tensor equals

$$\sigma = \sigma_{yy} + \sigma_{zz} = 2\,\sigma_{yy} \quad (26)$$

Equation (23) presents the required relation between the concentration and stress fields for the flat plate sample used in our experiments.
Let us analyse shortly the three contributions in (23) in a separate way: 1). The first term on the right hand side, proportional to the concentration difference Δc_i, is of local character. It expresses simply the stress arising in each volume element due to the concentration change, deviating from the stress free conditions. The proportionality to the partial molar volume is the obvious illustration of the proportionality between the strain and stress in the elastic region of deformation. The term discussed will contribute to the gradient of the chemical potential (driving force in diffusion) in a similar local way as the gradient of concentration in stress free conditions. Its final influence is expressed by an additional term to the thermodynamic correction due to non - ideality of the solution considered (see as example eq. (31)). 2). The second term on the right hand side of (23) has the physical meaning of the average concentration for which - if homogeneously distributed over the total plate - the stress would vanish. For such a case the first term (Δc_i) would exactly compensate the second in each volume element of the plate. To make its influence for the stress distribution more clear, let us assume a symmetrical concentration distribution in respect to the center plane of the plate ($x = \frac{L}{2}$). For such symmetry the third term in (23) vanishes and therefore the stress distribution is reduced to

$$\sigma_{yy} = \sigma_{zz} = -\frac{V_i Y}{3}\{\Delta c_i - \frac{1}{L}\int_o^L \Delta c_i dx\} \quad (27)$$

The realization of this symmetrical concentration distribution can be achieved by imposing identical initial and boundary conditions on the surfaces x=0 and x=L (see Fig.1)(if initially the same symmetry conditions for the plate are fulfiled). Let us now assume that for t=0 Δc_i=const>0 for x=0 and x=L. In other words we start a diffusion process from both (outer and inner) walls of the tube. At the entrance surfaces (x=0 and x=L) the stress will be equal (for t=0)

$$\sigma_{yy} = \sigma_{zz} = - \frac{V_i Y}{3} \Delta c_i \qquad (28)$$

as the integral in (27) will not contribute to the stresses (for t=0 only). Later on the stresses at these surfaces will also decrease as function of time due to the always positive contribution of the integral in (27). Keeping the boundary conditions

$$\Delta c_i = \text{const} \quad \text{for} \quad x=0 \quad \text{and} \quad x=L \quad \text{for} \quad t>0 \qquad (29)$$

we continue a diffusion process in which locally both terms of (27) will contribute to the time course of stresses. The numerical value of the integral will continuously increase, reaching at the diffusion equilibrium the final value of Δc_i. At this time, characterized by a new homogeneous concentration distribution in the plate, the stress will vanish, according to (27). In between there will be always a distance symmetrical from both walls of the tube (time dependant!) where the stress will be zero, just separating the outer ring with negative (tensile forces) stress from the internal slab with positive (compressive forces) stress. But the importance of the integral in (27) goes beyond the above interpretation: Namely it represents the first non - local contribution to the chemical potential of the mobile interstitial: Even in volume elements which the concentration changes did not reach by the diffusion kinetics the stress has a non - vanishing value. This is evident from (27) as for all x, characterized by $\Delta c_i = -c_i^0$, the stress does not vanish as long as the integral differs from the initial concentration. It is a compressive stress, equal for the whole internal slab which is not changed by the progressing diffusion front. This gradientless distribution of the compressive stress means that the integral of (27) will not contribute to the driving force of diffusion. Equation (27) makes clear that in a general sense no direct one to one relation exists between the concentration and stress distributions as long as the integral is time dependant. The same lack of correspondance exists between the stress distribution and the gradient of concentration, again as long as the integral in (27) is time dependant. The only condition for the realization of a direct connection between the concentration and stress fields is to make the integral in (27) time independant. It can be realized by a one step support of the plate by the additional amount of the diffusing component - for instance by a one step deposition of the interstitial at both surfaces of the tube (this can be easily carried through if a solid interstitial is involved). In this case the integral in (27) will be time independant, representing the constant average concentration of the diffusing interstitial. Such condition is different from that formulated by (29).

3). For our purpose the most crucial contribution to the stress field results from the second right hand side integral in (23). As it was already mentioned shortly above, this integral exhibits a non - vanishing value only for a non - symmetrical concentration distribution in respect to the middle surface of the plate ($x=\frac{L}{2}$). Such a non - symmetrical concentration course creates non - symmetrical stress fields which can mechanically result in a bending of the plate. This bending can be compared with the Gorsky effect where such bending may be imposed from outside. In the tube applied in our experiments (Lewis et al. 1983, 1988; Baranowski 1989) an externaly imposed Gorsky effect would be difficult to realize. Some qualitative illustrations of such bending caused by diffusion may be found (Lewis, Kandasamy and Baranowski 1988; Baranowski and Lewis 1989). The integral discussed will be the source of non - local contributions to the chemical potential of the hydrogen particles (influence on the equilibrium properties) as well as on the gradient of chemical potential (influence on the kinetic properties). Contrary to the integral in (27), the second integral in (23) will not vanish when formulating the gradient of stress. The same integral is responsible for the courses presented on Fig.3 and 4. The quantitative details will be discussed below. Now let us go back to equations derived in 2). and formulate some of them for the one - dimensional case only, in accordance with the schema presented on Fig.1. Thus the flow of the interstitial component (17) will simplify to

$$J_i = - D_i \left\{ \left(1+\frac{\partial \ln f_i}{\partial \ln c_i}\right) \frac{\partial c_i}{\partial x} - \frac{V_i c_i}{RT} \frac{\partial \sigma}{\partial x} \right\} \tag{30}$$

and taking (23) and (26) into account we derive

$$\frac{\partial \sigma}{\partial x} = - \frac{2}{3} V_i Y \left\{ \frac{\partial c_i}{\partial x} - \frac{12}{L^3} \int_o^L \Delta c_i (x-\frac{L}{2}) dx \right\} \tag{31}$$

Combining (30) with (31) we get

$$J_i = - D_i \left\{ \left(1+\frac{\partial \ln f_i}{\partial \ln c}\right) + \frac{2}{3} \frac{V_i^2 c_i Y}{RT} \right\} \frac{\partial c_i}{\partial x} +$$

$$+ \frac{8 V_i^2 c_i Y D_i}{RT\, L^3} \int_o^L \Delta c_i \,(x-\frac{L}{2})\, dx \tag{32}$$

Let us point out here a remarkable difference between the interstitial component flow in stress - free conditions - as expressed for instance in ideal solutions by (15) and the last equation (32) derived for a flat plate in the presence of self - stresses induced internally by the diffusion process. Two main differences are to be distinguished:

1). On the right hand side of (29) the first term, proportional to the concentration gradient, includes besides the correction for the non - ideality a second always positive term, being proportional to the square of the partial volume and to other always positive quantities. This additional term, due to the stresses present, enhances the diffusion process. It is proportional to the concentration of the interstitial component. The first term on the right hand side of (32), being proportional to the concentration gradient, is of local, Fickian character. The influence of the stress field is limited to the correction mentioned above.

2). The second term on the right hand side of (32) has a different character: As the integral is, for a given concentration distribution in the plate, a simple number, this term will induce a flow of the interstitial component everywhere where exists a non - vanishing concentration of this component. This flow is thus of non - local character. It arises in all volume elements of the plate with not vanishing concentration, independent of the existance or non - existance of a concentration gradient, or in more clear formulation, it exists even in elements to which the concentration changes due to Fickian diffusion did still not reach. This non - local character of the consequences of stresses was demonstrated in the experiments in metal - hydrogen systems, as shown schematically on Fig.3 and 4.

In terms of this discussion (32) can be presented as

$$J_i = J_{il} + J_{inl} \tag{33}$$

with the contributions

$$J_{il} = - D_i \left\{ \left(1+\frac{\partial \ln f_i}{\partial \ln c_i}\right) + \frac{2}{3} \frac{V_i^2 c_i Y}{RT} \right\} \frac{\partial c_i}{\partial x} \tag{34}$$

as the local, Fickian part

and

$$J_{inl} = \frac{8V_i^2 c_i Y D_i}{RT\, L^3} \int_0^L \Delta c_i \left(x-\frac{L}{2}\right) dx \tag{35}$$

as the non - local part of the total diffusion flow.

To make the characteristic feature of the non - local part of the diffusion flow more clear, let us remark that for a given time we can reformulate (35) as follows

$$J_{inl} = A\, c_i \qquad \text{(for a given t)} \tag{36}$$

where the constant A is a function of time only, but in terms of previously formulated approximations it is not a function of the geometrical coordinate. Equation (36) expresses the proportionality of the non - local diffusion flow to the local concentration of the mobile component. Thus the constant A in (36) has the physical meaning of mobility being universal for all volume elements of the plate.

Now we are able to express the balance equation for the mobile interstitial in a one - dimensional problem of a flat plate by introducing (32) into (18). The result is

$$\frac{\partial c_i}{\partial t} = D_i \left\{ \left(1 + \frac{\partial \ln f_i}{\partial \ln c_i}\right) + \frac{2}{3} \frac{V_i^2 c_i Y}{RT} \right\} \frac{\partial^2 c_i}{\partial x^2} + \frac{2}{3} \frac{D_i V_i^2 Y}{RT} \left(\frac{\partial c_i}{\partial x}\right)^2 -$$

$$- \left\{ \frac{8V_i^2 D_i Y}{L^3\, RT} \int_0^L \Delta c_i \left(x-\frac{L}{2}\right) dx \right\} \frac{\partial c_i}{\partial x} \tag{37}$$

This is the most basic equation for the problem considered, expressing the concentration as a function of time and geometrical coordinate. The complexities due to the creation of self - stresses are evident: Compared to the classical II. Fick's law - that is the first term on the right hand side without the contribution proportional to Y - two new terms are added. These terms change radically the mathematical character of the balance equation: Instead of the classical linear second order partial differential equation of parabolic character, we obtained a non - linear integro - differential equation.

The influence of self - stresses, as expressed by (23), is manifested by three additional terms in the balance equation (37): 1). The term proportional to the second derivative of concentration is simply the consequence of diffusion enhancing term in (32), which was added to the thermodynamic correction due to the non - ideality. 2). The non - linear term, proportional to the square of the concentration gradient, will probable be negligible if the concentration gradient is small - or could be eventually taken as an average constant term -. On the other hand it could be a source of interesting non - linear phenomena, including oscillations and more complex dissipative structures. Such an expectation could be of special real value if the stress field induced could exhibit time dependance. The introduction of external oscillations can be easily realized by a time dependant electrolytic supply pulses of constant current or alternative current of different frequencies of hydrogen at the outer or inner wall of the tube or by periodic changes of the hydrogen pressure in the surrounding gas volumes, whereby correlated different time scales could be imposed to the outer and inner wall. The importance of the non - linear term discussed may be of special value in cases where the concentration gradient is space independent. In such situations the term proportional to the second derivative of concentration in (37) disappears and the significance of the non - linear term will grow. Such situations may arise in times close to stationarity. 3). The term including the integral in (37) is due to the asymmetry of the stress field, as it was considered in details during the discussion after eq. (23) (point 3). in this discussion). Its origin is the non - local term in the flow equation (32), expressed separately in (35). Taking the equations (35-37) into account, we can express the term discussed in the following equivalent forms

$$\frac{\partial c_i}{\partial x} \left\{ - \frac{8V_i D_i Y}{L^3 RT} \int_o^L \Delta C_i \; (x-\frac{L}{2})dx \right\} = \frac{\delta J_{inl}}{\delta x} = A \frac{\partial c_i}{\partial x} \qquad (38)$$

This is the only term in the balance equation (37) which depends - through the integral - on the concentration distribution in the whole plate. Such an overall dependance contrasts the local character of the balance equation.

The interaction between the fields of stresses and concentration is of feed back character: The growth of concentration gradients due to diffusion induces a stress field which simultaneously influences the concentration field due to the diffusion flow created by the stresses. This interconnection between both fields has a further complicating consequence - namely it is responsible for the time dependant boundary conditions of our problem. Let us explain it in some details in connection to direct experimental conditions. As example we take the situation presented on Fig.3. Initially both sides of the metallic tube is kept in contact with the some pressure of gaseous hydrogen. At time t=0 the pressure at the outer wall of the tube is increased instantaneously from the initial value $p^o_{H_2}$ to a certain new value $p^*_{H_2}$ what invokes pressure changes inside the tube in course of time presented on Fig.3. At time t=0 the chemical potential of hydrogen at the entrance (outer) surface of the tube is determined by the hydrogen pressure in the gas phase what leads to a certain hydrogen concentration in the bulk layer attached to the surface. Assuming a high activity of the metallic surface in respect to the adsorption and dissociation of gaseous hydrogen molecules, we can expect a quick establishment of a thermodynamic equilibrium between gaseous hydrogen and the mentioned bulk layer. At the very beginning of the absorption process we can assume that the stress has no contribution to the chemical potential of hydrogen absorbed in the first atomic layers of the bulk phase. But as the absorption process develops, a stress field is growing up giving a well defined contribution to the chemical potential of the mobile hydrogen particles at the surface layer being in direct contact with gaseous hydrogen. Thus the maintaince of the equilibrium condition requires a continuous change of the hydrogen concentration at the surface layer because of the time changes of the stress contribution. This means that a time independant chemical potential of hydrogen in the gaseous phase, what is easily to be realized by a time independent hydrogen pressure, does not mean a time constancy of the hydrogen concentration in the contacting metallic layer.

Thus even if forgetting further influence of the stress field, the boundary conditions to be applied for the solution of the traditional II. Fick's law are hardly to be maintained on a time independant level. The exact realization of this condition would require a time dependant hydrogen pressure in the gaseous phase, being adjusted to the time changes of the stress at the contacting surface. Of course similar questions arise concerning the inner wall of the tube. We shall come back to this discussion later when considering some numerical estimations for short times of the experiment. Some more detailed remarks can be found elsewhere (Baranowski 1989).

An exact analytical solution of the balance equation (37) is a hard task. Just above we discussed shortly the difficulties in the formulation of boundary conditions in terms of concentration. Initial conditions are much easier to be fixed in an unique way, similarly to the classical problem of diffusion in stress - free conditions.

Let us shortly consider the problem of stationarity for the balance equation (37). The partial differential equation degenerates in this case to an ordinary differential equation, leaving the concentration as a function of the x - variable only:

$$\frac{\partial c_i}{\partial t} = 0 \qquad \text{for} \quad 0 \le x \le L \tag{39}$$

$$D_i \left\{\left(1+\frac{\partial \ln f_i}{\partial \ln c_i}\right) + \frac{2}{3}\,\frac{V_i^2 c_i Y}{RT}\right\} \frac{d^2 c_i}{dx^2} + \frac{2}{3}\,\frac{D_i V_i^2 Y}{RT}\left(\frac{dc_i}{dx}\right)^2 - \left\{\frac{8 V_i^2 D_i Y}{L^3 RT}\int_o^L \Delta c_i \left(x-\frac{L}{2}\right) dx\right\}\frac{dc_i}{dx} = 0 \tag{40}$$

For homogeneous distributions of the mobile interstitial $(\frac{dc_i}{dx}=0)$ the last equation is fulfiled in a trivial way. In classical (stress - free) permeation experiments a stationary flow through a flat membrane requires

$$\frac{dc_i}{dx} = \text{const} = \frac{\overline{\Delta c}_i}{L} \tag{41}$$

where $\overline{\Delta c}_i$ denotes the concentration difference between the outer and the inner surface of the plate. (40) is equivalent with a linear dependance of the concentration on the coordinate. Introducing such a dependance into (39) one can easily prove the fulfilment of this equation. On the other hand a linear dependance of the concentration in a flat plate is identical with vanishing stresses. To check this conclusion, we have to introduce into (23) a linear concentration course. Thus an alternative way for looking at (40) for the fulfilment of (39) would be the reduction of the modified II. Fick's law (19) to the classical version by cancellation of the terms including the gradient of the stress.

It is interesting to remark that in terms of (33) the fulfilment of the balance equation in stationary conditions (40) for linear concentration distributions does not mean a separate fulfilment for the local and non - local parts of the diffusion flow (33):

$$\mathrm{div}\, J_i = \mathrm{div}\, J_{il} + \mathrm{div}\, J_{inl} \tag{42}$$

but, comparing (34) and (35) with (40) it is clear that the participation of both parts of (42) is necessary.

In terms of this discussion it is obvious that the disappearance of stresses reduces the equation for the diffusion flow (32) in such stationary conditions to the classical, Fick's expression:

$$J_i = - D_i \left(1 + \frac{\partial \ln f_i}{\partial \ln c_i}\right) \frac{dc_i}{dx} \tag{43}$$

If a quadratic, third power dependance or exponential dependance of concentration on x is assumed, the stationary balance equation (40) is not fulfiled. Such concentration courses are equivalent with non - vanishing stresses what again can be proved by making use of the equation for the stress field (23.

Here the following more general problem may be taken up: Is the vanishing of stresses the very condition for stationarity? [First of all we have to be more specific, limiting the above question to systems with mobile interstitials only as cause of the stresses, that is leaving out of consideration cases of frozen (metastable) diffusion equilibria where gradients of concentration and accompanying stress fields can be maintained for infinite times (pseudo - stationary conditions)].

Let us discuss this question from the point of view of the minimal entropy production as criterion for stationarity in the linear region of non - equilibrium thermodynamics, known sometimes as Prigogine's principle (Prigogine 1947). For our purpose we have to consider the stationarity behavior of the diffusion flow formulated in the equation set (32) - (35). The local entropy production is proportional to the square of the diffusion flow. We have to examine if the complete reduction of stresses reduces the entropy production. This tendency is clearly realized in the local contribution to the diffusion flow (34) as the second part (proportional to V_i^2) is always positive. Thus for the "local" diffusion flow the disappearance of stresses has as consequence the reduction of the entropy production. Now the "non - local" contribution to the diffusion flow, as expressed by (35), has to be examined. As the term before the integral is always positive, we are left with the analysis of the sign of the integral itself. Let us write it down in two separate parts:

$$\int_0^L \Delta c_i \left(x-\frac{L}{2}\right) dx = \int_0^L \Delta c_i x dx - \frac{L}{2} \int_0^L \Delta c_i dx \tag{44}$$

Two extreme cases are to be distinguished for the initial conditions of a homogeneous concentration c_i^o of the interstitial: a). The absorption of additional interstitial in the plate, during which the following inequality is valid

$$\Delta c_i = c_i - c_i^o \geq 0 \qquad \text{for} \quad 0 \leq x \leq L \tag{45a}$$

and b). the desorption of the interstitial from the plate, characterized by

$$\Delta c_i = c_i - c_i^o \leq 0 \qquad \text{for} \quad 0 \leq x \leq L \tag{45b}$$

To be more unique in the discussion, and in accordance with the experimental conditions realized (Baranowski 1989), we assume that for x=0 the boundary conditions are only changed and maintained in more or less on time - independant values during the ab- and desorption processes. Having this settled, we find for case a). (absorption)that the first integral on the right hand side is always positive and the second always negative:

For case a). at all times:

$$\int_o^L \Delta c_i x dx > 0; \qquad - \frac{L}{2} \int_o^L \Delta c_i dx < 0 \tag{46}$$

For short times the second integral prevails and we have

$$\int_o^L \Delta c_i \left(x-\frac{L}{2}\right) dx < 0 \tag{47}$$

Inequalities (46) result from the fact that small initial penetration depths of the diffusion wave make the second integral of (44) larger numerically than the first one due to the factor $\frac{L}{2}$ before the second integral. But increased diffusion times make the numerical contribution of the first integral in (44) more and more significant, giving at some time instant the equality

$$\int_o^L \Delta c_i x dx = \frac{L}{2} \int_o^L \Delta c_i dx \tag{48}$$

what is equivalent to the vanishing of integral (47) and as consequence to the dissappearance of the non - local part of the diffusion flow. In other words the previous negative non - local diffusion flow goes over to a positive one. Such an inversion is equivalent with a conversion from the initial weakening of the total diffusion flow (33) to its strengthening in the following times (The local part of the diffusion flow is always positive in case a).). Thus the approach to the stationary state is characterized by a positive contribution of the non - local (stress induced) part of the diffusion flow. Therefore if stresses are reduced at the stationary state then a clear consequence of this will be a reduction of the entropy production. The above analysis leads us to the conclusion that a stress - free stationary state is surely (in case a). - effective absorption) characterized by a lower entropy production than a state with non - vanishing contributions of stresses to the diffusion flow.

For case b). at all times:

$$\int_o^L \Delta c_i x dx < 0; \qquad -\frac{L}{2}\int_o^L \Delta c_i dx > 0 \tag{49}$$

Here the non - local part of the diffusion flow will be positive in the first desorption period (thus as in case a). It will weaken the local part (always negative in the desorption process - case b).). In time course an inversion point, similar to that formulated by (48), will be met and the approach to stationarity will result in a strengthening of the total (negative) diffusion flow. Thus the dissappearance of stresses in stationary conditions will cause a reduction of the entropy production. Therefore one can conclude that a stress - free stationary state is surely in favour of the minimum entropy production principle as criterion of stationarity. One can find additional arguments supporting this conclusion: Following the diffusion kinetics and the accompanying changes of the stresses, it is easy to conclude that with growing time the intensity of the stress fields is clearly going down after passing a certain maximum, if starting from a stress free initial condition. Energetically is the creation of stresses equivalent with an increase of potential energy in the elastic media. Therefore one can expect that the approach to stationarity, as a state of a small deviation from thermodynamic equilibrium will reduce, as far as possible, the potential energy created by the diffusion of the mobile interstitial. Looking on the other hand on the structure of the total flow equation (32), it seems convincing that a minimum of the entropy production is in the simplest way to be expected by self reduction of the two terms due to stresses. It is hardly to expect that a combination of the classical Fick's term with the stress contributions could be an alternative approach.

The above conclusion about the stress - free condition for the stationary states contradicts the previous qualitative statements (Kandasamy 1988) about the non - linear stationary course of concentration. But first of all one has to distinguish two different stationary states: The linear increase of pressure in the metallic tube (Lewis et al 1988)implies a time dependant boundary condition at the inner tube wall, due to the changes of the hydrogen pressure in the surrounding gaseous phase. Such a stationary state does not imply the time constancy of concentration in each volume element, as formulated in (39). Therefore the difference between our discussion, based on (39), and the stationarity condition, probably discussed by Kandasamy (Kandasamy 1988) may be simply due to terminological confusion.

5). Comparison with experiments:

A systematic application of the balance equation derived (37) to experimental results, shortly characterized in 3). requires a solution of this equation with the appropriate initial and boundary conditions. The normal sequence would be first a general solution of (37) and later its adaptation to the experimentally realized conditions. The complex mathematical structure of this equation makes an exact analytical solution a rather difficult task. Little help can be expected from solutions of the celebrated Boltzmann integro - differential equation, used in kinetic theory of gases and electronic transport phenomena in metals (Chapman and Cowling 1951, Ziman 1969). There exists a quite formal similarity only between (37) and Boltzmann equation: Namely both are balance equations - (37) is the balance equation for the mobile interstitial embedded in an elastic matrix and Boltzmann equation represents the balance equation of the one - particle distribution function. In fact (37) could be derived from a Boltzmann equation if formulated for the interstitial component in the elastic matrix in a similar way as it is done for the mass balance equation of the gas particles. But a principal difference exists between the typical local character of Boltzmann equation and the non - local character of (37). Furtheron the integral in the Boltzmann equation plays the role of the source term for the one - particle distribution function whereby in (37) it represents the non - local part of the balance equation of the interstitial. In (37) the integral plays the role of one of the coupling terms between the fields of stresses and concentration and it is partially responsible for feed - back character of this balance equation. Due to the above differences little help can be expected for the solution of (37) from known solutions of the Boltzmann equation.

As a realistic proposition one could assume the solution of classical Fick's law - that is the state of stress free conditions - as a zeroth approximation, treating further, as additional contribution, the stress field evaluated on this basis. So far in literature one can find only examples of evaluations of stress fields, basing on concentration fields, taken over from stress - free solutions of balance equation (37) (Prussin 1961, Li 1978). One of the results of these calculations was the manifestation of the non - local character of the stress fields developed (Li 1978).

The most interesting results of hydrogen diffusion in $Pd_{0.8}Pt_{0.2}$ alloy in the presence of self - stresses are summarized in Fig.3 and 4. The time course of the hydrogen pressure inside the metallic tube reflects - to a certain degree - the time behavior of the total hydrogen diffusion flow, as given by (32), at the surface x=L.

Serious requirements are to be fulfiled if this is true. Pressure changes in the tube result due a hydrogen exchange between the internal (inner) surface of the tube and the gaseous phase. Such an exchange may exhibit a considerable delay if the surface activity is not sufficient in respect to the recombination process of hydrogen particles at the surface. Only monoatomic hydrogen particles (most probably as electron - screened protons) are present inside the metallic matrix what follows not only from the size of avalaible interstitials but also from the fulfilment of Sievert's law (Baranowski 1978). After penetrating the interphase from the bulk to the surface layer a recombination to diatomic atoms has to be followed before a release to the gaseous phase can take place. If this recombination process is slow, a gap of the hydrogen chemical potential can exist between the surface phase and the contacting phase of gaseous hydrogen. In such situations the time course of the gaseous hydrogen pressure will not monitor the bulk diffusion flow of hydrogen for x=L. To avoid the above mentioned gap of the hydrogen chemical potential - and thus to make the pressure course identical with the hydrogen flow - a special activation procedure (electrodeposition of a Pd-black layer) was always applied (Lewis et at. 1983, Baranowski 1989). The best method for checking the efficiency of the activation procedure is the determination of the hydrogen diffusion coefficient for a hydrogen free membrane (practically stress free conditions during the permeation process and no non - local contribution (35) due to $c_i=0$). If the evaluated diffusion coefficient is smaller than that determined with active surfaces, the activation perfomed is doubtful and should be repeated.

Being sure that the time course of the hydrogen pressure can be identified with the diffusion flow (32), the most elegant description of Fig.3 and 4 would require the theoretical evaluation of this flow as a function of time, basing on the concentration distribution of hydrogen in the plate (for $0 \leq x \leq L$). As we discussed above such on ambitious task could be only realized by the exact solution of the balance equation (37). As long as this is not known, we have to look for other possibilities of a comparison between theory and experiment. First of all let us interpret Fig.3 and 4 in terms of the flow equation (32) in a qualitative way: The initial conditions for the curves presented on Fig.3 and 4 is the homogeneous (stress - free) distribution of hydrogen in the plate. At a certain time instant (t=0) the boundary conditions at the outer wall of the tube (x=0) are changed, either by an instanteneous increase of the hydrogen activity (Fig.3) or a decrease of its activity (Fig.4). Let us analyse in some details the course presented on Fig.3: Taking the flow equation (32) into account, we can initially disregard its local part J_{il}, according to (34) as at the inner wall of the tube no concentration gradient exists ($\frac{dc_i}{dx}=0$ at x=L for t=0). Thus what counts for the initial period is the non - local part of (32), as given explicitly by (35). This non - local part is initially negative, because $\Delta c_i>0$ and the integral is negative, as it was discussed in details above (eq. (46) and (47)). Therefore the initial decrease of hydrogen pressure in the tube follows from (32). For much longer times the non - local part of (32) decreases, passes through zero and becomes later positive. Simultaneously the Fickian, local part of (32)(see (34)) plays for the exchange of hydrogen at the internal tube wall an increasing role giving finally, together with the contribution of the non - local part of the flow, the expected classical time course of the hydrogen pressure. Thus the shape of the curve, presented on Fig.3, is due to the superiority of the non - local part of the hydrogen flow in the initial absorption period and the conformity of the local and non - local parts later on. The appearance of a minimum in the internal hydrogen pressure is simply caused by an overlapping of both tendencies.

An explanation of Fig.4 (desorption process) is possible along similar lines, taking only the change of boundary conditions into account: At the time t=0 the hydrogen activity at the outer tube wall (x=0) is instantaneously reduced, leading to $\Delta c_i<0$. This results in a positive value of the non - local part of the hydrogen flow which prevails in the initial desorption. It leads to an increase of the hydrogen pressure in the tube. Later on the influence of the local (Fickian) part of the diffusion flow prevails, amplified by the sign change of the non - local part (see (45), (46) and the discussion). As a natural sequence Fig.4 exhibits a maximal hydrogen pressure in the tube after which the normal, Fickian behavior follows.

The above discussion proves that the non - local phenomena - especially the initial unexpected directions of the hydrogen flows - can be explained by the theoretical approach presented in 4).

Let us go now a step further, looking for a quantitative comparison between the experimental results and the theoretical expectation. So far this is possible only for short initial times of the ab- or desorption process. Under such circumstances one can neglect the contribution of the Fickian (local) part of the total diffusion flow (32) and estimate only the non - local part of this flow as being dominant for the experimental results recorded. In terms of (32) this means that we approximate for the time interval considered:

$$\left(\frac{dc_i}{dx}\right) \cong 0 \qquad \text{for} \quad z = L \tag{50}$$

what is equivalent with the approximation (33):

$$J_i = J_{il} + J_{inl} \cong J_{inl} \qquad \text{for} \quad z = L \tag{51}$$

For how long times these approximations can be considered as reasonable? In physical terms, the diffusion wave starting at x=0, due to the instantaneous boundary conditions change for t=0 should be as narrow as possible. In diffusion kinetics a useful criterion for such conditions in the inequality

$$\frac{D_i t}{L^2} << 1 \tag{52}$$

where t is the time counted from the instanteneous change of the boundary conditions (t=0) and D_i and L have the same meaning as used above. For such short time our plate can be treated as a semi - infinite medium because the concentration changes due to Fickian diffusion are far from the inner wall of the tube. The inequality (52) in mind, we can take over the known solution of Fick's equation for the boundary conditions considered (Crank 1956)

$$\frac{c_i - c_i^*}{c_i^o - c_i^*} = \text{erf} \frac{x}{2(D_i t)^{1/2}} \tag{53}$$

where erf means the error function and c_i^* denotes the time independant concentration of the interstitial for x=0 and t>0. If $c_i^* > c_i^o$ an absorption process takes place, whereby the inverse inequality $c_i^* < c_i^o$ will characterize a desorption of the interstitial from the plate, through the outer wall (x=0). With the concentration course (53) we can evaluate the integral in the non - local part of the overall diffusion flow J_{inl} (35). First of all let us remark that the following approximation holds

$$\int_o^L \Delta c_i \left(x-\frac{L}{2}\right) dx \approx -\frac{L}{2} \int_o^L \Delta c_i \, dx \tag{54}$$

This is justified as the region of x for which the integral exhibits non - vanishing values is small compared to $^L/_2$ due to the inequality (52). On the other hand the approximate integral (54) expresses the time changes of the mass of the interstitial, ab- or desorbed from the plate in the time interval considered. The introduction of (53) into (54) leads to

$$\frac{L}{2}\int_o^L \Delta c_i dx = L(c_i^* - c_i^o)\left(\frac{D_i t}{\pi}\right)^{1/2} \tag{55}$$

Thus the change of the interstitial content (positive or negative) is proportional to $t^{1/2}$ in the initial time period of the ab- or desorption process.

Let us remark here that the procedure carried through in the evaluation of integral (54) is equivalent to decoupling of the concentration and stress fields. In reality the course of the concentration as given by (53) is not exact because it is influenced by the developing stress field. In other words, the application of a stress free Fick's equation for the evaluation of the concentration field is an approximation. Such a procedure was extensively applied in the past (Prussin 1961, Li 1978) without stressing its non - exact character. But this has to be remembered in all further applications of (55).

The approximate evaluation of the integral (54) makes it possible to express explicitly (also in an approximate way) the stress field, the stress gradient and furtheron the non - local part of the hydrogen diffusion flow.

One derives the equation for the stress when introducing (54) and (55) into (23), giving

$$\sigma_{yy}=\sigma_{zz}=-\frac{V_iY}{3}\left\{\Delta c_i+\left(\frac{12x}{L^2}-\frac{8}{L}\right)(c_i^*-c_i^o)\left(\frac{D_it}{\pi}\right)^{1/2}\right\} \tag{56}$$

Like in the time course of the interstitial content (55) we find a $t^{1/2}$ dependance for the non - local part of the original expression for the stresses (23). Equation (56) is not complete as c_i in the term Δc_i should be consequently replaced by (53), leading finally to

$$\sigma_{yy}=\sigma_{zz}=\frac{-V_iY}{3}(c_i^*-c_i^o)\left[1-\mathrm{erf}\frac{x}{2(D_it)^{1/2}}+\frac{12x}{L^2}-\frac{8}{L}\left(\frac{D_it}{\pi}\right)^{1/2}\right] \tag{57}$$

For expressing the hydrogen flow in the initial period of ab- or desorption process we need the gradient of the stress which follows from (56) as equal to

$$\frac{\partial\sigma}{\partial x}=2\frac{\partial\sigma_{yy}}{\partial x}=-\frac{2}{3}V_iY\left[\frac{\partial c_i}{\partial x}+\left\{\frac{12}{L^2}(c_i^*-c_i^o)\left(\frac{D_it}{\pi}\right)^{1/2}\right\}\right] \tag{58}$$

Thus introducing (58) into (30) we derive

$$J_i=-D_i\left\{\left(1+\frac{\partial \ln f_i}{\partial \ln c_i}\right)+\frac{2V_ic_iY}{3RT}\right\}\frac{\partial c_i}{\partial x}-\frac{8V_i^2c_iYD_i}{RT\,L^2}(c_i^*-c_i^o)\left(\frac{D_it}{\pi}\right)^{1/2} \tag{59}$$

terms of (35) if it easy to recognize that the non - local part of the ...sion flow equals

$$J_{inl}=\frac{-8V_i^2c_iYD_i}{RT\,L^2}(c_i^*-c_i^o)\left(\frac{D_it}{\pi}\right)^{1/2} \tag{60}$$

In the experiments described in 3). the pressure of hydrogen inside the metallic tube was recorded as a function of time. This pressure course reflects the exchange of hydrogen between the inner wall of the tube (x=L) and the neighbouring gaseous phase. When starting the ab- or desorption process of hydrogen to or from the bulk metallic phase by changing the boundary conditions at the outer tube wall (x=0), there exists no concentration gradient at the inner wall (x=L). Therefore the local part of the hydrogen flow (59) can be neglected and for short times we are left with the non - local part (60) only. Assuming a very active surface of the inner tube wall, we can relate the flow (60) directly to the pressure change in the tube volume $P_i(H_2)$

$$\frac{dP_i(H_2)}{dt} = \frac{1}{2}\,\frac{RTS'}{V}\,J_{inl} \tag{61}$$

where S is the surface of the inner wall and V the volume of the tube (including the dead volume of the pressure gauge). $\frac{1}{2}$ in (61) takes into account the obvious fact that J_{inl} is expressed in moles of H whereby the pressure in the tube concerns the H_2 molecules. With proper units on the right hand side of (60) the flow J_{inl} will give the number of moles of H penetrating through the surface unit (m^2) in a time unit (sec). Deriving (61) we made use of the equation of state far an ideal gas.

Combining (61) with (60) and integrating in respect to time we obtain the following expression for the initial time course of the hydrogen pressure inside the tube

$$P_i(H_2) = P_i^o\,(H_2) - \frac{8}{3}\,\frac{V_i^{\,2}YD_i^{\,3/2}S}{\pi^{1/2}\,V\,L^2}\,c_i^o\,(c_i^*-c_i^o)\,t^{3/2} \tag{62}$$

Thus the hydrogen pressure in the tube should be - in the initial period of the process - proportional to time in power $^3/2$. This time dependance can be treated as a first quantitative criterion for the experimental verification of the theory developed in 4).. Plotting the pressure in the tube as a function of $t^{3/2}$ one should expect a linear course with the characteristic slope given by

$$\frac{dP_i(H_2)}{dt^{3/2}} = -\frac{8}{3}\,\frac{V_i^{\,2}YD_i^{\,3/2}S}{\pi^{1/2}VL^2}\,c_i^o\,(c_i^*-c_i^o) \tag{63}$$

A comparison of the numerical values of the slope, as evaluated from (63) and determined experimentaly can be treated as the most detailed comparison between theory and experiment.

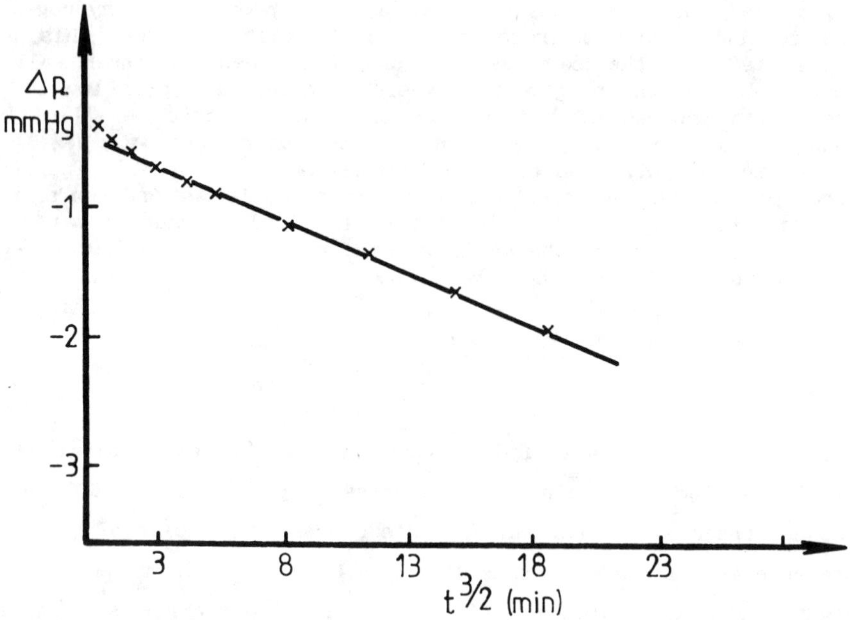

Fig.5. $\Delta p = p_i(H_2) - p_i^o(H_2)$ (in mm Hg) as a function of $t^{3/2}$ (min $^{3/2}$) measuredinside the tube during an electrolytic cathodic charging at 256K (I=200 mA, electrolyte solution HCl + CH_3OH + thiourea).

Fig.5 presents an example of an experimental curve plotted in the variable set $P_i(H_2)$ and $t^{3/2}$. We notice the fulfilment of a linear course as it is forseen by (62). As it was underlined during the derivation of (62), the course exemplified on Fig.5 is realized for the initial period only in which a negligible gradient of concentration exists near the inner wall of the tube (x=L). The slope of the straight line in Fig.5 equals

$$\frac{dP_i(H_2)}{dt^{3/2}} \cong 2 \times 10^{-2} \frac{Pa}{sec^{3/2}} \qquad (64)$$

Let us compare this numerical value with that calculated from (63): As the experimental details in which Fig.5 was realized are not allowing a quantitative fixation of c_i^o and c_i^* (due to the not well known initial and boundary conditions, mainly caused by the electrolytic charging), we can only estimate the order of magnitude of c_i^o if all other quantities in (63) are introduced in numbers. The following values were taken over: V_i

x 10^{-6} $\frac{m^3}{mole\ H}$ (Baranowski, Majchrzak and Flanagan 1971) Y = 3.05 x

aken for the alloy $Pd_{81}Pt_{19}$ as additive from the values for

$D_i = 4.0 \times 10^{-12} \frac{m^2}{sec}$ (Baranowski 1989) S, V and L for the

equal to 1.84 x $10^{-3} m^2$; $2.44 \times 10^{-5} m^3$ and $2.5 \times 10^{-4} m$. With these numbers we obtain from (64) and (63)

$$c_i^o \;(c_i^* - c_i^o) \cong 1.6 \times 10^6 \; \frac{mol^2}{m^6} \tag{65}$$

This leads to a mean value of hydrogen concentration in the alloy considered of the order of magnitude 10^3 mol/m^3 what is equivalent in atomic ratio

$$\frac{n_H}{n_{Pd_{81}Pt_{19}}} \cong 10^{-2} \tag{66}$$

The number in (66) follows from (65) if one takes into account - again from simple additive rule of the densities of pure Pd and Pt - that 1.132 x 10^5 moles of the alloy considered are occupying the volume of 1 m^3. It is a plausible number for the alloy considered (Lewis 1989).

To be more specific in a numerical comparison of (63) with experiment we have to analyse a situation in which the two concentrations in this equation (c_i^o and c_i^*) can be reduced to one. This can be realized if the outer wall of the tube can be maintained at a negligible hydrogen concentration ($c_i^* \cong c_i^o$). One can assume that such a requirement is fulfiled by an intense pumping off of the hydrogen volume surrounding initially the outer wall of the tube. The time course of pressure follows here the curve presented on Fig.4. For our purpose only the initial pressure increase in the tube has to be considered Fig.6 presents such an example

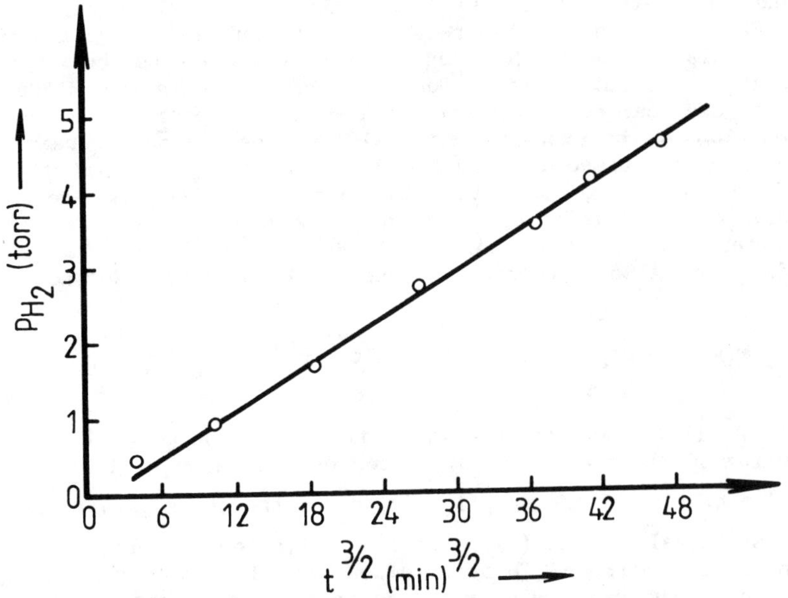

Fig.6. Pressure increase as a function of $t^3/2$ for a desorption process (intense pumping off of the hydrogen volume surrounding the outer tube wall) at 273K. Initial equilibrium pressure of gaseou hydrogen 680 Torr.

As in Fig.5 we find here the requirement of (63) fulfiled. The slope equals 3.8×10^{-2} Pa $\sec^{-3/2}$. Comparing this number with the right hand side of (63) whereby for the diffusion coefficient the value $5.9 \times 10^{-12} m^2 sec^{-1}$ was taken (Baranowski 1989) (all other numerical values were the same as above discussing Fig.5) we receive for c_i^o the value 1.54×10^3 mol m^{-3} what is equivalent to the atomic ratio $^{n_H}/_{n_{PdPt}}$ of about 0.014. This is the right order of magnitude but the number itself is lower around 4 times as compared with the measured concentration at the initial hydrogen pressure of 680 Torr (Lewis 1989) and seems also too low as compared with the data calculated from stationary permeation results at room temperature (Baranowski 1989). Which reasons can be suspected as standing behind this discrepancy? Besides the experimental errors of the individual quantities appearing in (63), following further reasons are probable: 1). It is difficult to prove how far the assumption $c_i^*=0$ at $x=0$ for $t\geq0$ is fulfiled. If the activity of the surface is not sufficient for the exchange of the hydrogen particles with the gas phase, a considerable error may be introduced. Assuming the real initial concentration of hydrogen as equal to 6×10^3 mol m^{-3} the c_i^* value should be 5.6×10^3 mol m^{-3} in order to fit the calculated value of effective c_i^{o2} equal to 2.37×10^6 mol^2 m^{-6}. In this case the surface concentration during the desorption would be only about 7% lower than the hydrogen concentration in the bulk phase. 2). Much more involved may be the surface at the inner wall of the tube. Here only small differences of the hydrogen chemical potential are responsible for the exchange between the gaseous phase and the surface of the tube. If here the surface activity is not sufficient, the exchange rate may be easily lower than the maximal value calculated from (62). According to this relation an instantaneous equilibrium should always exist between the surface and the gaseous bulk phase. Any deviation from this equilibrium in terms of existing differences of hydrogen chemical potential between the phases mentioned would result in a reduction of the exchange rate. 3). The most important reason for the discrepancy discussed is of quite different character. In order to make it clear, let us go back to (56), expressing the time course of the stresses in the initial time period of the ab- or desorption process. At the outer wall of the tube (x=0) we get, assuming time independent boundary condition in respect to concentration at this surface, from (56)

$$\sigma(x=0) = 2\sigma_{yy}(x=0) = -\frac{2V_i Y}{3}\{(c_i^* - c_i^o)[1 - \frac{8}{L}(\frac{D_i t}{\pi})^{1/2}]\} \qquad (67)$$

First of all the maximal stress at the outer surface will occur at the beginning of the process (t=0). Later on its value will be a decreasing …unction of time. For desorption $(c_i^* < c_i^o)$ the stress σ will be positive, … is equivalent (see (2)) with a negative contribution to the chemical …al of hydrogen. This reduction acts in favor of a delay of the …rocess bacause it replaces in a certain sense part of hydrogen … Assuming at t=0, $c_i^*=0$ and $c_i^o \cong 10^3$ mol m^{-3}, we calculate …s value of about 350 MPa if for V_i and Y the previously

193

mentioned numbers are applied. In terms of the chemical potential of hydrogen it is equivalent to about 600 J mol^{-1} what corresponds to concentration changes of the order 1 mol/m^3. Such a value is negligible as compared with the assumed concentration of hydrogen in the bulk phase which is three orders of magnitude higher. We conclude that the stresses arising at the outer tube wall can be neglected in our experimental conditions. The same conclusion is valid for the absorption process.

At the inner wall of the tube (x=L) (56) leads to

$$\sigma(x=L) = 2\,\sigma_{yy}(x=L) = -\frac{8V_iY}{3L}\,(c_i^*-c_i^o)\left(\frac{D_it}{\pi}\right)^{1/2} \tag{68}$$

At the start of the experiment (t=0) the inner tube wall is stress free but later on the stress increases with $t^{1/2}$. During a desorption process ($c_i^o > c_i^*$) the stress will be positive leading, according to (2), to a decrease of the hydrogen chemical potential. From this point of view an absorption process should follow at the inner tube wall what is opposite to the direction of the non - local diffusion flow, induced by the arising stress field in the bulk phase (see (60) and (62)). Inversely an absorption process at x=0 will create a negative stress at the inner wall (x=L), as can be seen from (68), and in consequence of this a positive change of the hydrogen chemical potential, invoking a release of hydrogen from the surface to the inner bulk phase. In consequence of this an increase of the hydrogen pressure in the tube should follow, being opposite to the pressure reduction caused by the non - local part of the hydrogen flow (see (62) and (63)). We conclude that, both from (67) and (68), a weakening of the non - local contribution should follow what surely can be a reason for a non - quantitative agreement between the experiments and the calculations based on (62) and (63). First of all as long as the influence of the stresses (68) prevails, a behavior schematically presented on Fig.7 can be expected.

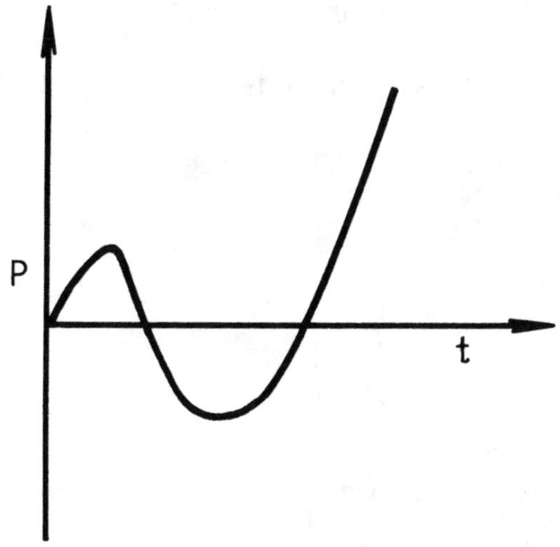

Fig.7. Expected time course of the hydrogen pressure inside the tube during an absorption process if the non - local flow (60) is preceded by the influence of the stress induced flow at the inner wall (68).

In such cases a more complex behavior could be observed than previously presented on Fig.3. Similarly to Fig.7 a desorption process - as shown generally by Fig.4 - should exhibit a minimum in the pressure course before the influence of the non - local diffusion flow with the corresponding maximum of the pressure will follow. It is worthwhile to mention that recently curves following the patterns of Fig.7 were observed during absorption of hydrogen in $Pd_{77}Ag_{23}$ tubes (Kandasamy, Lewis and Tang 1989). It remains now to explain which conditions are decidive for the realisation of Fig.3 or Fig.7? For this purpose we have to treat the exchange between the inner tube wall and the gas phase in a more detailed way. In the most simple approach we can assume that the exchange rate mentioned J_{sb} is proportional to the difference of the chemical potential of hydrogen at the inner surface (x=L) and in the surrounding gas phase.

$$J_{sb} = a\,[\mu_H(L) - \frac{1}{2}\,\mu_{H_2}(g)] \tag{69}$$

where the index sb denotes the exchange between the surface (s) (x=L) and the gaseous bulk phase (b). $\mu_H(L)$ and $\mu_{H_2}(g)$ denote the chemical potential of hydrogen at the inner surface of the tube and in the surrounding gaseous phase. When deriving the equations (61) - (63) we identified the non - local diffusion flow (60) with (69)

$$J_{inl} = J_{sb} = a\,[\mu_H(L) - \frac{1}{2}\,\mu_{H_2}(g)] =$$

$$= - \frac{8V_i^2 c_i Y D_i}{RTL^2} \times (c_i^* - c_i^o)\left(\frac{D_i t}{\pi}\right)^{1/2} \tag{70}$$

On the other hand the chemical potential of the hydrogen particles at the inner surface equals, in terms of the symbols used in (70)

$$\mu_H(L) = \mu_H^o + RT\,\ln c_i(L) - V_i\sigma\,(L) \tag{71}$$

if we neglect for simplicity the corresponding activity coefficient (see (1)). At the beginning of the ab- or desorption process (t=0) the equilibrium condition holds

$$\mu_H^o\,(L) = \frac{1}{2}\,\mu_{H_2}(g) = \mu_H^o + RT\,\ln c_i^o =$$

$$= \frac{1}{2}\,(\mu_{H_2}^o + RT \quad \ln P_i^o\,(H_2)) \tag{72}$$

As no stress field exists and the hydrogen pressure inside the tube corresponds to that used in (62) for t=0. For t>0 the stress in (71) has to be replaced by (68), giving

$$\mu_H(L) = \mu_H^o + RT\,\ln c_i(L) + \frac{8V_i^2 Y}{3L}\,(c_i^* - c_i^o)\left(\frac{D_i t}{\pi}\right)^{1/2} \tag{73}$$

Thus the time course of $\mu(L)$ will be determined by two terms: 1). One dependant on the concentration at the surface ($RT\ln c_i(L)$) and influenced directly by the non - local diffusion flow (60). 2). By the time course of the stress at the surface, given in (73) by the term proportional to $t^{1/2}$. As we discussed it already above, the two terms contribute in opposite directions to $\mu_H(L)$ - thus the effective exchange flow (69) may follow either the direction of the in - or outflow at the outer tube wall (x=0) or may go in the opposite direction. It is clear that for the beginning the influence of the surface stress will dominate as this part is transmitted with sound velocity from the outer to the inner wall. The contribution from the non - local diffusion, governed partially by diffusion kinetics, will therefore be manifested later. In this way the sequence of events shown on Fig.7 seems plausible. What still has to be taken into account is the contribution of the active Pd layer electrolytically deposed on the tube wall in order to increase the "a" coefficient in (69) that is to enhance the exchange rate between the surface and the gaseous phase. In the case of $Pd_{81}Pt_{19}$ alloy (Baranowski 1989) as presented on Fig.3 - Fig.6 a large difference in the hydrogen content exists between the bulk alloy and the Pd electrodeposed layer of Pd. Thus the quick impulse of the hydrogen flow caused by the stress changes of the surface (the small maximum in the first absorption period on Fig.7) may easily be sucked off by the Pd layer without giving an immidiate evidence in the pressure change. But if the difference in the hydrogen content between the Pd surface layer and the bulk alloy is much smaller, like it is the case in the $Pd_{77}Ag_{23}$ alloy, the first out - or input of hydrogen will manifest itself in the time course of the gaseous hydrogen pressure (see Fig.7). The reasoning presented will be especially valid if the electrodeposed Pd layer does not adhere strongly enough to the alloy surface making a direct transmission of the stress developed at the alloy surface to the contact layer between the gas phase and the contacting surface of the Pd layer not possible. Such a situation will surely be valid for thick electrodeposed Pd layers.Because a detailed discussion would require more knowledge about the Pd layers used, it is not desirable to develop here a quantitative approach.

6).General outlook: The time courses of hydrogen pressure, as presented on Fig.3 - 6, are qualitatively different from the classical permeation, as shown on Fig.2. A serious consequence of this difference may be a considerable error in the evaluated diffusion coefficient if the time lag method is applied (Lewis et al. 1983, Baranowski 1989, Kandasamy et al. 1989). Such a remark may be valid not only for the diffusion of interstitials. In more general terms each diffusion problem in elastic media requires a careful examination of a possible influence of self - stresses on the diffusion characteristics. The complex dependance of stress fields on the geometrical and boundary conditions of the problem considered makes such an examination even more urgent and desirable. On the other hand the development of stress fields during diffusion in solids is a common feature as long as volume changes are accompanying the mixing process. Cases of mixtures with vanishing partial volumes are rather artificial. Besides examples in which Gorsky (Volkl and Alefeld 1978) or Stoney (Cermak et al 1985, Kufudakis et al 1989) effects were applied for determination of diffusion coefficients, the influence of self - stresses was practically never considered in connection with experiments. It is easily to imagine that due to this neglectance several data measured in the presence of stresses are to be treated with caution and probably at least some of them are to be reconsidered. Let us remark that in our treatment the maximal stress fields, which can be created, were described. In reality a part of them is reduced due to annihilation at grain boundaries, dislocations or simply by plastic flow. Therefore in comparisons with experiments this possible source of discrepancy has to be seriously taken into account. From this point of view one has always to expect that the experimental results will be lower than that calculated for an ideal elastic medium. Our numerical comparisons presented above confirm this conclusion. A consequent incorporation of stresses into the diffusion characteristics in solids is in its very beginning and a wide field of activity lies surely before us.

References:

Baranowski B., Majchrzak S. and Flanagan T.B. 1971. The volume increase of fcc metals and alloys due to interstitial hydrogen over a wide range of hydrogen contents. J. Phys. F: Metal Phys. 1: 258-261.

Baranowski B. 1975. Nichtgleichgewichtsthermodynamik in der physikalischen Chemie. Leipzig, VEB Deutscher Verlag fur Grundstoffindustrie.

Baranowski B. 1978. Metal - hydrogen systems at high hydrogen pressures. In Hydrogen in Metals II eds. G.Alefeld and J.Volkl, 157-200. Berlin, Springer Verlag.

Baranowski B. 1989. Stress - induced diffusion in hydrogen permeation through $Pd_{81}Pt_{19}$ membranes. J. Less-Common Met. 154: 329-353.

Baranowski B., Lewis F.A., 1989. Non - local, non - Fickian diffusion of hydrogen in metals. Ber. Bunsenges. Phys. Chem. 93: 1225-1227.

Barrer R.M. 1956. Diffusion in and through Solids. Oxford, University Press.

Chapman S. and Cowling T.G. 1951. Mathematical Theory of Non - uniform Gases. Cambridge University Press, Cambridge.

Crank J. 1956. The Mathematics of Diffusion. Oxford, Oxford University Press.

Cermak J., Gardavska G., Kufudakis A. and Lejcek P. 1985. Diffusion - elastic effect - its relation to the Gorsky effect and its application to measurements of diffusivity of hydrogen in the crystal lattice of nickel with trapping sites. Z. Phys. Chem. (N.F.) 145: 239-250.

Fitts D.D. 1962. Non - equilibrium Thermodynamics. New York. Mc Graw - Hill.

Gibbs J.W. 1906. Scientific Papers. Vol.1, Longman, London.

Gorsky W.S. 1935. Theorie der elastischen Nachwirkung in ungeordneten Mischkristallen (elastische Nachwirkung zweiter Art). Phys. Z. der Sowjetunion 8: 457-471.

Groot de, S.R. and Mazur P. 1962. Non - equilibrium thermodynamics. Amsterdam, North - Holland Publ. Company.

Hirth J.P. 1977. In Stress corrosion cracking and hydrogen embrittlement in iron base alloys. ed. R.W.Staehle et al. NACE, Houston.

Kandasamy K. 1988. Hydrogen concentration distribution in elastic membranes during diffusion of hydrogen. Scripta metall. 22: 479-481.

Kandasamy K., Lewis F.A. and Tong X.Q. 1989. Lattice strain dependant uphill diffusion of hydrogen. Submitted to the Proc. of 4. International Conference of the Effects of Hydrogen on Material Behavior. Jackson Lake Lodge, Wyoming, U.S.A.

Kirchheim R. 1986. Interaction of hydrogen with external stress fields. Acta metall. 34: 37-42.

Kufudakis A., Cermak J. and Lewis F.A. 1989. Measurements of hydrogen diffusion coefficient in the α - phase concentration range of the Pd/H system by the diffusion - elastic technique. Z. Phys. Chem. (N.F.) 164: 1013-1018.

Larche F.C. and Cahn J.W. 1973. A linear theory of thermomechanical

equilibrium of solids under stress. Acta metall. 21: 1051-1063.
Larche F.C. and Cahn .W. 1978. Thermochemical equilibrium of multiphase solids under stress. Acta metall. 26: 1579-1589.
Larche F.C. and Cahn J.W. 1982. The effect of self - stresses on diffusion in solids. Acta metall. 30: 1835-1845.
Larche F.C. and Cahn J.W. 1985. The interaction of composition and stress in crystalline solids. Acta metall. 33: 331-357.
Larche F.C. 1988. Thermodynamics of stresses solids. In Solid State Phenomena Vol.3-4: p.205.
Lewis F.A., Magennis J.P., Mc Kee S.G. and Ssebuwufu P.J.M. 1983. Hydrogen chemical potentials and diffusion coefficients in hydrogen diffusion membranes. Nature (London) 306: 673-675.
Baranowski B. 1987. Uphill diffusion effects induced by self - stresses during hydrogen diffusion through metallic membranes. J. Less-Common Met. 134: L27-31.
Lewis F.A., Kandasamy K., Baranowski B. 1988. The "uphill" diffusion of hydrogen. Platinum Met. Rev. 32: 22-26.
Lewis F.A. (private communication).
Li J.C.M., Oriani R.A. and Darken L.S. 1966. The thermodynamics of stressed solids. Z. Phys. Chem. (N.F.) 49: 271-290.
Li J.C.M. 1978. Physical chemistry of some microstructural phenomena. Metall. Trans. A 9: 1353-1380.
Prigogine I. 1947. Etude Thermodynamique des Phenomenes irreversibles. Editions Desoer, Liege.
Prussin S. 1961. Generation and distribution of dislocations by solute diffusion. J. Appl. Phys. 32: 1876-1881.
Stephenson G.B. 1988. Deformation during interdiffusion. Acta metall. 36: 2663-2683.
Timoshenko S. 1934. In Theory of Elasticity. I. ed., New York, Mc Graw - Hill Book Company.
Volkl J. and Alefeld G. 1978. Diffusion of Hydrogen in Metals. In Hydrogen in Metals I. eds. G. Alefeld and J. Volkl, 321-348. Berlin, Springer Verlag.
Volkl J. and Alefeld G. 1979. Anelasticity due to long - range diffusion. Z. Phys. Chem. (N.F.) 114: 123-139.
Wipf H. 1978. Electro - and Thermotransport of Hydrogen in Metals. In Hydrogen in Metals II. eds. G. Alefeld and J. Volkl, 273-301. Berlin, Springer Verlag.
Ziman J.M. 1969. Principles of the Theory of Solids. Cambridge University Press, Cambridge.

A Variational Approach to Transient Heat Transfer Theory

B. D. Vujanović
Faculty of Technical Sciences
University of Novi Sad
V. Vlahovića 3, 21000 Novi Sad, Yugoslavia

ABSTRACT

In this study we have analized a class of variational principles of the Hamilton's type whose stationarity conditions leads to the transient heat conduction phenomena with a finite speed of thermal propagation. Our effort has been focused to employ these variational principles (in the sense of the direct methods of variational calculus) as a tool for finding approximate solutions of the Fourier's type transient thermal processes in which the speed of thermal propagation is infinite.

1. INTRODUCTION

It is well known from the classical analytical mechanics and many other branches of physics, that the most general description of the behaviour of the dynamical systems can be accomplished by means of the variational principles. Especially, the Hamilton (integral) variational principle, when correctly stated enables one:

a) To derive correct differential equations of a dynamical process together with an appropriate number of natural boundary conditions.

b) To obtain conservation laws and to study the dynamical symmetry by the examination of invariancy of the Hamilton's action integral with respect to finite and infinitesimal transformation of coordinates, velocities, time etc. of the system.

c) To apply numerous direct methods of variational calculus for obtaining the exact and approximate solutions of linear and nonlinear boundary value problems.

However, the possibility of employing these important possibilities is restricted only to those physical systems whose behaviour can be described completely by means of a monogenic functional, called Lagrangian function, or simply Lagrangian, from which all physical manifestations of the dynamical system can be uniquely derived.

Traditionally, the Hamilton's variational principle of a physical process described by means of a field function $T=T(t,x,y,z)$ can be stated as the problem of finding the stationary value of the definite integral:

$$I(T) = \int_{t_0}^{t_1} \int_V L(t,x,y,z,T_x,T_y,T_z,T_t)\, dVdt \qquad (1.1)$$

where L denotes the suitably selected Lagrangian function depending on time t, Cartesian coordinates x, y and z and its first partial derivates: $T_t=\partial T/\partial t$, $T_x=\partial T/\partial x$, etc. It is to be understood that the time interval (t_0,t_1) in which the evolution of the physical process goes on is specified. Also is specified the fixed volume V and the surface S enclosing V in which the process is taking place.

In the Hamiltonian method (for more details, see for example Gelfand and Fomin (1963)), the actual physical process developing between any two fixed instants (t_0,t_1) and in a given portion of space V, is compared with a slightly varied process. To calculate the variation of the action integral (1.1), we shall suppose that the field function T(t,x,y,z) is specified on the surface S and for $t=t_0$ and $t=t_1$. Thus, this function is not subject to the variations on the spaceous and temporal boundaries. Let us introduce the variation of the field function T in the form

$$\delta T = \bar{T}(t,x,y,z) - T(t,x,y,z) = \varepsilon\eta(t,x,y,z) \qquad (1.2)$$

where $\bar{T}(t,x,y,z)$ denotes an infinitesimally different function from T(t,x,y,z) and ε is a small constant. The variation of the action integral I(T) is the difference

$$\delta I = I(\bar{T}) - I(T) = \\ = \int_{t_0}^{t_1} \int_V [L(t,x,y,z,T + \varepsilon\eta, T_x + \varepsilon\eta_x, T_y + \varepsilon\eta_y, T_z + \varepsilon\eta_z) - \\ - L(t,x,y,z,T,T_x,T_y,T_z)]\, dVdt \qquad (1.3)$$

Expanding the difference appearing in the integrand into a Taylor series in powers of ε, retaining the terms of the first order, performing the partial integration and applying the divergence theorem, we find

$$\delta I = \int_{t_0}^{t_1} \int_V \left[\frac{\partial L}{\partial T} - \frac{\partial}{\partial t}\frac{\partial L}{\partial T_t} - \frac{\partial}{\partial x}\frac{\partial L}{\partial T_x} - \frac{\partial}{\partial y}\frac{\partial L}{\partial T_y} - \right. \\ \left. - \frac{\partial}{\partial z}\frac{\partial L}{\partial T_z} \right] \varepsilon\eta\, dVdt + \int_V \varepsilon \frac{\partial L}{\partial T_t} \eta \Big|_{t_0}^{t_1} dV + \\ + \int_{t_0}^{t_1} \int_S \left(\frac{\partial L}{\partial T_x} dydz + \frac{\partial L}{\partial T_y} dxdz + \frac{\partial L}{\partial T_z} dxdy\right) \varepsilon\eta\Big|_S dt \qquad (1.4)$$

As mentioned above, the field function T(t,x,y,z) is completely specified on the spaceous and temporal boundaries, namely

$$\eta(t_0,x,y,z) = \eta(t_1,x,y,z) = 0, \text{ everywhere in V including S} \qquad (1.5)$$

$$\eta(t,x,y,z)\big|_S = 0\ , \text{ for every } t,\ t_0 \leq t \leq t_1 \qquad (1.6)$$

appart from these restrictions on the boundaries, the function η is arbitrary.

The necessary condition for I(T) to be stationary is that $\delta I=0$, for all possible admissible variations satisfying the boundary terms (1.5) and (1.6). From (1.4) it follows that the action integral I will be stationary for

$$\frac{\partial L}{\partial T} - \frac{\partial}{\partial t}\frac{\partial L}{\partial T_t} - \frac{\partial}{\partial x}\frac{\partial L}{\partial T_x} - \frac{\partial}{\partial y}\frac{\partial L}{\partial T_y} - \frac{\partial}{\partial z}\frac{\partial L}{\partial T_z} = 0 \qquad (1.7)$$

which is known as the Euler-Lagrange equation.

As an example of the special interest, we consider the generalized heat conduction equation with the finite speed of the thermal disturbance:

$$\tau \frac{\partial^2 T}{\partial t^2} + \frac{\partial T}{\partial t} - \alpha\nabla^2 T = 0 \qquad (1.8)$$

where $T=T(t,x,y,z)$ is the temperature, ∇^2 is the Laplacian operator which stands for: $\partial^2/\partial x^2+\partial^2/\partial y^2+\partial^2/\partial z^2$, α denotes the thermal diffusivity, and τ represents the material thermal relaxation time, which is a given constant for a particular material. Such an equation is partial hyperbolic differential equation and has a finite speed V of the thermal disturbance, equal to

$$V = (\alpha/\tau)^{1/2} \qquad (1.9)$$

For a large variety of initial and boundary conditions, the equation (1.8) has the solutions which take the form of waves propagating through the medium at the constant velocity V, while decreasing exponentially with time. (For more details about the physical model reflecting this equation see, for example, Eckert and Drake (1972) and the literature cited therein. Note, that the solutions of equation (1.8) has been found for variety of initial and boundary conditions by Baumeister and Hamill (1969). Also, the phase transition problems with the finite speed of thermal propagation have been studied recently by Sadd and Didlake (1977) and Socio and Gualtieri (1983)).

It was demonstrated (Vujanovic 1971) that the equation (1.8) can be expressed in terms of the Hamilton's variational principle. Consider the action integral (1.1) with the Lagrangian function:

$$L = \left\{ \frac{\tau}{2}\left(\frac{\partial T}{\partial t}\right)^2 - \frac{\alpha}{2}\left[\left(\frac{\partial T}{\partial x}\right)^2 + \left(\frac{\partial T}{\partial y}\right)^2 + \left(\frac{\partial T}{\partial z}\right)^2\right]\right\} e^{t/\tau} \qquad (1.10)$$

It is easy to demonstrate, by repeating the same procedure described by the equations (1.2)-(1.6) that the Euler-Lagrange equation (1.7) will generate the correct differential equation (1.8). Thus, the variational principle governing the hyperbolic heat transfer theory is fully established.

In this report we intend to employ the variational principle just described, as a vehicle for obtaining approximate solutions of the classical (Fourier's) transient heat conduction problems, in which the material relaxation time τ tends to zero and the speed of the thermal propagation is infinite.

2. VARIATIONAL PRINCIPLE WITH VANISHING PARAMETER

In order to amploy the utilities mentioned in a)-c) of the Introduction, many attempts have been made to describe classical (Fourier's) transient linear and nonlinear heat transfer problems by means of a correctly stated variational principle of Hamilton's type. Unfortunately, even for the linear case, the classical Fourier's parabolic differential equation

$$\frac{\partial T}{\partial t} - \alpha \nabla^2 T = 0 \tag{2.1}$$

cannot be derived from a variational principle of the Hamilton's type, since there not exists a Lagrangian function whose Euler-Lagrange equation in the form (1.7) will generate the equation (2.1).

It is clear from (1.8), that for $\tau \to 0$, this equation becomes the Fourier's (2.1). Therefore, in order to employ the action integral of the generalized heat conduction theory in the realm of the Fourier's theory, we should obey the following rules:

i Use the action integral in the form (1.10) whose structure can, if necessary, be adapted with extra boundary terms, in accordance to the problem in question.

ii When the process of variation is effectuated and the variational equation $\delta I = 0$ is formed, the resulting equation should be devided by the common factor $e^{t/\tau}$.

iii In the Euler-Lagrange equation obtained in this way, put $\tau \to 0$.

We note, that the use of the variational principle defined in this way, which we usually refere as the variational principle with vanishing parameter, will be employed as a starting point for finding approximate solutions of transient heat conduction problems of the Fourier's type.

The variational principle with vanishing parameter can be easily applied to the nonlinear heat conduction problems, in which thermophysical parameters are temperature dependent (Vujanović 1971), (Vujanovic, Djukic 1972), (Stokic 1973), (Kozdoba 1975). For this case, equation (2.1) is replaced by

$$c(T)\frac{\partial T}{\partial t} = \frac{\partial}{\partial x}\left[k(T)\frac{\partial T}{\partial x}\right] + \frac{\partial}{\partial y}\left[k(T)\frac{\partial T}{\partial y}\right] + \frac{\partial}{\partial z}\left[k(T)\frac{\partial T}{\partial z}\right] \tag{2.2}$$

where $c=c(T)$ denotes the temperature-dependent volumetric heat capacity, i.e., the product of density $\rho(T)$ and heat capacity $c_0(T)$ at constant pressure, and $k=k(T)$ stands for the thermal conductivity. This equation can be derived from the Hamilton's principle whose Lagrangian function is of the form

$$L = \left\{\frac{\tau}{2}\, c(T)\, k(T)\left(\frac{\partial T}{\partial t}\right)^2 - \frac{1}{2}\, k^2(T)\left[\left(\frac{\partial T}{\partial x}\right)^2 + \left(\frac{\partial T}{\partial y}\right)^2 + \left(\frac{\partial T}{\partial z}\right)^2\right]\right\} e^{t/\tau} \tag{2.3}$$

Supposing that the temperature field is completely specified on the temporal and spaceous boundaries, applying the Euler-Lagrange equation (1.7), we obtain, after deviding by $e^{t/\tau}$:

$$-c(T)\frac{\partial T}{\partial t} + \frac{\partial}{\partial x}\left[k(T)\frac{\partial T}{\partial x}\right] + \frac{\partial}{\partial y}\left[k(T)\frac{\partial T}{\partial y}\right] + \frac{\partial}{\partial z}\left[k(T)\frac{\partial T}{\partial z}\right] =$$

$$= \frac{\tau}{k(T)}\left\{\frac{\partial}{\partial t}\left[c(T)k(T)\frac{\partial T}{\partial t}\right] - \right.$$

$$\left. - \frac{1}{2}\frac{\partial}{\partial T}\left[c(T)k(T)\right]\left(\frac{\partial T}{\partial t}\right)^2\right\} \tag{2.4}$$

Therefore, by letting $\tau \to 0$, the nonlinear equation (2.2) is obtained.

As suggested by Biot (see Biot (1970) and the references cited therein) by introducing the heat flow vector

$$\underset{\sim}{H}(t,x,y,z) = H_1(t,x,y,z)\underset{\sim}{i} + H_2(t,x,y,z)\underset{\sim}{j} + H_3(t,x,y,z)\underset{\sim}{k} \tag{2.5}$$

we are able to replace the scalar heat conduction equation (2.1) with the equivalent system of the first order

$$\text{div } \underset{\sim}{H} = -cT \text{ , (conservation of energy)} \tag{2.6}$$

$$\frac{\partial \underset{\sim}{H}}{\partial t} = -k \text{ grad } T \text{ , (the law of heat conduction)} \tag{2.7}$$

where $\underset{\sim}{i}$, $\underset{\sim}{j}$ and $\underset{\sim}{k}$ are the unit vectors of the Cartesian axes x, y and z. H_1, H_2 and H_3 are the components of the heat flow vector in the directions x, y and z and c and k are constants. This system can be written more explicitely as

$$\frac{\partial H_1}{\partial x} + \frac{\partial H_2}{\partial y} + \frac{\partial H_3}{\partial z} = -cT \tag{2.8}$$

and

$$\frac{\partial H_1}{\partial t} = -k\frac{\partial T}{\partial x} \text{ , } \frac{\partial H_2}{\partial t} = -k\frac{\partial T}{\partial y} \text{ , } \frac{\partial H_3}{\partial t} = -k\frac{\partial T}{\partial z} \tag{2.9}$$

It is to be noted that equations (2.6) and (2.7) play the central role in a variational non-Hamiltonian formulation of transient heat conduction problems to which we usually refere as the Biot's variational principle. Note also, that if one takes the divergence of (2.7) and uses (2.6), one gets by elimination of $\underset{\sim}{H}$

$$c\frac{\partial T}{\partial t} = k \text{ div(grad } T) \tag{2.10}$$

However, this equation with $\alpha = k/c$ is identical with the linear heat conduction equation (2.1).

It is of interest to note that the Biot's "canonical" equations (2.8) and (2.9) can be derived (in the sense of the variational principle with vanishing parameter), from the Hamilton's principle whose Lagrangian function is

$$L = \left\{\frac{\tau c}{2}\sum\left(\frac{\partial H_i}{\partial t}\right)^2 + \frac{1}{2}kc^2T^2 + \frac{1}{2}kc\left[\left(\frac{\partial H_1}{\partial x}T - \frac{\partial T}{\partial x}H_1\right) + \right.\right.$$

$$\left.\left. + \left(\frac{\partial H_2}{\partial y}T - \frac{\partial T}{\partial y}H_2\right) + \left(\frac{\partial H_3}{\partial z}T - \frac{\partial T}{\partial z}H_3\right)\right]\right\} e^{t/\tau} \tag{2.11}$$

Here, the Lagrangian function depends on four field functions T, H_1, H_2 and H_3 thus, the Euler-Lagrange's equations are of the form (1.7) and also

$$\frac{\partial L}{\partial H_i} - \frac{\partial}{\partial t}\frac{\partial L}{\partial H_{it}} - \frac{\partial}{\partial x}\frac{\partial L}{\partial H_{ix}} - \frac{\partial}{\partial y}\frac{\partial L}{\partial H_{iy}} - \frac{\partial}{\partial z}\frac{\partial L}{\partial H_{iz}} = 0 , \qquad (2.12)$$

$(i=1,2,3)$

Entering with (2.11) into these equations one obtains from (1.7) the equation (2.8) and from (2.12), after dividing by $e^{t/\tau}$, we find

$$\begin{aligned} \tau \frac{\partial^2 H_1}{\partial t^2} + \frac{\partial H_1}{\partial t} &= -k \frac{\partial T}{\partial x} \\ \tau \frac{\partial^2 H_2}{\partial t^2} + \frac{\partial H_2}{\partial t} &= -k \frac{\partial T}{\partial y} \\ \tau \frac{\partial^2 H_3}{\partial t^2} + \frac{\partial H_3}{\partial t} &= -k \frac{\partial T}{\partial z} \end{aligned} \qquad (2.13)$$

Note that the equations (2.8) and (2.13) had been introduced (in a different form) by Luikov (1967) and in the form presented here they are given also by Mikhailov and Glazunov (1985). However, the exact variational character of these equations, demonstrated here, has not been reported so far.

For the case of nonlinear heat conduction in which the volumetric heat capacity is temperature dependent i.e., $c=c(T)$, the foregoing considerations can be easily adapted.

Let us introduce following auxiliary functions

$$h(T) = \int_0^T c(\xi)d\xi \text{ , (the heat content)} \qquad (2.14)$$

and

$$F(T) = \int_0^T h(\xi)d\xi \qquad (2.15)$$

Consider the Lagrangian function

$$L = \left(\frac{\tau}{2}\left[\sum_i \left(\frac{\partial H_i}{\partial t}\right)^2 + F(T) + \frac{k}{2}\left[\left(\frac{\partial H_1}{\partial x} T - \frac{\partial T}{\partial x} H_1\right) + \right.\right.\right.$$
$$\left.\left.\left. + \left(\frac{\partial H_2}{\partial y} T - \frac{\partial T}{\partial y} H_2\right) + \left(\frac{\partial H_3}{\partial z} T - \frac{\partial T}{\partial z} H_3\right)\right]\right]\right) e^{t/\tau} \qquad (2.16)$$

It is easy to verify that the Euler-Lagrange equations (1.7) and (2.12) will generate the equations (2.13) and, instead of (2.8), we obtain

$$\frac{\partial H_1}{\partial z} + \frac{\partial H_2}{\partial y} + \frac{\partial H_3}{\partial z} = -h(T) \qquad (2.17)$$

Letting $\tau \to 0$, we are led to the equations which can be written in vectorial form

$$\operatorname{div} \underset{\sim}{H} = -h(T) \qquad (2.18)$$

$$\frac{\partial \underset{\sim}{H}}{\partial t} = -k \operatorname{grad}T \qquad (2.19)$$

where the heat conductivity k is assumed to be a given constant.

The differential equations thus obtained are identical with the equations given by Biot (1970) which form the basis for the formulation of the Biot's variational principle.

3. RELATED SUBJECTS AND MODIFICATIONS

Variational principle describing the hyperbolic nature of the propagation of heat, whose form for the linear conduction is specified by the Lagrangian function (1.1), has been the subject of a variety of attempts to extend its form to a wider area of dissipative and irreversible physical systems.

First, it was shown (Djukic 1971), that by introducing into Lagrangian function an exponential multiplicative term which contains a parameter similar to the material relaxation time, the general equations of the boundary layer theory of an incompressible transient fluid flow can be obtained, by employing the rules of vanishing parameter principle. This generalization, by employing the direct methods, has been used afterwards in the study of the transient laminar boundary layer theory of incompressible fluids (Vujanovic, Djukic, Pavlovic 1973), boundary layer theory of an incompressible power-law non-Newtonian fluids (Djukic 1971), (Djukic, Vujanovic 1975), (Djukic, Vujanovic, Tatic 1973), boundary layer theory of conducting fluids (Vujanovic, Strauss, Djukic 1972) and some problems arising in magneto-hydrodynamics (Boričić, Nikodijević 1984).

Second, in a series of papers, Sieniutycz (1977,1978,1979,1984) has employed the Lagrangian functions similar to those given by (1.11), having exponential multiplicative terms which are of considerably complicated structure, but containing also relaxation parameters. He formulated variationally the general equations of irreversible thermodynamics which one can refer as "hyperbolic irreversible thermodynamics". The Lagrangian functions generating the differential equations describing uncoupled heat, mass and momentum transport in incompressible media are considered. Attention is also paid to the study of Lagrangian functions which generate wave equations that govern heat conduction as well as simultaneous heat and mass transfer in moving media in Euler representation.

Third, since the general rules for constructing the Lagrangian function for a given irreversible process (containing the relaxation parameters) are not known, several rules for finding this function has been suggested recently.

In order to express the single differential equation

$$F(t,x,T,T_t,T_x,T_{tx},T_{tt},T_{xx}) = 0 \ , \tag{3.1}$$

in variational way, Atanackovic and Jones (1974) formulated a variational principle whose Lagrangian function is of the form

$$L = m^3 T_{ttt}\, e^{t/m}\, F(t,x,T,T_t,T_x,T_{tx},T_{tt},T_{xx}) \tag{3.2}$$

Since the Lagrangian function contains the derivatives higher than first, the Euler-Lagrange equation is given by

$$\frac{\partial L}{\partial T} - \frac{\partial}{\partial t}\frac{\partial L}{\partial T_t} - \frac{\partial}{\partial x}\frac{\partial L}{\partial T_x} - \frac{\partial}{\partial x}\frac{\partial L}{\partial T_x} + \frac{\partial^2}{\partial t^2}\frac{\partial L}{\partial T_{tt}} + \frac{\partial^2}{\partial t\partial x}\frac{\partial L}{\partial T_{tx}} + \\ + \frac{\partial^2}{\partial x^2}\frac{\partial L}{\partial T_{xx}} - \frac{\partial^3}{\partial t^3}\frac{\partial L}{\partial T_{ttt}} = 0 \tag{3.3}$$

Entering with (3.2) into (3.3), dividing the resulting equation by $e^{t/m}$, we arrive to the equation of the form

$$F + m(....) = 0 \tag{3.4}$$

where the neglected terms in the last equation are proportional to m or of higher powers of m. By letting $m \to 0$, one obtains the differential equation (3.1). Atanackovic and Jones have used this vanishing parameter variational principle in the study of heat conduction problems with the infinite speed of thermal propagation.

Third, considerably simpler and wider form of Lagrangian functions for general irreversible thermodynamic system have been proposed by Mikhailov and Glazunov (1985).

Their rule for constructing Lagrangian functions of any system of partial differential equations of a multi-component irreversible physical system can be formed as

$$L = \frac{m}{2} \sum (\text{Differential equation of the process})^2 e^{t/m} \tag{3.5}$$

where m, as in the case of Atanackovic-Jones variational formulation, is a constant parameter which tends to zero after the variational process has been accomplished. Note, that in contrast to the variational principles discussed in the Sections 1 and 2, where the material relaxation time has clear physical meaning, in the last two variational approaches, the vanishing relaxation parameter m has not any obvious physical interpretation.

4. A BRIEF OUTLINE OF THE METHOD OF PARTIAL INTEGRATION

In this section we shall briefly describe the direct method usually refered as the method of pertial integration, or Kantorowich method, which can be easily and efficiently applied for solving approximately a variety of transient linear and nonlinear heat transfer problems. As already mentioned, we will focus our attention to solve the classical (i.e. Fourier's) class of problems for which the velocity of thermal propagation is infinite. Thus, we have to employ the variational principles with vanishing parameter.

The general rules for applying this method can be traced as follows.

a) If we are using the variational principle which describes the heat conduction equation of the type (2.1) or (2.2), we usually select an approximate temperature profile in the form

$$T = T(t,x,y,z,q_1(t),q_2(t),\dots,q_n(t)) \tag{4.1}$$

which, besides the independent variables t, x, y and z depends also upon a finite number of, so called, generalized coordinates $q_i(t)$, $(i=1,2,\dots,n)$. They are unknown functions of time and should be considered as mutually independent. These functions are introduced in such a way to describe rationally the temperature field of the particular problem. Also, the selection of q's should be guided by the tendency to satisfy maximal number of initial and boundary conditions of the problem in question. Frequently, we are able to satisfy all prescribed initial and boundary conditions by a proper choice of (4.1).

Introducing (4.1) into the action integral (1.1), for a given form of the Lagrangian function, and integrating with respect to the spaceous coordinates (whose boundaires are given), one obtains a reduced form of the action integral

$$I = \int_{t_0}^{t_1} \bar{L}(t,q_1,\ldots,q_n,\dot{q}_1,\ldots,\dot{q}_n)dt \tag{4.2}$$

The functions $q_i(t)$ can be selected such that the Euler-Lagrange equations for the reduced variational problems are satisfied, i.e.

$$\frac{d}{dt}\frac{\partial\bar{L}}{\partial\dot{q}_i} - \frac{\partial\bar{L}}{\partial q_i} = 0 \ , \ (i=1,\ldots,n) \tag{4.3}$$

However, due to the particular form of the Lagrangian functions, the differential equations thus obtained have, as a rule, the following structure

$$[K_i(t,q_1,\ldots,q_n,\dot{q}_1,\ldots,\dot{q}_n) + \\ + \tau P_i(t,q_1,\ldots,q_n,\dot{q}_1,\ldots,\dot{q}_n,\ddot{q}_1,\ldots,\ddot{q}_n)]e^{t/\tau} = 0 \tag{4.4}$$

employing rules of the variational principles with vanishing parameters i-iii given in the Section 2, we arrive to the system of ordinary equations of the first order $K_i=0$, which should be integrated subject to the given set of initial conditions $q_i(0)$.

b) If we decide to study heat conduction problems by means of the Biot's canonical equations (2.6) or (2.7) and their equivalent Hamiltonian formulation based on the Lagrangian functions (2.11) or (2.16), we proceed as follows.

First, we select again a trial temperature field in the form (4.1) and integrate the differential equation (2.6) or (2.18). Thus, we obtain the heat flow vector

$$\underset{\sim}{H} = \underset{\sim}{H}(t,x,y,z,q_1,\ldots,q_n) \tag{4.5}$$

which satisfies these equations identically. It should be noted that the integration of these equation for two and three dimensional problems can be a very difficult task. However, for one dimensional cases the vector $\underset{\sim}{H}$ has just one component and can be easily found by quadratures. It should be pointed out that the request for finding the heat flow vector $\underset{\sim}{H}$ in the form (4.5), by integration of (2.6) or (2.18) is one of the essential steps in applying the Biot's variational principle (see, Biot 1970). Entering with (4.1) and (4.5) into the action integral whose Lagrangian function is given by (2.11) for linear problems, or into (2.16) for nonlinear ones, then integrating over the specified volume, we arrive at a reduced form of the action integral, whose form again is of the type (4.2).

5. ILLUSTRATIVE EXAMPLES

In this section we will demonstrate by means of several concrete examples the general features of application of the variational principles with vanishing parameter. Particulary we will consider the variational principles whose Lagrangian functions are given by (1.10), (2.11) and (3.5). Note that the variational principle with vanishing parameter for Lagrangian functions of the form (1.10) or (2.3) has been applied in numerous practical situations involving, practically all classes of boundary conditions used in engineering. (See, for example, Vujanovic (1971), Vujanovic, Strauss (1971), Vujanovic, Djukic (1972), Vujanovic, Jones (1989)).

5.1. Surface Radiation Problem

Let us consider a semi-infinite body which is initially heated to a constant temperature T_0. The body looses heat energy in such a way that the free surface, at x=0, radiates thermal energy into the ambient with zero temperature. The body is thermally insulated along its length and the thermo-physical parameters are assumed to be constant.

Mathematically, the problem can be stated as follows. Find the solution of the heat transfer equation

$$c \frac{\partial T}{\partial t} = k \frac{\partial^2 T}{\partial x^2} \tag{5.1}$$

subject to the initial condition

$$T(0,x) = T_0 = \text{const.}, \text{ for } x \geqq 0 \tag{5.2}$$

The nonlinear boundary condition at the free surface can be expressed in the form

$$k \frac{\partial T}{\partial x} = hT^m \text{ , at } x = 0, \text{ for } t > 0 \tag{5.3}$$

where c, k, h and m are given constant parameters.

In order to describe this problem in a variational form, we introduce the action integral

$$I(T) = \int_0^\infty \int_0^{q_2} \left[\frac{\tau c}{2} \left(\frac{\partial T}{\partial t}\right)^2 - \frac{k}{2} \left(\frac{\partial T}{\partial x}\right)^2\right] e^{t/\tau} \, dxdt -$$

$$- \int_0^\infty \frac{hT^{m+1}(t,0)}{m+1} e^{t/\tau} \, dt \tag{5.4}$$

where the upper limit q_2 depends on the selected temperature profile.

To verify that this is the correct variational principle we need to calculate the variation of (5.4). Thus, we find

$$\delta I(T) = \varepsilon \int_0^\infty \int_0^{q_2} \left(\tau c \frac{\partial^2 T}{\partial t^2} + c \frac{\partial T}{\partial t} - k \frac{\partial^2 T}{\partial x^2}\right) e^{t/\tau} \, \eta(t,x) dxdt$$

$$+ \varepsilon \int_0^\infty \left[\left(k \frac{\partial T}{\partial x} - hT^m\right)_{x=0} \eta(t,0) - k \frac{\partial T(q_2,t)}{\partial x} \eta(t,q_2)\right] e^{t/\tau} dt$$

$$+ \varepsilon \int_0^{q_2} \tau c \frac{\partial T}{\partial t} \eta(t,x)\Big|_0^\infty e^{t/\tau} \, dx \tag{5.5}$$

Since the temperature at the boundary x=0 is not specified, the variation $\eta(t,0)$ is arbitrary. Supposing the logical boundary condition

$$\frac{\partial T(q_2,t)}{\partial x} = 0 \tag{5.6}$$

and noting that the temperature at t=0 and t=∞ is fully specified, it follows from (5.5) that the variational equation $\delta I(T)=0$, leads to the hyperbolic heat conduction equation

$$\tau c \frac{\partial^2 T}{\partial t^2} + c \frac{\partial T}{\partial t} - k \frac{\partial^2 T}{\partial x^2} = 0 \tag{5.7}$$

and the boundary condition (5.3).

Let us suppose that the temperature field can be described by the trial function

$$T = T_o - (T_o - q_1)(1 - \frac{x}{q_2})^3 \tag{5.8}$$

where $q_1(t)$ and $q_2(t)$ are the generalized coordinates denoting the surface temperature at x=0, and the depth of thermal penetration respectively.

Combining (5.3) and (5.8) we have following algebraic relation between the generalized coordinates

$$\Lambda = \frac{3k(T_o - q_1)}{q_2} - hq_1^m = 0 \tag{5.9}$$

Substituting (5.8) into the action integral (5.4), integrating with respect to x from 0 to q_2, one obtains

$$I = \int_0^\infty L(t,\tau,q_1,q_2,\dot{q}_1,\dot{q}_2)dt =$$

$$= \int_0^\infty \left(\frac{\tau c}{2}\left[\frac{1}{7}\dot{q}_1^2 q_2 - \frac{1}{7}\dot{q}_1\dot{q}_2(T_o - q_1) + \frac{3}{35}(T_o - q_1)^2 \frac{\dot{q}_2^2}{q_2}\right]\right.$$

$$\left. - \frac{9}{10} k \frac{(T_o - q_1)^2}{q_2} - \frac{h}{m+1} q_1^{m+1}\right) e^{t/\tau} dt \tag{5.10}$$

Before applying the Euler-Lagrange equations, we note that due to the relation (5.9), the generalized coordinates q_1 and q_2 are not mutually independent. Therefore, (see, for example, Gelfand, Fomin (1963)), introducing a Lagrangian multiplier λ, we form a new Lagrangian function

$$\bar{L} = L + \lambda\Lambda e^{t/\tau} =$$

$$= \left(\frac{\tau c}{2}\left[\frac{1}{7}\dot{q}_1^2 q_2 - \frac{1}{7}\dot{q}_1\dot{q}_2(T_o - q_1) + \frac{3}{35}(T_o - q_1)^2 \frac{q_2^2}{q_2}\right] - \right.$$

$$- \frac{9}{10} k \frac{(T_o - q_1)^2}{q_2} - \frac{h}{m+1} q_1^{m+1} + \lambda\left[3k \frac{T_o - q_1}{q_2} - \right.$$

$$\left.\left. - hq_1^m\right]\right) e^{t/\tau} , \tag{5.11}$$

and treat the coordinates q_1 and q_2 as independent.

Substituting (5.11) into (4.3), deviding by $e^{t/\tau}$, letting $\tau \to 0$, we obtain

$$c\left[\frac{1}{7}\dot{q}_1 q_2 - \frac{1}{14}\dot{q}_2(T_o - q_1)\right] - \frac{9}{5}k\frac{T_o - q_1}{q_2} +$$

$$+ hq_1^m + \lambda\left(\frac{3k}{q_2} + hmq_1^{m-1}\right) = 0 \qquad (5.12)$$

$$-\frac{c}{14}\dot{q}_1 + \frac{3}{35}c(T_o - q_1)\frac{\dot{q}_2}{q_2} - \frac{9}{10}k\frac{T_o - q_1}{q_2^2} + 3\lambda\frac{k}{q_2^2} = 0 \qquad (5.13)$$

Therefore, we have obtained three equations (5.12), (5.13) and (5.9) for finding q_1, q_2 and λ.

Eliminating λ from these equations, and introducing dimensionless coordinates in the form

$$X_1 = 1 - q_1/T_o\ ,\quad X_2 = q_2 hT_o^{m-1}/k\ ,\quad \bar{t} = (hT_o^{m-1}/k)^2\alpha t\ ,$$

$$\alpha = k/c \qquad (5.14)$$

we obtain from (5.12) and (5.13)

$$\left[\frac{3}{14}X_2 + \frac{1}{42}m(1 - X_1)^{m-1}X_2^2\right]\dot{X}_1 + \left[\frac{11}{70}X_1 + \right.$$

$$\left. + \frac{1}{35}mX_1(1 - X_1)^{m-1}X_2\right]\dot{X}_2 + \frac{9}{10}\frac{X_1}{X_2} -$$

$$- (1 - X_1)^m - \frac{3}{10}m(1 - X_1)^{m-1}X_1^2 = 0 \qquad (5.15)$$

This equation should be considered together with the algebraic relation (5.9) whose form in new variables is

$$3\frac{X_1}{X_2} - (1 - X_1)^m = 0 \qquad (5.16)$$

Note, that an overdot in equation (5.15) denotes the differentiation with respect to the dimensionless time $\bar{t}$. Note also, that the system (5.15) and (5.16) should be integrated subject to the given initial conditions

$$X_1(0) = 0\ ,\quad X_2(0) = 0 \qquad (5.17)$$

Numerical integration of this nonlinear system can be easily accomplished for various values of the radiating parameter m. However, for the sake of simplicity, we will analyze only the asymptotic behaviour of the solution for small and long times of the dimensionless time $\bar{t}$.

A simple asymptotic analysis reveals that for small times, the solution of (5.15) and (5.16) is independent of the nonlinear parameter m and can be represented as

$$\left.\begin{aligned} X_1 &= 1.121\ \bar{t}^{1/2} \\ X_2 &= 3.363\ \bar{t}^{1/2} \end{aligned}\right\} \text{ for } \bar{t}\to 0 \qquad (5.18)$$

For large values of $\bar{t}$, the solution depends on the parameter m and takes the form

$$X_1 = 1 - \left(\frac{0.337}{\bar{t}}\right)^{1/2m} \qquad \text{for } \bar{t}\to\infty \qquad (5.19)$$

$$X_2 = 5.170\ \bar{t}^{1/2}$$

As reported by Lardner (1963), (who has analyzed the same problem by using the Biot's variational principle), the exact asymptotic solutions for short and long times for m=4, i.e. for the case of the black-body radiation are

$$X_{1(exact)} = 1.128\ \bar{t}^{1/2}, \quad \text{for } \bar{t}\to\infty$$

$$X_{1(exact)} = 1-0.864\ \bar{t}^{-1/8}, \quad \text{for } \bar{t}\to\infty \qquad (5.20)$$

From (5.18) and (5.19), for m=4, we find

$$X_{1(approx.)} = 1.121\ \bar{t}^{1/2}, \quad \text{for } \bar{t}\to 0$$

$$X_{1(approx.)} = 1-0.873\ \bar{t}^{-1/8}, \quad \text{for } \bar{t}\to\infty \qquad (5.21)$$

Thus, it is evident that the approximate solution obtained by the help of variational principle with vanishing parameter compares very good with the exact solution, and the percentage error does not exceeds 1%.

5.2. Heating of a Slab whose Surface Temperature is Given Function of Time

In this example we study the heating of a semi-infinite slab initially at zero temperature whose face at x=0, is subject to the power-law boundary temperature

$$T(t,0) = f(t) = at^{n/2} \qquad (5.22)$$

where a is given constant and n is a positive integer. We shall suppose that the thermo-physical parameters are constant, and the temperature profile is taken to be

$$T = f(t)\left(1 - \frac{x}{q}\right)^3 \qquad (5.23)$$

where q=q(t) denotes the depth of thermal penetration into the body.

For this one dimensional case, the heat conduction equations (2.6) and (2.7) become

$$\frac{\partial H}{\partial x} = -cT \qquad (5.24)$$

$$\frac{\partial H}{\partial t} = -k\ \frac{\partial T}{\partial x} \qquad (5.25)$$

Using (5.23) and integrating (5.24) the heat flow function is found to be

$$H = \frac{1}{4}\ cf(t)\ q\left(1 - \frac{x}{q}\right)^4 + P(t) \qquad (5.26)$$

where P(t) is an arbitrary function of time. To find P(t), we substitute the last equation into (5.25):

$$-k\frac{\partial T}{\partial x} = \frac{cfnq}{8t}\left(1-\frac{x}{q}\right)^4 + \frac{cf\dot{q}}{4}\left(1-\frac{x}{q}\right)^3 +$$

$$+ cf\dot{q}\,\frac{x}{q}\left(1-\frac{x}{q}\right)^3 + \dot{P}(t) \qquad (5.27)$$

Thus, we conclude from this relation, that under the condition

$$-k\left.\frac{\partial T}{\partial x}\right|_{x=0} = \frac{cfnq}{8t} + \frac{cf\dot{q}}{4}\,, \qquad (5.28)$$

the function P is equal to zero.

Using the Lagrangian function (2.11) for one dimensional case, we consider the action integral in the form

$$I = \int_0^\infty\int_0^q \left[\frac{\tau}{2}\left(\frac{\partial H}{\partial t}\right)^2 + \frac{1}{2}\,kcT^2 - \frac{1}{2}\,k\left(\frac{\partial T}{\partial x}\,H - \frac{\partial H}{\partial x}\,T\right)\right]e^{t/\tau}dxdt +$$

$$+ \int_0^\infty \frac{1}{2}\,kTH\Big|_{x=0}\, e^{t/\tau}dt \qquad (5.29)$$

It is easy to demonstrate that the stationarity condition $\delta I=0$, (in which the functions T and H are subject to variations), will generate correct differential equations (5.24) and (5.25).

By substitution of (5.26), (for P=0) and (5.23) into (5.29), integrating over the space integral, one arrives to the following equation

$$I = \int_0^\infty \bar{L}(t,\tau,q,\dot{q})dt$$

with

$$\bar{L} = \left[\frac{1}{2}\,\tau c^3 f^2\left(\frac{1}{56}\,\dot{q}q^2 + \frac{1}{96}\,\frac{q^2\dot{q}n}{t}\right) + \frac{5}{28}\,kc^2f^2q\right]e^{t/\tau} \qquad (5.30)$$

Applying the Euler-Lagrange equation for the coordinate q, deviding the resulting equation by $e^{t/\tau}$, and letting $\tau\to 0$, one obtains the ordinary differential equation of the first order

$$\frac{1}{56}\,q\dot{q} + \frac{1}{192}\,\frac{q^2 n}{t} = \frac{5}{28}\,\alpha \qquad (5.31)$$

whose solution, subject to the initial condition q(0)=0, is found to be

$$q = \left(\frac{240}{12+7n}\right)^{1/2}(\alpha t)^{1/2} \qquad (5.32)$$

The incident heat flux at the surface x=0, can be calculated from (5.28):

$$Q_{s(approx.)} = \left(-k\left.\frac{\partial T}{\partial x}\right|_{x=0}\frac{\sqrt{\alpha t}}{kf}\right) = \frac{n+1}{8}\sqrt{\frac{240}{12+7n}} \qquad (5.33)$$

The exact heat flux (Carlslaw, Jaeger, 1959) is

$$Q_{s(exact)} = \frac{\Gamma(1+\frac{n}{2})}{\Gamma(\frac{n+1}{2})} \qquad (5.34)$$

where the symbol Γ denotes the gamma function.

Note that the approximate solution (5.33) compares very good with the exact solution (5.34). The largest percentage error is about 2% for the range of the expontential parameter n between 0 to 7.

5.3. A Melting Problem

As the last illustrative example we shall consider the melting of a semi-infinite solid body initially at the melting temperature θ_p, whose surface x=0 is raised suddenly to the temperature θ_0 and held there for a time $t \geq 0$. Supposing the temperature distribution only in the liquid phase, then the depth of the thermal penetration is identical with the location of the melting line x=s(t).

Introducing the dimensionless temperature: $T=(\theta-\theta_0)/(\theta_0-\theta_p)$ and supposing a liquid of constant thermo-physical properties, the governing differential equation is

$$\frac{\partial T}{\partial t} = \alpha \frac{\partial^2 T}{\partial x^2} \quad , \quad (\alpha=k/\rho c_0=k/c) \tag{5.35}$$

The boundary condition at x=0 is

$$T(0,t) = 1 \text{ , for } t>0 \tag{5.36}$$

The following two conditions must be satisfied at the interface:

$$T(t,s) = 0 \tag{5.37}$$

and

$$-\frac{\partial T}{\partial x} = \frac{L}{c(\theta_0 - \theta_p)} \frac{ds}{dt} \text{ , for } x=s \tag{5.38}$$

where L denotes the coefficient of the latent heat of melting.

Let us represent the heat distribution in the liquid phase in the following form (Vujanovic, Bačlić (1976), (Vujanovic, Jones 1989)

$$T = A(\frac{x}{s} - 1) + (A + 1)(\frac{x}{s} - 1)^2 \tag{5.39}$$

where A is an unknown constant. Note that the boundary conditions (5.36) and (5.37) are identically satisfied, while the melting boundary condition (5.38) remains to be matched in the course of analysis.

In order to recast the differential equation (5.35) in variational form, we consider the action integral

$$I = \int_0^\infty \int_0^s \frac{m}{2} (\frac{\partial T}{\partial t} - \alpha \frac{\partial^2 T}{\partial x^2})^2 e^{t/m} dxdt \tag{5.40}$$

where the integrand (the Lagrangian function) is formed in accordance to the rule given by (3.5). Note that in contrast to the examples considered previously, the parameter m does not have in this case a sound physical interpretation.

Upon substituting (5.39) into (5.40), integrating with respect to x from 0 to s, we find

$$I = \int_0^{\infty} \bar{L}(t,m,A,s,\dot{s})dt$$

where

$$\bar{L} = \frac{m}{2}\left[\frac{1}{15}(2A^2 - A + 2)\frac{\dot{s}^2}{s} + \frac{2}{3}(A^2 - A - 2)\frac{\dot{s}}{s^2} + 4(A + 1)^2 \frac{1}{s^3}\right]e^{t/m} \tag{5.41}$$

Applying the Euler-Lagrange equation for the coordinate s, deviding by $e^{t/m}$, and letting $\tau \to 0$, one arrives to the equation

$$2s\dot{s} = -10\,\frac{A^2 - A - 2}{2A^2 - A + 2} \tag{5.42}$$

which could be easily integrated subject to the initial condition $s(0)=0$. On the other hand, by substituting (5.39) into (5.38), we are led to another differential equation for s:

$$2s\dot{s} = -A\mu \tag{5.43}$$

where

$$\mu = \frac{2(\theta_0 - \theta_p)c}{L} \tag{5.44}$$

Equating the right-hand sides of (5.42) and (5.43), we obtain the following algebraic equation for A:

$$2\mu A^3 - (\mu + 10)A^2 + (2\mu + 10)A + 20 = 0 \tag{5.45}$$

From (5.43), we find the position of the melting line

$$s = \sqrt{-\mu A}\,(\alpha t)^{1/2} \tag{5.46}$$

The relevant negative roots of the equation (5.45) are calculated for $0.5 \leq \mu \leq 3$, and the position of the melting line, for various values of μ is presented in the Table 5.1. The exact solution found by Carlslaw-Jaeger (1959), is also presented in the same Table. It is seen, that the agreement between the exact and approximate solutions is very good for the whole range of melting parameter μ.

TABLE 5.1. The position of the melt line $s/(\alpha t)^{1/2}$ for various values of the melting parameter μ

μ	Exact	Approximate
0.5	0.6802	0.6806
1.0	0.9296	0.9315
1.5	1.1042	1.1082
2.0	1.2401	1.2466
2.5	1.3517	1.3608
3.0	1.4464	1.4580

Note, that the same problem was discussed by means of the Gauss' principle of least constraint by Vujanovic and Bačlić (1976). The identical results are obtained as presented here.

6. DISCUSSIONS

In this study we have analyzed a class of variational principles of the Hamilton's type, which describe the transient heat conduction phenomena with a finite speed of thermal propagation, conditionally named "hyperbolic process". Our effort has been focused to employ these variational principles as a tool for finding the approximate solutions of the Fourier's type heat conduction problems, i.e. for the cases in which the speed of the thermal propagation is infinite or, equivalently, when the material relaxation time tends to zero (named here "parabolic process").

Our considerations have been based on the premise that, as the relaxation parameter tends to zero, the solution of the hyperbolic process tends to the solution of corresponding parabolic, (Fourier's) process. The authors believes, that this premise can be confirmed rigorously by using the mathematical theory of the singular perturbation analysis. Instead, we have simply relied upon numerous practical examples performed by many researchers (see, for example, Baumeister and Hamill (1969)), where it have been demonstrated that in each particular case, the solution of hyperbolic heat conduction equation tends to the solution of Fourier's equation when the material relaxation time tends to zero. Several examples of the engineering significance, stemming from the classical Fourier's theory, have been analyzed.

REFERENCES

Atanackovic, T.M., Jones, E.E. (1974). A Generalized Variational Formulation. J.Math.Anal.and Appl. 48: 672-676.

Baumeister, K.J., Hamill, T.D. (1969). Hyperbolic Heat-Conduction Equation - A Solution for the Semi-Infinite Body Problem. Journal of Heat Transfer 91: 543-548.

Biot, M.A. (1970). Variational Principles in Heat Transfer (A Unified Lagrangian Analysis of Dissipative Phenomena), Oxford, Clarendon Press.

Boričić, Z., Nikodijević, D. (1984). Application of the Variational Method to the Study of the Magneto-Hydrodynamics Boundary Layers on the Axial-Symmetric Solids by the Fluids with Variable Conductivity (in Serbian), Receuil des Travaux de l'Institut Mathematique, Neuvelle serie 4: 12-21.

Carlslaw, H.S., Jaeger, J.C. (1959). Conduction of Heat in Solids, 2nd ed. Oxford, Clarendon Press.

Djukic, Dj.S. (1971). Stationary Plane Laminar Boundary Layer of Non-Newtonian Fluid with the Power-Law Rheological Equation of State (in Serbian). Doctoral thesis. Dept. of Mechanics, Univ. of Belgrade.

Djukic, Dj.S., Vujanovic, B.D., Tatic, N. (1973). On Two Variational Methods for Obtaining Solutions to Transport Problems. Chem.Eng.J. 5: 145-152.

Djukic, Dj.S., Vujanovic, B.D. (1975). A Variational Principle for the Two-Dimensional Bondary-Layer Flow of Non-Newtonian Power-Law Fluids. Rheologica Acta 14: 881-890.

Eckert, E.R.G., Drake, R.M.Jr. (1972). Analysis of Heat and Mass Transfer, New York, McGraw-Hill.

Gelfand, I.M., Fomin, S.V. (1963). Calculus of Variations. Englewood Cliffs, N.J.

Kozdoba, L.A. (1975). Methods of Solution of the Nonlinear Problems in Heat Transfer (in Russian), "Nauka", Moscow.

Lardner, T.J. (1963). Biot's Variational Principle in Heat Conduction. AIAA Journal 1: 196-206.
Luikov, A.V. (1967). Theory of Heat Transfer. "Vishaja Shkola", Moscow.
Mikhailov, Yu.A., Glazunov, Yu.T. (1985). Variational Methods in the Nonlinear Heat and Mass Transfer (in Russian), "Znanie", Riga.
Sadd, M.H., Didlake, J. (1977). Non-Fourier Melting of a Semi-Infinite Solid. Journal of Heat Transfer, Trans. ASME 99: 25-28.
Sieniutycz, S. (1977). The Variational Principles of Classical Type for Non-Coupled Non-Stationary Irreversible Transport Processes with Convective Motion and Relaxation. Int.J.Heat Mass Transfer 20: 1221-1231.
Sieniutycz, S. (1978). Zasady Wariacyjne Dla Falowych Rownan Przenoszenia Ciepla I Masy W Osrodkach Ruchomych, Inzynieria Chemiczna 8: 649-668.
Sieniutycz, S. (1979). The Wave Equations for Simultaneous Heat and Mass Transfer in Moving Media-Structure Testing, Time-Space Transformations and Variational Approach. Int.J.Heat Mass Transfer 22: 585-599.
Sieniutycz, S. (1984). Variational Approach to Extended Irreversible Thermodynamics of Heat ana Mass Transfer. J.Non-Equlib. Thermodyn. 9: 61-70.
Socio de L.M., Gualtieri, G. (1983). A Hyperbolic Stefan Problem. Quarterly of Applied Mathematics 43: 253-259.
Stokić, D.B. (1973). Radiation of Solids with Thermally Changeable Conductivity. Publications de l'Institut Mathematique 16 (30): 159-168.
Vujanovic, B.D. (1971). An Approach to Linear and Nonlinear Heat Transfer Problem Using a Lagrangian. AIAA Journal 9: 131-135.
Vujanovic, B.D., Strauss, A.M. (1971). Heat Transfer with Nonlinear Boundary Conditions Via a Variational Principle. AIAA Journal 9: 327-331.
Vujanovic, B.D., Djukic, Dj.S. (1972). On One Variational Principle of Hamilton's Type for Nonlinear Heat Transfer Problem. Int. J.Heat Mass Transfer 15: 1111-1123.
Vujanovic, B.D., Strauss, A.M., Djukic, Dj.S. (1972). A Variational Problem for a Power Law Non-Newtonian Conducting Fluid. Ingenieur-Archiv 41: 381-386.
Vujanovic, B.D., Djukic, Dj.S., Pavlovic, M. (1973). A Variational Principle for the Laminar Boundary Layer Theory. Bollettino U. M.I. 4: 377-391.
Vujanovic, B.D. (1976). Applications of Analytical Mechanics to Nonconservative Field Theory. Ren.Sem.Mat.Univers.Torino 35: 88-95.
Vujanovic, B.D., Bačlić, B. (1976). Applications of Gauss' Principle of Least Constraint to the Nonlinear Heat-Transfer Problem. Int.J.Heat Mass Transfer 19: 721-730.
Vujanovic, B.D., Jones, S.E. (1989). Variational Methods in Nonconservative Phenomena, Academic Press, Boston.

Some Problems in Nonequilibrium Thermomechanics

G. C. Sih
Institute of Fracture and Solid Mechanics
Lehigh University, Bethlehem, Pennsylvania 18015, USA

ABSTRACT

Nonequilibrium phenomena are characterized by the rate at which a process takes place such that the state traversed by a system cannot be described in terms of physical parameters representing the system as a whole. Their description becomes realistic only if the thermal changes are synchronized with motion of the mass elements. Those ambiguities and inconsistencies encountered in classical mechanics would thus be bypassed. Employed will be the isoenergy density theory for characterizing the nonequilibrium/irreversible behavior of gas, liquid and solid in addition to phase transformation. A complete account of laminar to turbulent flow including transition can be made. Viscous sublayer correction to the velocity profile obtained from Prandtl's mixing length theory is no longer necessary. The location and instance at which fluid separates from emersed solid object can be predicted. Gas, when compressed in a confined region, condenses nonuniformly. Temperature in uniaxial tensile specimens remained below the ambient condition even though the stress and strain are well into the nonlinear regime. Concepts and notions established from equilibrium theories should be scrutinized so that they will not be misapplied to explain nonequilibrium phenomena.

INTRODUCTION

Physical processes are, by nature, nonequilibrium as evolution is made possible by continuous changes. Nonequilibrium theories have followed two schools of thought in the literature. *Statistical and kinetic* theories reposed on transport equations for the particles. They have developed only for particular molecular models and special cases of irreversible processes. These models probe deeper into the physics of matters and provide numerical data for the coefficients in the phenomenological relations. They do not lend themselves to engineering applications where the combined effects of geometry, material and external disturbance are accounted for explicitly. The formalisms of nonequilibrium *thermodynamics* theories (Onsager 1931; Prigogine 1955; DeGroot and Mazur 1962 and Truesdell and Toupin 1960) are vastly different from the molecular models; they are based on a priori assumptions. Thermal change and motion of mass elements are not considered as a single intrinsic operation. That is, the additional concept of heat was needed to determine the temperature. The linear relation between the change of the entropy *exchanged* between the system and surroundings and that *produced* in the system leaves out the possibility that the heat exchanged and produced could interact. Nonequilibrium invokes changeability such that the states traversed by a system cannot be described in terms of thermodynamic coordinates referring to the system as a whole. In other words, coefficients in the equations of state or constitutive relations vary from location to location and they depend on the external disturbance. Application of the same state of equation or constitutive relation to all mass elements in a system cannot be justified for nonequilibrium processes.

From the viewpoint of unification, it would be satisfying if the forms of motion arising from mechanical, electromagnetic, nuclear, etc., can be described in space and time by a single theory, not only qualitatively but also quantitatively. Since motion and thermal change are inherent in all physical processes, their mutual interaction must be accounted for in the development of a model that unifies all phenomena in the way of the classical field theories. To this end, the isoenergy space concept (Sih 1985; 1988) was introduced. Such a space is occupied by elements whose size remains finite and depends on the local rotation and deformation without restrictions. The orientation and size of these elements are determined at each location and time step according to the history of prescribed disturbance from the condition that the same energy is transmitted across all surfaces of the element. Hence, the name *isoenergy element* is used to emphasize this special property. Unique correspondence of energy state between uniaxial and multiaxial stress/strain state are made possible without loss in generality. Only a brief summary of the isoenergy density theory will be given for the sake of reference. The objective of this communication is to emphasize the application of the theory to some previously unsolved and/or unexplained problems in engineering.

ISOENERGY DENSITY THEORY

Unlike the classical field theories in which the change of volume with surface, say $\Delta V/\Delta A$, is assumed to vanish in the limit, this quantity will be kept finite and is assumed to have three components denoted by V_j (j = 1,2,3). The surface energy density vector also possesses three components S_j (j = 1,2,3) which are related to the volume energy density W as

$$S_j = V_j \cdot W, \quad j = 1,2,3 \tag{1}$$

Equation (1) is referred to a system of rectangular Cartesian coordinates ξ_i (i = 1,2,3) and implies the exchange of surface and volume energy density.

Isoenergy Element. An isoenergy surface possesses the property that the transmission of energy on such a surface is directional independent. If ξ_i (i = 1, 2,3) are the current state coordinates, then the condition

$$S_1 = S_2 = S_3 \rightarrow S \tag{2}$$

determines the position of the element that the same S prevails on all orthogonal surfaces. On the isoenergy surface, the scalar W has a unique form (Sih 1985; 1988)

$$W = \iint \lambda V \mathrm{d}e\mathrm{d}e \tag{3}$$

in which λ served as a weighting function that defines the current isoenergy density state[1] and the limits of integration; it can depend on e.

Equations (1) and (2) imply that

$$V_1 = V_2 = V_3 \rightarrow V \tag{4}$$

The finiteness of V makes the theory nonlocal. General expressions for the rate change of volume with surface V can be found (Sih 1988) where deformation, rotation and change in element size are combined into one operation. In equation (3), e stands for any one of the nine displacement gradients

[1]A different form of λ would be chosen for evaluating energy density for the nuclear or electromagnetic field.

$$e_{ij} = \frac{\partial u_i}{\partial \xi_j}, \quad i,j = 1,2,3 \tag{5}$$

Thermal/Mechanical Interaction. Mass elements deform and change temperature simultaneously. Interdependence of these two effects cannot be assumed as a priori but must be determined (Sih 1985; 1988):

$$\frac{\Delta\Theta}{\Theta} = -\lambda V \frac{\Delta e}{\Delta \mathcal{D}/\Delta e}$$

in which Θ represents the nonequilibrium temperature in contrast to T in classical thermodynamics that applies only to equilibrium states. The dissipation energy density $\mathcal{D}$ is determined on the isoenergy density plane; it is a positive definite quantity:

$$d\mathcal{D} = dW - dA, \quad \mathcal{D} \geq 0; \ \frac{d\mathcal{D}}{dt} \geq 0 \tag{7}$$

with A being the available energy. Equation (7) is equivalent to the conservation of energy. The dissipated and available energy are mutually exclusive. Because Θ can be determined from V, e and $\mathcal{D}$ without heat, the isoenergy density formulation represents a fundamental departure from thermodynamics. Irreversibility is reflected through the H-function:

$$dH = -\frac{d\mathcal{D}}{\Theta} \tag{8}$$

The negative sign indicates work is done on the system. The change of H can be positive, zero or negative. It is not related to the Boltzmann H-theorem in statistical dynamics for dilute gases.

Isostress and Isostrain. Since an isoenergy density state is completely defined by (V,e) as W in equation (3), there is no need for the stress quantity. A theory can be formulated using the concept of force alone would be sufficient. For the sake of familiarity, however, the isostress τ referred to isoenergy density plane will be defined:

$$\tau = \int \lambda V de \tag{9}$$

as λ depends on e as mentioned earlier. A one-to-one correspondence between (V,e) and (τ,e) is invoked by equation (9) and is illustrated in Figures 1(a) and 1(b). Variations of λ with e have the same qualitative form as V versus e. A unique feature of the isoenergy density theory can be stated as follows:

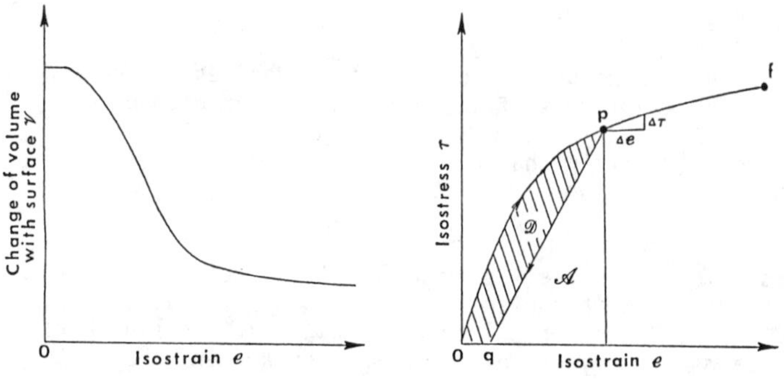

(a) $\mathcal{V}$ versus e (b) τ versus e

Figure 1. Equivalence of slope on isostress and isostrain curve to change of volume with surface.

The state of an isoenergy element defined by (V,e) or (τ,e) is determined for each load step and time increment without a preknowledge of the so-referred to "constitutive relations" in classical field theory.

Only the relation between τ and e for the reference state of a material is required. Once the loading steps are specified, adjustment on (τ,e) is made for each load step at every location in the system. A new curve opf as shown in Figure 1(b) is found for each element. Unloading, say from p to q, is also determined analytically so that both $\mathcal{D}$ and A can be found. In view of equation (9), equation (6) also takes the form

$$\frac{\Delta\Theta}{\Theta} = -\frac{\Delta\tau\Delta e}{\Delta\mathcal{D}} \tag{10}$$

This illustrates how Θ is found for each increment change of Δτ and Δe, Figure 1(b).

Governing Equations. As in the classical field theories, the governing equations can be derived by application of the conservation of linear and angular momentum and conservation of energy. Nonequilibrium, nonlocal, irreversible, and large/finite deformation effects are also included in addition to having finite element size. This necessitates the distinction between the Cauchy stress $\underset{\sim}{\sigma}$ and isostress $\underset{\sim}{\tau}$ which are related as

$$\tau_{ij} = \sigma_{ij} + \rho(\ddot{u}_i - h_i)V_j \tag{11}$$

where u_i are the displacement components. The quantities $\rho\ddot{u}_i$ and ρh_i represent the inertia and body force, respectively, while ρ is the mass density and dot stands for differentiation with time. The isostress tensor $\underset{\sim}{\tau}$ is not symmetric on account of the nonlocal character of the deformation. The conservation of energy applied to a system with volume Λ enclosed by surface Σ yields

$$\int_\Lambda \rho h_i \dot{u}_i dV + \int_\Lambda \overset{n}{T}_i \dot{u}_i dA = \int_\Lambda \rho\ddot{u}_i\dot{u}_i dV + \frac{d}{dt}\int_\Lambda \mathcal{W} dV \tag{12}$$

With n_i being the components of the outward unit normal vector, the tractions are given by

$$\overset{n}{T}_i = \tau_{ji}n_j, \quad \tau_{ij} \neq \tau_{ji} \tag{13}$$

Derivation of equation (12) also reveals that

$$\mathcal{W} = \tau_{ij}\dot{e}_{ij} \tag{14}$$

Make note that $\mathcal{W}$ is not an elastic potential. It applies to an irreversible and dissipative process as implied by equations (7) and (8). The equations of equilibrium are

$$\frac{\partial\tau_{ji}}{\partial\xi_j} + \rho h_i = \rho\ddot{u}_i \tag{15}$$

Existence and Boundness. The existence of the isoenergy density function $\mathcal{W}$ has been proved (Sih 1988). Only the theorem will be given as follows:

There exists an isoenergy density function $\mathcal{W}$ that can be obtained by integrating λV twice with respect to any one of the nine displacement gradients $\partial u_i/\partial\xi_j$ (i,j = 1,2,3) where V is the change of volume with surface for a given λ.

Mathematical equivalence of the foregoing statement is that

$$W = \int \tau_{11} de_{11} = \int \tau_{12} de_{12} = \ldots = \int \tau_{23} de_{23} \rightarrow \int \tau de \tag{16}$$

This means that any one of nine pairs (τ_{11},e_{11}), $(\tau_{12},e_{12}),\ldots,(\tau_{23},e_{23})$ or (τ,e) can be used to yield the same W. Hence, the uniaxial isostress τ versus isostrain e plot provides a general representation of the energy state even though the other stress components τ_{ij} are also acting on the isoenergy element.

Because nonequilibrium implies nonuniqueness, uniqueness proof can be provided only for equilibrium states (Sih 1988):

Isoenergy equilibrium states (V,e) or (τ,e) are unique for positive definite isoenergy density function W, and the satisfaction of equilibrium and continuity.

The nonequilibrium fluctuation of isoenergy states can only be bounded by their neighboring equilibrium states. Boundness and limit have been rigorously established and can be summarized (Sih 1988):

Fluctuations of nonequilibrium/irreversible isoenergy states are bounded by equilibrium/irreversible isoenergy states and they tend to definite limits depending on the initial step.

What can be deceiving is that the outward appearance of some of the governing equations in the isoenergy density theory may be similar to those in ordinary mechanics; the underlying principles and assumptions are completely different. The departure begins from the start in equation (1) where the exchange of S and V does not take place in ordinary mechanics.

NONUNIFORM CONDENSATION OF COMPRESSED GAS

The isoenergy density theory (Sih 1985; 1988) can, in general, be applied to analyze the nonequilibrium/irreversible behavior of gas, liquid or solid. Such versatility can be illustrated by first solving a nontrivial moving boundary problem concerning a gas compressed at a prescribed rate. The pressure and temperature are allowed to increase nonuniformly until condensation occurs. Classical thermodynamics cannot be used because the process is rate dependent.

Response of CO_2 Gas. Consider the region R which is 25.4 mm by 127 mm and occupied by CO_2 gas as shown in Figure 2. The gas is compressed at a displacement

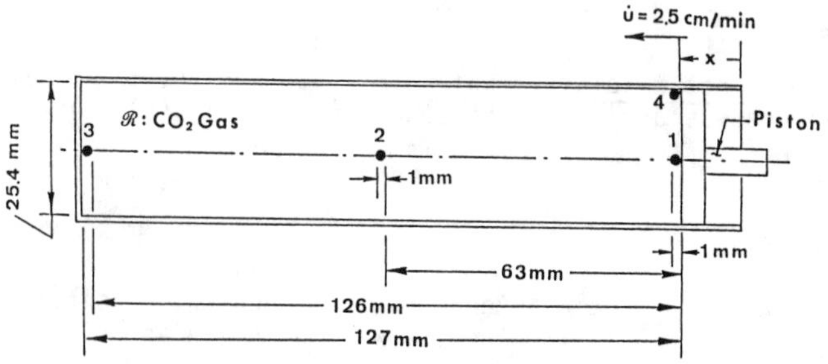

Figure 2. Carbon dioxide (CO_2) gas compressed at a rate of $\dot{u}$ = 2.5 cm/min.

rate of $\dot{u}$ = 2.5 cm/min. The van der Waals relation

$$(P + \frac{a}{v^2})(v-b) = RT \tag{17}$$

can be used to approximate the *reference* equation of the state for CO_2. Equation (17) holds reasonably well for the liquid and vapor phase and near and above the critical point. The three constants a, b and R can be determined from the critical values of $P_c = 73.87 \times 10^5$ Pa, $v_c = 94$ cm^3/g mole and $T_c = 304.2°K$ for CO_2. Instead of plotting the pressure P as a function of the volume V where v = V/m is the specific volume with m being the mass, reference will be made to the isostress τ which, for a gas, equals to P and isostrain e defined as dV/V. Application of equation (17) gives the reference state behavior of CO_2 in Figure 3.

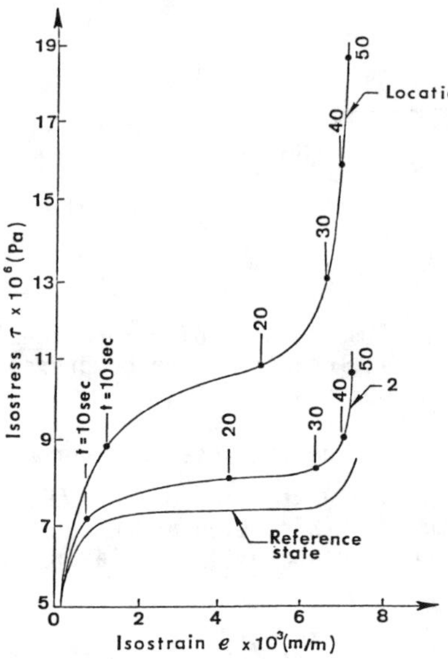

Figure 3. Isostress and isostrain response for CO_2 gas.

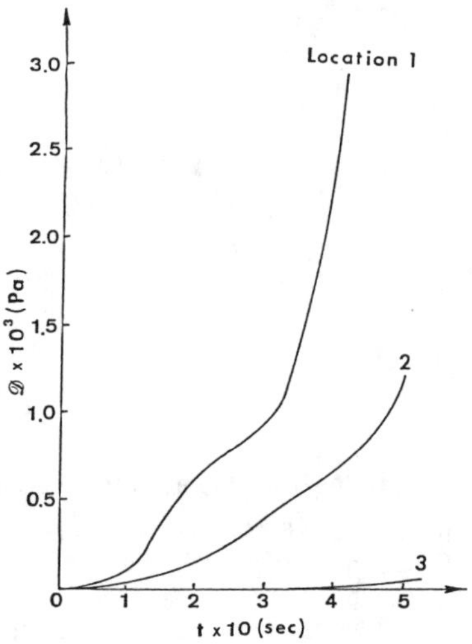

Figure 4. Time history of energy dissipation at locations 1, 2 and 3.

Following the method of solution described (Sih 1988), five time steps t = 10, 20,...,50 sec are taken to obtain (V^1,e^1), (V^2,e^2),...,(V^5,e^5) for every mass element in the system. Displayed in Figure 3 are the τ versus e relations for locations 1 and 2 which are located next to the piston and near the middle of the region R in Figure 2. Both curves differed significantly from the reference state and show that the local deformation rate at 1 is indeed much higher than that at 2. The behavior of the CO_2 gas when compressed would be different throughout R and cannot be characterized by one set of constants a, b and R.

Dissipation and Temperature Change. Unloading is carried at each time step by applying a negative displacement rate. This determines the path pq and $\mathcal{D}$ as shown in Figure 1(b). The time history of $\mathcal{D}$ at locations 1, 2 and 3 up to 50 sec is shown graphically in Figure 4. As it is expected, energy dissipated next to the piston is much higher than that in the mid-section. The gas at location 3 being farthest away from the piston is hardly disturbed and dissipates little energy even at the end of 50 sec.

Once τ, e and $\mathcal{D}$ are known, the temperature distribution Θ in the gas can be determined from equation (10). Figure 5 shows that for more than 20 sec, Θ at

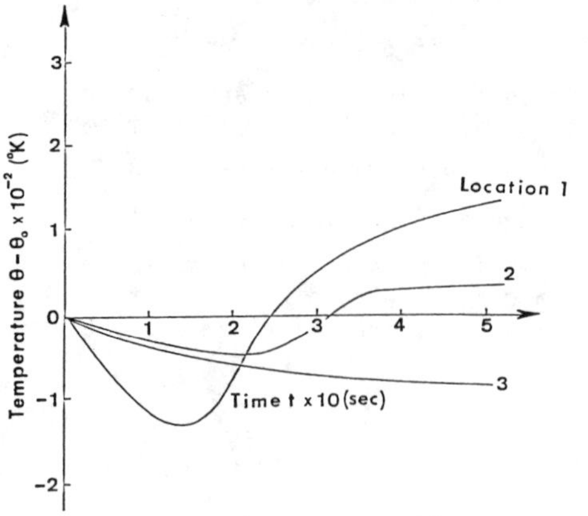

Figure 5. Time history of temperature at locations 1, 2 and 3.

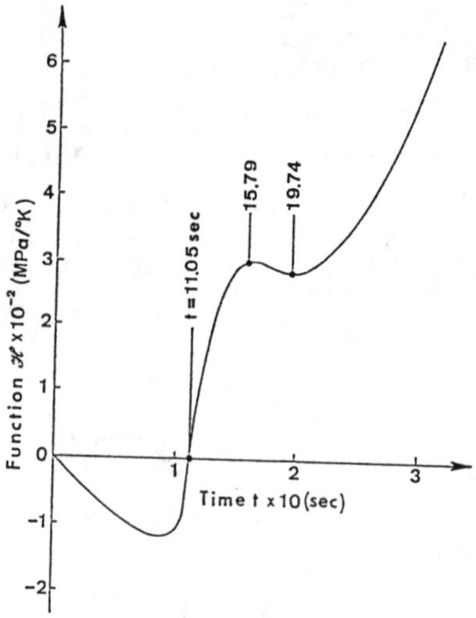

Figure 6. Time history of H-function at location 1 for compressed CO_2 gas.

location 1 is below the ambient temperature Θ_0. This cooling/heating phenomenon is reminiscence of that which occurred in the uniaxial extension of the SAFC-40R steel (Sih and Tzou 1987) and of the 6061-T6 aluminum (Sih, Lieu and Chao 1987). Location 2 experiences the same behavior with Θ rising above Θ_0 at a later time while Θ at 3 remained below Θ_0 up to 50 sec.

Because of cooling/heating, the H-function in equation (8) also undergoes a sign change. A temporary restoration of order is predicted locally when H increases negatively. This occurred at location 1 for less than the interval of the first time step, i.e., 10 sec as the gas responded to the initial movement of the piston. Thereafter, H increases with time which signifies increase in the disorder. An inflection in the H-curve is found between t ≃ 16 and 20 sec. Phase transition[2] is predicted (Sih 1987) where the gas at location 1 may transform to liquid at this time. Additional information on the latent heat is required.

Phase Transformation. The strain rate of energy dissipation density $\Delta\mathcal{D}/\Delta e$ in equation (6) plays the same role as the latent heat $\Delta Q/\Delta V$ in classical thermodynamics for determining phase transformation. The relation

$$\frac{\Delta\mathcal{D}}{\Delta e} \rightarrow \frac{\Delta Q}{\Delta V} = T\,\frac{\Delta S}{\Delta V} \qquad (18)$$

holds only in the limit when $\mathcal{D} \rightarrow Q$. Here, ΔQ is the heat required per unit mass of substance in changing the thermodynamic state from temperature T to T+ΔT and ΔV is the change of volume for a unit mass. The corresponding change in entropy

[2]A change in microstructure ahead of a crack in a 1020 steel specimen was predicted from the inflection point on the H-curve. This result was verified experimentally (Sih, Tzou and Michopoulos 1987).

is ΔS. The basic difference between $\Delta \mathcal{D}/\Delta e$ and $\Delta Q/\Delta V$ is that the former accounts for rate effect in phase transformation and the latter does not.

For the CO_2 gas, the latent heat for phase transition can be estimated from the classical thermodynamics treatment that gives $\Delta Q/\Delta V = 6.0244 \times 10^5$ Pa. This value intersects on the $\Delta \mathcal{D}/\Delta e$ versus time curves in Figure 7 at approximately t = 24.3 sec for location 1 and t = 31.8 sec for location 2. The curve for location 3 is so far below threshold that no phase transformation is predicted. A slightly earlier time of t ≈ 20 sec is estimated from the occurrence of an

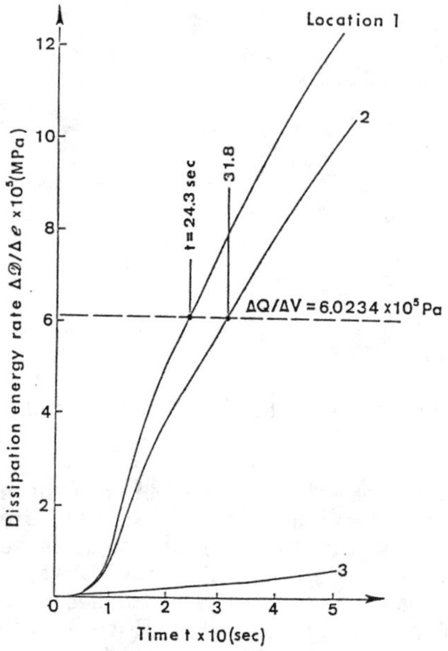

Figure 7. Strain rate of energy dissipation density as a function of time for compressed CO_2 gas.

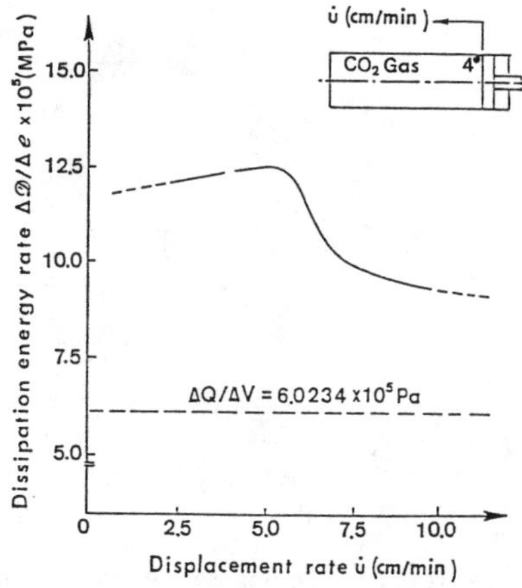

Figure 8. Rate effect on condensation at location 4 for compressed CO_2 gas.

inflection on the H-curve in Figure 6. This would correspond to a lower $\Delta \mathcal{D}/\Delta e$ for phase change. The rate at which the gas is compressed makes a difference. Latent heat required to cause phase transition depends on $\dot{u}$. Figure 8 illustrates this influence for location 4 at the reentrant corner which experiences the highest local energy concentration. The strain rate of energy dissipation does not increase monotonically with the rate of compression; it increased only slightly up to $\dot{u} \simeq 5.42$ cm/min and then dropped considerably. Minimization of energy dissipation for nonequilibrium processes is not apparent. Calculations such as those in Figure 8 at location 4 can be done for all mass elements and the average can be used to determine $\dot{u}$ that corresponds to minimum strain rate of energy dissipation.

Summarized in Figures 9(a) to 9(e) are, respectively, the nonuniform condensation pattern for t = 10,20,...,50 sec. These predictions are made by solving a moving boundary problem. The finite element grid pattern is updated after each time step. The distances travelled by the piston are given in Figures 9. Condensation first appears at the reentrant corners where the piston slides against the wall. This is shown in Figure 9(a) for t = 10 sec. At t = 20 sec, the pis-

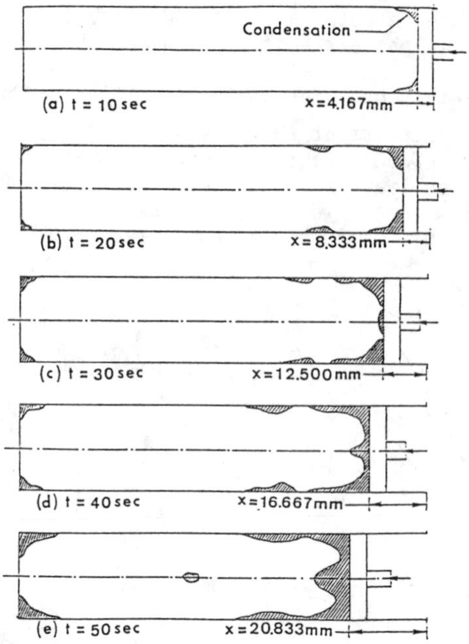

Figure 9. Time history of nonuniform condensation pattern for compressed CO_2 gas.

ton travelled almost twice as far with x = 4.167 mm while condensation occurred also at the far end of the reentrant corners in addition to those isolated along the wall, Figure 9(b). Gas near the center of the piston began to condense between t = 20 and 30 sec. This is shown in Figure 9(c) where the liquid region has spread over most of the piston surface. Coalescence of the disconnected portions of the liquid phase along the wall near the piston takes place in Figures 9(d) and 9(e) for t = 40 and 50 sec. Predicted is also an island of condensed gas in Figure 9(e). This result is by no means obvious and is subject to experimental verification.

Laminar/Turbulent Flow Between Parallel Planes. The over-a-century postulate of *no slippage* at a liquid/solid interface in viscous fluid mechanics was seriously questioned when computational failure occurred at Weissenberg number of order of unity (Panar and Denn 1986). An overestimation of the intensification of local energy did not agree with experimental observation. This conflict of physical reality that a real fluid can slip casts doubt on the validity of the Navier-Stokes equations when applied to explain boundary layer phenomena. The Prandtl's mixing-length approach requires a priori assumption to distinguish the difference between the velocity distribution in the turbulent and laminar region, creating a so-called transition zone. Such empiricisms do not arise in the isoenergy density theory since the complete velocity field can be obtained without making assumptions on the flow characteristics. This will be done for the flow between parallel planes.

Reference State. The one-dimensional laminar flow solution based on Newton's law of viscosity will be used as the reference state for the problem of flow passing between two parallel planes as shown in Figure 10. The fluid with a dynamic viscosity μ occupies the region R which is 2 m by 10 m and is symmetric about the x-axis. In the one-dimensional case of steady-state and homogeneous flow, the governing equation

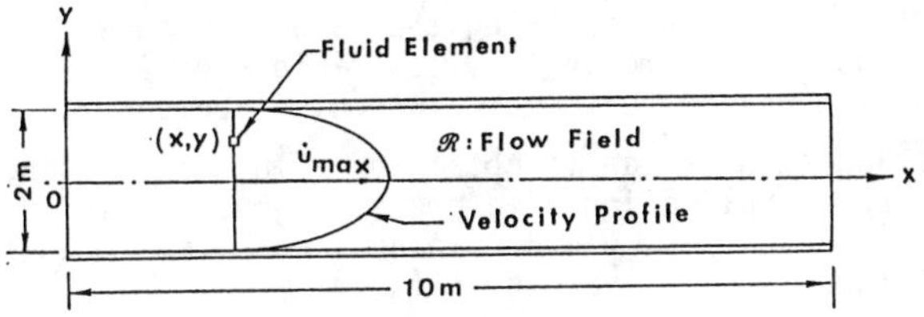

Figure 10. Schematic of flow between parallel planes.

$$\dot{u}_x(y) = \frac{1}{2\mu}\frac{dP}{dx}(1-y^2) \tag{19}$$

satisfies the conditions

$$\dot{u}_x = 0;\ y = 1,\ \frac{d\dot{u}_x}{dy} = 0;\ y = 0 \tag{20}$$

Equation (19) is assumed to be valid only as a reference state in the limit of vanishing pressure gradient. Once a gradient of pressure is applied across $\mathcal{R}$ in Figure 10, the flow field adopts a two-dimensional character where both $\dot{u}_x$ and $\dot{u}_y$ will be present.

The baseline of comparison for velocity distribution in turbulent flow corresponds to the Prandtl mixing length velocity expression (Daily and Herleman 1966):

$$\frac{\dot{u}_x - \dot{u}_{max}}{\dot{u}_*} = \frac{1}{\kappa}\log\left(\frac{y}{y_o}\right) \tag{21}$$

where κ is referred to as a universal constant in turbulent flow and $y_o = 1$ m for the problem in Figure 10. The shear stress velocity $\dot{u}_*$ is given by

$$\dot{u}_* = \sqrt{\frac{\sigma^o_{xy}}{\rho}} \tag{22}$$

The mass density is ρ and the shear stress at the wall $y = \pm 1$ is denoted by σ^o_{xy}. Experimental data (Nikuradse 1933) on fluid flow velocity are generally collated by three different expressions. The normalized quantity

$$\frac{\dot{u}}{\dot{u}_*} = \frac{\dot{u}_* y}{\nu},\quad \frac{\dot{u}_* y}{\nu} < 5 \tag{23}$$

applies to the laminar *sublayer* while

$$\frac{\dot{u}}{\dot{u}_*} = 11.5\log\left(\frac{\dot{u}_* y}{\nu}\right) - 3.05,\quad 5 < \frac{\dot{u}_* y}{\nu} < 30 \tag{24}$$

is for the *buffer* zone. For the *turbulent* core, $\dot{u}/\dot{u}_*$ takes the form

$$\frac{\dot{u}}{\dot{u}_*} = 5.75\log\left(\frac{\dot{u}_* y}{\nu}\right) + 5.5,\quad \frac{\dot{u}_* y}{\nu} > 30 \tag{25}$$

In equations (23) to (25), ν is the kinematic viscosity which is equal to μ/ρ.

Fluid Inhomogeneity. Instead of applying the same μ in equation (19) or ν in equations (23) to (25) everywhere in the flow field, the isoenergy density theory will be employed to determine the complete displacement field u_x and u_y from which the stresses and strains can be obtained. Fluid inhomogeneity is established immediately upon application of a pressure gradient across the section at x = 0 and x = 10 m. That is, μ or ν can no longer be taken as constants as they are different for different values of (x,y).

Let a uniform pressure P be specified at x = 0 while a zero P is maintained at x = 10 m. A pressure gradient of $\Delta P/\Delta x = P/10$ Pa/m is thus created. Time rate effect will also be analyzed for $\Delta\dot{P}/\Delta x$ = 0.0919, 0.1379 and 0.2758 Pa/m sec. They will be referred to simply as $\dot{P}$ = 0.9193, 1.379 and 2.758 Pa/sec. To begin with, the isoenergy density plane for each fluid element will be determined by application of equation (4) for the two-dimensional flow field involving only V_1 and V_2. The initial slope of the isostress and isostrain curve which is equivalent to specifying[3] the dynamic viscosity in Newtonian flow is taken to be $d\tau/de = 0.896 \times 10^{-1}$ Pa. Different time steps can then be taken to obtain (V^1,e^1), (V^2,e^2), etc., for each fluid element. Plotted in Figure 11 are data for the isostress τ and isostrain e of elements at the entrance x = 0 and y = 0, 0.5 y_0 and 1.0 y_0 for $\dot{P}$ = 0.9193 Pa/sec. Both τ and e are negative over the cross-section except near the wall y/y_0 = 1.0 where they are positive. This feature is also observed at x = 5.0 m where τ is decreased by two orders of magnitude, Figure 12. These data correspond to t = 15 and 30 sec. Use is made of the Lagrangian interpolation scheme for the values of τ and e from 0 to 30 sec. Inhomogeneity of the fluid can be exhibited by computing for the variations of σ_{xy} with $\dot{\gamma}_{xy}$. A time interval Δt = 0.3 sec is used to compute for

$$\dot{\gamma}_{xy} = \frac{1}{2\Delta t}[(\gamma_{xy})_{t+\Delta t} - (\gamma_{xy})_{t-\Delta t}] \tag{26}$$

The results for x = 5.0 m and y = 0.0, 0.5 and 1.0 m are given in Figure 13 show that σ_{xy} varies nonlinearly with $\dot{\gamma}_{xy}$. Interpreting the ratio $\sigma_{xy}/\dot{\gamma}_{xy}$ as the apparent dynamic viscosity μ^*, Figure 14 shows that the classical assumption of a constant μ holds near the entrance at t = 30 sec for x = 0; $y/y_0 < 0.6$. Large variations of μ^* with y/y_0 take place further downstream for x = 5.0 and 10.0 m.

Values of the apparent dynamic viscosity are near those for the SAE 30 oil (Pao 1967) whose dynamic viscosity coefficient μ = 0.383 Pa·sec at 20°C and its density is ρ = 824.63 kg/m^3.

Velocity Distribution. Comparison of the isoenergy density solution with experiment (Nikuradse 1933; Pao 1967) can be best illustrated using the velocity data. The data in Figure 15 corresponds to the variations of the normalized distance y/y_0 with the velocity defect $(\dot{u}_{max}-\dot{u})/\dot{u}_*$ for $\dot{P}$ = 0.9193 Pa/sec and t = 30 sec. The shear stress velocity is computed from equation (22) for the SAE 30 oil. Shown by the open circles are results obtained from the Prandtl mixing-length equation (21) for $\kappa = 1/3$. They fit the experimental data represented by the plus signs reasonably well except in the laminar sublayer within which the dotted straight line obtained from equation (23) is used and did not give a good correlation. Predictions made from the isoenergy density theory are given

[3]The difference is that μ is defined as the ratio of shear stress σ_{xy} and shear strain rate $\dot{\gamma}_{xy}$ while e is the shear displacement gradient in the isoenergy density theory.

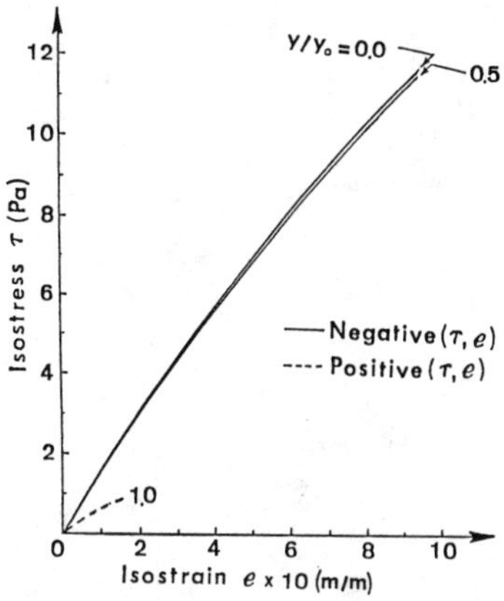

Figure 11. Variations of isostress with isostrain at x = 0 with different y for $\dot{P}$ = 0.9193 Pa/sec.

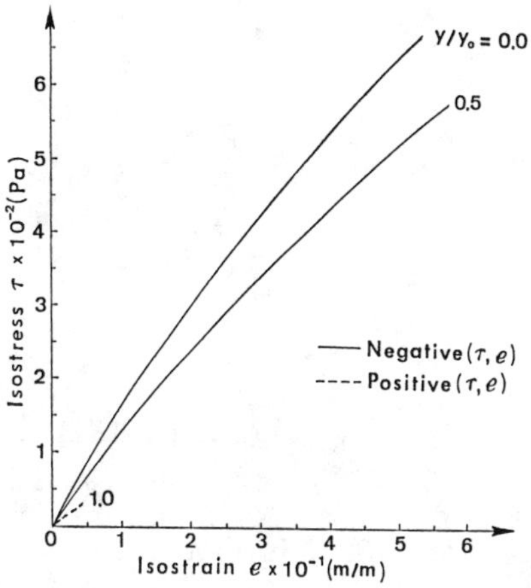

Figure 12. Variations of isostress with isostrain for fluid elements at x = 5.0 m and different y for $\dot{P}$ = 0.9193 Pa/sec.

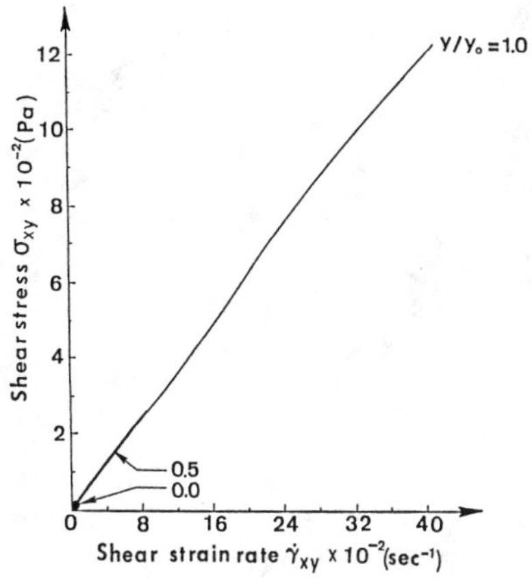

Figure 13. Shear stress versus shear strain rate at x = 5.0 m and different y for $\dot{P}$ = 0.9193 Pa/sec.

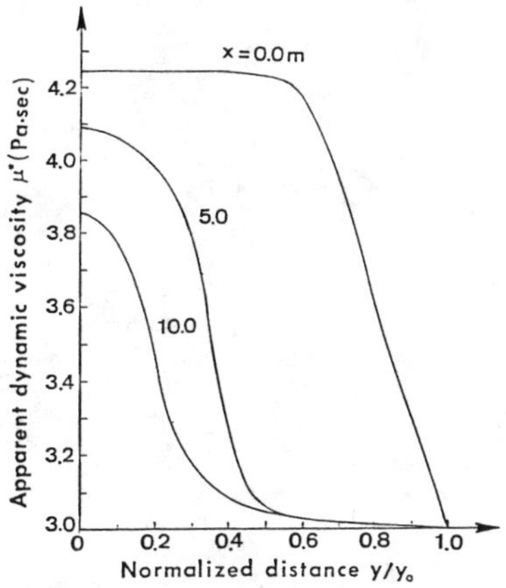

Figure 14. Variations of apparent dynamic viscosity flow field at t = 30 sec for $\dot{P}$ = 0.9193 Pa/sec.

by the completely-solid circles for x = 5.00 m and partially-solid circles for x = 4.85 m; they embraced the measured data. The agreement is not conclusive because the experiments (Nikuradse 1933; Pao 1967) made no reference to the location or distance x at which $\dot{u}$ is measured while the theoretical results apply to those at approximately half-way from the entrance. Figure 16 provides an alternate representation of the data by plotting the logarithmic term with argument $\dot{u}_* y/\nu$ against $\dot{u}/\dot{u}_*$. Three sets of data for x = 4.60, 4.65 and the average from x = 3.50 to 8.33 m are found from the isoenergy density theory. They followed a gradually rising trend in contrast to the disconnected straight line segments obtained from equations (23), (24) and (25) for the laminar sublayer, buffer zone and turbulent core, respectively.

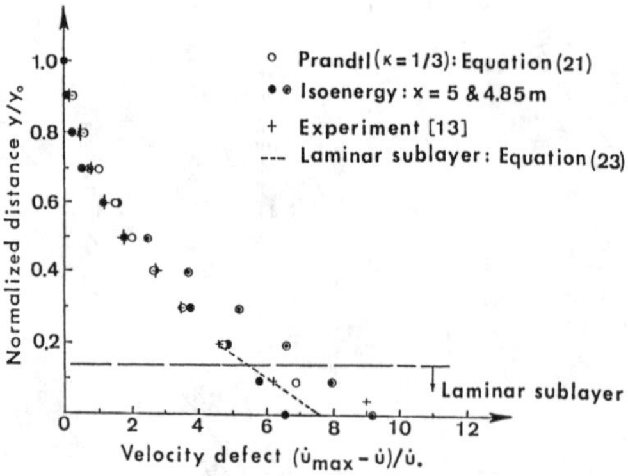

Figure 15. Distribution of velocity defect at x = 4.85 and 5.00 m for $\dot{P}$ = 0.9193 Pa/sec and t = 30 sec.

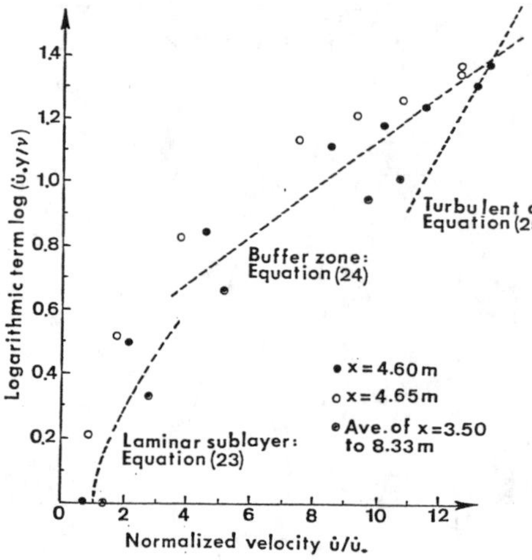

Figure 16. Comparison of isoenergy density solution at x = 4.60 and 4.65 m with logarithmic velocity distribution for $\dot{P}$ = 0.9193 Pa/sec and t = 30 sec.

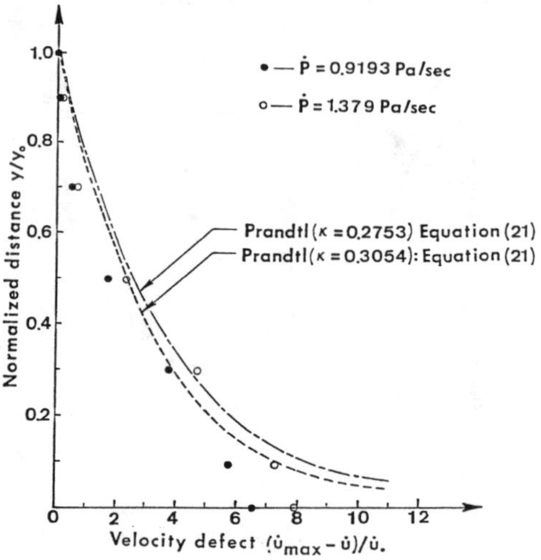

Figure 17. Effect of pressure gradient rate at x = 5.0 m for t = 30 sec.

Exhibited in Figure 17 is the influence of pressure gradient rate on velocity distribution for t = 30 sec and x = 5.0 m. Largest deviation in the velocity defect occurred at y/y_0 = 0 as $\dot{P}$ is increased from 0.9193 to 1.399 Pa/sec. According to the isoenergy density solution, the parameter κ in equation (21) is flow rate dependent; it decreased from 0.2753 to 0.3054 as $\dot{P}$ is increased. Further increase of $\dot{P}$ up to 2.758 Pa/sec is made in Figure 18. The term $\log(\dot{u}_* y/\nu)$ increased with $\dot{u}/\dot{u}_{max}$ at a much more rapid rate at t = 10 sec than the case in Figure 16 which corresponds to a lower flow rate. No trend could be established from the Prandtl mixing length equation.

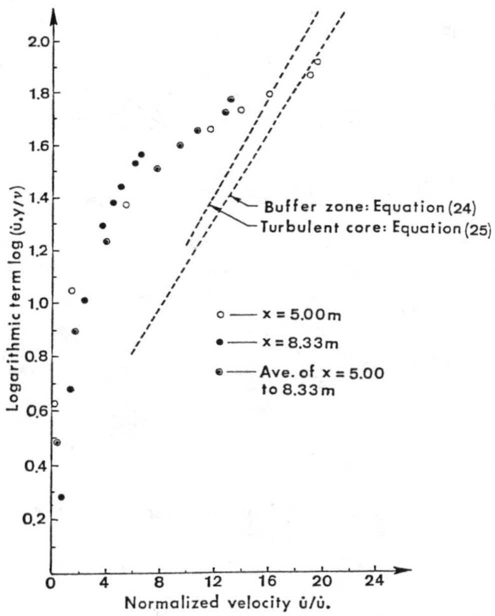

Figure 18. Comparison of isoenergy density solution at x = 5.00 and 8.33 m with logarithmic velocity distribution for $\dot{P}$ = 2.758 Pa/sec and t = 10 sec.

High Rate Flow Characteristics. Classical fluid mechanics makes no reference to the change of applied pressure gradient rate that obviously can greatly affect the laminar/turbulent characteristics of the flow. Further illustration on the influence of $\dot{P}$ on the flow field will be made. An order of magnitude increase on $\dot{P}$ to 13.790 Pa/sec leads to a dramatic change in the time variation of the stresses and strains. Following the method of solution in [6], the isostresses τ and isostrains e are found at different locations of (x,y) for t = 4, 8 and 12 sec.

Near the entrance at x = 0.50 m, both τ and e are negative and they increase significantly with time for y/y_0 up to 0.9. Near the wall at $y/y_0 \simeq 1$, τ and e dropped suddenly and became slightly positive. These features are displayed in Figures 19 and 20. Of interest are also the development of the velocity profile as the fluid traverses across the full length of the region *R* in Figure 10. The series of velocity profiles would be different for each time instance. Figure 21 shows $\dot{u}/\dot{u}_{max}$ as a function of y/y_0 and x for t = 12 sec. An oscillatory behavior is observed; the velocity curve at x = 0.5 m bulged and then deflated at x = 5.0 m. This process is repeated at x = 6.5 and 8.0 m but not as pronounced. At x = 6.5 m, the velocity profile is nearly parabolic. The dotted curve connected by the solid circles is the commonly known laminar solution. As the flow proceeds toward the exit that is maintained at zero pressure, the velocity field acquires a more erratic behavior because of the sudden change in pressure drop. Figure 22(a) shows a reversal of $\dot{u}/\dot{u}_{max}$ at $y/y_0 \simeq 0.675$ m. At x = 10 m, the flow changed direction twice. The presence of strong vortices are thus predicted at the exit. While their magnitude can be computed, it is, however, more pertinent to present results on the energy dissipation in the fluid associated with temperature changes.

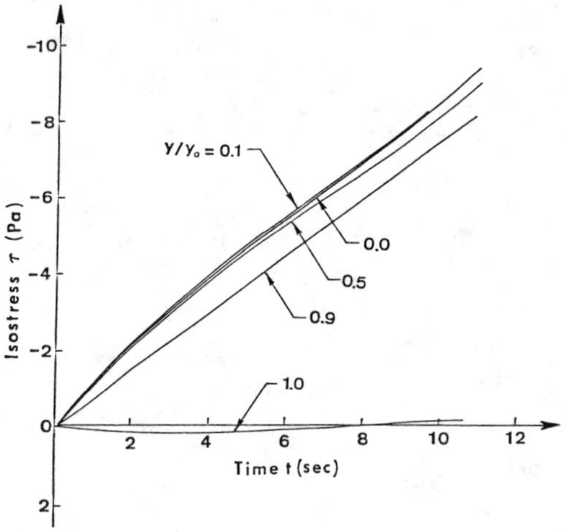

Figure 19. Time history of negative isostress for $\dot{P}$ = 13.790 Pa/sec at x = 0.5 m.

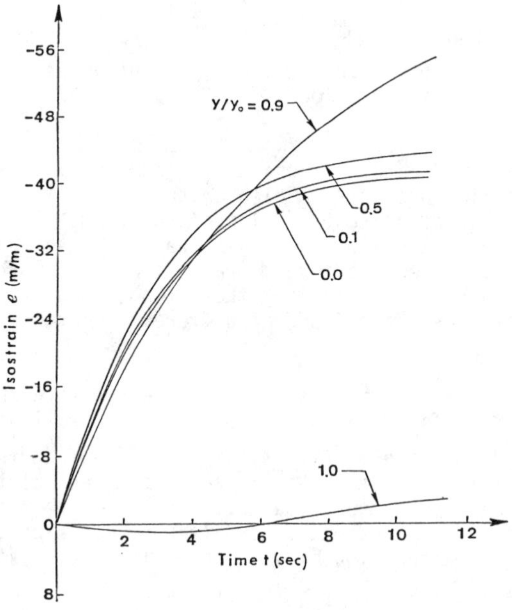

Figure 20. Time history of negative isostrain for $\dot{P}$ = 13.790 Pa/sec at x = 0.5 m.

The unloading procedure with reference to Figure 1(b) applies equally well to fluid elements and it can be used to determine the distribution of dissipated energy. As it is expected, the largest energy dissipation occurs near the wall at $y/y_0 \simeq 0.9$ as the flow is developed. This is shown in Figure 23 for x = 0.5 m and time t > 6 sec. The fluid near the mid-section at $y/y_0 \simeq 0$ dissipated

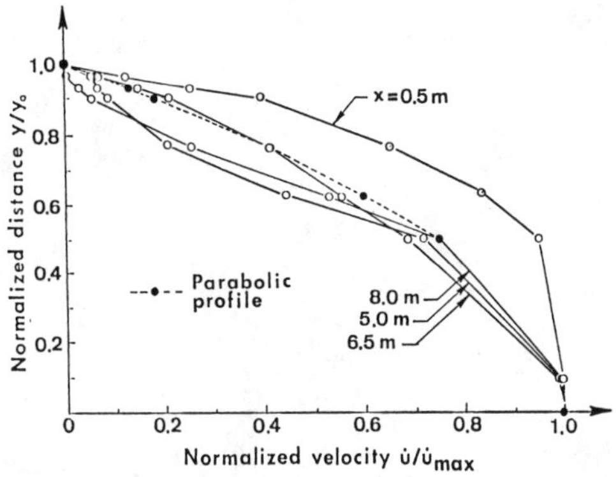

Figure 21. Distribution of normalized velocity for $\dot{P}$ = 13.790 Pa/sec at t = 12 sec.

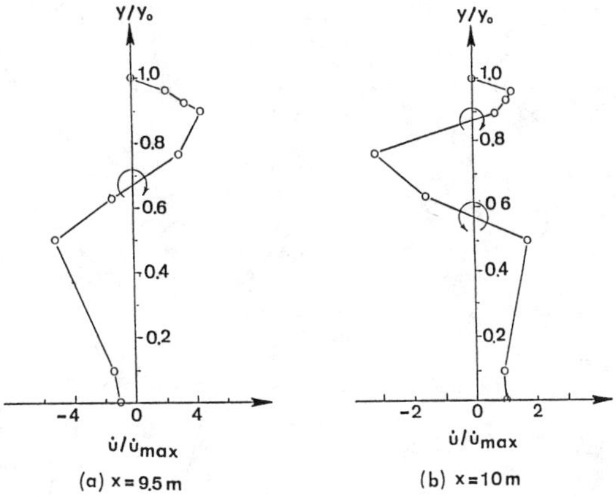

Figure 22. Distribution of normalized velocity at x = 9.5 and 10 m for $\dot{P}$ = 13.790 Pa/sec and t = 12 sec.

the least amount of energy except for the initial period t < 3 sec where the flow is highly unstabilized. The phenomenon of cooling/heating that is now well-known in the uniaxial extensions of solids (Sih and Tzou 1987; Sih, Lieu and Chao 1987) is also predicted in fluid flow. For the elements at x = 0.5 m and y/y_0 up to 0.9 m, Figure 24 shows that the temperature Θ dropped below the ambient condition; it rises about Θ_0 only after some lapse in time. The condition $\Theta = \Theta_0$ for the curve $y/y_0 = 0$ occurred at $t \simeq 4$ sec. Fluctuation of the temperature dies down quickly near the wall. In fact, the temperature Θ at the wall rised slightly above Θ_0 and then dropped below ambient. Heating preceded cooling instead of cooling and heating. The oscillatory character of thermal changes in the fluid elements cannot be neglected for it is an inherent part of laminar/turbulent flow.

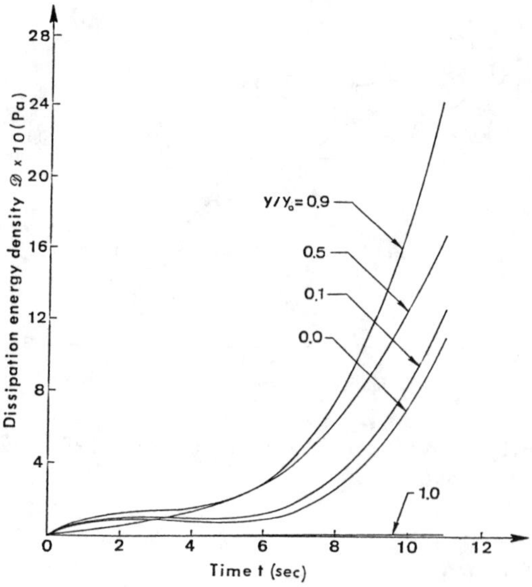

Figure 23. Time history of dissipation energy density for $\dot{P}$ = 13.790 Pa/sec at x = 0.5 m.

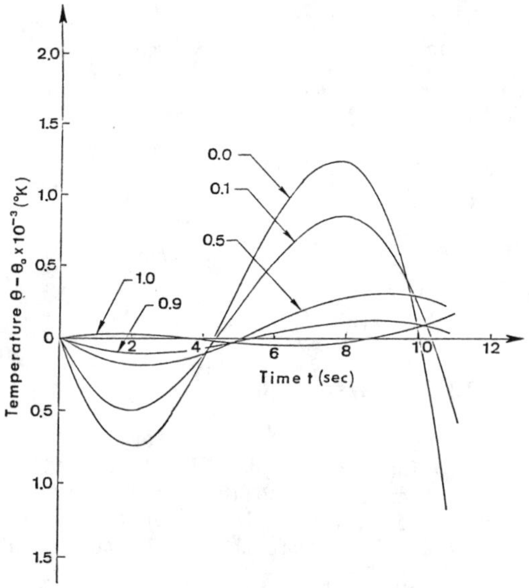

Figure 24. Time history of temperature change for $\dot{P}$ = 13.790 Pa/sec at x = 0.5 m.

FLUID/SOLID SEPARATION

The characteristics of flow around solid bodies have been studied extensively in the literature (Schlichting 1968; Goldstein 1965; Prandtl and Tiejens 1934; Chang 1970). Local instability in the form of bubble creation or complete separation of the fluid from solid would occur as the flow speed is gradually increased. These phenomena cannot be adequately explained by classical fluid mechanics on account of the highly nonequilibrium behavior of the fluid, particularly at the instance just before separation. Predictions of flow separation based on the solution of the flow field prior to instability have been made (Sih 1980). The location of flow separation was assumed to coincide with the maximum of the local increase in the potential energy density function. That is, the local momentum loss makes it impossible for the material volume to overcome the steep potential energy barrier and hence separation would prevail. Surprisingly enough, this criterion, when incorporated into the inviscid flow solution, gave excellent results for those situations where separation occurred very suddenly. Separation points at angles ±125.26° from the forward stagnation point of a stationary circular cylinder in a uniform flow field were predicted (Sih 1980). This agreed closely with known experiments (Goldstein 1965) that reported separation for angles between ±122° and ±130° from the forward stagnation point. Shifting of the separation points for the rotating cylinder also agreed with observation (Prandtl and Tiejens 1934). A thin viscous boundary layer next to the solid apparently has little or no influence on fluid instability at large. The simplicity of the method (Sih 1980) provides an easy means of locating the separation of flow around complex two- and three-dimensional solid objects.

Emersed Circular Cylinder. Instability initiates from the sudden release of local energy. As the fluid impinges on the solid, energy exchange takes place across the fluid/solid interface. Energy transmitted into the solid at the forward stagnation point may not be negligible and may influence the downstream condition at instability. Barring from such an effect, the analysis would consider a uniform flow of $\dot{u}_0$ = 178.8 m/sec passing a deformable circular cylinder in a region designated as R_1, Figure 25. The fluid is water and the solid oc-

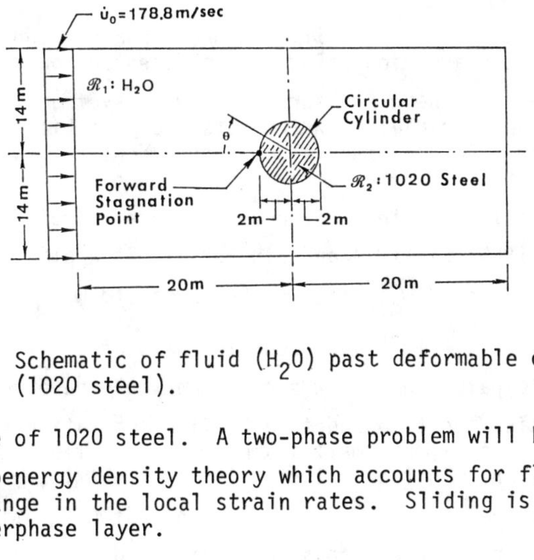

Figure 25. Schematic of fluid (H_2O) past deformable circular cylinder (1020 steel).

cupies R_2 is made of 1020 steel. A two-phase problem will be solved by application of the isoenergy density theory which accounts for fluid viscosity, nonlinearity and change in the local strain rates. Sliding is permitted in the solid/liquid interphase layer.

The values of $d\tau/de$ for the reference state of R_1 and R_2 are, respectively, 1.070×10^3 MPa and 2.068×10^5 MPa. Since the rate change of volume with surface for the liquid is much larger than that for the solid, the initial values of $V°$ equal to 10^2 and 1 will be taken to distinguish this difference. For a two-dimensional problem, V_3 in the direction normal to the flow can be fixed; it is assumed to be 200 for the liquid and 2 for the solid. The above information suffices to start the calculation for (V^1, e^1), (V^2, e^2), etc., in both the solid and fluid elements.

Slippage and Separation. Fluid slippage along the surface of the solid cylinder occurs owing to the discontinuity in the local circumferential velocity $\dot{u}_\theta$. Figure 26 displays the results for t = 2 x 10^{-2} sec at which time $\dot{u}_\theta$ reached a

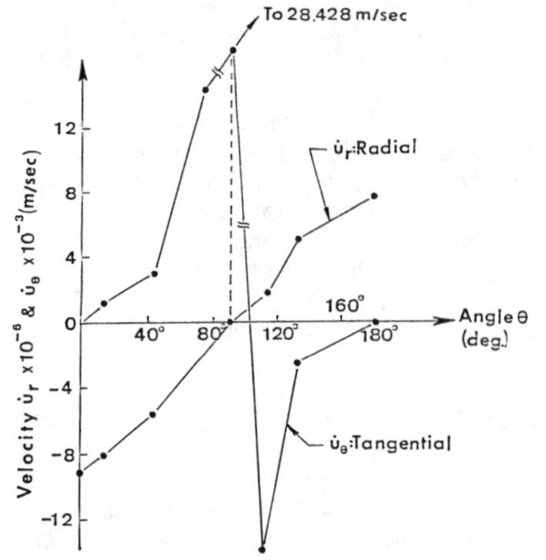

Figure 26. Angular variations of radial and tangential velocity components near cylinder surface at t = 2 x 10^{-2} sec.

maximum of 28.428 m/sec at θ ≃ 90°. It then drops sharply and reverses direction. Near the forward stagnation point, $\dot{u}_r$ is directed toward the cylinder. At about θ ≃ 90°, the fluid near the surface flows away from the solid. This change in the direction of radial flow suggests the possibility of flow separation. Indeed, the variations of the volume energy density $\mathcal{W}$ with θ in Figure 27 shows that a maximum of the minimum $\mathcal{W}$ occurs at θ ≃ 125°. The significance of $\mathcal{W}_{min}^{max}$ in relation to failure initiation by fracture in solids has been well documented (Sih 1972-1983; Sih 1974 and Sih 1984). It applies also to the onset of flow separation should $\mathcal{W}_{min}^{max}$ attain the threshold $\mathcal{W}_c$. Since the value of $\mathcal{W}_{min}^{max}$ at a later time t = 4 x 10^{-2} sec occurring also at θ ≃ 125° decreased in amplitude, it is anticipated that flow separation would have occurred at t = 2 x 10^{-2} sec. The precise location and time of flow separation would be difficult to verify experimentally as small asymmetry of the flow field can greatly disturb the condition that initiates instability. Preliminary calculations show that $\mathcal{W}_{min}^{max}$ oscillates near θ ≃ 125°. The range of θ and time within which flow separation could initiate can thus be established.

Vortex Formation. The shedding of vortices off the cylinder has been the subject of many past discussions in fluid mechanics. Once the fluid has physically separated from the cylinder, the wake region must then be modelled accordingly. Without getting into such details, the tendency for vortex formation can be seen from the velocity field at t = 6 x 10^{-2} sec. Using the values of

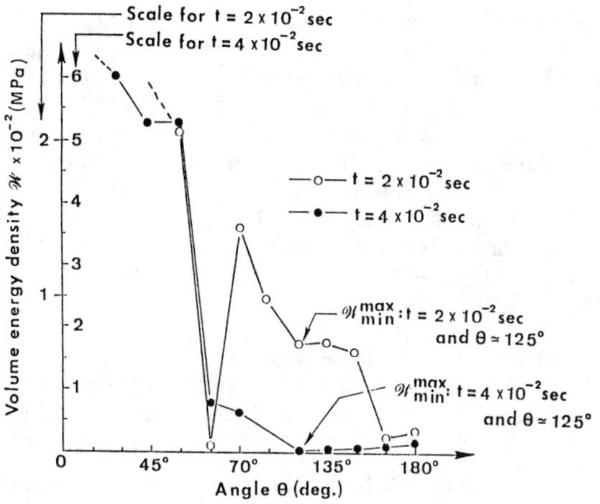

Figure 27. Angular variations of volume energy density for t = 2 x 10^{-2} and 4 x 10^{-2} sec.

$\dot{u}_r$ and $\dot{u}_\theta$, the vortices in Figure 28 can be computed and they are scattered behind the cylinder. The asymmetry pattern can be easily obtained by referring the data to a horizontal line tilted with an angle of attack Θ_0 oriented less than 10° from the geometric center line.

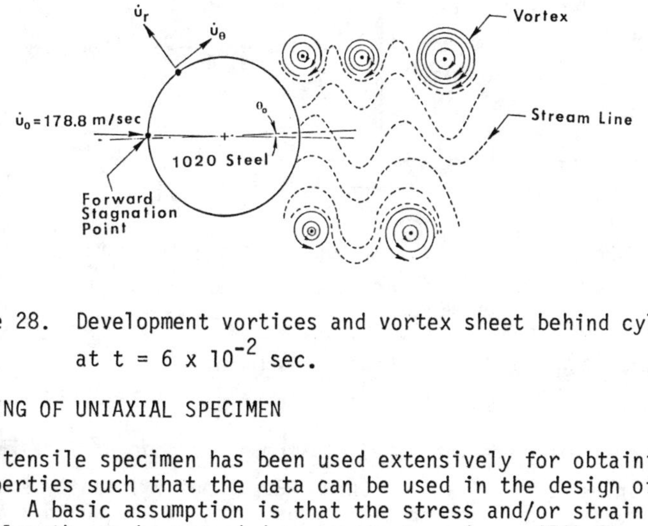

Figure 28. Development vortices and vortex sheet behind cylinder at t = 6 x 10^{-2} sec.

COOLING/HEATING OF UNIAXIAL SPECIMEN

The uniaxial tensile specimen has been used extensively for obtaining basic mechanical properties such that the data can be used in the design of more complex systems. A basic assumption is that the stress and/or strain state within certain gage length can be regarded as constant or in equilibrium. Such an idealization would encounter considerable difficulties in a wide range of loading rates, specimen sizes and environmental conditions. For instance, specimens tested under varying temperature conditions are more sensitive to change in strain rate, geometry and/or size. Classical thermo-mechanical theories[4] (Cernocky and Krempl 1981) have not succeeded in characterizing such behavior mainly because they are unable to synchronize the thermal change with mechanical deformation. Temperature or its gradient on the solid surface is generally specified rather than derived according to the test conditions. The prescription of constant temperature on the solid surface cannot be physically realized because there would always be a layer of solid/air interface within which appreciable change in heat transfer takes place. Thermal fluctuation on the specimen surface can, in turn, affect the bulk behavior of the uniaxial specimen and interpretation of the data.

Thermally and Mechanically Induced Deformation. Without making a priori assumption on the temperatures or temperature gradients distribution that prevail on the uniaxial tensile specimen, the far field temperature is maintained uniform at a prescribed value to be denoted by Θ_0. The temperature changes and displacements of the mass elements in both the solid and its surrounding will be found. A schematic of the two-phase system is given in Figure 29 in which the

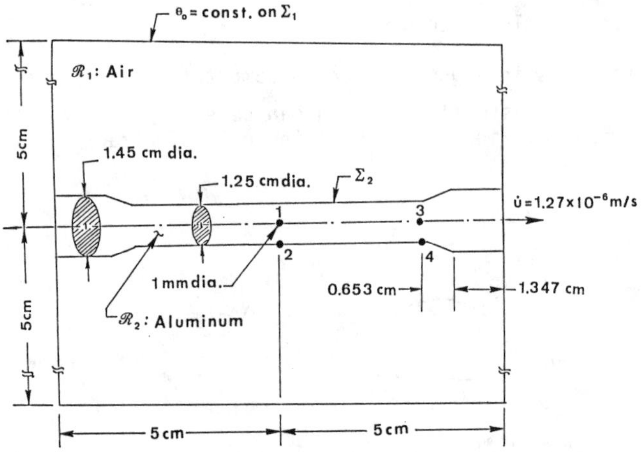

Figure 29. Schematic of aluminum tensile specimen in air environment with temperature Θ_0.

circular cylinder occupies the region R_2 enclosed by Σ_2 is engulfed by the air in region R_1 with boundary Σ_1 on which Θ_0 is maintained constant. The solid specimen is stretched at a displacement rate of $\dot{u} = 1.27 \times 10^{-6}$ m/sec.

The influence of four different far field temperatures Θ_0 = 25°, 75°, 125° and 175°C will be investigated (Sih and Chou 1989). A sufficiently small load increment load is first taken so that the properties of the solid and air can be

[4]Contrived into the equilibrium theory of thermoviscoplasticity (Cernocky and Krempl 1981) is the nonequilibrium cooling/heating phenomenon which invalidates the work in principle. This can be further evidenced from the wrong deduction made by the same theory (Cernocky and Krempl 1981) that torsional loading produced no reversal of heat flow. Experiments (Lieu 1986) at Lehigh University showed that the effect of cooling/heating is just as pronounced in torsion as in extension.

assumed to be known and thus preassigned. All subsequent responses will be determined individually for each solid and fluid element and time step. Table 1

Table 1. Reference Properties of 6061 Aluminum and Air.

Temperature Θ_0 (°C)	Initial Slope $d\tau/de \times 10^5$ (Pa)	
	6061 Aℓ	Air
25°	8.167×10^5	2.0
75°	7.500	2.5
125°	6.500	3.0
175°	5.000	3.5

gives the initial values of $d\tau/de$ for the 6061 aluminum and air at different values of Θ_0. Following the procedure discussed (Sih 1988), results will be obtained for t = 120,240,...,720 sec.

Local and Global Behavior. The nonequilibrium stress and strain behavior can be seen from the difference between the local and global response as illustrated in Figures 30 and 31 for Θ_0 = 25°C. Referring to the four locations 1,

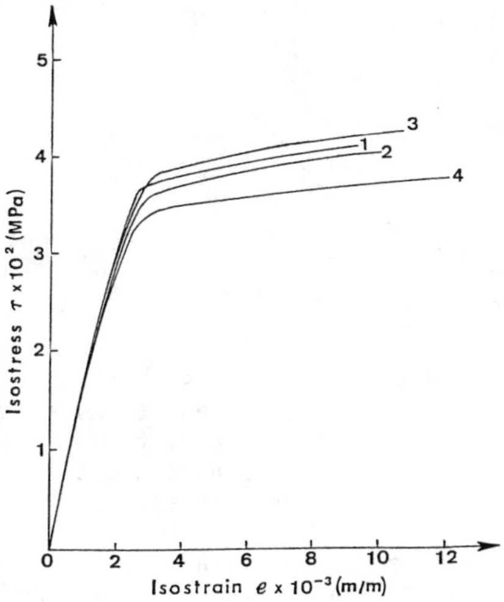

Figure 30. Isostress versus isostrain for ambient temperature of 25°C.

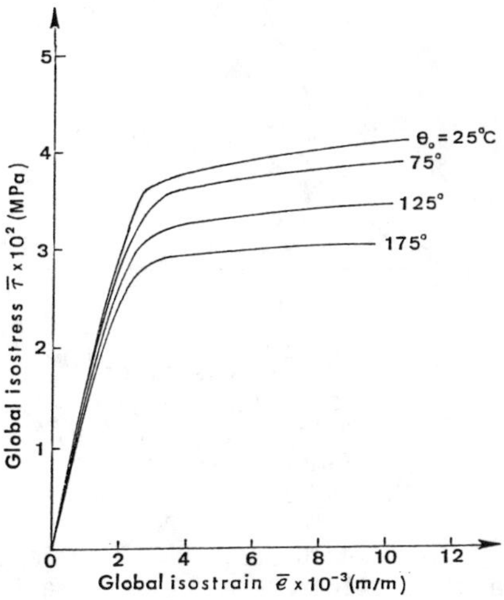

Figure 31. Global isostress-isostrain response for different ambient temperatures.

2, 3 and 4 in Figure 29, it is seen that the isostress and isostrain curves differed in the nonlinear range. Along the centroidal axis of cylinder where dilatational effect is more pronounced, the elements would experience a higher strain rate. This is indicated by curves 1 and 3 in contrast to curves 2 and 4 for elements lying on the cylinder surface which is more readily influenced by distortion. The highest local strain rate occurs at 3 being closest to the end of load application. Curves such as those in Figure 30 can also be found for

Θ_0 = 75°, 125° and 175°C. The average values of τ and e for all the solid elements in region R_2 are shown graphically in Figure 31. They are given as $\bar{\tau}$ and $\bar{e}$. The curve for Θ_0 = 25° in Figure 31 lies almost exactly in between those labelled as 1 and 2 in Figure 30 such that (τ,e) at 1 and 3 are above average and 2 and 4 are below average. The same interpretation holds for Θ_0 = 75°, 125° and 175°C. A gradual decrease in the strain rate is seen as the temperature surrounding the stretched solid is increased.

As a result of the difference between loading and unloading, energy is dissipated and the amount can be computed as $\mathcal{D}$ defined in equation (7) for each element. Figure 32 gives the time history of the dissipation energy density at 1, 2, 3 and 4 for Θ_0 = 25°C. All the curves rised slowly at first and then more

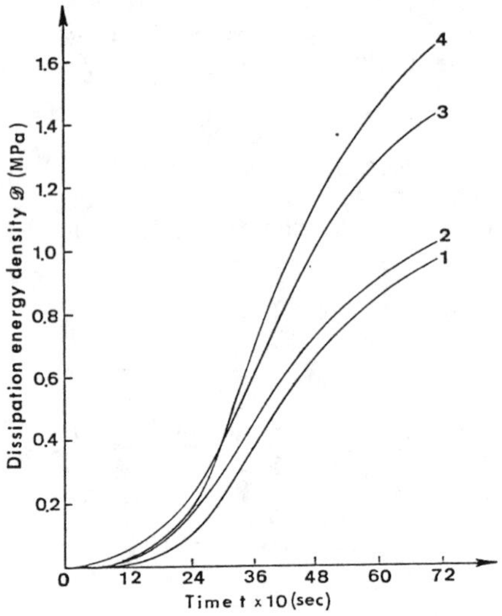

Figure 32. Variations of dissipation energy density with time for ambient temperature of 25°C.

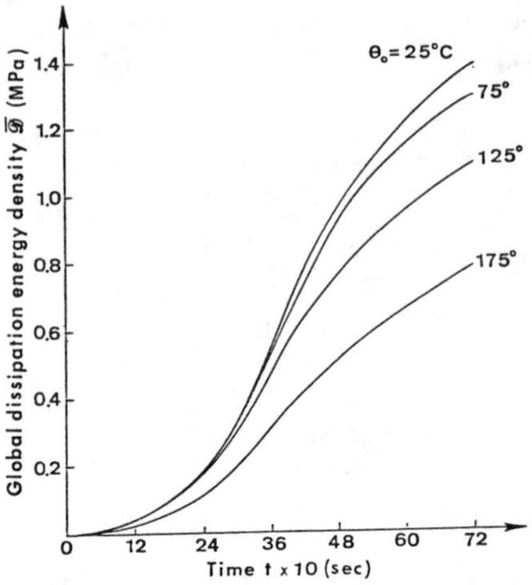

Figure 33. Global dissipation energy density as a function of time for different ambient temperatures.

rapidly for t > 240 sec. Much more energy is lost at 3 and 4 near the end than in the mid-section where 1 and 2 are located. The global average $\bar{\mathcal{D}}$ curve for Θ_0 = 25°C in Figure 33 lies just below curve 3 in Figure 32. This implies that the mid-section elements dissipate less energy on the average and store more available energy to do damage. Significant decrease in dissipation energy is also seen if the temperature of the surrounding environment is increased. Figure 33 shows that an appreciable decrease in $\bar{\mathcal{D}}$ occurs for all time at Θ_0 = 175°C.

The data in Figures 30 to 33 inclusive can be used with the help of equation (10) to obtain the time history of temperature distribution. The familiar cooling/heating effect is again observed in Figure 34 at locations 1, 2, 3 and 4 for Θ_0 = 25°C. Location 3 first returned to the ambient condition followed by location 2 while 1 and 4 lagged behind. Even for t > 700 sec when the stress and strain are well into the nonlinear regime as shown in Figure 30, the temperatures at 1 and 4 are still below 25°C. The classical theory of thermoplasticity would have predicted a rise in temperature above Θ_0 immediately after

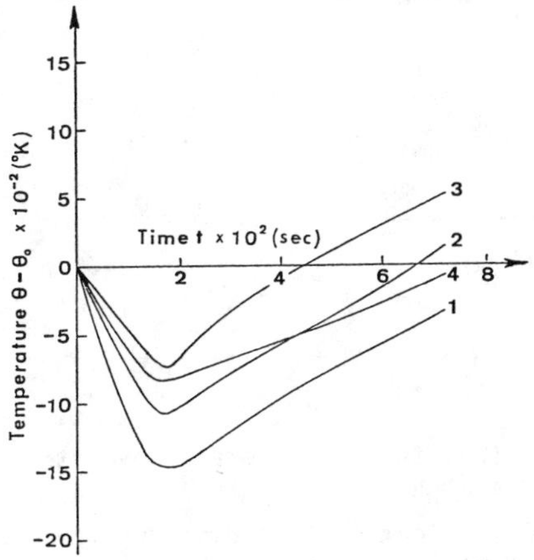

Figure 34. Change of temperature with time at locations 1, 2, 3 and 4 for Θ_0 = 25°C.

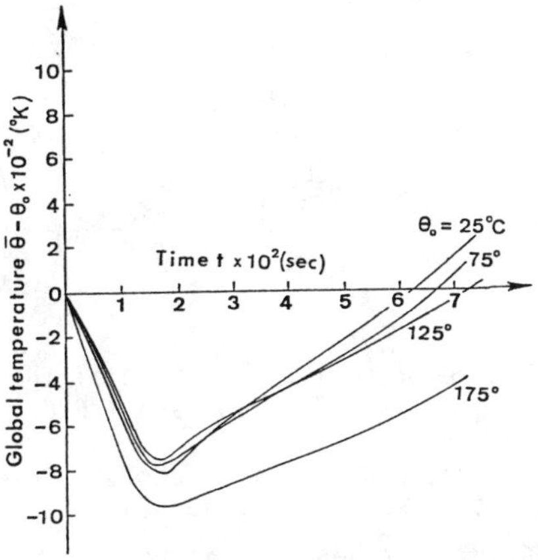

Figure 35. Global time history of temperature change in solid for different ambient temperatures.

the stress and strain behaved nonlinearly. Variations of the temperature $\bar{\Theta}$ averaged from all elements with time are given by the curves in Figure 35. Note that the ambient temperature Θ_0 is different for each curve. What should be emphasized is that $\bar{\Theta}$ in the solid not only varied continuously with time but also changed sign after more than ten (10) minutes of loading. Unless this time lag, which has been verified by experiments (Sih and Tzou 1987; Sih, Lieu and Chao 1987) is properly accounted for, thermal changes and mechanical deformation in the cylinder would not be synchronized. A gross misrepresentation of data would result if the curves in Figure 31 were used to determine the temperature dependency of constitutive coefficients in classical thermo-mechanical theories. The ambient temperatures Θ_0 = 25°, 75°, 125° and 175°C are not representative of the changing thermal conditions in the specimen. To reiterate, nonequilibrium thermal states in the uniaxial specimen change continuously with time for the entire load history; their reference to a single temperature would be meaningless and serve no useful purpose.

Displayed in Figure 36 are the values of $\bar{H}$ plotted against the absolute temperature Θ for different Θ_0. A different scale of Θ is used for each Θ_0 as given in Table 2 in which Θ^4 in °K is equal to the ambient temperature Θ_0 in °C. The

Table 2. Absolute Temperature Scale Θ for Different Θ_0.

Ambient Θ_0 (°C)	Temperature Scale Θ (°K)				
	Θ^1	Θ^2	Θ^3	Θ^4	Θ^5
25°	298.04	298.08	298.12	298.16	298.20
75°	348.04	308.08	348.12	348.16	348.20
125°	398.04	398.08	398.12	398.16	398.20
175°	448.04	448.08	448.12	448.16	448.20

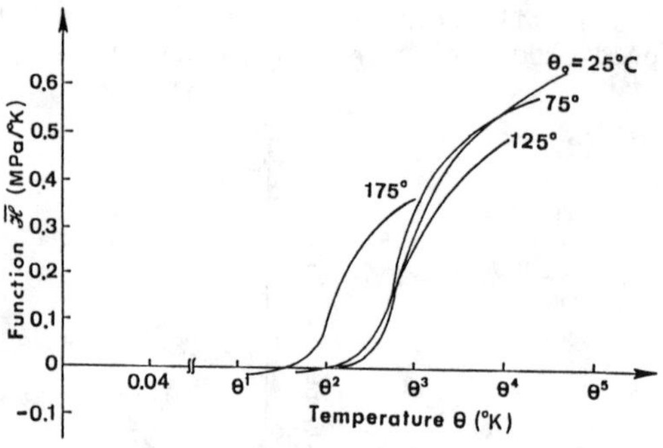

Figure 36. Global average of the function $\bar{H}$ over 720 sec for Θ_0 = 25°, 75°, 125° and 175°C.

data correspond to the *H*-function in equation (8); they are averaged over the entire time span of 720 sec for all the solid elements in R_2. As a general trend, $\bar{H}$ increased with increasing Θ with the implication that disorder tends to increase with increasing temperature. As mentioned earlier, the *H*-function can be positive, zero and negative.

Pseudo-Nonequilibrium Thermodynamics. Non-equilibrium thermal/mechanical uniaxial data are constantly used although their interpretation is not always clear. In connection with the pseudo-nonequilibrium thermodynamics based on the hypothesis of Clausius, the "reversible" and "irreversible" portions of a process can be divided such that the total entropy variation of a system can be written as

$$dS = d_r S + d_i S \tag{27}$$

The portion

$$d_r S = \frac{d_r Q}{T} \tag{28}$$

is the external (or reversible) entropy variation belonging to the heat $d_r Q$ exchanged in a reversible way with the surroundings and T is the absolute temperature. Produced inside the system by irreversible processes is the positive variation

$$d_i S \geq 0 \tag{29}$$

which is zero in the reversible or equilibrium case. The external supply of entropy $d_r S$ may be positive, zero or negative but the internal production of entropy $d_i S$ is positive definite. In the case of equality in equation (29), equation (27) gives the second law for reversible processes:

$$dS = \frac{d_r Q}{T} \tag{30}$$

The inequality in equation (29) leads to the inequality

$$dS > \frac{d_r Q}{T} \qquad (31)$$

for irreversible processes.

The complete form of equation (27) must be preserved for the thermodynamics of irreversible processes because the generation of entropy $d_i S$ during the evolution of a physical process may not be zero. Linear independence of $d_r S$ and $d_i S$ as expressed by equation (27), however, need not always hold for an irreversible process.

The nonequilibrium temperatures when averaged over the total time span of 720 sec may be as the pseudo-equilibrium data and used to compute $d_r S$ in equation (28). Substituting the expression

$$d_r Q = C_p dT \qquad (32)$$

into equation (28) and integrating, the result is

$$S_T = C_p \log(\frac{T}{T_o}) + S_o \qquad (33)$$

The equilibrium specific heat coefficient C_p at constant pressure for the 6061 aluminum is 870 J/kg°K. Its variation with temperature is not significant for solids in the temperature range considered. In equation (33), S_o is referred to T_o = 250°K. The results for $\Delta_r S$ defined as the difference of S_r and S_o at 250°K are understood to be global average of all elements as in the case of an equilibrium system.

The data in Table 3 is relevant to the Third Law of Thermodynamics that emphasize the unattainability of absolute zero. It is particularly relevant to the

Table 3. Entropy Change Based on Pseudo-Equilibrium Temperature Data.

Ambient Θ_o (°K)	Pseudo-Equilibrium Temperature $\bar{\Theta}-\Theta_o$ (°K)	Pseudo-Equilibrium Temperature $\bar{T}$ (°K)	Entropy Change $\Delta_r S$ (J/kg°K)
298.16	-0.03254	298.127	153.17
348.16	-0.03610	348.124	228.06
398.16	-0.03838	398.122	404.81
448.16	-0.07007	448.090	507.67

physics of very low temperatures. Shown by the dotted curves A and B in Figure 37 are the entropy decay for two hypothetical processes. Curve A approached T = 0 with $\Delta_r S \neq 0$ and curve B with $\Delta_r S = 0$. The solid curve represents the change in entropy states in an aluminum specimen stretched at $\dot{u} = 1.27 \times 10^{-6}$ m/sec. Extrapolation of this curve to intersect the T-axis yield the lowest temperature which is far from zero. The end temperature of any physical process cannot be absolute zero. While the curve in Figure 37 would follow a different path if $\dot{u}$ is altered, it should not intersect the ordinate.

Temperature Decay. Displayed in Figures 38 and 39 are the temperature Θ^* averaged over approximately 700 secs which varied as a function of r from the specimen wall. The radius of the cylinder is r = 0.625 cm which coincides with the

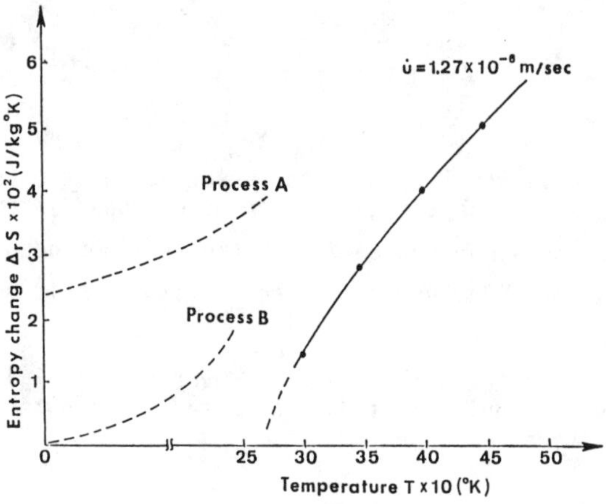

Figure 37. Entropy versus temperature for aluminum specimen stretched at $\dot{u} = 1.27 \times 10^{-6}$ m/sec.

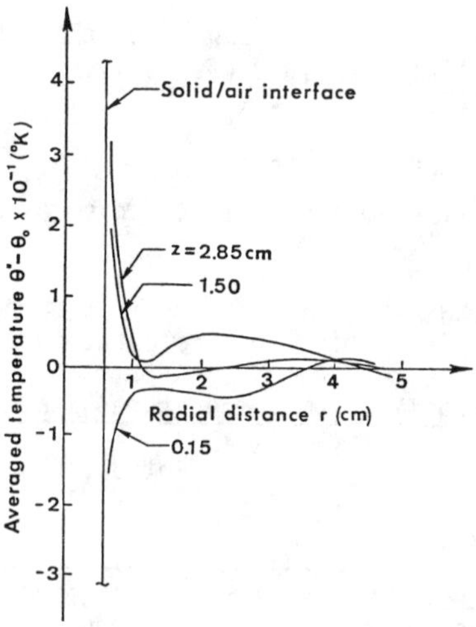

Figure 38. Average temperature decay in air from solid wall at $\Theta_0 = 25°C$.

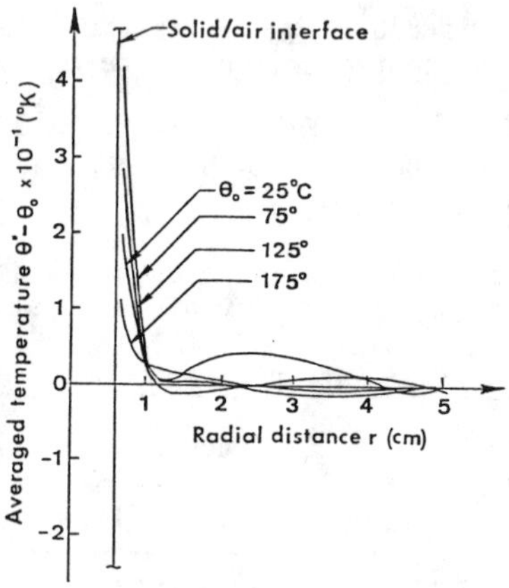

Figure 39. Average temperature decay in air at z = 1.5 cm for different ambient temperatures.

solid/air interface. Referring to Figure 11, z = 0 corresponds to the mid-plane of the cylinder while z = 3.0 cm is the half gage length that spans the distance between points 1 and 3 or 2 and 4. Note that the average temperature difference $\Theta^*-\Theta_0$ for $\Theta_0 = 25°C$ deviates significantly from the ambient condition near the wall; it oscillates with the distance r with decreasing amplitude. All the curves tend to the ambient conditions for r greater than one-half of the specimen gage length, i.e., r > 3.0 cm for all z. Figure 39 gives a plot of $\Theta^*-\Theta_0$

versus r for the four (4) different ambient temperatures of Θ_0 = 25°, 75°, 125° and 175°C. Again, all the curves attained their highest values near the solid wall and eventually decay to Θ_0 for sufficiently large r.

CONCLUDING REMARKS

A great deal of care should be exercised in distinguishing and/or defining physical processes with reference to their equilibrium or nonequilibrium states. The former infers to systems in which all parts possess the *same* physical properties, both locally and globally. Nonequilibrium arise on account of disturbance in the interior of a system or between a system and its surroundings. The local stress, strain, temperature and energy density vary from one part of a system to another at each time instance. The continuously changing nature of the constituents cannot be described by referring to the system as a whole unless reference is made to some averages, statistically or otherwise. Such averages on space and time are not always clearly stated in the correlation of experimental data, especially when their interpretation relies on theories that are valid only when the system is in equilibrium. Classical continuum mechanics theories are particularly vulnerable when applied to explain experimental data for *open* systems whose states are, by nature, nonequilibrium. Mathematically enforced or idealized thermodynamic processes for *isolated* systems have little or no practical application.

A system is always disturbed by its surroundings through contact, let it be a solid/gas or liquid/solid interface. The highly unstable state of affairs at these interphases are generally bypassed in mechanics for they are specified as boundary conditions or interfaces across which the displacements and/or stresses are assumed to be continuous. Strictly speaking, all systems consist at least two-phase. The compressed gas interacts with the solid wall of the container. Resistance of the solid exerted on the passing fluid creates turbulence or instability in the form of separation. Mechanical and thermal behavior of the uniaxial specimen can be highly influenced by the surrounding air temperature. These examples have been discussed in connection with predictions made from the isoenergy density theory that emphasizes the exchange of surface and volume energy. Even though this interchange of energy is well-known in crystal nucleation theory (Shewmon 1969), it is not commonly known that this effect has been excluded from the theories of continuum mechanics in the process of shrinking the element size where the rate change of volume with surface vanished in the limit. Reconciliation of thermodynamics and continuum mechanics (Truesdell and Toupin 1960) has failed to recognize the real cause for the decoupling of thermal and mechanical effects.

Description of nonequilibrium phenomena necessitates the simultaneous account of size, time and temperature such that the subdivision of solids, liquids and gases will be made with due consideration given to time and the corresponding local energy dissipation reflected via temperature changes. The time scale for observing microscopic events would be much smaller than that for macroscopic events but much larger than that at the atomic level. Temperature fluctuations in a microelement obviously will differ from those in a macroelement which are the averages of many microelements. Scaling of size/time/temperature from the atomic to the microscopic and the macroscopic is automatically done in the isoenergy density theory. This can be seen from equation (6) in which $\mathcal{V}$ and e can be determined for any elements size, say with linear dimension of 10^{-8} cm to 10^{-2} cm inclusive, the dissipation energy $\mathcal{D}$ and temperature Θ follow accordingly. Hence, events at the atomic, microscopic and macroscopic level can be interrelated with consistency free of ambiguity. The progressive damage characteristics of uniaxial tensile and compressive specimens (Sih and Chao 1989) were discussed in connection with the scaling of size/time/temperature from the macroscopic to the microscopic and the atomic. Phenomena such as turbulence, eddies and waves at the different scale level are also easily quantified. The highly transitory predictions may appear disorderly as in Brownian motion, but they are deterministic. To be kept in mind is that science deals with the bookkeeping of relative changes; it should not and does not delve into the unattainable "absolute". Ask not for turbulence in an absolute sense, but the onset of turbulence or change from one state to another is deterministic and is all that science is capable of providing.

REFERENCES

Cernocky, E. P. and E. Krempl. 1981. A theory of thermoviscoplasticity for uniaxial mechanical and thermal loading. Journal de Mechanique Appliquee 5: 293-321.

Cernocky, E. P. and E. Krempl. 1981. A coupled, isotropic theory of thermoviscoplasticity based on total strain and overstress and its prediction in monotonic torsional loading. Journal of Thermal Stresses 4: 69-82.

Chang, P. K. 1970. Separation of Flow: Pergamon Press.

Daily, J. W. and D. R. F. Herleman. 1966. Fluid Dynamics: 232-240, Addison Wesley.

DeGroot, B. R. and P. Mazur. 1962. Nonequilibrium Thermodynamics: Amsterdam, North Holland Publishing Company.

Goldstein, S. 1965. Modern Developments in Fluid Dynamics: Dover.

Lieu, F. L. 1986. Experiments on cooling/heating of aluminum tube under torsion, Institute of Fracture and Solid Mechanics, Lehigh University.

Nikuradse, J. 1933. Stromungsgesetze in Ranben Robren, VDI-Forschungsheft 361; Beilage zu Forschung auf dem Gebiete der Ingenieur-wesen, Ausgabe B, Band IV.

Onsager, L. 1931. Reciprocal relations in irreversible processes I, Phys. Rev. 37: 405-426 and Reciprocal relations in irreversible processes II, 38: 2265-2279.

Panar, M. and M. M. Denn. 1986. High-performance polymers. Mechanical Engineering: 108: 72-95.

Pao, R. H. F. 1967. Fluid Dynamics: Charles E. Merrill Books, Inc. 326-328.

Prandtl, L. and O. G. Tiejens. 1934. Applied Hydro- and Aero-Mechanics: Dover.

Prigogine, L. 1955. Introduction to Thermodynamics of Irreversible Processes: Springfield.

Schlichting, H. 1968. Boundary-Layer Theory: McGraw-Hill.

Shewmon, P. G. 1969. Transformations in Metals: McGraw-Hill.

Sih, G. C. 1972-1983. Introductory chapters of Mechanics of Fracture I to VII, ed. G. C. Sih. The Hague: Martinus Nijhoff Publishers.

Sih, G. C. 1974. Some basic problems in fracture mechanics and new concepts. Journal of Engineering Fracture Mechanics 5: 365-377.

Sih, G. C. 1980. Phenomena of instability: fracture mechanics and flow separation, Naval Research Reviews 32: 30-42.

Sih, G. C. 1984. Fracture mechanics of engineering structural components. In Fracture Mechanics Methodology, eds. G. C. Sih and L. O. Faria, 35-101. The Netherlands: Martinus Nijhoff Publishers.

Sih, G. C. 1985. Mechanics and physics of energy density and rate of change of volume with surface. Journal of Theoretical and Applied Fracture Mechanics 4: 157-173.

Sih, G. C. 1987. Thermal/mechanical interaction associated with the micromechanisms of material behavior, Institute of Fracture and Solid Mechanics Monograph, Library of Congress No. 87-080715.

Sih, G. C. 1988. Thermomechanics of solids: nonequilibrium and irreversibility. Journal of Theoretical and Applied Fracture Mechanics 9: 175-198.

Sih, G. C. and C. K. Chao. 1989. Scaling of size/time/temperature - Part 1: Progressive damage in uniaxial tensile specimen and Part 2: Progressive damage in uniaxial compressive specimen. Journal of Theoretical and Applied Fracture Mechanics 12: 93-119.

Sih, G. C. and D. M. Chou. 1989. Nonequilibrium thermal/mechanical response of 6061 aluminum alloy at elevated temperature. Journal of Theoretical and Applied Fracture Mechanics 12: 19-31.

Sih, G. C., F. L. Lieu and C. K. Chao. 1987. Thermal/mechanical damage of 6061-T6 aluminum tensile specimen. Journal of Theoretical and Applied Fracture Mechanics 7: 67-78.

Sih, G. C. and D. Y. Tzou. 1987. Irreversibility and damage of SAFC-40R steel specimen in uniaxial tension. Journal of Theoretical and Applied Fracture Mechanics 7: 23-30.

Sih, G. C., D. Y. Tzou and J. G. Michopoulos. 1987. Secondary temperature fluctuation in cracked 1020 steel specimen loaded monotonically. Journal of Theoretical and Applied Fracture Mechanics 7: 79-87.

Truesdell, C. and R. A. Toupin. 1960. The Classical Field Theory, Handbuch der Physik, 3rd ed., Berlin/Gottingen/Heidelberg: Springer.

A Lagrangian Formulation of Chemical Reaction Dynamics Far from Equilibrium

J. S. Shiner
Physiologisches Institut, Universität Bern
Bühlplatz 5, CH-3012 Bern, Switzerland

ABSTRACT
A Lagrangian formulation of the dynamics of systems of chemical reactions on the macroscopic, phenomenological level of mass action kinetics is developed for transients and stationary states arbitrarily far from thermodynamic equilibrium. The Lagrangian approach represents a unification of the dynamics of systems of chemical reactions with those of most other sorts of processes, for which Lagrangian formulations are known, and simplifies the treatment of systems where chemical reactions are coupled to other sorts of processes. For chemical reactions the Lagrangian is the appropriate (equilibrium) thermodynamic potential, e.g. the Gibbs energy under conditions of constant temperature and pressure. The key to the formulation is the definition of appropriate "resistances" for chemical reactions in a dissipation function of the Rayleigh form in terms of the state of the chemical system, so that mass action kinetics can be recovered. The Lagrangian formulation points out in a natural way an algebraic symmetry obeyed by systems of chemical reactions at stationary states arbitrarily far from thermodynamic equilibrium, which is a more general case of the Onsager reciprocity shown near equilibrium. When inertial-inductive effects are absent, as in the case for chemical reactions on a macroscopic, phenomenological level, the Lagrangian approach can be recast as a variational formulation which reduces to the principle of minimum entropy production near equilibrium. It also leads naturally to a network representation of systems of chemical reactions, which allows one to treat them using the established methods of network analysis.

INTRODUCTION

Chemistry has historically presented thermodynamics with problems. Not until Josiah Willard Gibbs (1931) introduced the abstract concept of the chemical potential in the last century could chemical aspects be fully incorporated into equilibrium thermodynamics. Several decades then passed before DeDonder (1928) recognized that the affinity of a chemical reaction is its net thermodynamic driving force. Onsager (1931) illustrated his derivation of the general class of reciprocal relations for irreversible processes which bear his name with an analogy to a system of three chemical reactions, but Onsager reciprocity, the symmetry he found for stationary states near equilibrium in what are now known as the phenomenological equations, is valid for systems of chemical reactions in general only near equilibrium, whereas for many other sorts of processes the symmetry holds without the near equilibrium restriction, at least for the ideal case (Prigogine 1967, Glansdorff and Prigogine 1971, Keizer 1977, Nicolis and Prigogine 1977). In the network theory of thermodynamics arbitrarily far from equilibrium chemical reactions have presented such problems that formulations have been developed using the so-called "kinetic forces" (Peusner 1982, Shiner 1984), which at least allow one to recover the dynamics of the reactions, although the actual thermodynamic information is lost. The difficulties with chemical reactions are also seen in the differences between the dynamics of chemical reactions and the dynamics of other sorts of processes. The dynamics of many types of processes, mechanical or electrical ones for example, are governed by a common formalism (Goldstein 1950, MacFarlane 1970); indeed their equations of motion take an analogous form, at least in the ideal case. On the other hand, the equations of motion for chemical reactions seem to take a different form, and the formalism common to other types of dynamic processes is apparently not valid for chemical reactions. For ideal mechanical systems or electrical networks the flow through a dissipative element is proportional to the force across the element, but for chemical reactions the dissipative flows (reaction rates) are proportional to the conjugate thermodynamic forces (reaction affinities) only in the near equilibrium regime (Prigogine 1967, Nicolis and Prigogine 1977). For transport processes the range of validity of the local equilibrium assumption of nonequilibrium thermodynamics is coincident with the range of linear behavior, but for chemical reactions the validity of the local equilibrium assumption is independent of whether the reactions operate in the linear near equilibrium regime or not (Prigogine 1967, Nicolis and Prigogine 1977).

It will be shown how the dynamics of systems of chemical reactions on the macroscopic, phenomenological level of chemical kinetics may be reformulated so that the equations of motion are given by the formalism common to other sorts of dissipative processes (Shiner, 1990). The formalism is expressed in terms of Lagrangian dynamics extended to allow for dissipative effects via a Rayleigh dissipation function (Goldstein 1950). To be able to define the Lagrangian and the dissipation function it is necessary to assume the local equilibrium condition. Note, however, that this condition is independent of whether the reactions operate in the near equilibrium regime or far from equilibrium, as already mentioned. The key to the Lagrangian formulation of the dynamics of systems of chemical reactions will be to allow the resistances in the dissipation function to depend on the state of the system, as has been done by Grabert et al. (1983) and Sieniutycz (1987a) and Shiner (1987). It is fair to ask why such a formulation is desired; after all, phenomenological chemical kinetics are well established empirically, and, as Marion (1965) has noted for classical mechanics, a Lagrangian formulation of chemical kinetics must be an *a posteriori* formulation, since it must simply yield the equations of motion in the well known form of chemical kinetics.

The first reason for a Lagrangian formulation of chemical reaction kinetics is that such a formulation represents a unification of the dynamics of chemical processes with those of many other sorts of processes. The Lagrangian approach has long been known to be valid for mechanical systems, electrical networks and fluid mechanics, to name a few examples, and it has been clear that these processes are governed by a common underlying formalism. By reformulating chemical kinetics in terms of a Lagrangian it is immediately seen that the dynamics of chemical reactions can be written in the same form as those of other sorts of processes and are also governed by the same underlying formalism. Of course, for chemical kinetics on the macroscopic, phenomenological level there are no inertial-inductive effects and the classical Lagrangian approach for conservative systems must be extended to allow for dissipative effects with the help of a dissipation function. However, these conditions may also arise for the other sorts of processes described by a Lagrangian approach (Goldstein, 1950). The reformulation should not only achieve a unification of the dynamics of chemical reactions with those of other processes in some simple sense; it should also simplify the treatment of systems where chemical reactions are coupled to other sorts of dissipative processes, mechanochemical and electrochemical systems for example, even if these processes do show inertial-inductive effects, since only one formalism will be needed for the dynamics of all processes. The treatment of such sorts of coupled systems has been complicated in the past in that one has had to first treat chemical aspects in terms of chemical reaction kinetics, then treat the other sorts of processes in terms of another formalism appropriate for these other sorts and then integrate the two different formalisms. This is not only cumbersome; it has also meant that chemical aspects on the one hand and mechanical or electrical aspects on the other hand have often not been treated on an equal footing. Muscle contraction is a good example. The only available physically correct formulation of this phenomenon which considers both chemical and mechanical aspects was developed more than a decade ago (Hill 1974, 1975) and places emphasis heavily on the chemistry. One first considers all possible chemical states of muscle and writes down the chemical evolution equations for the possible chemical reactions between these states, taking into account the effects that the mechanics must have on the chemical rate constants according to equilibrium thermodynamics. Mechanical motion appears only in the time rates of change of the various chemical states of muscle, which are written in the form of continuity equations involving the velocity of contraction of the muscle itself; nowhere does one see an equation of motion for the mechanics. While this approach is correct as far as it goes, it is not completely satisfying. It would be desirable to formulate the problem in such a way that one had equations of motion for both the chemical aspects (chemical kinetics) and for the mechanical aspects and that the coupling between the two sorts of processes occurred naturally. With the Lagrangian approach to chemical reaction dynamics recently developed, it should be possible to do exactly this. Furthermore, the treatment of muscle contraction should be simplified by using the Lagrangian approach, since the use of generalized coordinates would obviate the need to distinguish formally between the chemistry and the mechanics.

The second reason for the Lagrangian approach to chemical reaction dynamics is that it may also be presented as a variational formulation which reduces to the Onsager-Rayleigh principle of least dissipation of energy (Onsager 1931) in the near equilibrium regime, or equivalently to the theorem of minimum entropy production (Prigogine 1967, Nicolis and Prigogine 1977). It thus leads to an extension of these results from the linear near equilibrium regime to states arbitrarily far from thermodynamic equilibrium. The extension is also valid for other processes without inertial-inductive effects and therefore for systems where chemical reactions are coupled to other sorts of processes as long as these other processes do not show any inertial-inductive effects; if they do, then the extension is no longer valid. Several variational principles, including Onsager's, are already known for the dynamics of dissipative systems (Keizer 1977). These are, however, for the most part restricted to the near equilibrium regime (Onsager 1931, Machlup and Onsager, 1953, Onsager and Machlup 1953) or to the neighborhood of a stable stationary state (not too near a critical point) far from equilibrium (Keizer 1977, 1979a, 1979b). On the other hand, the formulation presented here is valid arbitrarily far from equilibrium, for both stationary states and transient conditions, and regardless of the stability of the stationary state. Previously known variational principles are also based for the most part on studies of the behaviors of fluctuations (Onsager 1931, Machlup and Onsager, 1953, Onsager and Machlup 1953, Prigogine 1967, Glansdorff and Prigogine 1971, Nicolis and Prigogine 1977); in contrast, the approach here is macroscopic since it is based on the phenomenology of chemical kinetics. An exception to these other variational results is the work of Sieniutycz (1987a), who has used the field approach to develop a Lagrangian for chemical reaction dynamics and has obtained some of the results presented here.

The Lagrangian formulation has also yielded a new symmetry for systems of chemical reactions far from equilibrium (Shiner 1987). Onsager reciprocity, or symmetry, for ideal processes is expressed by the phenomenological equations of nonequilibrium thermodynamics, which relate the thermodynamic forces X_i to their conjugate flows J_i at the stationary state (Onsager 1931, Prigogine 1967, Glansdorff and Prigogine 1971, Nicolis and Prigogine 1977):

$$X_i = \sum_j \mathcal{R}_{ij} J_j \,. \tag{1.1}$$

These relations are symmetric in the generalized or phenomenological resistances $\mathcal{R}_{ij}$:

$$\mathcal{R}_{ij} = \mathcal{R}_{ji} \,, \tag{1.2}$$

or, equivalently, since the $\mathcal{R}_{ij}$ are constants:

$$(\partial X_i/\partial J_j) = (\partial X_j/\partial J_i) \,. \tag{1.3}$$

For chemical reactions, however, previous approaches had yielded symmetry in the phenomenological equations in general only for stationary states close to equilibrium. The Lagrangian formulation of the dynamics of systems of chemical reactions leads naturally to stationary state phenomenological equations of the form of equation (1.1), and these equations display symmetry in the matrix of phenomenological resistances, as in equation (1.2) (Shiner 1987). In general this symmetry is only an algebraic one, and differential symmetry of Onsager reciprocity is not found except near equilibrium and for certain other special cases. To distinguish this property from Onsager reciprocity, it was called algebraic symmetry.

The final reason for applying the Lagrangian formalism to systems of chemical reactions is that it leads naturally to a network representation of systems of chemical reactions. Network thermodynamics was the first form of nonequilibrium thermodynamics able to treat chemical reactions far from equilibrium (Peusner 1970, Oster, Perelson and Katchalsky 1973, Wyatt 1978). Indeed, the Lagrangian formalism for chemical reactions could have been developed by first noting the form which the formalism takes for electrical networks and then giving the various elements of electrical networks chemical interpretations. The symmetry property in equation (1.2) is a necessary condition for a network representation. The network representation of systems of chemical reactions is valuable in that it makes it possible to apply the many well known results for networks and methods for network analysis to systems of chemical reactions. It should be distinguished between the the Lagrangian approach to the dynamics of systems of chemical reactions under discussion here which uses the true thermodynamic forces for the reactions, the affinities, and its network representation on the one hand, and, an earlier approach (Shiner 1984), which uses the so-called "kinetic forces" for the reactions, and its corresponding network "thermodynamics" (Peusner 1982, Shiner 1984, Peusner et al. 1985) on the other hand. The Lagrangian formulation here not only yields correct results dynamically but also energetically and thermodynamically; the approach using the kinetic forces is based only on dynamic analogies, but its "Lagrangian" does not have dimensions of energy, nor its "dissipation function" those of power. Thus, it does not recast chemical kinetics into a form in which the equations of motion are given by the formalism common to other sorts of dissipative processes. Furthermore, it is restricted to (pseudo) first-order reaction-diffusion systems.

In this paper only systems of chemical reactions under conditions of constant temperature and pressure will be considered. Volume changes and heat flow will also be neglected. This paper is organized as follows. In the next section the systems of chemical reactions under consideration, their thermodynamic representation and the resistance for a chemical reaction will be defined. The Lagrangian formulation for systems of chemical reactions will then be presented. In the following section the alternate variational formulation and the extension of the principle of minimum entropy production valid at stationary states arbitrarily far from thermodynamic equilibrium are derived. It is then proven that algebraic symmetry holds for all stationary states, arbitrarily far from thermodynamic equilibrium and regardless of the stability of the stationary state. Examples of the Lagrangian formulation and the network representation for several systems of chemical reactions are discussed in the following section. Finally the results and directions for their future extension are summarized.

DEFINITIONS

We consider a chemical system of N species at constant temperature and pressure, and disregard any volume changes and heat flow which might occur. n_i is defined as the number of moles of species i, and μ_i, its chemical potential. To insure that the chemical potentials μ_i are meaningful it is necessary to assume the validity of the local equilibrium condition; note, however, that this assumption is in no way incompatible with the system's being arbitrarily far from equilibrium. According to the Brussels school (Prigogine 1967, Nicolis and Prigogine 1977) the assumption of local equilibrium is valid under far from equilibrium[1] conditions if the chemical reactions occur on a time scale much slower than the time scale of the elastic collisions within the reaction system. This is the case under the conditions under which chemical reactions are usually studied.

The N species in the system undergo $\mathcal{N}$ reactions j:

$$\sum_{i=1}^{N} \nu_{+ij}\,[i] \underset{k_{-j}}{\overset{k_j}{\longleftrightarrow}} \sum_{i=1}^{N} \nu_{-ij}\,[i]\,. \tag{2.1}$$

[1]The definition of "far from equilibrium" that we use here is simply the condition that at least some of the reactions are not operating under the near equilibrium conditions, where all thermodynamics forces X_i are much less than the thermal energy RT, where R is the gas constant.

ν_{+ij} and ν_{-ij} are, respectively, the forward and backward stoichiometric coefficients for species i in reaction j; k_j and k_{-j} are the corresponding rate constants. The approach we are taking here is valid for any form of rate law for reactions (1) as long as it is consistent with the requirements of thermodynamic equilibrium (Corso 1983). [See equation (2.4d) below]. However, for simplicity we will restrict this paper to the case of the law of mass action , where the rate $\dot{\xi}_j$ of the jth reaction is given by

$$\dot{\xi}_j = k_j \prod_{i=1}^{N} a_i^{\nu_{+ij}} - k_{-j} \prod_{i=1}^{N} a_i^{\nu_{-ij}} . \tag{2.2}$$

a_i is the relative activity of species i,

$$a_i = \exp[(\mu_i - \mu_i^o)/RT] , \tag{2.3}$$

where μ_i^o is its chemical potential in some standard state. T is temperature and R, the gas constant.[2]

The thermodynamic force driving reaction j is its affinity (DeDonder 1928), defined as

$$A_j \equiv \sum_{i=1}^{N} (\nu_{+ij} - \nu_{-ij})\mu_i \tag{2.4a}$$

$$= \sum_{i=1}^{N} (\nu_{+ij} - \nu_{-ij})(\mu_i^o + RT \ln a_i) \tag{2.4b}$$

$$= RT\ln \mid (k_j \prod_{i=1}^{N} a_i^{\nu_{+ij}})/(k_{-j} \prod_{i=1}^{N} a_i^{\nu_{-ij}}) \mid . \tag{2.4c}$$

The last equality results from the identity

$$\sum_{i=1}^{N} (\nu_{+ij} - \nu_{-ij})\mu_i^o = RT\ln \mid k_j/k_{-j} \mid , \tag{2.4d}$$

2The law of mass action is usually written in terms of concentrations instead of activities. Activities are used here since we wish to derive a thermodynamic formulation. To a certain extent, it is a choice between incorporating any effects of nonideality into the activities or into the rate constants. If the reader objects to using activities, he may consider only the ideal case, where activities can be replaced with concentrations.

which follows since both $\dot{\xi}_j$ and A_j must vanish at thermodynamic equilibrium.

Chemical reactions such as those in equation (2.1) are dissipative processes, like the flow of current through an electrical resistance or mechanical motion against friction. For a dissipative process in general one defines a resistance R_j through the relation $X_j = R_j J_j$, where X_j is a generalized force (electrical potential or mechanical force) and J_j is the conjugate flow (electrical current or mechanical motion). For the chemical reactions (2.1) this becomes $A_j = R_j \dot{\xi}_j$. Therefore, using equations (2.2) and (2.4c) we take

$$R_j \equiv \frac{RT\ln \left| \left(k_j \prod_{i=1}^{N} a_i^{\nu_{+ij}}\right) / \left(k_{-j} \prod_{i=1}^{N} a_i^{\nu_{-ij}}\right) \right|}{k_j \prod_{i=1}^{N} a_i^{\nu_{+ij}} - k_{-j} \prod_{i=1}^{N} a_i^{\nu_{-ij}}} \tag{2.5}$$

as the definition of the resistance of the jth reaction. Grabert et al. (1983) have used a similar definition in their study of fluctuations in chemical reactions in ideal mixtures, as have Wyatt in formulating the network thermodynamics of chemical reactions and Sieniutycz (1987a) in his work on a field approach to a Lagrangian formulation of chemical reactions. This definition is actually the key to the Lagrangian formulation; the other definitions in this section follow from the standard thermodynamics of chemical reactions.

The time rate of change of the number of moles of species i due to reaction j is

$$\dot{n}_{ij} = (\nu_{-ij} - \nu_{+ij})\, \dot{\xi}_j \tag{2.6a}$$

so that the total time rate of change of n_i due to all $\mathcal{N}$ reactions is

$$\dot{n}_{iReac} = \sum_{j=1}^{\mathcal{N}} \dot{n}_{ij} = \sum_{j=1}^{\mathcal{N}} (\nu_{-ij} - \nu_{+ij})\, \dot{\xi}_j \,. \tag{2.6b}$$

The system of chemical reactions described up to this point is closed with respect to each chemical species; the description therefore would be valid only for systems at or relaxing to thermodynamic equilibrium. We would, however, like to include nonequilibrium stationary states and relaxation to them in our treatment. To accomplish this we must open the system of chemical reactions to some of the species involved and include sources for these species to maintain the system at nonequilibrium stationary states or to drive the system toward such states. To this end we allow for N' (< N) ideal external sources of chemical potential μ_i^{ex} ; the conjugate number of moles is n_i^{ex}. The source i will then hold the chemical potential of species i at the value μ_i^{ex} by varying the flow through the source. A system with such sources which would be in principle experimentally realizable could be constructed by placing the system of chemical reactions in a membrane infinitely permeable to those species whose values are to be held at the values μ_i^{ex} but completely impermeable to the other species and maintaining the volume of the reservoir surrounding the membrane much larger than the volume within the membrane. Then the permeable species would instantaneously diffuse in and out of the membrane to hold their chemical potentials at the corresponding values in the reservoir surrounding the membrane, without altering the μ_i^{ex} in the reservoir. This situation is entirely analogous to that of ideal voltage sources in electrical networks. Other examples of plausible methods of realizing such sources can be found in Ross et al. (1988).

The extensive variables defining the system of chemical reactions -- the n_i, the n_i^{ex}, and the advancements of reaction $\xi_j = \int \dot{\xi}_j \, dt$ -- are not independent, of course, but connected by conservation of mass relations, one for each internal species. For those internal species for which there are no external sources (i > N'), the total change in the number of moles of species i is simply that due to the chemical reactions, and from equation (2.6b) we obtain the conservation of mass relations

$$f_i = \dot{n}_i - \sum_{j=1}^{\mathcal{N}} (\nu_{-ij} - \nu_{+ij})\, \dot{\xi}_j = 0, \; N' + 1 \le i \le N. \tag{2.7a}$$

For those species for which there are external sources an additional term appears in the mass balance equation due to flow through the source: $\dot{n}_i = \dot{n}_{iReac} + \dot{n}_i^{ex}$. This equation expresses the control of the number of moles of species i by the source i so that the chemical potential of species i remains at the value of the chemical potential of the source; i.e., just the right amount of flow $\dot{n}_i^{ex}$ occurs through the source i to counterbalance any change in μ_i due to the chemical reactions. (See also the discussion above of ways of constructing the sources.) The conservation of mass relations for the controlled species are thus

$$f_i = \dot{n}_i - \sum_{j=1}^{\mathcal{N}} (\nu_{-ij} - \nu_{+ij})\, \dot{\xi}_j - \dot{n}_i^{ex} = 0, \; 1 \le i \le N'. \tag{2.7b}$$

THE LAGRANGIAN FORMULATION

The formulation of the dynamics of dissipative systems which we use here is the classical Lagrangian formulation for conservative systems extended to allow for dissipative effects by a Rayleigh dissipation function (Goldstein 1950). The equations of motion are given by

$$\frac{d}{dt}\frac{\partial \mathcal{L}}{\partial \dot{q}} - \frac{\partial \mathcal{L}}{\partial q} + \frac{\partial \mathcal{F}}{\partial \dot{q}} + \sum_i \lambda_i \frac{\partial f_i}{\partial \dot{q}} = 0 \, . \tag{3.1}$$

$\mathcal{L}$ is the Lagrangian, defined as the difference in the kinetic (E) and potential (U) energies ($\mathcal{L} \equiv E - U$); $\mathcal{F}$ is the Rayleigh dissipation function; q is a generalized coordinate, i.e. a variable used to describe the system; $\dot{q} \equiv dq/dt$ is a generalized velocity; the f_i are equations of constraint among the $\dot{q}$'s (or q's), all written so that

$f_i \equiv 0$; and the λ_i are the corresponding undetermined multipliers; these arise when all the generalized coordinates q are not independent of another.[3]

Since we are concerned in this paper with a macroscopic description of the dynamics of systems of chemical reactions on the phenomenological level of deterministic chemical kinetics, where there are no inertial-inductive effects, the kinetic energy term vanishes from the Lagrangian. We also consider only the conditions of constant temperature and pressure, where chemical reactions are usually studied; the Gibbs (free) energy G rather than the potential energy is then the appropriate thermodynamic potential[4], and the Lagrangian becomes

$$\mathcal{L} = - G. \tag{3.2a}$$

There are two contributions to G here. The first is due to the N "internal" species undergoing the chemical reactions:

$$G_{IN} = \sum_{i=1}^{N} n_i \mu_i ; \tag{3.2b}$$

the second is due to the N' external sources:

$$G_{EX} = \sum_{i=1}^{N'} n_i^{ex} \mu_i^{ex} . \tag{3.2c}$$

The total Gibbs energy is $G = G_{IN} - G_{EX}$. Note that G_{EX} makes a negative contribution since G is the Gibbs energy of the system whereas G_{EX} refers to the external sources.[5] The Lagrangian for the system can now be expressed as

$$\mathcal{L} = - \sum_{i=1}^{N} n_i \mu_i + \sum_{i=1}^{N'} n_i^{ex} \mu_i^{ex} . \tag{3.2d}$$

---

[3]Normally one writes the equations of constraint in terms of the q's and not the $\dot{q}$'s; the partial derivatives of the equations of constraint in equation (3.1) are then carried out with respect to the q's. It is simpler to use the $\dot{q}$'s here, however, since we then do not have to introduce initial conditions. If one prefers to use the q's, he would simply integrate the equations of constraint we use here over time.

[4]For any other choice of independent variables the appropriate potential function can be found by Legendre transformations; see (Callen 1960).

[5]The situation is exactly the same as for electrical networks. There ideal voltage sources would also make a negative contribution to the energy of the network. This follows since by convention the current flow through the voltage sources is taken to flow out of the source, whereas for internal network components, current flows into the component. We adopt the same convention here: $\dot{n}_i$ flows into species i, but $\dot{n}_i^{ex}$ flows out of source i. Otherwise $\dot{n}_i^{ex}$ would have the opposite sign in the mass conservation relations in equation (2.7b).

We write the Rayleigh dissipation function here as

$$\mathcal{F} = \frac{1}{2} \sum_{j=1}^{\mathcal{N}} R_j \dot{\xi}_j^2 \qquad (3.3)$$

where the resistances R_j for the reactions are given by equation (2.5). Equation (3.3) has the usual form $1/2\ \Sigma\ R_j \dot{q}_j^2$, where $\dot{q}_j$ is a dissipative flow and R_j the conjugate resistance.

The generalized coordinates defining the system of chemical reactions are the n_i, the n_i^{ex}, and the advancements of reaction ξ_j, which are connected by the conservation of mass relations in equations (2.7) in the previous section, one for each internal species. In the Lagrangian terminology these relations are the equations of constraint, which we note here again.

$$f_i = \dot{n}_i - \sum_{j=1}^{\mathcal{N}} (\nu_{-ij} - \nu_{+ij}) \dot{\xi}_j - \dot{n}_i^{ex} = 0,\ 1 \le i \le N'. \qquad (3.4a)$$

$$f_i = \dot{n}_i - \sum_{j=1}^{\mathcal{N}} (\nu_{-ij} - \nu_{+ij}) \dot{\xi}_j = 0,\ N' + 1 \le i \le N. \qquad (3.4b)$$

The Lagrangian (3.2d), the dissipation function (3.3) along with the definitions of the R_j (2.5), and the equations of constraint (3.4a,b) complete the Lagrangian description of systems of chemical reactions. Carrying out the operations indicated in equation (3.1), we find

for $q = n_i$: $\qquad \mu_i + \lambda_i = 0,\ 1 \le i \le N; \qquad (3.5a)$

for $q = n_i^{ex}$: $\qquad -\mu_i^{ex} - \lambda_i = 0,\ 1 \le i \le N'; \qquad (3.5b)$

and for $q = \xi_j$: $\qquad R_j \dot{\xi}_j - \sum_{i=1}^{N} \lambda_i (\nu_{-ij} - \nu_{+ij}) = 0,\ 1 \le j \le \mathcal{N}. \qquad (3.5c)$

Equations (3.5a) and (3.5b) imply that

$$\mu_i = \mu_i^{ex},\ 1 \le i \le N'; \qquad (3.6)$$

i.e., the chemical potentials of the internal species controlled by external sources are held at the values of the chemical potentials of the corresponding sources. Upon substituting from (3.5a) into (3.5c) for the λ_i, we find

$$R_j \dot{\xi}_j = \sum_{i=1}^{N} (\nu_{+ij} - \nu_{-ij}) \mu_i,\ 1 \le j \le \mathcal{N}. \qquad (3.7)$$

The right hand side of equation (3.7) is just the driving force for the jth chemical reaction, its affinity [see equation (2.4a)]. Recalling the definition of R_j in equation (2.5), we obtain the usual form of the rate of a chemical reaction

$$\dot{\xi}_j = k_j \prod_{i=1}^{N} a_i^{\nu_{+ij}} - k_{-j} \prod_{i=1}^{N} a_i^{\nu_{-ij}} , \tag{3.8}$$

where we have used the identity

$$\sum_{i=1}^{N} (\nu_{+ij} - \nu_{-ij}) \mu_i^{\circ} = RT\ln | k_j/k_{-j} | \tag{2.4d}$$

again.

We see that the Lagrangian formulation does indeed lead to the correct equations of motion for a system of chemical reactions. The chemical potentials of the internal species controlled by external sources are held at the values of the corresponding sources [equation (3.6)], and equation (3.7) is the law of mass action for reaction j. This result is guaranteed because we could define the chemical resistances R_j so that $R_j \dot{\xi}_j = A_j$ and so that the R_j are independent of the rates of reaction in the system. Otherwise, the term $\partial \mathcal{F}/\partial \dot{q}$ in equation (3.1) would not yield $R_j \dot{\xi}_j$ for $q = \xi_j$ and the equations of motion obtained from equation (3.1) would not be correct for the law of mass action.

THE VARIATIONAL FORMULATION

The Lagrangian formulation of the previous section may also be presented as a variational problem in a more transparent manner. This also allows a comparison with previous work on variational principles for nonequilibrium dissipative systems (Onsager 1931, Keizer 1977, Keizer 1979a). We first note that with the help of the equations of constraint (3.4) the time variation of the Gibbs energy (3.2d) may be written

$$\begin{aligned} \dot{G} &= \sum_{i=1}^{N} \dot{n}_i \mu_i - \sum_{i=1}^{N'} \dot{n}_i^{ex} \mu_i^{ex} \\ &= \sum_{j=1}^{\mathcal{N}} [\sum_{i=1}^{N} (\nu_{-ij} - \nu_{+ij}) \mu_i] \dot{\xi}_j + \sum_{i=1}^{N'} (\mu_i - \mu_i^{ex}) \dot{n}_i^{ex} . \end{aligned} \tag{4.1}$$

Now we set the variation of the sum, $\dot{G} + \mathcal{F}$, with respect to the flows in the system equal to zero, where $\mathcal{F}$ is given by equation (3.3) with the resistances from equation (2.5). Since the flows $\dot{n}_i$ have already been eliminated through the equations of constraint, only the remaining flows $\dot{\xi}_j$ and $\dot{n}_i^{ex}$ are to be varied. For the $\dot{n}_i^{ex}$ we find

$$\partial(\dot{G} + \mathcal{F})/\partial\dot{n}_i^{ex} = \mu_i - \mu_i^{ex} = 0 \, , \; 1 \le i \le N' \, , \tag{4.2}$$

and for the $\dot{\xi}_j$

$$\partial(\dot{G} + \mathcal{F})/\partial\dot{\xi}_j = \sum_{i=1}^{N} (\nu_{-ij} - \nu_{+ij})\, \mu_i + R_j\dot{\xi}_j = 0 \, , \; 1 \le j \le \mathcal{N}. \tag{4.3}$$

Equation (4.2) is just equation (3.6) again, and equation (4.3) is identical to (3.7), which leads to the proper expressions for the rates of reaction (2.2) when the definition of the R_j in equation (2.5) is taken into account. Thus the Lagrangian approach is equivalent to the variational formulation

$$\delta_{\dot{q}}(\dot{G} + \mathcal{F}) = 0 \tag{4.4}$$

where $\delta_{\dot{q}}$ implies that the variation is taken only with respect to the flows.

Now let us assume for the moment that $\delta_{\dot{q}}(\dot{G} + \mathcal{F})$ is the total variation of $(\dot{G} + \mathcal{F})$; equation (4.4) then states that $(\dot{G} + \mathcal{F})$ is an extremum. Let us further assume that the second-order variation with respect to the flows

$$\delta^2_{\dot{q}\dot{q}}(\dot{G} + \mathcal{F}) = \delta^2_{\dot{q}\dot{q}}\mathcal{F} = \sum_{j=1}^{\mathcal{N}} R_j(d\dot{\xi}_j)^2 \ge 0 \tag{4.5}$$

is also the total second-order variation of $(\dot{G} + \mathcal{F})$; then the extremum is a minimum, since the R_j are all ≥ 0. In other words, the equations of motion of a system of chemical reactions are given by the condition that the sum of the rate of increase of the Gibbs energy and the dissipation function is a minimum.

In the near equilibrium regime studied by Onsager (1931) in his original paper the R_j are considered constant and the rate of increase of the entropy of the system, S, is taken to be a linear functional of the flows in the system. Recalling that the only processes we are considering here are chemical reactions under isothermal conditions, we note that the rate of increase of the entropy of the system of chemical reactions is $S = -\dot{G}/T$.[6] In addition, the dissipation function used by Onsager is $\phi = \mathcal{F}/T$.

Thus, under these Onsager conditions the first- and second-order variations of $(\dot{G} + \mathcal{F})$ with respect to the flows are the total variations, and the previous paragraph is simply a restatement for chemical reactions of sections 5 and 6 of Onsager's original work. Following Onsager further we note that at the stationary state

---

[6]In the notation used here Onsager's S and S* would be S_{IN} and $-S_{EX}$.

$$\dot{G}_{IN} = \sum_{i=1}^{N} \dot{n}_i \mu_i \tag{4.6}$$

[see equation (3.2b)] vanishes since all $\dot{n}_i = 0$ by definition at the stationary state. Therefore, equation (4.4) becomes [see equations (3.2c,d)]

$$\delta_{\dot{q}}(- \dot{G}_{EX} + \mathcal{F}) = 0 ,$$
$$- \dot{G}_{EX} + \mathcal{F} = \text{minimum}. \tag{4.7}$$

If furthermore the flows on the boundary of the system are taken constant, this becomes

$$\delta_{\dot{q}} \mathcal{F} = 0, \; \mathcal{F} = \text{minimum}. \tag{4.8}$$

Equation (4.8) is just Rayleigh's principle of least dissipation of energy, as stated by Onsager (1931) for the near equilibrium regime. The last two paragraphs are also simply a statement of the well known theorem of minimum entropy production (Prigogine 1967, Nicolis and Prigogine 1977) in the near equilibrium regime.

Now let us return to the more general situation arbitrarily far from equilibrium where the variations with respect to the flows are not the total variations of $(\dot{G} + \mathcal{F})$. This follows in general since G depends on the μ_i, as do the R_j [(equation 2.5)] in $\mathcal{F}$ (through the a_i), and since the μ_i are functions of the n_i. Or, since we have used the equations of constraint to eliminate the $\dot{n}_i$, we can also eliminate the n_i through the integrated equations of constraint and view the μ_i as functions of the n_i^{ex} and the ξ_j. The system of reactions is then completely defined in a "phase space" with $2(\mathcal{N} + N')$ dimensions ξ_j, n_i^{ex}, $\dot{\xi}_j$, $\dot{n}_i^{ex}$. Along the hypersurfaces of constant ξ_j and n_i^{ex}, all variations of $(\dot{G} + \mathcal{F})$ except those with respect to the flows vanish of course. Actually, the situation is not so restrictive. For $\delta_{\dot{q}}(\dot{G} + \mathcal{F})$ and $\delta^2_{\dot{q}\dot{q}}(\dot{G} + \mathcal{F})$ to be the total first- and second-order variations of $(\dot{G} + \mathcal{F})$, it is not necessary that all ξ_j and n_i^{ex} be constant but only that all n_i be constant. From the equations of constraint (2.7a,b), this condition is met by

$$\sum_{j=1}^{\mathcal{N}} (\nu_{-ij} - \nu_{+ij})\, d\xi_j + dn_i^{ex} = 0, \; 1 \le i \le N' ;$$
$$\sum_{j=1}^{\mathcal{N}} (\nu_{-ij} - \nu_{+ij})\, d\xi_j = 0, \; N' + 1 \le i \le N. \tag{4.9}$$

We may now state the alternative variational formulation corresponding to the Lagrangian approach of the previous section as follows: along the hypersurfaces defined by equation (4.9), the sum of the rate of increase of the Gibbs energy and the dissipation function is a minimum:

$$\delta_{\dot{q}}(\dot{G} + \mathcal{F}) = 0 \,, \qquad G + \mathcal{F} = \text{minimum.} \tag{4.10}$$

Along these hypersurfaces equations (4.7) and (4.8) are also valid under appropriate conditions. Thus the variational formulation presented here and the corresponding Lagrangian approach of the previous section represent an extension of Onsager's work to systems of chemical reactions arbitrarily far from equilibrium, with the constraint that the variations are only to be taken with respect to the flows in the system.

SYMMETRY

The equations of motion for a system of chemical reactions as recovered from the Lagrangian formulation naturally lead to symmetric relations in the stationary state relation between the phenomenological fluxes of the reactions and the conjugate thermodynamic forces (Shiner 1987). To demonstrate this let us first introduce a vector notation for the variables and parameters of the system, where a superscript T denotes the transpose:

$$\begin{aligned}
\dot{n} &= (\dot{n}_1,\dot{n}_2,\ldots,\dot{n}_N)^T \\
\dot{n}^{ex} &= (\dot{n}_1^{ex},\dot{n}_2^{ex},\ldots,\dot{n}_{N'}^{ex})^T \\
\mu &= (\mu_1,\mu_2,\ldots,\mu_N)^T \\
\mu^{ex} &= (\mu_1^{ex},\mu_2^{ex},\ldots,\mu_{N'}^{ex})^T \\
\mu^{in} &= (\mu_{N'+1},\mu_{N'+2},\ldots,\mu_N)^T \\
\dot{\xi} &= (\dot{\xi}_1,\dot{\xi}_2,\ldots,\dot{\xi}_{\mathcal{N}})^T \,.
\end{aligned} \tag{5.1}$$

$\dot{n}$, $\dot{n}^{ex}$, μ, μ^{ex}, μ^{in}, and $\dot{\xi}$ are the column vectors of the rates of change of the number of moles of species i, the flow through the N' external sources, the chemical potentials of the species i, the chemical potentials of the N' external sources, the chemical potentials of the species not controlled by external sources, and the $\mathcal{N}$ rates of reaction. Let R be the ($\mathcal{N}$ X $\mathcal{N}$) diagonal matrix of the resistances of the reactions with elements

$$R_{ij} = \delta_{ij}R_i, \tag{5.2}$$

where δ_{ij} is the Kronecker delta; and let ν be the (N X $\mathcal{N}$) matrix of net stoichiometric coefficients with elements

$$\nu_{ij} = \nu_{-ij} - \nu_{+ij} \, . \tag{5.3}$$

The equations of motion and the mass balance equations for the system of chemical reactions can now be written

$$R\dot{\xi} = -\nu^{T}\mu, \tag{5.4}$$

$$\dot{n} = \nu\dot{\xi} + \begin{pmatrix} \dot{n}^{ex} \\ --- \\ 0 \end{pmatrix} \tag{5.5}$$

where 0 is a column vector with all elements equal to zero. At the stationary state all $\dot{n}_i$ vanish by definition and the mass balance equations become

$$\nu\dot{\xi} = -\begin{pmatrix} \dot{n}^{ex} \\ --- \\ 0 \end{pmatrix}. \tag{5.6}$$

All of the $\dot{\xi}_j$ are thus not independent at the stationary state. We arbitrarily take the last (N - N') $\dot{\xi}_j$ to be the dependent reactions rates[7] and partition $\dot{\xi}$ into a vector of independent reaction rates $\dot{\xi}_I$ and a vector of dependent rates $\dot{\xi}_D$:

$$\dot{\xi} = \begin{pmatrix} \dot{\xi}_I \\ --- \\ \dot{\xi}_D \end{pmatrix} \tag{5.7}$$

where

$$\dot{\xi}_I = (\dot{\xi}_1, \dot{\xi}_2, \ldots, \dot{\xi}_{\mathcal{N} - N + N'})^T$$

$$\dot{\xi}_D = (\dot{\xi}_{\mathcal{N} - N + N' + 1}, \dot{\xi}_{\mathcal{N} - N + N' + 2}, \ldots, \dot{\xi}_{\mathcal{N}})^T \, .$$

---

[7]It is assumed that N > N - N'; otherwise, the stationary state conditions can be fulfilled only when all $\dot{\xi}_j$ and all $\dot{n}_i^{ex}$ vanish, i.e. at thermodynamic equilibrium. Similarly, we assume that $\mathcal{N} \leq N \Rightarrow \mathcal{N} - N + N' \leq N'$, which insures that there are more external sources than independent reaction rates in $\dot{\xi}_I$ as defined here. In the other case, when there are more reaction rates in $\dot{\xi}_I$ than there are external sources, all of the $\dot{\xi}_j$ in $\dot{\xi}_I$ are not truly independent, and we should choose a smaller set of independent reaction rates. Since the demonstration of symmetry in this latter case follows the same lines as that in the former case, it will not be presented here.

We also partition ν as

$$\nu = \left(\begin{array}{c|c} \nu_{II} & \nu_{ID} \\ \hline \nu_{DI} & \nu_{DD} \end{array}\right), \tag{5.8}$$

where

$$\nu_{II} = \begin{pmatrix} \nu_{11} & \cdots & \nu_{1\,\mathcal{N}-N+N'} \\ \vdots & & \vdots \\ \nu_{N'1} & \cdots & \nu_{N'\,\mathcal{N}-N+N'} \end{pmatrix},$$

$$\nu_{ID} = \begin{pmatrix} \nu_{1\,\mathcal{N}-N+N'+1} & \cdots & \nu_{1\mathcal{N}} \\ \vdots & & \vdots \\ \nu_{N'\mathcal{N}-N+N'+1} & \cdots & \nu_{N'\mathcal{N}} \end{pmatrix},$$

$$\nu_{DI} = \begin{pmatrix} \nu_{N'+1\,1} & \cdots & \nu_{N'+1\,\mathcal{N}-N+N'} \\ \vdots & & \vdots \\ \nu_{N1} & \cdots & \nu_{N\,\mathcal{N}-N+N'} \end{pmatrix},$$

$$\nu_{DD} = \begin{pmatrix} \nu_{N'+1\,\mathcal{N}-N+N'+1} & \cdots & \nu_{N'+1\,\mathcal{N}} \\ \vdots & & \vdots \\ \nu_{N\,\mathcal{N}-N+N'+1} & \cdots & \nu_{N\mathcal{N}} \end{pmatrix}.$$

The stationary state condition can now be written

$$\left(\begin{array}{c|c} \nu_{II} & \nu_{ID} \\ \hline \nu_{DI} & \nu_{DD} \end{array}\right)\begin{pmatrix} \dot{\xi}_I \\ \hline \dot{\xi}_D \end{pmatrix} = -\begin{pmatrix} \dot{n}^{ex} \\ \hline 0 \end{pmatrix} \tag{5.9a}$$

or

$$\nu_{II}\dot{\xi}_I + \nu_{ID}\dot{\xi}_D = -\dot{n}^{ex}, \tag{5.9b}$$

$$\nu_{DI}\dot{\xi}_I + \nu_{DD}\dot{\xi}_D = 0. \tag{5.9c}$$

The equations of motion can now also be partitioned:

$$\left(\begin{array}{c|c} R_{II} & R_{ID} \\ \hline R_{DI} & R_{DD} \end{array}\right)\left(\begin{array}{c} \dot{\xi}_I \\ \hline \dot{\xi}_D \end{array}\right) = \left(\begin{array}{c|c} R_{II} & 0 \\ \hline 0 & R_{DD} \end{array}\right)\left(\begin{array}{c} \dot{\xi}_I \\ \hline \dot{\xi}_D \end{array}\right)$$

$$= -\left(\begin{array}{c|c} \nu_{II}^T & \nu_{DI}^T \\ \hline \nu_{ID}^T & \nu_{DD}^T \end{array}\right)\left(\begin{array}{c} \mu^{ex} \\ \hline \mu^{in} \end{array}\right) \tag{5.10a}$$

or

$$R_{II}\dot{\xi}_I = -\nu_{II}^T \mu^{ex} - \nu_{DI}^T \mu^{in}, \tag{5.10b}$$

$$R_{DD}\dot{\xi}_D = -\nu_{ID}^T \mu^{ex} - \nu_{DD}^T \mu^{in} \tag{5.10c}$$

where R has also been partitioned in the following manner

$$R = \left(\begin{array}{c|c} R_{II} & R_{ID} \\ \hline R_{DI} & R_{DD} \end{array}\right) \tag{5.11}$$

into

$$R_{II} = \begin{pmatrix} R_1 & \cdots & \cdots & 0 \\ \cdot & \ddots & \cdot & \cdot \\ \cdot & \cdot & \ddots & \cdot \\ 0 & \cdots & \cdots & R_{\mathcal{N}-N+N'} \end{pmatrix},$$

$$R_{ID} = R_{DI} = 0,$$

$$R_{DD} = \begin{pmatrix} R_{\mathcal{N}-N+N'+1} & \cdots & \cdots & 0 \\ \cdot & \ddots & \cdot & \cdot \\ \cdot & \cdot & \ddots & \cdot \\ 0 & \cdots & \cdots & R_{\mathcal{N}} \end{pmatrix}.$$

Here 0 is a matrix whose elements all vanish identically.

We now eliminate the dependent reaction rates using equation (5.9c)[8]:

---

[8]We assume that ν_{DD} is not singular.

$$\dot{\xi}_D = -\nu_{DD}^{-1}\nu_{DI}\dot{\xi}_I . \tag{5.12}$$

We can now also eliminate the chemical potentials of the species not controlled by external sources μ^{in} by first substituting equation (5.12) into equation (5.10c):

$$-R_{DD}\nu_{DD}^{-1}\nu_{DI}\dot{\xi}_I = -\nu_{ID}^{T}\mu^{ex} - \nu_{DD}^{T}\mu^{in}$$

or

$$\mu^{in} = \left(\nu_{DD}^{T}\right)^{-1}\left(R_{DD}\nu_{DD}^{-1}\nu_{DI}\dot{\xi}_I - \nu_{ID}^{T}\mu^{ex}\right). \tag{5.13}$$

Substituting this result into equation (5.10b) we have finally

$$R_{II}\dot{\xi}_I = -\nu_{II}^{T}\mu^{ex} - \nu_{DI}^{T}\left(\nu_{DD}^{T}\right)^{-1}\left(R_{DD}\nu_{DD}^{-1}\nu_{DI}\dot{\xi}_I - \nu_{ID}^{T}\mu^{ex}\right)$$

or (5.14)

$$\left[R_{II} + \nu_{DI}^{T}\left(\nu_{DD}^{T}\right)^{-1}R_{DD}\nu_{DD}^{-1}\nu_{DI}\right]\dot{\xi}_I = \left[-\nu_{II}^{T} + \nu_{DI}^{T}\left(\nu_{DD}^{T}\right)^{-1}\nu_{ID}^{T}\right]\mu^{ex} .$$

These are the phenomenological relations expressing the stationary state relations between the independent reaction rates $\dot{\xi}_j$, $1 \le j \le \mathcal{N} - N + N'$, and the chemical potentials of the external sources μ_i^{ex}, $1 \le i \le N'$. It can be written in the standard form

$$\mathcal{R}J = X, \tag{5.15}$$

where $J = \dot{\xi}_I$ is the vector of phenomenological flows,

$$X = \left[-\nu_{II}^{T} + \nu_{DI}^{T}\left(\nu_{DD}^{T}\right)^{-1}\nu_{ID}^{T}\right]\mu^{ex} \tag{5.16}$$

is the vector of conjugate forces, and the matrix of phenomenological resistances is given by

$$\mathcal{R} = R_{II} + \nu_{DI}^{T}\left(\nu_{DD}^{T}\right)^{-1}R_{DD}\nu_{DD}^{-1}\nu_{DI} . \tag{5.17}$$

It can easily be seen that $\mathcal{R}$ is symmetric. First note that R_{II} and R_{DD} are diagonal and therefore symmetric, and that

$$\left(\nu_{DD}^{T}\right)^{-1} = \left(\nu_{DD}^{-1}\right)^{T} \tag{5.18}$$

since ν_{DD} is a square matrix. Therefore,

$$\nu_{DI}^{T}\left(\nu_{DD}^{T}\right)^{-1} = \nu_{DI}^{T}\left(\nu_{DD}^{-1}\right)^{T} = \left(\nu_{DD}^{-1}\nu_{DI}\right)^{T}, \tag{5.19}$$

and

$$\nu_{DI}^{T}\left(\nu_{DD}^{T}\right)^{-1} R_{DD}\, \nu_{DD}^{-1}\, \nu_{DI} \tag{5.20}$$

is symmetric, since $B^{T}CB$ is symmetric if C is symmetric. Therefore $\mathcal{R}$ is a symmetric matrix and $\mathcal{R}_{ij} = \mathcal{R}_{ji}$.

Thus the phenomenological equations for systems of chemical reactions are symmetric in an algebraic sense, even though they contain the full dynamics of the system of chemical reactions. However, they do not display Onsager reciprocity

$$(\partial X_i/\partial J_j) = (\partial X_j/\partial J_i) \tag{5.21}$$

in general, since the $\mathcal{R}_{ij}$ depend on the phenomenological forces X_i through the μ_i^{ex}. In some special cases the differential symmetry of Onsager reciprocity is obeyed, of course; among these special cases is thermodynamic equilibrium. It has been remarked that the property of algebraic symmetry is trivial, since the equations of motion for the chemical reactions as obtained from the Lagrangian formulation, written here in vector form in equation (5.4), are of the form of the phenomenological equations (5.15)[9], and the matrix of resistances for the chemical reactions R is diagonal; thus the reactions are uncoupled. However, equation (5.4) involves all of the rates of reaction, not just the independent ones. As DeGroot and Mazur (1962) noted, a system of coupled reactions, only some of which are independent, may show coupling when the dependent rates are eliminated. The coupling is apparent in the equations of constraint (5.5), written as mass balance equations. One encounters an analogous situation in the study of electrical networks. For each resistance R, the current I flowing through the resistance is related to the voltage V across it by V = IR. However, when various resistances are connected together in a network, one finds that not all of the currents are independent due to Kirchkoff's current law at the nodes. This point will be illustrated in some of the examples in the next section.

The property of differential symmetry [equation (5.21)] is often sought because of the simplicity offered by linear, symmetric equations; indeed much work has been done on finding special cases where the phenomenological equations for systems of chemical reactions are linear and/or symmetric in the sense of Onsager reciprocity far from equilibrium (Rothschild et al. 1980, Essig and Caplan 1981, Stucki et al. 1983); some biologists have even argued for the existence of such cases on the basis of Darwinian evolution (Stucki et al. 1983). Nonetheless any differential symmetry found was not a general property of stationary states far from equilibrium, as the algebraic symmetry is.

---

[9]The right hand side of equation (5.4) is just the vector of the affinities, the thermodynamic forces for the reactions.

Phenomenological equations of the form (5.15), but with constant resistances, have also been found to be valid for chemical reactions in the neighborhood of stationary states far from equilibrium if the thermodynamic forces and fluxes are replaced with their perturbations around the stationary state (Prigogine 1967, Glansdorff and Prigogine 1971, Keizer 1977, Nicolis and Prigogine 1977). These linear equations do not in general possess a symmetry property, however, and furthermore are restricted to the neighborhood of stable stationary states. The equations (5.15) here are valid for any stationary states, regardless of their stability, and contain the full dynamic behavior of the system of chemical reaction, not just its linear behavior around the stationary states.

EXAMPLES AND NETWORK REPRESENTATION

In this section several examples not given elsewhere are presented. The examples will serve to illustrate the application of the Lagrangian formulation to systems of chemical reactions and the algebraic symmetry property derived in the last section. The connection between the Lagrangian approach and the network thermodynamics of chemical reactions (Peusner 1970, Oster et al. 1973, Wyatt 1978) will be emphasized by showing the network for some of the examples and demonstrating how particular results may be obtained either from the Lagrangian or the network approach. We will only use the linear graph form of representation of the networks, not the bond graph representation.

The Reaction: m A $\longleftrightarrow$ p B

The first example is very simple; it will allow us to introduce the symbols used for the network representation. Consider the single chemical reaction

$$m\,A \longleftrightarrow p\,B$$

where m molecules of species A react to form p molecules of species B. The Lagrangian, dissipation function and equations of constraint are

$$\mathcal{L} = -\,n_A\mu_A - n_B\mu_B\,;$$

$$\mathcal{F} = R_1\dot{\xi}_1{}^2/2\,;$$

$$f_A = \dot{n}_A + m\dot{\xi}_1 = 0\,; \tag{6.1}$$

$$f_B = \dot{n}_B - p\,\dot{\xi}_1 = 0\,;$$

where

$$R_1 = \frac{RT\ln\left|(k_1a_A^m)/(k_{-1}a_B^p)\right|}{k_1a_A^m - k_{-1}a_B^p}\,. \tag{6.2}$$

Carrying out the operations to find the equations of motion, we find

$$\mu_A + \lambda_A = 0; \quad \mu_B + \lambda_B = 0; \quad R_1\dot{\xi}_1 + m\,\lambda_A - p\,\lambda_B = 0. \tag{6.3}$$

Solving the first two of these for the λ's and substituting the result into the third equation we have

$$R_1\dot{\xi}_1 = m\,\mu_A - p\,\mu_B = 0 \tag{6.4}$$

which gives the correct equation for the rate of reaction when the definition of R_1 is taken into account.

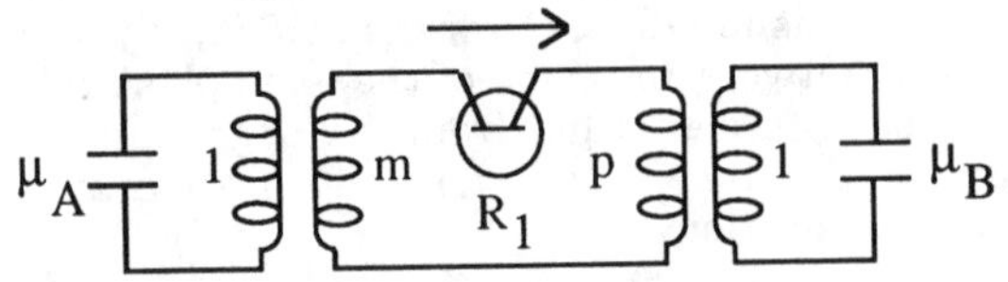

Figure 1. The reaction mA ↔ pB.

The network for this example is shown in Figure 1. Each of the chemical species A and B is represented by the symbol normally used to represent a capacitor in an electrical network. This choice is made since the separation of charge on a capacitor and the increase in mole number of a chemical species both represent passive energy storage. Each "chemical capacitor" is labelled with the chemical potential μ_i of the species it represents. The mole number n_i of this species is considered by convention to flow into the capacitor towards "chemical ground". The symbol used for the resistance of a chemical reaction is a slightly modified version of the symbol introduced by Wyatt (1978); it should remind one of the symbol used for a field-effect transistor (FET) in electrical networks. This choice is motivated by the qualitative similarity between the constitutive relation for the "chemical resistance" and that for an FET. The only change made in the symbol used here from that introduced by Wyatt is that its connection to ground is not shown explicitly. "Current flow" through the resistance, i.e. the rate of the reaction, is taken to be positive in the direction of the arrow. Since the net stoichiometric coefficients of both species A and B in the reaction are not 1, the ideal transformers shown must be included (Wyatt 1978); their sole effect is to multiply the flow of the corresponding species by a constant factor. It is easy to verify that network analysis of this simple case yields the same results as the Lagrangian formulation.

<u>Enzyme Catalysis with Two States</u>

The second example consists of three reactions among four species:

Reaction 1: e + s ⟷ es

Reaction 2: es ⟷ e + p

Reaction 3: s $\leftrightarrow$ p

The first two reactions represent the simplest possible model of catalysis by enzymes; the catalyzed reaction is the isomeric conversion of s into p. The free enzyme e binds the substrate s in the first reaction; in the second the "enzyme-substrate complex" es releases the product p. For good measure the uncatalyzed conversion of s into p is included in reaction 3. In models of enzyme kinetics it is usually assumed that the activities of substrate and product are held constant, so that in general a nonequilibrium stationary state is achieved. For this reason we include external sources for s and p.

The Lagrangian, dissipation function and equations of constraint are:

$$\mathcal{L} = - n_s\mu_s - n_p\mu_p - n_e\mu_e - n_{es}\mu_{es} + n_s^{ex}\,\mu_s^{ex} + n_p^{ex}\,\mu_p^{ex}\,;$$

$$\mathcal{F} = (R_1\dot{\xi}_1^{\,2} + R_2\dot{\xi}_2^{\,2} + R_3\dot{\xi}_3^{\,2})/2\,;$$

$$\begin{aligned}
f_s &= \dot{n}_s + \dot{\xi}_1 + \dot{\xi}_3 - \dot{n}_s^{ex} = 0\,;\\
f_p &= \dot{n}_p - \dot{\xi}_2 - \dot{\xi}_3 - \dot{n}_p^{ex} = 0\,;\\
f_e &= \dot{n}_e + \dot{\xi}_1 - \dot{\xi}_2 = 0\,;\\
f_{es} &= \dot{n}_{es} - \dot{\xi}_1 + \dot{\xi}_2 = 0\,;
\end{aligned} \tag{6.5}$$

where

$$\begin{aligned}
R_1 &= \frac{RT\ln|(k_1a_ea_s)/(k_{-1}a_{es})|}{k_1a_ea_s - k_{-1}a_{es}}\,;\\
R_2 &= \frac{RT\ln|(k_2a_{es})/(k_{-2}a_ea_p)|}{k_2a_{es} - k_{-2}a_ea_p}\,;\\
R_3 &= \frac{RT\ln|(k_3a_s)/(k_{-3}a_p)|}{k_3a_s - k_{-3}a_p}\,.
\end{aligned} \tag{6.6}$$

After carrying out the Lagrangian operations according to equation (3.1) we have

$$\begin{aligned}
&\mu_s = \mu_s^{ex}; \quad \mu_p = \mu_p^{ex};\\
&R_1\dot{\xi}_1 = \mu_e + \mu_s^{ex} - \mu_{es}; \quad R_2\dot{\xi}_2 = \mu_{es} - \mu_e - \mu_p^{ex}; \quad R_3\dot{\xi}_3 = \mu_s^{ex} - \mu_p^{ex}.
\end{aligned} \tag{6.7}$$

Using the last two equations of constraint, we find at the stationary state:

$$\dot{n}_e = \dot{n}_{es} = 0 \Longrightarrow \dot{\xi}_1 = \dot{\xi}_2 \equiv \dot{\xi}_C \tag{6.8}$$

is the net rate of the catalysis of s into p. Using the first two of the equations for $R_i\dot{\xi}_i$ to eliminate μ_e and μ_{es}, we have

$$(R_1 + R_2)\,\dot{\xi}_C = \mu_s^{ex} - \mu_p^{ex}. \tag{6.9}$$

Since $\dot{\xi}_3$ is the rate of the uncatalyzed conversion of s into p, the total rate of conversion of s into p is

$$\dot{\xi}_T = \dot{\xi}_C + \dot{\xi}_3 = \left(\frac{1}{R_1 + R_2} + \frac{1}{R_3}\right)\left(\mu_s^{ex} - \mu_p^{ex}\right). \tag{6.10}$$

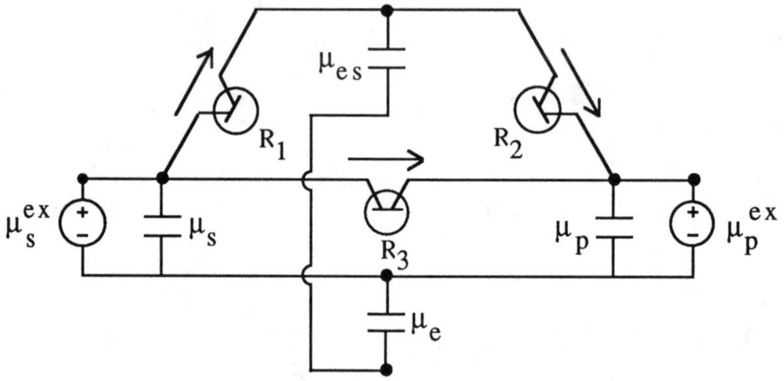

Figure 2. The network representation of enzyme catalysis with two enzyme states.

The network for this reaction scheme is shown in Figure 2. Since the stoichiometric coefficients of all species in all of the reactions are 1, there are no mutual inductors. The external sources μ_i^{ex} for the species s and p are represented by the same symbol used to represent ideal voltage sources in electrical networks. As is conventional in electrical networks for the current leaving a voltage source, the mole number n_i^{ex} corresponding to the source μ_i^{ex} is assumed to flow out of the source. If we would now interpret this figure as an electrical network, we would see that the equations of constraint are simply Kirchkoff's current law at the nodes, and that the equations of the form

$$\mu_i = \mu_i^{ex} \text{ and } R_3\dot{\xi}_3 = \mu_s^{ex} - \mu_p^{ex} \tag{6.11}$$

are expressions of Kirchkoff's voltage law around the loops of the network. These analogies hold in general for systems of chemical reactions. Indeed they allow the network representation of systems of chemical reactions **and** their Lagrangian formulation.

The network representation can simplify the analysis of chemical reactions significantly. Regarding Figure 2 as an electrical network again, we can analyze the stationary state by simply disregarding the capacitors. The network then looks like the one in Figure 3. In this simple case we can easily obtain the stationary state equations by inspection.

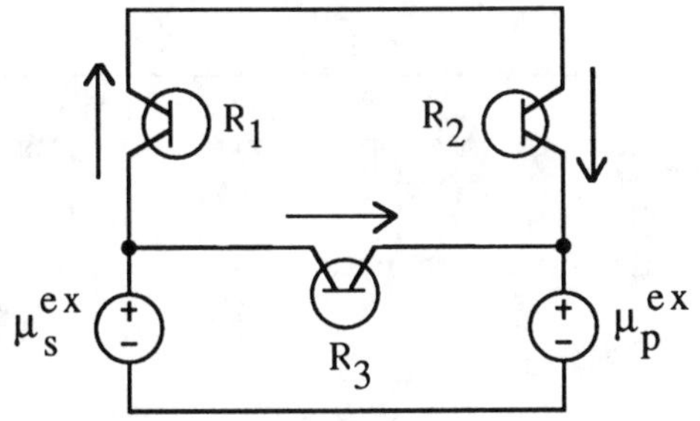

Figure 3. The network representation of enzyme catalysis with two enzyme states with capacitors removed for analysis of the stationary state.

Enzyme Catalysis with Three States

A more accurate model for the enzymatic catalysis of the conversion of s into p is given by the following set of reaction:

Reaction 1: e + s $\leftrightarrow$ es

Reaction 2: es $\leftrightarrow$ ep

Reaction 3: ep $\leftrightarrow$ e + p

Reaction 4: s $\leftrightarrow$ p

Here we have not only an enzyme-substrate complex es but also an enzyme-product complex ep. The enzymatic cycle has three states instead of the two of the simpler model above.

Since the two- and three-state models of enzyme catalysis are similar, however, I will not present the Lagrangian formulation for this example, but only the network representation as shown in Figure 4.

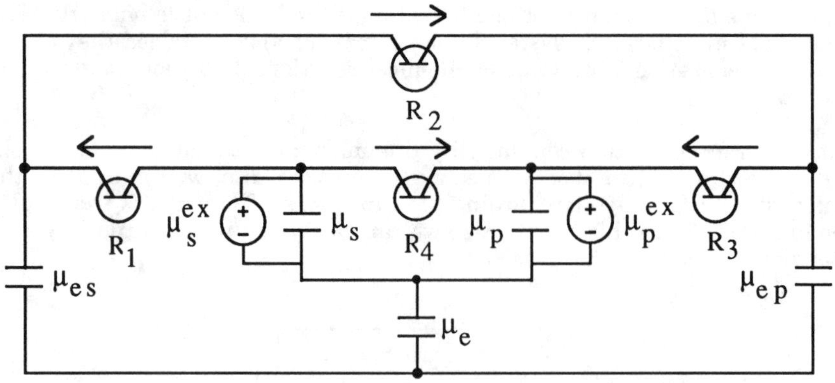

Figure 4. The network representation of enzyme catalysis with three states.

At the stationary state this network reduces to that shown in Figure 5.

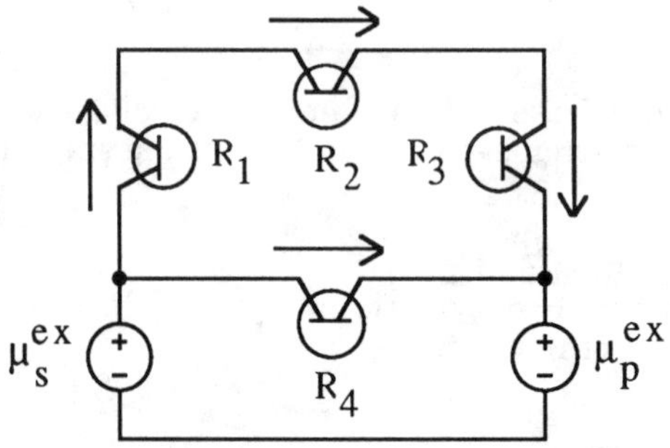

Figure 5. The network representation of enzyme catalysis with three enzyme states with capacitors removed for analysis of the stationary state.

From this figure we can immediately write the stationary state relations between the thermodynamic driving force and the rates of reaction:

$$R_4\dot{\xi}_4 = \mu_s^{ex} - \mu_p^{ex}; \quad (R_1 + R_2 + R_3)\,\dot{\xi}_C = \mu_s^{ex} - \mu_p^{ex},$$
$$\dot{\xi}_T = \dot{\xi}_C + \dot{\xi}_4 = \left(\frac{1}{R_1 + R_2 + R_3} + \frac{1}{R_4}\right)\left(\mu_s^{ex} - \mu_p^{ex}\right) \tag{6.12}$$

where $\dot{\xi}_C$ is the net rate of catalysis, and $\dot{\xi}_T$ is the total rate of conversion of s into p. The reader is encouraged to verify that these equations also follow from the Lagrangian formulation.

An Elementary Model for Coupled, Carrier Mediated Membrane Transport

Now we turn to an example which shows coupling between two independent reactions at the stationary state. The reactions are

Reaction 1: e + a $\leftrightarrow$ ea

Reaction 2: ea + b $\leftrightarrow$ eab

Reaction 3: e + a' $\leftrightarrow$ e'a'

Reaction 4: e'a' + b' $\leftrightarrow$ e'a'b'

Reaction 5: ea $\leftrightarrow$ e'a'

Reaction 6: eab $\leftrightarrow$ e'a'b'

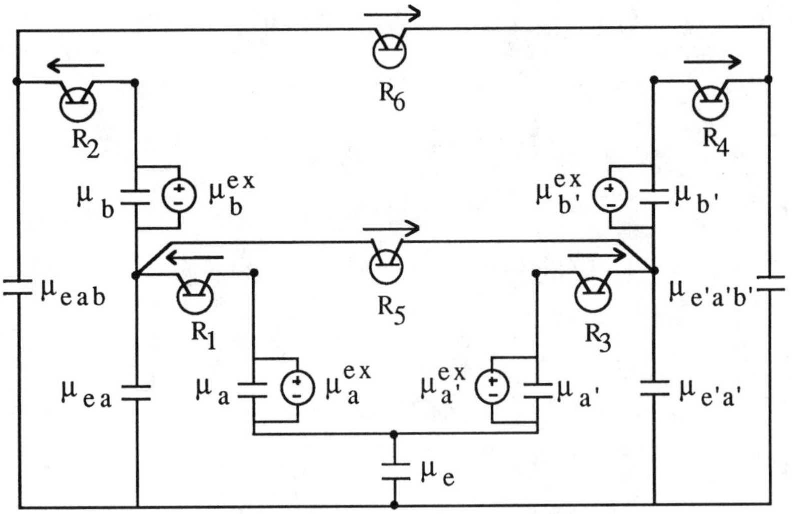

Figure 6. The network representation of coupled transport.

The network is shown in Figure 6. These reactions can be interpreted to represent carrier mediated transport across a membrane (Hill 1977), as is often seen in biological systems. e is a carrier molecule, which can diffuse rapidly across the membrane, so that the activity of e is the same on both sides. Species a and b, on the other hand, are completely impermeable to the membrane; the same chemical species on the other side of the membrane are designated a' and b'. Although a is impermeable to the membrane, its complex with the carrier molecule, ea, can cross the membrane to the other side, where it is designated e'a'. b and b' cannot form a complex with the carrier molecule e alone, but can form complexes with the ea and e'a' complexes on either side of the membrane. eab can also diffuse across the membrane to become e'a'b'. That only one species e is considered and not two, e and e', is equivalent to the assumption that e can diffuse very rapidly across the membrane in comparison with the other permeable species, so that its activities on both sides of the membrane are equal. The diffusion of the carrier complexes, reactions 5 and 6, are treated here as first-order reactions with permeabilities as rate constants.

The Lagrangian and the dissipation function are easy to write down and will not be shown here. The equations of constraint are

$$
\begin{aligned}
f_a &= \dot{n}_a + \dot{\xi}_1 - \dot{n}_a^{ex} = 0 ; \\
f_b &= \dot{n}_b + \dot{\xi}_2 - \dot{n}_b^{ex} = 0 ; \\
f_{a'} &= \dot{n}_{a'} + \dot{\xi}_3 - \dot{n}_{a'}^{ex} = 0 ; \\
f_{b'} &= \dot{n}_{b'} + \dot{\xi}_4 - \dot{n}_{b'}^{ex} = 0 ; \\
f_e &= \dot{n}_e + \dot{\xi}_1 + \dot{\xi}_3 = 0 ; \\
f_{ea} &= \dot{n}_{ea} - \dot{\xi}_1 + \dot{\xi}_2 + \dot{\xi}_5 = 0 ; \\
f_{eab} &= \dot{n}_{eab} - \dot{\xi}_2 + \dot{\xi}_6 = 0 ; \\
f_{e'a'} &= \dot{n}_{e'a'} - \dot{\xi}_3 + \dot{\xi}_4 - \dot{\xi}_5 = 0 ; \\
f_{e'a'b'} &= \dot{n}_{e'a'b'} - \dot{\xi}_4 - \dot{\xi}_6 = 0 .
\end{aligned}
\tag{6.13}
$$

From the Lagrangian formulation the equations of motion are

$$
\begin{aligned}
R_1\dot{\xi}_1 &= \mu_a^{ex} + \mu_e - \mu_{ea} ; \\
R_2\dot{\xi}_2 &= \mu_b^{ex} + \mu_{ea} - \mu_{eab} ; \\
R_3\dot{\xi}_3 &= \mu_{a'}^{ex} + \mu_e - \mu_{e'a'} ; \\
R_4\dot{\xi}_4 &= \mu_{b'}^{ex} + \mu_{e'a'} - \mu_{e'a'b'} ; \\
R_5\dot{\xi}_5 &= \mu_{ea} - \mu_{e'a'} ; \\
R_6\dot{\xi}_6 &= \mu_{eab} - \mu_{e'a'b'} .
\end{aligned}
\tag{6.14}
$$

where we have already substituted the values of the external forces for species a, b, a', b'. At the stationary state the last five of the equations of constraint yield

$$\begin{pmatrix} \dot{\xi}_3 \\ \dot{\xi}_4 \\ \dot{\xi}_5 \\ \dot{\xi}_6 \end{pmatrix} = \begin{pmatrix} -1 & 0 \\ 1 & -1 \\ 0 & 2 \\ 0 & -1 \end{pmatrix} \begin{pmatrix} \dot{\xi}_1 \\ \dot{\xi}_2 \end{pmatrix}. \tag{6.15}$$

Substituting into the equations of motion for the last four reactions, we have

$$\begin{aligned} -R_3\dot{\xi}_1 &= \mu_{a'}^{ex} + \mu_e - \mu_{e'a'}; \\ -R_4\dot{\xi}_2 &= \mu_{b'}^{ex} + \mu_{e'a'} - \mu_{e'a'b'}; \\ R_5(\dot{\xi}_1 - \dot{\xi}_2) &= \mu_{ea} - \mu_{e'a'}; \\ R_6\dot{\xi}_2 &= \mu_{eab} - \mu_{e'a'b'}; \end{aligned} \tag{6.16a}$$

or

$$\begin{aligned} \mu_e - \mu_{ea} &= -\mu_{a'}^{ex} - R_5(\dot{\xi}_1 - \dot{\xi}_2), \\ \mu_{ea} - \mu_{eab} &= -\mu_{b'}^{ex} - R_4\dot{\xi}_2 + R_5(\dot{\xi}_1 - \dot{\xi}_2) - R_6\dot{\xi}_2 \end{aligned} \tag{6.16b}$$

Finally, substituting these results into the equations of motion for the first two reactions, we find

$$\begin{aligned} \mu_a^{ex} - \mu_{a'}^{ex} &= (R_1 + R_3 + R_5)\,\dot{\xi}_1 - R_5\dot{\xi}_2 \\ \mu_b^{ex} - \mu_{b'}^{ex} &= -R_5\dot{\xi}_1 + (R_2 + R_4 + R_5 + R_6)\,\dot{\xi}_2 \end{aligned} \tag{6.17}$$

or

$$\begin{pmatrix} X_1 \\ X_2 \end{pmatrix} = \begin{pmatrix} \mathcal{R}_{11} & \mathcal{R}_{12} \\ \mathcal{R}_{21} & \mathcal{R}_{22} \end{pmatrix} \begin{pmatrix} J_1 \\ J_2 \end{pmatrix}. \tag{6.18}$$

These are the stationary state phenomenological equations for the coupled transport of a and b. The left hand side of the first one, $X_1 = \mu_a^{ex} - \mu_{a'}^{ex}$, is the thermodynamic driving force for the transport of a from one side of the membrane to the other, and the left hand side of the second equation, $X_2 = \mu_b^{ex} - \mu_{b'}^{ex}$, is the thermodynamic force for the transport of b. $J_1 = \dot{\xi}_1$ is the net rate of disappearance of a from the unprimed side of the membrane, which equals the net rate of appearance of a' on the primed side of the membrane, and is therefore the net rate of transport of a, the thermodynamic flow conjugate to X_1. Similarly, $J_2 = \dot{\xi}_2$ is the net rate of transport of species b, the flow conjugate to the thermodynamic force X_2. The phenomenological coupling resistance is symmetric: $\mathcal{R}_{12} = \mathcal{R}_{21} = -R_5$. Why this is so and why the coupling resistance is negative is immediately apparent when one inspects the network with capacitances removed for the stationary state (Figure 7).

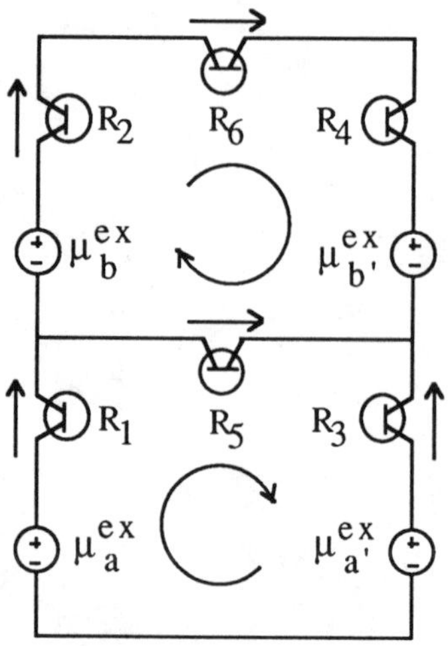

Figure 7. The network representation of coupled transport with capacitors removed for analysis of the stationary state.

On performing a loop analysis, one sees immediately that J_1 and J_2 flow in opposite directions through resistance R_5. Physically, $\mathcal{R}_{12}$ is negative since reaction 5 represents the diffusion of the carrier complexed with a alone across the membrane. When this complex diffuses, the carrier is no longer available to complex with b and transport it across the membrane.

The Brusselator

As a last example, we consider the Brusselator (Prigogine and Nicolis 1971), a prototypical model for oscillating chemical reactions, as modified by Wyatt (1978) for an analysis using network thermodynamics. The reactions are

Reaction 1: $A \leftrightarrow X$

Reaction 2: $B + X \leftrightarrow Y + D$

Reaction 3: $X \leftrightarrow E$

Reaction 4: $2X + Y \leftrightarrow 3X$

The network will not be presented here, since Wyatt (1978) has shown it in detail. (See his Figures 12, 13 and 14.) The activities of species A, B, D, E are assumed to be controlled by external sources, whereas X and Y are internal species. The overall reaction due to reactions 1 and 3 is A $\longleftrightarrow$ E, and that due to reactions 2 and 4 is B $\longleftrightarrow$ D. The Lagrangian formulation leads to the following equations for the reaction rates:

$$\begin{aligned} R_1\dot{\xi}_1 &= \mu_A^{ex} - \mu_X ; \\ R_2\dot{\xi}_2 &= \mu_B^{ex} - \mu_D^{ex} + \mu_X - \mu_Y ; \\ R_3\dot{\xi}_3 &= -\mu_E^{ex} + \mu_X ; \\ R_4\dot{\xi}_4 &= -\mu_X + \mu_Y . \end{aligned} \tag{6.19}$$

Particularly interesting is the resistance for the autocatalytic reaction 4:

$$R_3 = \frac{RT \ln |(k_4 a_Y)/(k_{-4} a_X)|}{k_4 a_X^2 a_Y - k_{-4} a_X^3} . \tag{6.20}$$

The autocatalyis expresses itself only in the denominator, not in the numerator, which is just the affinity of the reaction. At the stationary state, defined by $\dot{n}_X = \dot{n}_Y = 0$, the equations of constraint lead to $\dot{\xi}_1 = \dot{\xi}_3$ and $\dot{\xi}_2 = \dot{\xi}_4$. Using this and the last two of the rate equations to eliminate μ_X and μ_Y from the first two of the rate equations, we find the phenomenological equations at the stationary state:

$$\begin{aligned} (R_1 + R_3)\,\dot{\xi}_1 &= \mu_A^{ex} - \mu_E^{ex} \\ (R_2 + R_4)\,\dot{\xi}_2 &= \mu_B^{ex} - \mu_D^{ex} . \end{aligned} \tag{6.21}$$

Surprisingly, these equations show no coupling, even though oscillations in X and Y may occur under appropriate conditions. Of course, when this happens, the stationary state as defined here is unstable. The instability and the coupling responsible for the oscillations is not apparent when the phenomenological equations at the stationary state are written in this form. They find their expression in the resistances themselves, particularly in R_4 for the autocatalytic reaction.

DISCUSSION

I have presented a Lagrangian formulation of the dynamics of systems of chemical reactions under the local equilibrium condition, extending the classical Lagrangian approach through a Rayleigh dissipation function to describe the dissipative aspects of the reactions. The Lagrangian formulation is equivalent to a variational approach stating that the variation of the sum of the (time) rate of change of the Gibbs free energy and the dissipation function with respect to the flows in the system is a minimum. This approach leads to an extension of the Onsager principle of least energy dissipation to the far from equilibrium regime. I have further shown that the stationary state phenomenological equations for systems of chemical reactions display an algebraic symmetry property far from equilibrium regardless of the stability of the stationary state and given several examples of the relation between the Lagrangian formulation and the network thermodynamics of systems of chemical reactions.

Building on earlier work on variational principles for nonequilibrium coupled heat and mass transfer (Sieniutycz 1980, 1981, 1984,1987b, 1988), Sieniutycz (1987a) has derived similar results for reaction-diffusion systems from an action functional, which may be written as the integral over space and time of

$$(\mathcal{L}_{cons} + \mathcal{L}_{diss}) \exp(t/\tau). \tag{7.1}$$

$\mathcal{L}_{cons}$ is the classical conservative Lagrangian, here the difference between the sum of the kinetic energies of the diffusing species K^d and the Gibbs free energy:

$$\mathcal{L}_{cons} = K^d - G. \tag{7.2}$$

$\mathcal{L}_{diss}$, which one might call the dissipative Lagrangian, is simply the product of τ and the dissipation function $\mathcal{F}$ [equation (3.3)] I have used in this paper, with the resistances defined as in equation (2.5):

$$\mathcal{L}_{diss} = \tau \mathcal{F}. \tag{7.3}$$

τ is a relaxation time which "characterizes the average time between collisions." The exponential factor in the functional takes into account the (thermodynamic) irreversibility of time. When τ is small and the rates of reaction are not too large, conditions which insure the validity of the assumption of local equilibrium, Sieniutycz recovers the dynamics of mass action kinetics, and also finds the extension of the principle of minimum dissipation valid far from equilibrium found in this paper, as he earlier found an extension for coupled heat and mass transfer (Sieniutycz 1980, 1984). He argues that the extension should have the same status as Onsager's principle near equilibrium, even though only the variations with respect to the flows of the system are considered, i.e. the state of the system is not varied, since the state is also not varied in the Onsager treatment. Sieniutycz's approach would seem to be more general than mine. It includes inertial terms, does not demand the assumption of local equilibrium, results from the extremization of an action functional, as does the classical Lagrangian formulation, and is not limited to chemical reaction kinetics (Sieniutycz 1980, 1981, 1984,1987b, 1988). In addition, Sieniutycz and Berry (1989) have recently shown that the conservation laws for energy and momentum obtained from Sieniutycz's approach are in agreement with those obtained from statistical mechanics. It should be noted, however, that in the general case his formulation results in extra terms in the rate equations for the chemical reactions, which lead him to define extended reaction affinities related to "nonequilibrium chemical potentials". It is not known whether these are the proper extensions of the equilibrium concepts of chemical potential and reaction affinities valid when the local equilibrium assumption no longer holds.

The results of Sieniutycz and those presented here differ from previous ones on variational principles for chemical reactions. These are limited to the near equilibrium regime (Onsager 1931, Machlup and Onsager 1953, Onsager and Machlup 1953) or the neighborhood of a stable stationary state (Keizer 1977, 1978, 1979a, 1979b) and are based on the studies of fluctuations for the most part (Onsager 1931, Machlup and Onsager 1953, Onsager and Machlup 1953, Prigogine 1967, Glansdorff and Prigogine 1971, Nicolis and Prigogine 1977). The present approach is valid arbitrarily far from equilibrium and is macroscopic in the sense that it is based on the phenomenology of chemical kinetics. The work here should also be distinguished from the Lagrangian approach to first-order reaction systems (Shiner 1984) based on the "kinetic forces" for chemical reactions (Peusner 1982, Shiner 1984). There one notes that in the ideal case the rate of a first order reaction can be written in the form

$$k_{ij}c_i - k_{ji}c_j = \frac{1}{R_{ij}}\left(\frac{n_i}{C_i} - \frac{n_j}{C_j}\right), \tag{7.4}$$

where the k's are the first-order rate constants, the c's concentrations, and

$$k_{ij}c_i = \frac{n_i}{R_{ij}C_i}, \quad k_{ji}c_j = \frac{n_j}{R_{ij}C_j}, \tag{7.5}$$

although there is some controversy as to how R_{ij} and the C's should be interpreted (Peusner 1982, Shiner 1984, Peusner et al. 1985). The analogy with an electrical network is clear when one makes the identifications R - resistor, C - capacitor, and n - charge. The driving force, the kinetic force, for the reaction is then

$$\left(\frac{n_i}{C_i} - \frac{n_j}{C_j}\right). \tag{7.6}$$

For chemical reactions, however, the kinetic force is not the thermodynamic force conjugate to the rate of the reaction, and the formulations using the kinetic forces lack any thermodynamic or energetic meaning, although they do reproduce the dynamics of first-order reaction systems correctly. Furthermore, these formulations cannot be extended to reactions of order greater than one.

The kinetic forces are often used in formulations and applications of the network thermodynamics of systems of chemical reactions and of reaction-diffusion systems (Peusner 1982, Shiner 1984, Peusner et al. 1985). The applications often go further and choose whatever representation of chemical kinetics is most convenient for developing analogies to electrical networks (Wyatt et al. 1980, May and Mikulecky 1982). The choice of representation can then simplify the network analysis and simulation of the dynamics considerably. Unfortunately, the form of the chemical kinetics chosen often has no direct relation to thermodynamics. In these cases one would better refer to network simulations or network analogies, as Wyatt et al (1980) do, than network thermodynamics; the latter term could then be reserved for the truly thermodynamic network formulations of chemical reactions (Peusner 1970, Oster et al. 1973, Wyatt 1978).

Finally the question arises as to the significance of the algebraic symmetry proven for the phenomenological equations for systems of chemical reactions at stationary states arbitrarily far from equilibrium. The more powerful and restrictive property of Onsager reciprocity is valid only near equilibrium and for certain special cases. Should one view the differential symmetry of Onsager reciprocity as the important one and the algebraic symmetry as a weak correlate, or is the algebraic symmetry the general property and differential symmetry a special case valid in general only near equilibrium? One can argue for the latter case, since the algebraic property is the one which allows a network representation of systems of chemical reactions, electrical networks and mechanical systems. It would thus seem to be a general property of dynamic systems.

ACKNOWLEDGEMENTS
This research was supported in part by grant no. 3.033-0.87 from the Swiss National Science Foundation. I wish to thank Joel Keizer and Bernard Isaak for many helpful discussions.

REFERENCES & FOOTNOTES

Callen, H.B. 1960. Thermodynamics. New York: Wiley.
Corso, P.L. 1983. Thermodynamic and kinetic descriptions of equilibrium. J. Phys. Chem. 87: 2416-2419.
DeDonder, Th. 1928. L'Affinite. Paris: Gauthier-Villars [cited in Prigogine (1967)].
DeGroot, S.R. and P. Mazur. 1962. Nonequilibrium Thermodynamics, chpt. 10. Amsterdam: North-Holland.
Essig, A. and S.R. Caplan. 1981. Active transport: conditions for linearity and symmetry far from equilibrium. Proc. Natl. Acad. Sci. USA 78: 1647-1651.
Gibbs, J.W. 1931. On the equilibrium of heterogeneous substances. In The Collected Works of J. Willard Gibbs, 55-353. New York, Longmans-Green. (Originally published in Trans. Conn. Acad. III: 108-248, 1875-1876 and 343-524, 1877-1878.)
Glansdorff, P. and I. Prigogine. 1971. Thermodynamic Theory of Structure, Stability and Fluctuations. New York: Wiley-Interscience.
Goldstein, H. 1950. Classical Mechanics. Cambridge: Addison-Wesley.
Grabert, H., P. Hänggi and I. Oppenheim. Fluctuations in reversible chemical reactions. 1983. Physica 117A: 300-316.
Hill, T.L. 1974. Theoretical formalism for the sliding filament model of contraction of striated muscle. Part I. Prog. Biophys. Mol. Biol. 28: 267-340.
Hill, T.L. 1975. Theoretical formalism for the sliding filament model of contraction of striated muscle. Part II. Prog. Biophys. Mol. Biol. 29: 105-159.
Hill, T.L. 1977. Free Energy Transduction in Biology. New York: Academic.
Keizer, J. 1977. Variational principles in nonequilibrium thermodynamics. Biosystems 8: 219-226.
Keizer, J. 1978. Thermodynamics at nonequilibrium steady states. J. Chem. Phys. 69: 2609-2620.
Keizer, J. 1979a. Nonequilibrium thermodynamics and the stability of states far from equilibrium. Accts. Chem. Res. 12: 243-249.
Keizer, J. 1979b. Thermodynamics of nonequilibrium processes. In Pattern Formation by Dynamic Systems and Pattern Recognition, ed. H. Haken, 266-277. New York: Springer.
MacFarlane, A.G.J. 1970. Dynamical System Models. London: Harrap.
Machlup, S. and L. Onsager. 1953. Fluctuations and irreversible processes. II. systems with kinetic energy. Phys. Rev. 91:1512-1515.
Marion, J.B. 1965. Classical Dynamics of Particles and Systems. New York, Academic Press.
May, J.M. and D.C. Mikulecky. 1982. The simple model of adipocyte hexose transport: kinetic features, effect of insulin and network thermodynamic computer simulations. J. Biol. Chem. 257:11601-11608.

Nicolis, G. and I. Prigogine. 1977. Self-Organization in Nonequilibrium Systems. New York: Wiley-Interscience.
Onsager, L. 1931. Reciprocal relations in irreversible processes. I. Phys. Rev 37: 405-426.
Onsager, L. and S. Machlup. 1953. Fluctuations and irreversible processes. Phys. Rev. 91:1505-1512.
Oster, G.F., A. Perelson and A. Katchalsky. 1973. Network thermodynamics: dynamic modelling of biophysical systems. Q. Rev. Biophys. 6:1-134.
Peusner, L. 1970. PhD. Dissertation, Harvard University. Principles of Network Thermodynamics: Theory and Biophysical Applications. Lincoln, Mass.: Entropy Ltd. (1987).
Peusner, L. 1982. Global reaction: diffusion coupling and reciprocity in linear asymmetric kinetic networks. J. Chem. Phys. 77: 5500-5507.
Peusner, L., D.C. Mikulecky, B. Bunow and S.R. Caplan. 1985. A network thermodynamic approach to Hill and King-Altman reaction-diffusion kinetics. J. Chem. Phys. 83: 5559-5566.
Prigogine, I. 1967. Thermodynamics of Irreversible Processes. New York: Wiley-Interscience.
Prigogine, I. and G. Nicolis. 1971. Biological order, structure and instabilities. Quart. Rev. Biophys. 4:107-.
Ross, J., K.L.C. Hunt and P.M. Hunt. 1988. Thermodynamics far from equilibrium: reactions with multiple stationary states. J. Chem. Phys. 88: 2719-2729.
Rothschild, K.J., S.A. Ellias, A. Essig and H.E. Stanley. 1980. Nonequilibrium linear behavior of biological systems: existence of enzyme-mediated multidimensional inflection points. Biophys. J. 30: 209-230.
Shiner, J.S. 1984. A dissipative Lagrangian formulation of the network thermodynamics of (pseudo-) first-order reaction-diffusion systems. J. Chem. Phys. 81:1455-1465.
Shiner, J.S. 1987. Algebraic symmetry in chemical reaction systems at stationary states arbitrarily far from thermodynamic equilibrium. J. Chem. Phys. 87:1089-1094.
Shiner, J.S. 1990. A formulation of chemical reaction dynamics applicable to biological systems. In preparation.
Sieniutycz, S. 1980. The variational principle replacing the principle of minimum entropy production for coupled non-stationary heat and mass transfer processes with convective motion and relaxation. Int. J. Heat Mass Transfer 23:1183-1193.
Sieniutycz, S. 1981. Action functionals for linear wave dissipative systems with coupled heat and mass transfer. Phys. Lett. 84A:98-102.
Sieniutycz, S. 1984. Variational approach to extended irreversible thermodynamics of heat and mass transfer. J. Non-Equilib. Thermodyn. 9:61-70.
Sieniutycz, S. 1987a. From a least action principle to mass action law and extended affinity. Chem. Eng. Sci. 42:2697-2711.
Sieniutycz, S. 1987b. Variational approach to the fundamental equations of heat, mass, and momentum transport in strongly unsteady-state processes. I. General action functionals. Int. Chem. Eng. 27:545-555.
Sieniutycz, S. 1988. Variational expressions underlying the equations for the transport of heat, mass, and momentum in highly unsteady-state processes. II. Examples of application of the theory. Int. Chem. Eng. 28:353-361.
Sieniutycz, S. and R.S. Berry. 1989. Conservation laws from Hamilton's principle for nonlocal thermodynamic equilibrium fluids with heat flow. Phys. Rev. A 40:348-361.
Stucki, J.W., M. Compiani and S.R. Caplan. 1983. Efficiency of energy conversion in model biological pumps. Optimization by linear nonequilibrium thermodynamic relations. Biophys. Chem. 18: 101-109.

Wyatt, J.L. 1978. Network representation of reaction-diffusion systems far from equilibrium. Comput. Programs Biomed. 8: 180-195.
Wyatt, J.L., D.C. Mikulecky and J.A. DeSimone. 1980. Network modelling of reaction-diffusion systems and their numerical solution using SPICE. Chem. Eng. Sci. 35: 2115-2128.

Consistency of Chemical Kinetics with Thermodynamics

S. Lengyel
Central Research Institute for Chemistry
Hugarian Academy of Sciences
H-1025 Pusztaszeri ut 59-67. Budapest, Hungary

ABSTRACT

A review of the contribution of the Hungarian group of nonequilibrium thermodynamics to the study of consistency of chemical kinetics with thermodynamics. Stoichiometrically dependent and independent elementary chemical reactions are defined and their entropy production is given in bilinear forms. The exact form of the nonlinear constitutive equation is derived for stoichiometrically independent chemical reactions and satisfaction of the Gyarmati-Li generalized reciprocity relations is shown in multireaction chemical kinetic systems. Evolution of reversible and irreversible chemical reactions is described in closed adiabatically insulated systems. The governing principle of dissipative processes is applied to chemical kinetic systems and the kinetic mass action law is derived from the dissipation potentials of suitable form. By this derivation the kinetic mass action law is integrated into nonequilibrium thermodynamics. It is shown that kinetics of chemical reactions is consistent with thermodynamics. Accordingly, the old conception which identified affinity divided by the temperature with the thermodynamic force driving a chemical reaction must be refused. Instead, the affinity must be replaced by the partial Gibbs free energy of the ensemble of the reactants of the reaction. The corresponding flux (rate) potential, however, is a function of both the partial Gibbs free energy of the ensemble of the reactants and the partial Gibbs free energy of the products of the reaction as well. The theory is used to the solution of a chemical problem which others tried but were unable to solve. Some problems arising in the study of the consistency of chemical kinetics with thermodynamics are discussed.

1. INTRODUCTION

The relationship between velocity of chemical reactions and heat phenomena accompanying chemical reactions is a very old topic examined by many authors in the past. The origins go back to times when neither of the two branches of science thermodynamics and chemical kinetics were developed. Later on, the relationship between phenomenological chemical kinetics and modern thermodynamics became the subject of a great number of studies. Some recent papers give analyses and lists of the relevant literature (Lengyel and Gyarmati 1981a). The present article is written to give a review of the contribution of the Hungarian group of nonequilibrium thermodynamics to the theory of the relationship between thermodynamics of irreversible processes and phenomenological reaction kinetics, and to publish some recent results.

2. DEPENDENT AND INDEPENDENT CHEMICAL REACTIONS AND THEIR ENTROPY PRODUCTION

We assume that in a continuous medium out of equilibrium the elementary chemical reactions

$$\sum \nu'_{kt} B_k \longrightarrow \sum \nu''_{kt} B_k \qquad (t=1,2,\ldots,S) \qquad (2.1)$$

involving the chemical components $B_1,\ldots,B_k,\ldots,B_K$ proceed at the rates

$$\vec{J}_1,\ldots,\vec{J}_t,\ldots,\vec{J}_S \qquad (2.2)$$

in all volume elements of the continuum. In Eq. (2.1), summation has to be carried out over all reactants of reaction t on the left hand side and over all products of the same reaction on the right hand side. The reaction rates (2.2) are space and time dependent fields. By the complete set of reactions (2.1) entropy is produced in unit volume of a volume element at the rate

$$\sigma = \sum_{t=1}^{S} \vec{J}_t[(A'_t/T)-(A''_t/T)] \ , \qquad (2.3)$$

since the amount of substance of component B_k per unit volume is changed at the rate

$$\sigma_k = \sum_{t=1}^{S} (\nu''_{kt}-\nu'_{kt})\vec{J}_t \qquad (k=1,2,\ldots,K) \qquad (2.4)$$

and

$$\Gamma_k = -\mu_k/T \qquad (k=1,2,\ldots,K) \qquad (2.5)$$

is the corresponding intensive variable and, finally,

$$\sigma = \sum_{k=1}^{K} \Gamma_k \sigma_k \qquad (2.6)$$

(see Section 6).

In Eq. (2.3)

$$A'_t = \sum \nu'_{kt} \mu_k \qquad (2.7)$$

is the partial value of the Gibbs free energy belonging to the ensemble of the reactants of reaction t and

$$A''_t = \sum \nu''_{kt} \mu_k \qquad (2.8)$$

is the partial value of the Gibbs free energy of the ensemble of the products of the same reaction. The symbols μ and T denote chemical potentials of the chemical components and temperature, respectively.

If the rank of the stoichiometric matrices $[\nu'_{kt}]$ and $[\nu''_{kt}]$ is $Q<S$, the complete set of reactions (2.1) will be linearly dependent and the maximum number of linearly independent chemical equations will be Q. Let the collection of reactions

$$\sum \nu'_{ku} B_k \longrightarrow \sum \nu''_{ku} B_k \qquad (u=1,2,\ldots,Q) \qquad (2.9)$$

be such an independent subset. In this case the relations

$$A'_t = \sum_{u=1}^{Q} \gamma_{tu} A'_u \;; \quad A''_t = \sum_{u=1}^{Q} \gamma_{tu} A''_u \qquad (2.10)$$

and

$$\sigma = \sum_{t=1}^{S} \vec{J}_t [(A'_t/T) - (A''_t/T)] = \sum_{u=1}^{Q} \vec{J}^{*}_u [(A'_u/T) - (A''_u/T)] \qquad (2.11)$$

will hold with

$$\vec{J}^{*}_{u} = \sum_{t=1}^{S} \gamma_{tu} \vec{J}_{t} \qquad (u=1,2,\ldots,Q) \; . \qquad (2.12)$$

The coefficients $\gamma_{tu} \gtreqless 0$ are small integers or zero. (For details see Lengyel and Gyarmati 1981b and Lengyel 1989b). According to Eq. (2.11), the entropy production by the complete set (2.1) of elementary chemical reactions is a bilinear form of S rates ($\vec{J}_t$) and S dependent forces $(A'_t/T)-(A''_t/T)$ or, alternatively, a bilinear form of Q rates ($\vec{J}^{*}_{u}$) and Q linearly independent forces $(A'_u/T)-(A''_u/T)$.

3. CONSTITUTIVE (PHENOMENOLOGICAL) EQUATIONS. RECIPROCITY RELATIONS

Constitutive equations relate thermodynamic fluxes (rates) to thermodynamic forces and vice versa. In the case of the complete set (2.1) of elementary chemical reactions the kinetic mass action law (Guldberg and Waage 1867)

$$\vec{J}_{t} = \vec{k}_{t} \prod_{k=1}^{K} c_{k}^{\nu'_{kt}} \qquad (t=1,2,\ldots,S) \qquad (3.1)$$

holds in ideal systems or

$$\vec{J}_{t} = \vec{k}_{t} \prod_{k=1}^{K} a_{k}^{\nu'_{kt}} \qquad (t=1,2,\ldots,S) \qquad (3.2)$$

in nonideal systems (Lengyel 1989b). In Eqs. (3.1) and (3.2) $\vec{k}_t$ is the rate coefficient of reaction t, c_k the concentration in amount of substance per unit volume and a_k the relative activity of component B_k. By substitution of

$$c_k/c^{\theta} = \exp[(\mu_k-\mu^{\theta}_k)/RT] \qquad (k=1,2,\ldots,K) \qquad (3.3)$$

and

$$a_k = \exp[(\mu_k-\mu^{\circ}_k)/RT] \; , \qquad (k=1,2,\ldots,K) \qquad (3.4)$$

respectively, Eqs. (3.1) and (3.2) will be transformed into the Marcelin-Kohnstamm-type (Marcelin 1910; Kohnstamm and Scheffer 1911) nonlinear constitutive equations

$$\vec{J}_{t} = \vec{\lambda}_{t} e^{A'_t/RT} \qquad (t=1,2,\ldots,S) \qquad (3.5)$$

where

$$\vec{\lambda}_t = \vec{k}_t \exp(-\sum \nu'_{kt}\mu^{\theta}_k/RT) \tag{3.6}$$

and R is the gas constant. Standard values are labeled by θ or o.

Since the complete set $A'_1,\ldots,A'_t,\ldots,A'_S$ is linearly dependent, nothing can be directly said about the interrelations of the partial derivatives

$$\partial\vec{J}_t/\partial A'_s \qquad (t,s=1,2,\ldots,S) \; . \tag{3.7}$$

The properties of the complete set

$$\vec{J}^*_1,\ldots,\vec{J}^*_u,\ldots,\vec{J}^*_Q \; , \tag{3.8}$$

however, are different, because the variables on which they depend (i.e. $A'_1,\ldots,A'_u,\ldots,A'_Q$) are independent. Substituting Eqs. (2.10) and (3.5) into (2.12) we obtain the explicit form

$$\vec{J}^*_u = \sum_{t=1}^{S} \gamma_{tu}\vec{\lambda}_t \exp(\sum_{v=1}^{Q} \gamma_{tv}A'_v/RT) \qquad (u=1,2,\ldots,Q) \tag{3.9}$$

for $\vec{J}^*_u$. In the case of these constitutive equations the general reciprocity relations

$$\partial\vec{J}^*_u/\partial A'_v = \partial\vec{J}^*_v/\partial A'_u \qquad (u,v=1,2,\ldots,Q) \tag{3.10}$$

as postulated by Gyarmati (Gyarmati 1961a; Gyarmati 1961b) and by Li (Li 1958) in general, and in the special case of chemical reactions by VAN Rysselberghe (VAN Rysselberghe 1962) are satisfied, since

$$\partial\vec{J}^*_u/\partial A'_r = \sum_{t=1}^{S} (\gamma_{tu}\gamma_{tr}/RT)\; \vec{\lambda}_t \exp(\sum_{v=1}^{Q} \gamma_{tv}A'_v/RT) = \partial\vec{J}^*_r/\partial A'_u \tag{3.11}$$

(see Lengyel 1988).

Among the S elementary chemical reactions (2.1) there will be several pairs of antagonistic processes. In other words, the products of reaction t will be identical with the reactants of reaction s and vice versa. In such cases we shall speak about a reversible chemical reaction and write

$$\sum_k \nu'_{kt}B_k \rightleftharpoons \sum_k \nu''_{kt}B_k \; . \tag{3.12}$$

The reversible chemical reaction will advance at the net rate

$$J_t = \vec{J}_t - \overleftarrow{J}_t \tag{3.13}$$

where $\overleftarrow{J}_t=\vec{J}_s$ denotes the rate of the backward reaction, i.e. of the reaction reverse to reaction t.

4. EVOLUTION OF NONEQUILIBRIUM SYSTEMS

If a thermodynamic system is out of equilibrium and accordingly, processes advance in it, its state variables and, consequently, fluxes (rates) and forces of the processes will continuously change with time t in all points r of the system. We speak, in a word, about evolution of the system, namely evolution in space and in time. If we restrict the analysis to the chemical reactions (2.1), we may describe the evolution in time in any volume element of the system by means of the coordinates

$$\vec{J}_1,\ldots,\vec{J}_t,\ldots,\vec{J}_S\ ;\ A'_1/T,\ldots,A'_t/T,\ldots,A'_S/T\ ;\ A''_1/T,\ldots,A''_t/T,\ldots,A''_S/T\ . \tag{4.1}$$

In the space of this coordinate system (namely (4.1)) the evolution will follow a trajectory. If the thermodynamic system is closed (no exchange of matter with the surroundings) and adiabatically insulated (exchange of heat with the surroundings prevented), it will approach step by step an equilibrium state where the rate of entropy production vanishes in all points r. The net rates J_t of the reversible chemical reactions will vanish too. The rates $\vec{J}_t$ and $\overleftarrow{J}_t$, however, will assume equal ($\vec{J}_t = \overleftarrow{J}_t$) but finite values. The forces (in equilibrium)

$$A'_t/T = A''_t/T \qquad (t=1,2,\ldots,S) \tag{4.2}$$

will assume finite values as well. Such an equilibrium is traditionally termed dynamic equilibrium. Irreversible chemical reactions do not permanently advance in the system. They transiently happen and the chemical species involved are not always present in the system; they temporarily occur then disappear and are absent in equilibrium. Hence, the rates $\vec{J}_s$ and forces A'_s/T and A''_s/T of the irreversible chemical reactions will vanish in equilibrium, provided that the chemical potentials of these species are related to standard values in infinitely diluted solutions. These standard values are known from usage in the physical chemistry of such solutions.

5. VARIATION PRINCIPLES IN THERMODYNAMICS

Thermodynamic systems evolve in the course of time. The coordinate system of the Cartesian components of the generalized fluxes and of the thermodynamic forces can be used to define the trajectories of this evolution. If elementary chemical reactions (2.1) are the only processes taking place in the system, the evolution will follow a trajectory prescribed by nature in the space of the coordinates (4.1).

Trajectory is a concept of theoretical mechanics where it was first used to specify the curve represented by the coordinates $q_r(t)$ $(r=1,2,\ldots,f)$ as functions of time t as parameter, i.e. the mechanical path in configuration space. In dynamics trajectory specifies the curve represented by the momenta $p_r(t)$ in momentum space (or by velocities $\dot{q}_r(t)$ in velocity space). See for example Mercier 1959. Extended to nonequilibrium thermodynamics trajectory specifies the curve represented by the flux components and force components as functions of time in the corresponding flux-force space. Like in dynamics, one of all imaginable trajectories is favored by nature in thermodynamics too. This trajectory can be singled out by a variation principle in dynamics and in thermodynamics as well.

According to the variation principle (Hamilton's principle in mass-point mechanics), it is possible to find a function (the Lagrange-function) for which any variation corresponding to independent variations of the coordinates and velocities, vanishes. The Lagrange function or simply Lagrangian accounts for interaction between particles.

In thermodynamics variation principles were conceived as least energy dissipation (Onsager 1931), minimum entropy production (valid for steady state processes alone) (Prigogine 1945; Prigogine 1947) and governing principle of dissipative processes (Gyarmati 1965; Gyarmati 1969; Gyarmati 1970). Among them the last one is the most general. In the governing principle of dissipative processes in continua the Lagrangian

$$L = \int_V (\sigma - \psi - \varphi)\, dV \qquad (5.1)$$

is used where the density σ of entropy production rate is the bilinear form of the scalar components of the fluxes and of the thermodynamic forces, ψ is the density of the flux potential, a function of the force components and φ is the density of the force potential, a function of the flux components. The functions σ, ψ and φ are space and time dependent fields. V is the volume of the system.

6. CHEMICAL REACTIONS AS LOCAL PROCESSES IN THE UNIFIED FIELD THEORY OF THERMODYNAMICS

Nonequilibrium thermodynamics of continuous media was elaborated in the 1940s and 1950s (Meixner 1941a; Meixner 1941b; Meixner 1942; Meixner 1943a; Meixner 1943b; Prigogine 1947; de Groot 1951) and rounded off by a unified field theory (Gyarmati 1970) and by the governing principle of dissipative processes (Gyarmati 1965; Gyarmati 1969; Gyarmati 1970) in the 1960s. According to Gyarmati's unified field theory of thermodynamics, in a continuous medium in a nonequilibrium state the densities $a_1(r,t),\dots,a_i(r,t),\dots,a_f(r,t)$ of the extensive quantities are space and time dependent fields. The extensive quantities are carried by <u>transport</u> <u>processes</u> with the current densities $J_1(r,t),\dots,J_i(r,t),\dots,J_f(r,t)$. These transport currents increase the extensive quantities in unit volume of a volume element by the inflow rates $-\nabla\cdot J_i$ $(i=1,2,\dots,f)$ (the symbol $\nabla\cdot$ is for divergence). Transport processes may be accompanied by mechanical, electromagnetic etc. work performed by internal forces (the pressure tensor) and/or by external (electromagnetic etc.) forces which are in cause and effect relation with the tensorial or vectorial currents. The corresponding extensive quantity (internal energy in the first place) will then be increased at the rate σ_i termed source density.

The entropy density s is a function of the complete set of the extensive densities: $s \equiv s(a_1,\dots,a_i,\dots,a_f)$ and so are the densities of other thermodynamic functions like enthalpy, Helmholtz free energy and Gibbs free energy. The intensive variables $\Gamma_1,\dots,\Gamma_i,\dots,\Gamma_f$ are the partial derivatives of the entropy density with respect to the densities of the extensive quantities and are space and time dependent fields as well.

In addition to the transport processes, <u>local processes</u> also take place in the continuum: chemical reactions in the first place. They increase the corresponding quantities (in the case of chemical reactions the amounts of substance of the chemical components per unit volume) at the rates $\sigma_k(r,t)$ $(k=1,2,\dots,K)$ termed source densities. As seen, we have to distinguish two kinds of processes: transport processes and local (non-transport) processes. The latter are often called purely dissipative processes.

Pertaining to the extensive quantities the balance equations

$$\partial a_i/\partial t = -\nabla\cdot J_i + \sigma_i \qquad (i=1,2,\dots,f) \tag{6.1}$$

hold. In words, the rate of increase of an extensive quantity is equal to the inflow $-\nabla\cdot J_i$ due to transport processes plus the rate of generation σ_i due to local or in some cases to transport processes (or to both).

Entropy is assumed to flow in a continuous medium with the current density

$$J_s = \sum_{i=1}^{f} \Gamma_i J_i \tag{6.2}$$

due to the transport processes and is assumed to be generated (produced) due both to transport processes and local processes. According to the entropy balance equation, the density σ_s of the rate of entropy production is equal to the entropy change rate

$$\partial s/\partial t = \sum_{i=1}^{f} \Gamma_i \partial a_i/\partial t = -\nabla\cdot\sum_{i=1}^{f} \Gamma_i J_i + \sum_{i=1}^{f} J_i\cdot\nabla\Gamma_i + \sum_{i=1}^{f} \Gamma_i \sigma_i \tag{6.3}$$

minus the rate of entropy inflow $-\nabla\cdot J_s$. Thus, we have the entropy production rate in unit volume

$$\sigma_s = \sum_{i=1}^{f} J_i\cdot\nabla\Gamma_i + \sum_{i=1}^{f} \Gamma_i \sigma_i \ . \tag{6.4}$$

The last term, among others, contains the entropy production by the purely dissipative processes. In Eqs. (6.3) and (6.4) the symbols ∇ and $\nabla\cdot$ are for gradient and divergence, respectively.

The elementary chemical reactions (2.1) are local (purely dissipative; non-transport) processes advancing at the rates (2.2). Let the complete set of the chemical reactions be assumed stoichiometrically independent. Chemical component B_k is produced by the reactions (2.1) at the rate (2.4) and the corresponding entropy production by the complete set of reactions (2.1) will be (2.3). The partial values A'_t and A''_t of the Gibbs free energy are defined by Eqs. (2.7) and (2.8).

For thermodynamic calculations pertaining to chemical kinetic systems, in addition to the thermodynamic relation (second law of thermodynamics)

$$\sigma = \sum_{t=1}^{S} \vec{J}_t [(A'_t/T) - (A''_t/T)] > 0 \qquad (6.5)$$

expressing positive entropy production by the complete set of the chemical reactions (2.1), also the equations

$$\vec{J}_t = \vec{k}_t \prod_{k=1}^{K} a_k^{\nu'_{kt}} = \vec{\lambda}_t e^{A'_t/RT} \qquad (t=1,2,\ldots,S) \qquad (6.6)$$

of chemical kinetics must be known. By the use of information on the thermodynamic and the kinetic equations and their coefficients the evolution of the system can be studied. In other words, in this theory for the thermodynamic examination of chemical kinetic systems the combination of the two branches of science thermodynamics and chemical kinetics is necessary.

7. SYNTHESIS OF THERMODYNAMICS AND CHEMICAL KINETICS

We may, however, raise this combination to a higher level, to the level of synthesis or in other words, integration of the two branches of science if we start with the most general variation principle of thermodynamics, namely with Gyarmati's governing principle of dissipative processes (Gyarmati 1965; Gyarmati 1969; Gyarmati 1970). According to the global form of this principle, the time and space evolution of the system (say a continuous medium of volume V in the time interval $t_0 \leq t < \infty$) is governed by the equation

$$\delta \int_V \int_{t=t_0}^{\infty} (\sigma - \psi - \varphi)\, dV dt = \int_V \int_{t=t_0}^{\infty} (\delta\sigma - \delta\psi - \delta\varphi)\, dV dt = 0 \qquad (7.1)$$

where δ denotes variation.

If we specialize this principle of general validity to the case of the elementary chemical reactions (2.1) (considered as stoichiometically independent) in a continuous medium, we shall have Eq. (2.3) for the density of entropy production rate whose variation is:

$$\delta\sigma = \sum_{t=1}^{S} \vec{J}_t \delta(A'_t/T) - \sum_{t=1}^{S} \vec{J}_t \delta(A''_t/T) + \sum_{t=1}^{S} [(A'_t/T) - (A''_t/T)]\delta\vec{J}_t . \tag{7.2}$$

The integration of phenomenological chemical kinetics into thermodynamics can be performed by the proper choice of the dissipation potentials ψ and φ. Let us postpone the specification of the force potential

$$\varphi \equiv \varphi(\vec{J}_1, \ldots, \vec{J}_t, \ldots, \vec{J}_S) . \tag{7.3}$$

For the flux (rate) potential ψ we postulate the function

$$\psi = 2R\sum_{t=1}^{S} \vec{\lambda}_t e^{A'_t/RT}\left(1-e^{A''_t/RT}\right). \tag{7.4}$$

Substituting the variations of σ, ψ and φ into Eq. (7.1) we obtain following specific form of the governing principle of dissipative processes

$$\int_V \int_{t=t_0}^{\infty} \Bigg\{ \sum_{t=1}^{S} \left[\vec{J}_t - 2\vec{\lambda}_t e^{A'_t/RT} + 2\vec{\lambda}_t e^{(A'_t+A''_t)/RT}\right] \delta(A'_t/T) -$$

$$- \sum_{t=1}^{S} \left[\vec{J}_t - 2\vec{\lambda}_t e^{(A'_t+A''_t)/RT}\right] \delta(A''_t/T) +$$

$$+ \sum_{t=1}^{S} \left[(A'_t/T) - (A''_t/T) - (\partial\varphi/\partial\vec{J}_t)\right] \delta\vec{J}_t \Bigg\} dVdt = 0 \quad . \tag{7.5}$$

Due to independence of the variations in Eq. (7.5) we arrive at the equations

$$\vec{J}_t = 2\vec{\lambda}_t e^{A'_t/RT} - 2\vec{\lambda}_t e^{(A'_t+A''_t)/RT} \qquad (t=1,2,\ldots,S) \tag{7.6}$$

and

$$\vec{J}_t = 2\vec{\lambda}_t e^{(A'_t+A''_t)/RT} \qquad (t=1,2,\ldots,S) \tag{7.7}$$

valid in any point of the system at any time. Their sum gives the kinetic mass action law

$$\vec{J}_t = \vec{\lambda}_t e^{A'_t/RT} = \vec{k}_t \prod_{k=1}^{K} a_k^{\nu'_{kt}} \qquad (t=1,2,\ldots,S) \tag{7.8}$$

according to which the reaction rate is a function of the partial value A'_t of the Gibbs free energy (or of the relative activities

a_k) of the ensemble of the reactants only and does not depend on the partial value A''_t of the Gibbs free energy (or the relative activities) of the ensemble of the reaction products. Thus, the postulated form of ψ covers the phenomenological theory of reaction kinetics.

Postulating now the function

$$\varphi = R\sum_{t=1}^{S} \vec{J}_t[\ln(2\vec{J}_t/\vec{\lambda}_t)-1] \qquad (7.9)$$

for the force potential, in addition to Eqs. (7.6) and (7.7) we obtain

$$(A'_t/T)-(A''_t/T) - R\ln(2\vec{J}_t/\vec{\lambda}_t) = 0 \qquad (7.10)$$

as the specified form of the expression in brackets of the last term of Eq. (7.5). Combination of Eq. (7.10) with the logarithmic form of Eq. (7.7), namely with

$$(A'_t/T)+(A''_t/T) = R\ln(\vec{J}_t/2\vec{\lambda}_t) \qquad (7.11)$$

leads to the inverse (logarithmic) form of Eq. (7.8), i.e. to the kinetic mass action law once again.

Summing up we may say that the specified functions (7.4) and (7.9) for the dissipation potentials cause the governing principle of dissipative processes to include the whole phenomenological theory of chemical kinetics. The same postulated functions for the dissipation potentials, on the other hand, cause the principle to be in accordance with the specified bilinear form (2.3) of the entropy production and with its variation (7.2).

It should be added here that it is not necessary to assume the complete set (2.1) of the elementary chemical reactions to be stoichiometrically independent. If the set (2.1) is linearly dependent, the subset (2.9) will be independent. In this case, using relations (2.10)-(2.12), instead of equations (7.2) and (7.4) we have the equations

$$\delta\sigma = \sum_{u=1}^{Q}\vec{J}^{*}_u\delta(A'_u/T)-\sum_{u=1}^{Q}\vec{J}^{*}_u\delta(A''_u/T)+\sum_{t=1}^{S}[(A'_t/T)-(A''_t/T)]\delta\vec{J}_t \qquad (7.12)$$

and

$$\psi = 2R\sum_{t=1}^{S}\vec{\lambda}_t e^{\sum_{v=1}^{Q}\gamma_{tv}A'_v/RT}\left(1-e^{\sum_{v=1}^{Q}\gamma_{tv}A''_v/RT}\right) \qquad (7.13)$$

respectively. For the specified local form of the governing principle of dissipative processes we shall obtain the equation

$$\sum_{u=1}^{Q}\left[\vec{J}_u^{*}-2\sum_{t=1}^{S}\gamma_{tu}\vec{\lambda}_t\exp\left(\sum_{v=1}^{Q}\gamma_{tv}A'_v/RT\right)+\right.$$

$$\left.+\ 2\sum_{t=1}^{S}\gamma_{tu}\vec{\lambda}_t\exp\left(\sum_{v=1}^{Q}\gamma_{tv}(A'_v+A''_v)/RT\right)\right]\delta(A'_u/T)\ -$$

$$-\sum_{u=1}^{Q}\left[\vec{J}_u^{*}-2\sum_{t=1}^{S}\gamma_{tu}\vec{\lambda}_t\exp\left(\sum_{v=1}^{Q}\gamma_{tv}(A'_v+A''_v)/RT\right)\right]\delta(A''_u/T)\ +$$

$$+\sum_{t=1}^{S}\left[(A'_t/T)-(A''_t/T)-R\ \ln(2\vec{J}_t/\vec{\lambda}_t)\right]\delta\vec{J}_t\ =\ 0 \qquad (7.14)$$

where all variations are linearly independent. It is clear that finally we obtain the same result as in the case in which the set (2.1) was assumed to be independent.

If the S elementary reactions (2.1) are coupled to S/2 reversible reactions, reactants and products will exchange their roles in the backward reactions with respect to those of the forward reactions. The reversible reactions will advance at the net rates $J_t = \vec{J}_t - \overleftarrow{J}_t$ $(t=1,2,\ldots,S/2)$ which will replace $\vec{J}_1,\ldots,\vec{J}_t,\ldots,\vec{J}_S$ in the bilinear form of the entropy production. Substitution of the corresponding functions for entropy production and dissipation potentials finally will lead to the constitutive equation

$$J_t = \lambda_t\left(e^{A'_t/RT} - e^{A''_t/RT}\right). \qquad (7.15)$$

This equation was heuristically postulated, independently and almost at the same time (December 5 and 24, respectively, in 1910) by Marcelin in Paris (Marcelin 1910) and Kohnstamm and Scheffer in Amsterdam (Kohnstamm and Scheffer 1911) who with this postulation made the first step on the way of development of thermodynamics of irreversible processes. Their equation, namely, was the first one that expressed the rate of a (irreversible) process in terms of quantities defined in thermostatics, i.e. in terms of the partial values of the Gibbs free energy of the ensemble of the reactants and of that of the products. In the development of thermodynamics of irreversible processes the roles of Onsager and Prigogine are generally known, renown and honored. The most powerful phenomenological thermodynamic theory, however, is Gyarmati's unified field theory (Gyarmati 1970) with the governing principle of dissipative processes (Gyarmati 1965; Gyarmati 1969) that by suitable choices of the dissipation potentials (Lengyel 1988; Lengyel 1989b) can be made a thermodynamic umbrella covering the phenomenological theory of chemical kinetics and as a review of the literature shows (Lengyel 1989b), other science branches like elasticity, viscous and plastic flow, heat conduction, electric conduction, diffusion etc. and their cross effects.

8. SOLUTION OF A CHEMICAL PROBLEM

As an example we shall treat a chemical kinetic system constructed by Bataille, Edelen and Kestin (Section 7 of Bataille, Edelen and

Kestin 1978). The system consists of two reversible reactions among the four species B_1, B_2, B_3 and B_4 :

$$[1] \quad B_1+2B_2 \rightleftharpoons B_3+2B_4 \tag{8.1}$$
$$[2] \quad B_1+2B_3+B_4 \rightleftharpoons 4B_2 \ . \tag{8.2}$$

In the system we actually have the four reactions

$$(1) \quad B_1+2B_2 \longrightarrow B_3+2B_4 \tag{8.3}$$
$$(2) \quad B_1+2B_3+B_4 \longrightarrow 4B_2 \tag{8.4}$$
$$(3) \quad B_3+2B_4 \longrightarrow B_1+2B_2 \tag{8.5}$$
$$(4) \quad 4B_2 \longrightarrow B_1+2B_3+B_4 \tag{8.6}$$

that are linearly independent, because the rank of the stoichiometric matrices $[\nu'_{kt}]$ and $[\nu''_{kt}]$ is equal to 4. The partial values of the Gibbs free energy of reactants (A'_t) and of products (A''_t) are the following

$$A'_1 = \mu_1+2\mu_2 \qquad A''_1 = \mu_3+2\mu_4 \tag{8.7}$$
$$A'_2 = \mu_1+2\mu_3+\mu_4 \qquad A''_2 = 4\mu_2 \tag{8.8}$$
$$A'_3 = \mu_3+2\mu_4 \qquad A''_3 = \mu_1+2\mu_2 \tag{8.9}$$
$$A'_4 = 4\mu_2 \qquad A''_4 = \mu_1+2\mu_3+\mu_4 \ . \tag{8.10}$$

The set A'_1, A'_2, A'_3, A'_4 is linearly independent. The set A''_1, A''_2, A''_3, A''_4 as well. The two sets are interrelated by the equations

$$A''_3 = A'_1 \ ; \ A''_4 = A'_2 \ ; \ A''_1 = A'_3 \ ; \ A''_2 = A'_4 \ . \tag{8.11}$$

The affinities of the reactions are

$$A_1 = A'_1-A''_1 = \mu_1+2\mu_2-\mu_3-2\mu_4 = -A_3 \tag{8.12}$$
$$A_2 = A'_2-A''_2 = \mu_1-4\mu_2+2\mu_3+\mu_4 = -A_4 \ , \tag{8.13}$$

according to the fact that reaction (3) is reverse to reaction (1) and reaction (4) reverse to reaction (2). Bataille, Edelen and Kestin have shown that the net rates

$$J_1 = \lambda_1\left(e^{A_1/RT}-e^{-A_1/RT}\right) \tag{8.14}$$

$$J_2 = \lambda_2\left(e^{A_2/RT}-e^{-A_2/RT}\right) \tag{8.15}$$

do not satisfy the reciprocity relation

$$\partial J_1/\partial(A_2/T) = \partial J_2/\partial(A_1/T) \ . \tag{8.16}$$

As we see, the affinities divided by the temperature were considered as independent forces in the two reversible reactions [1] and [2] whose net rates are represented by Eqs. (8.14) and (8.15).

In our theory we separately treat the four independent reactions (1)-(4) whose rates are

$$\vec{J}_1 = \vec{\lambda}_1 e^{A_1'/RT} \tag{8.17}$$

$$\vec{J}_2 = \vec{\lambda}_2 e^{A_2'/RT} \tag{8.18}$$

$$\vec{J}_3 = \vec{\lambda}_3 e^{A_3'/RT} \tag{8.19}$$

$$\vec{J}_4 = \vec{\lambda}_4 e^{A_4'/RT} \quad . \tag{8.20}$$

Since the equations for the variables A_1', A_2', A_3', A_4' are linearly independent, the reciprocity relations

$$\partial\vec{J}_r/\partial(A_s'/T) = \partial\vec{J}_s/\partial(A_r'/T) = 0 \qquad (r,s=1,2,3,4;\ r\neq s) \tag{8.21}$$

will be trivially satisfied provided that the variables A_t'/T $(t=1,2,3,4)$ are identified with the thermodynamic forces. This is in opposition to the general usage followed also by Bataille, Edelen and Kestin. Thus we see that the proper choice of the thermodynamic forces ensures satisfaction of the generalized reciprocity relations, as predicted by Lengyel and Gyarmati (Lengyel and Gyarmati 1981a).

The dissipation potentials ψ and φ can easily be composed corresponding to Eqs. (7.4) and (7.9) as

$$\psi = 2R\sum_{t=1}^{4}\vec{\lambda}_t\left[e^{A_t'/RT} - e^{(A_t'+A_t'')/RT}\right] \tag{8.22}$$

and

$$\varphi = R\sum_{t=1}^{4}\vec{J}_t\left[\ln(2\vec{J}_t/\vec{\lambda}_t)-1\right] \tag{8.23}$$

whose variation with respect to the independent forces and reaction rates combined with the variation of the entropy production leads to the equations

$$\vec{J}_t = 2\vec{\lambda}_t e^{A_t'/RT} - 2\vec{\lambda}_t e^{(A_t'+A_t'')/RT} \tag{8.24}$$

$$\vec{J}_t = 2\vec{\lambda}_t e^{(A_t'+A_t'')/RT} \tag{8.25}$$

and

$$(A_t'/T) - (A_t''/T) = R\ \ln(2\vec{J}_t/\vec{\lambda}_t) \tag{8.26}$$

where t=1,2,3,4. Finally, combination of Eqs. (8.24) and (8.25) gives the kinetic mass action law

$$\vec{J}_t = \vec{\lambda}_t e^{A'_t/RT} = \vec{k}_t \prod a_k^{\nu'_{kt}} \qquad (t=1,\ldots,4) \qquad (8.27)$$

and combination of Eqs. (8.25) and (8.26) the inverse (logarithmic) form

$$A'_t = RT \ln(\vec{J}_t/\vec{\lambda}_t) \qquad (t=1,\ldots,4) \qquad (8.28)$$

of the mass action law for all four of the independent reactions (1)-(4).

On the other hand, the authors having constructed the system of the two reversible reactions [1] and [2] (Bataille, Edelen and Kestin 1978), in addition to showing that the general reciprocity relations are not satisfied (provided that the affinities divided by the temperature are assumed to be the thermodynamic forces), calculated a rather complicated dissipation function ψ.

9. DISCUSSION OF SOME PROBLEMS OF PAST, PRESENT AND FUTURE

9.1 The past

During the past 35 years a series of authors postulated Onsager's reciprocity relations in the generalized form

$$\partial J_r/\partial X_s = \partial J_s/\partial X_r \qquad (r,s=1,2,\ldots,F) \qquad (9.1)$$

where the subindexed J and X denote Cartesian components of tensorial fluxes and thermodynamic forces, respectively. The tensorial order may be two, one or zero. A review and analysis of the attempts to prove Eqs. (9.1) or verify the satisfaction of these equations was given in 1981 (Lengyel and Gyarmati 1981a) and, restricted to a few of these authors in 1989 (Lengyel 1989b). In all cases it was assumed that the force components $X_1,\ldots,X_r,\ldots,X_s,\ldots,X_F$ were identical with the force components involved in the explicit expression (bilinear form)

$$\sigma = \sum_{r=1}^{F} J_r X_r \qquad (9.2)$$

of the entropy production rate. Particularly the thermodynamic force

$$X_t = (A'_t/T)-(A''_t/T) = A_t/T \qquad (t=1,2,\ldots,S) \qquad (9.3)$$

was assumed in the case of elementary chemical reaction t. In Eqs. (9.3) A_t is the affinity of reaction t. All attempts were unsuccessful including that of Lengyel and Gyarmati (Lengyel and Gyarmati 1981b; Lengyel and Gyarmati 1986). These authors, however, emphasized over and over again that the apparent inconsistency between thermodynamics and chemical kinetics must not be accepted. They continued to postulate Eqs. (9.1) and drew the conclusion that a new thermodynamic force must be sought which will ensure satisfaction of the generalized reciprocity relations.

Some authors examined the constitutive equations

$$J_t = \lambda_t \left(e^{A'_t/RT} - e^{A''_t/RT} \right) \qquad (t=1,2,\ldots,S) \qquad (9.4)$$

of a complete set of independent reversible chemical reactions (Bataille, Edelen and Kestin 1978; Lengyel and Gyarmati 1981b; Lengyel and Gyarmati 1986) and have found that the Onsager reciprocity relations were satisfied, but the generalized reciprocity relations were not.

Before showing below the reason for this frustration we recall that in Onsager's theory (Onsager 1931), generally considered as the foundation of the thermodynamics of irreversible processes, linear constitutive equations are assumed in the form

$$J_i = \sum_{k=1}^{n} L_{ik} X_k \qquad (i=1,2,\ldots,n) \qquad (9.5)$$

and in the inverse form

$$X_i = \sum_{k=1}^{n} R_{ik} J_k \qquad (i=1,2,\ldots,n) \qquad (9.6)$$

which are good approximations in the case of transport processes. In the linear theory reciprocity relations (9.1) are identical with the relations

$$L_{ik} = L_{ki} \quad ; \quad R_{ik} = R_{ki} \qquad (i,k=1,2,\ldots,n) \qquad (9.7)$$

bearing Onsager's name.

The authors referred to above, have considered, as already mentioned, the affinity divided by the temperature as the thermodynamic force (i.e. A_t/T) of the reaction

$$\sum \nu'_{kt} B_k \longrightarrow \sum \nu''_{kt} B_k \ . \qquad (9.8)$$

In other words, they have taken A_t/T $(t=1,2,\ldots,S)$ as independent variables. In opposition to this choice, in the application of the governing principle of dissipative processes to elementary chemical reactions A'_t/T and A''_t/T $(t=1,2,\ldots,S)$ are taken as independent variables. Using these independent variables and the relations $\vec{J}_s = \overleftarrow{J}_t$; $J_t = \vec{J}_t - \overleftarrow{J}_t$; $A'_s = A''_t$; $A''_s = A'_t$; $\vec{\lambda}_s = \overleftarrow{\lambda}_t = \vec{\lambda}_t = \lambda_t$, we shall arrive at the specified local form

$$\delta(\sigma-\psi-\varphi) = \sum_{t=1}^{S/2} \left[J_t - 2\lambda_t e^{A'_t/RT} + 4\lambda_t e^{(A'_t+A''_t)/RT} \right] \delta(A'_t/T) \ -$$

$$-\sum_{t=1}^{S/2}\left[J_t+2\lambda_t e^{A''_t/RT}-4\lambda_t e^{(A'_t+A''_t)/RT}\right]\delta(A''_t/T)\ +$$

$$+\sum_{t=1}^{S/2}\left[A_t/T-\partial\varphi/\partial J_t\right]\delta J_t\ =\ 0 \qquad (9.9)$$

of the governing principle if reactions t and s are coupled to a reversible reaction. Owing to the fact that the complete set of variables in Eq. (9.9) is linearly independent we obtain as the final result the equation

$$J_t = \vec{J}_t-\overleftarrow{J}_t = \lambda_t\left(e^{A'_t/RT}-e^{A''_t/RT}\right) \qquad (t=1,2,\ldots,S/2)\ , \qquad (9.10)$$

i.e. the Marcelin-Kohnstamm equation.

If, on the other hand, in the bilinear form

$$\sigma = \sum_{t=1}^{S}\vec{J}_t A_t/T \qquad (9.11)$$

$\vec{J}_1,\ldots,\vec{J}_t,\ldots,\vec{J}_S$ and $A_1/T,\ldots,A_t/T,\ldots,A_S/T$ are taken as independent variables and $\varphi \equiv \varphi(\vec{J}_1,\ldots,\vec{J}_t,\ldots,\vec{J}_S)$ and

$$\psi = R\sum_{t=1}^{S}\vec{\lambda}_t e^{A_t/RT} \qquad (9.12)$$

as dissipation potentials with $\vec{\lambda}_t$ not depending on the variables, the governing principle will take the local form

$$\sum_{t=1}^{S}\left[\vec{J}_t - \vec{\lambda}_t e^{A_t/RT}\right]\delta(A_t/T)+\sum_{t=1}^{S}\left[(A_t/T)-(\partial\varphi/\partial\vec{J}_t)\right]\delta\vec{J}_t = 0. \qquad (9.13)$$

Due to the (assumed) linear independence of the complete set of variables we shall obtain

$$\vec{J}_t = \vec{\lambda}_t e^{A_t/RT} = \vec{\lambda}_t e^{-A''_t/RT}\cdot e^{A'_t/RT} =$$

$$= \vec{K}_t e^{A'_t/RT} \qquad (t=1,2,\ldots,S) \qquad (9.14)$$

where

$$\vec{K}_t = \vec{\lambda}_t e^{-A''_t/RT} \qquad (t=1,2,\ldots,S). \qquad (9.15)$$

Equations (9.15) are at variance with the experimental fact that the rate coefficients of chemical reactions do not depend on the concentrations or relative activities of the chemical components of the system. We see now the thermodynamic reason why all attempts to prove the generalized reciprocity relations necessarily failed, provided that the affinity divided by the temperature was considered to be the thermodynamic force driving the chemical reaction. These attempts belong now to the past in the history of the topic.

Some authors (see for example Section 1 of Chapter V in Prigogine 1967) linearized Eqs. (9.4) by expanding in Taylor's series and omitting terms of higher than first degree. Thus, the linear constitutive equation

$$J_t = \frac{\lambda_t}{R}\left(\frac{A'_t}{T} - \frac{A''_t}{T}\right) = \frac{\lambda_t}{R}\frac{A_t}{T} \tag{9.16}$$

with affinity A_t as variable was obtained. Omission of terms of higher degrees in A'_t and A''_t, however, is not justified, because the quantities A'_t/RT and A''_t/RT may have rather large values even in equilibrium as we have seen in Section 4.

9.2 The present

Concerning the dissipation potentials for systems of chemical reactions the question may arise what is added by them to chemical kinetics? Is it not simply the kinetic mass action law rewritten in the form of a potential? The answers to these questions can be based on the comparison of the two potentials Eq. (7.4) and

$$\psi_0 = R\sum_{t=1}^{S}\vec{\lambda}_t e^{A'_t/RT} \tag{9.17}$$

in the case of the independent complete set of elementary reactions (2.1). While Eq. (9.17) is, indeed, not more than the kinetic mass action law rewritten in potential form, Eq. (7.4) together with Eq. (7.2) substituted into Eq. (7.1) expressing the governing principle of dissipative processes leads both to the kinetic mass action law and to the specified bilinear form of the entropy production by the given chemical reactions. In other words, while the rate potential Eq. (7.4) together with the governing principle performs the synthesis of chemical kinetics and thermodynamics, the simpler potential (9.17) does not play such a role.

Another important question may arise about the role of the generalized reciprocity relations. Is satisfaction of the general reciprocity relations (9.1) equivalent to complete consistency of chemical kinetics and thermodynamics? It is easy to show that the answer is no. In Eqs. (6.6), expressing the kinetic mass action law, only the variables $A'_1/T,\dots,A'_t/T,\dots,A'_S/T$ appear and nothing more can be proved than that the relations

$$\partial\vec{J}_r/\partial(A'_s/T) = \partial\vec{J}_s/\partial(A'_r/T) = 0 \qquad (r,s=1,2,\dots,S;\ r\neq s) \tag{9.18}$$

are (trivially) satisfied, provided that the reactions are linearly independent. On the other hand, from Eqs. (7.4) and (7.7) we see that also the relations

$$\partial\vec{J}_r/\partial(A''_s/T) = \partial\vec{J}_s/\partial(A''_r/T) \qquad (r,s=1,2,\dots,S;\ r\neq s) \tag{9.19}$$

must hold. But from the equations of the kinetic mass action law alone we shall not know anything about the variables $A''_1/T,\ldots,A''_t/T,\ldots,A''_s/T$ and their role. Their existence and their role will be clear only by application of the governing principle of dissipative processes and suitable choice of the dissipation potentials (7.4) and (7.9). Therefore, satisfaction of the general reciprocity relations (3.11) does not justify the conclusion that the kinetics of a complete set of elementary chemical reactions is consistent with thermodynamics.

9.3 The future

The evolution of chemical kinetic systems of elementary reactions described in Section 4 is restricted, however, to closed and adiabatically insulated systems. The treatment must eventually be extended to open systems, since such systems are extremely important in nature, i.e. in physical chemistry, biochemistry and biology etc.

10. REFERENCES

Bataille, J., D.G.B. Edelen and J. Kestin. 1978. Nonequilibrium thermodynamics of the nonlinear equations of chemical kinetics. J. Non-Equilibr. Thermodyn. 3: 153-168.

De-Groot, S.R. 1951. Thermodynamics of irreversible processes. Amsterdam:North-Holland.

Guldberg, C.M. and P. Waage. 1867. Etudes sur les Affinites Chimiques. Christiania. Brøgger et Christie.

Gyarmati, I. 1961a. On the phenomenological basis of irreversible thermodynamics I. Periodica Polytechn. Chem. Eng. 5: 219-243.

Gyarmati, I. 1961b. On the phenomenological basis of irreversible thermodynamics II. (On a possible nonlinear theory). Periodica Polytechn. Chem. Eng. 5: 321-339.

Gyarmati, I. 1965. On a general variational principle of nonequilibrium thermodynamics. Zh. Fiz. Khim. [Russ. J. Phys. Chem.] 39: 1489-1493.

Gyarmati, I. 1969. On the governing principle of dissipative processes and its extension to nonlinear problems. Ann. Phys. (Leipzig) 23: 353-378.

Gyarmati, I. 1970. Nonequilibrium thermodynamics. Field theory and variational principles. Berlin: Springer.

Kohnstamm Ph. and F.E.C. Scheffer 1911. Thermodynamic potential and velocities of reactions. Proc. Sect. Sci. K. ned. Akad. Wet. (Amsterdam) 13: 789-800.

Lengyel, S. 1988. Deduction of the Guldberg-Waage mass action law from Gyarmati's governing principle of dissipative processes. J. Chem. Phys. 88: 1617-1621.

Lengyel, S. 1989a. Chemical kinetics and thermodynamics. A history of their relationship. Computer Math. Applic. 17: 443-455.

Lengyel, S. 1989b. On the relationship between thermodynamics and chemical kinetics. Z. Phys. Chem. (Leipzig) 270: 577-589.

Lengyel, S. 1989c. Reversible chemical reactions in the unified field theory of thermodynamics. Acta Phys. Hung. 66: 1-.

Lengyel, S. and I. Gyarmati. 1981a. Nonlinear thermodynamical studies of homogeneous chemical kinetic systems. Periodica Polytechnica Chem. Eng. 25: 63-99.

Lengyel, S. and I. Gyarmati. 1981b. On the thermodynamics of elementary chemical reactions in homogeneous systems. J. Chem. Phys. 75: 2384-2389.
Lengyel, S. and I. Gyarmati. 1986. Constitutive equations and reciprocity relations of nonideal homogeneous closed chemical kinetic systems. Acta Chim. Hung. 122: 7-17.
Li, J.C.M. 1958. Thermodynamics of nonisothermal systems. The classical formulation. J. Chem. Phys. 29: 747-754.
Marcelin, R. 1910. Sur la mécanique des phénomènes irréversibles. C.R. Acad. Sci. 151: 1052-1054.
Meixner, J. 1941a. Zur Thermodynamik der Thermodiffusion. Ann. Phys. 39: 333-356.
Meixner, J. 1941b. Zur Teorie der elektrischen Transporterscheinungen im Magnetfeld. Ann. Phys. 40: 165-180.
Meixner, J. 1942. Reversible Bewegungen von Flüssigkeiten und Gasen. Ann. Phys. 41: 409-425.
Meixner, J. 1943a. Zur Thermodynamik der irreversiblen Prozesse in Gasen mit chemisch reagierenden, dissoziirenden und anregbaren Komponenten. Ann. Phys. 43: 244-270.
Meixner, J. 1943b. Zur Thermodynamik der irreversiblen Prozesse. Z. Phys. Chem. (B) 53: 235-263.
Mercier, A. 1959. Analytical and canonical formalism in physics. 10-14. Amsterdam: North-Holland.
Onsager, L. 1931. Reciprocal relations in irreversible processes. Phys. Rev. 37: 405-426. Phys. Rev. 38: 2265-2279.
Prigogine, I. 1945. Bull, Acad. r. Belg. Cl. Sci. 31: 600.,
Prigogine, I. 1947. Etudes thermodynamiques des phenomènes irreversibles. Liège: Desoer.
Prigogine, I. 1967. Introduction to thermodynamics of irreversible processes. 3rd ed. New York. London. Sydney. Interscience, Wiley.
VAN Rysselberghe, P. 1962. General Reciprocity Relation between the Rates and Affinities of Simultaneous Chemical Reactions. J. Chem. Phys. 36: 1329-1330.330.

Explosive, Conservative, and Dissipative Systems and Chemical Oscillators

Henrik Farkas and Zoltan Noszticzius
Institute of Physics
Technical University of Budapest
Budapest, H-1521, Hungary

Abstract

This paper reviews some of the results achieved by the authors and their coworkers in the field of mathematical modeling of chemical oscillators. First, after a brief historical introduction, the oscillatory Belousov-Zhabotinsky (BZ) reaction is discussed as a chemical background of the investigations. Then two different mathematical models of the BZ reaction, the Explodator and the Oregonator, are described as serial and parallel oscillatory networks. It is shown that serial type oscillatory networks can be derived from the famous Lotka-Volterra scheme. In the following, stability properties of two-dimensional generalized Lotka-Volterra (GLV) models are analyzed. It is proven that GLV models can be explosive, conservative, or dissipative depending on a parameter. The explosive-dissipative transition can go through critical Hopf bifurcation. In this case, the GLV model is conservative at the bifurcation point. The authors present integrals for conservative models and Liapunov functions for the explosive and dissipative models. Limit cycle oscillators can be constructed by adding a so called "limiting" reaction to an explosive model. Next, the problem of the "simplest" oscillating chemical model is investigated. A realistic chemical mechanism is built around an oscillator which is the simplest one from a mathematical point of view. The model is studied by the parametric representation method. Finally, some thermodynamic aspects of chemical oscillators are discussed. An analogy between the second law of thermodynamics for closed systems and an evolutionary criterion for open systems maintained far from equilibrium is suggested.

CONTENTS

1. INTRODUCTION

> *"These regions we may rightly call Mundus Novus a New World."*
> The Florentine banker and traveler, Amerigo Vespucci on America.

A decade ago when we started our journey through the strange new world of oscillating chemical reactions, that land was already known from the works of its famous discoverers and investigators like Lotka (1920), Volterra (1931), Bonhoffer (1948), Belousov (1951), Zhabotinsky (1964), Prigogine and Lefever (1968), Field, Kőrös, and Noyes (1972), and others. Nevertheless, we found that there were still problems in this area, interesting enough to deal with them. In this paper we want to summarize some of our results achieved by us and our coworkers while working on those problems.

The authors of this chapter have different backgrounds; they came from different fields. Ten years ago, when we started our common work on the Belousov-Zhabotinsky (BZ) reaction (Noszticzius and Farkas 1981) H. F., a physicist, was working on theoretical problems of non-equilibrium thermodynamics (Farkas 1968, 1975a, 1975b, 1978, 1980). At the same time Z. N., a chemist, was interested in experimental aspects of oscillating reactions (Noszticzius 1977, 1979a, 1979b, Noszticzius and Bódiss 1979, 1980,). We found this combination of different fields as a useful approach during the years of of our common research on the mechanism of the oscillatory BZ reaction (Noszticzius, Farkas and Schelly 1984a, 1984b, Farkas and Noszticzius 1985b, Farkas, Kertész, and Noszticzius 1986, Gáspár, Noszticzius, and Farkas 1987, Farkas, Noszticzius, Savage, and Schelly 1989) and related theoretical problems (Farkas and Noszticzius 1985a, 1987). Here, after giving a brief introduction to the chemical background, we will review mainly the theoretical results. The chemical introduction explains why we do think that consecutive autocatalytic reaction networks, like the famous Lotka-Volterra system, can play an important role in the BZ and other oscillating reactions. Then we discuss stability and bifurcation properties of some generalized Lotka-Volterra schemes. Specifically, we show that such schemes can display dissipative, conservative, or explosive behavior depending on the parameters. The dissipative-explosive transition takes place via critical Hopf bifurcation. At the critical parameter value - which separates the dissipative and explosive regions - the system is conservative. Integrals of the conservative systems (the integral curves are closed orbits) can serve as Liapunov functions for the related dissipative and explosive systems. Furthermore, we show that explosive systems can serve as "explodator cores" in constructing chemical oscillators of the limit cycle type. Our Explodator model of the BZ reaction (Noszticzius, Farkas and Schelly 1984a, 1984b) was constructed following this recipe: a limiting reaction was added to an explosive network to obtain a limit cycle oscillator.

Next, the problem of the "simplest" chemical oscillator is discussed. This oscillator uses the simplest base: the linear (harmonic) oscillator, and onto this base we built a "chemically reasonable" oscillator with four variables. Bifurcations for this oscillator are studied with the aid of parametric representation method (Gilmore 1981, Farkas, Gyökér and Wittmann 1989, 1990).

Finally, looking for relationship between thermodynamics and nonlinear dynamics, we suggest that for open chemical systems (oscillatory or not) the most general evolutionary criterion is that these systems are dissipative ones, and consequently, they evolve toward an asymptotic state.

2. Chemical oscillators

The Belousov-Zhabotinsky reaction

The BZ reaction is the oscillatory oxidation of an organic substrate (mostly malonic acid) by acidic bromate in the presence of a redox catalyst (mostly Ce^{3+}/Ce^{4+}) (Field 1985, Tyson 1985). The reaction was discovered by the Russian chemist Belousov (1951) and was further studied by his compatriot Zhabotinsky, a biophysicist (Zhabotinsky 1964, 1985). When it is placed into a CSTR ("continuously fed stirred tank reactor"), the BZ reaction is a prime example for an *open* chemical system *far from equilibrium* (Nicolis and

Prigogine 1977). The system exhibits several exotic phenomena characteristic for nonlinear dynamic systems: simple and complex oscillations, excit- ability, multistability and chaos (Maselko and Swinney 1986, Noszticzius, Wittmann, and Stirling 1985, 1987, De Kepper and Bar-Eli 1983, Hudson and Mankin 1981, Swinney and Roux 1984). Even if it is studied in batch, oscillations can easily occur for several hours. A qualitative mechanism was already suggested by Zhabotinsky himself (Vavilin and Zhabotinsky 1969), but the first quantitative mechanism for the BZ reaction is due to Field, Kőrös and Noyes (1972). Later on, Field and Noyes presented a skeletonized version of the FKN mechanism: this is the famous Oregonator model (Field and Noyes 1974). In the recent years, in the light of new experimental results, the details of the mechanism were modified, even new models have appeared: the Explodator (Noszticzius, Farkas and Schelly 1984), the revised (Noyes 1984, Tyson 1984), the modified (Field and Boyd 1985), and the amplified Oregonator (Ruoff and Noyes 1986), the Tandem Oscillator (Noszticzius, McCormick, Swinney, and Schelly 1987), the Radicalator (Försterling, Murányi, and Noszticzius 1990). We do not want to discuss here all the mechanistic problems of the BZ reaction which are still debated (Noszticzius 1979b, Ganapatisubramanian and Noyes 1982, Ruoff and Schwitters 1984, 1986, Varga and Kőrös 1986, Noszticzius and McCormick 1987, 1988, Noyes, Field, Försterling, Kőrös and Ruoff 1989, Hayes, Schmidt, and Meisel 1989, Ruoff and Vestvik 1989, Russo 1990). Instead, we want to emphasize some general features which are characteristic for oscillating chemical networks, and the BZ reaction will be taken as an illustrative example only.

Positive and negative feedback loops in chemical oscillators

For a qualitative explanation of the chemical oscillators it is comfortable to apply the feedback concept used extensively by Franck (1978), and recently by Epstein and coworkers (Luo and Epstein 1990). According to this interpretation, the oscillations arise from the antagonistic interactions of a relatively fast acting, unstabilizing positive feedback and a slower acting (delayed) negative feedback. In this picture, oscillations can be explained as follows. Even if the system is in a steady state, any perturbation will be amplified by the fast acting positive feedback. As a consequence, the system will start to move away rapidly from that unstable steady state. After a while, the delayed negative feedback overcomes that effect and forces the system back to the steady state. Because of the delay, however, an overshoot (or undershoot)(Rábai, Bazsa, and Beck 1979) will occur. That overshoot will be amplified by the positive feedback again and now the system leaves the steady state in the other direction. This sequence of events eventually leads to undamped oscillations in the system. Now let us see how chemical reactions can establish positive and negative feedback loops.

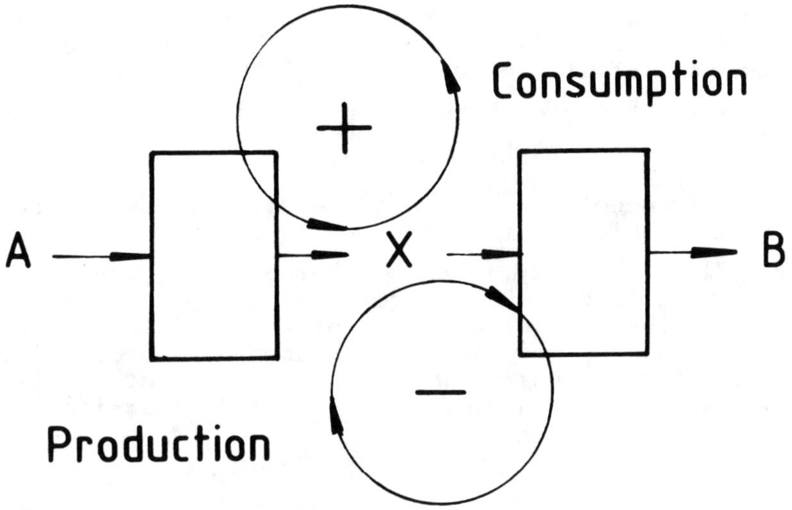

Figure 1. Antagonistic feedback loops

The positive feedback loop acts on the *production* of the intermediate X: the rate of its production r_p grows monotonically with increasing X (here we use X to denote the chemical species itself and its concentration as well). The simplest chemical manifestation of the positive feedback loop is an autocatalytic process like Process (T):

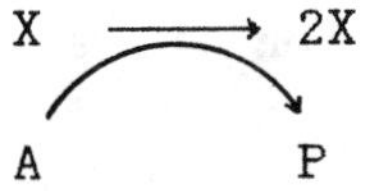

(T)

Here A is some starting material and P is some product of Process (T). In the case of BZ reaction (T) is the first autocatalytic process investigated by Thompson (1971) and Noyes, Field, and Thompson (1971):

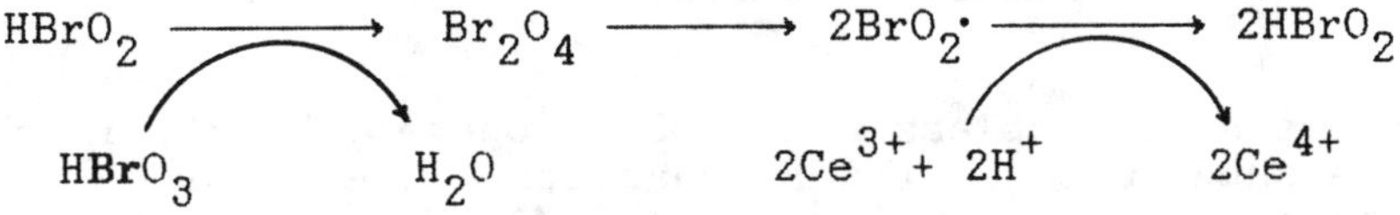

The delayed negative feedback loop acts on the *consumption* of X. An increase of X will result, after some delay, in an increased rate of its consumption r_c. Thus X will grow or decrease alternately, depending on the difference of the two rates:

$$\dot{X} = r_p - r_c$$

Naturally, $\dot{X}$ cannot depend on X exclusively, otherwise no oscillations would occur. (If r_c would depend solely on X then there would be no delay.) Other intermediates besides X and other reactions should play a role in the delayed negative feedback loop. First, consumption of X should take place in a chemical reaction, thus, there is another intermediate the so-called *control intermediate* which reacts with the *autocatalytic intermediate* X in the control reaction (C)

$$X + Y \longrightarrow \text{products} \qquad \text{(C)}$$

For example, Br^- can play the role of the control intermediate in the BZ reaction:

$$HBrO_2 + Br^- \xrightarrow{H^+} 2HOBr$$

Second, the control intermediate Y must not appear immediately (i.e. simultaneously with the autocatalytic intermediate X) because a delay in the negative feedback loop is necessary. Thus Z, a precursor of Y should appear at first and the control intermediate Y should be produced from Z in a subsequent reaction

$$Z \longrightarrow fY \qquad \text{(D)}$$

Here f is a stoichiometric factor.

In the case of the BZ reaction there are two candidates for the delay reaction (D). One of them which appears in the original Oregonator is (O5):

$$2Ce^{4+} + BrMA \longrightarrow fBr^- + \text{other products} \qquad \text{(O5)}$$

that is $Z=2Ce^{4+}$. Here BrMA denotes bromomalonic acid. Theoretical calculations (Clarke 1976) show that f should be around 1.

There are some problems with (O5), however. First, in the case of some substrates of the BZ reaction like oxalic acid, no brominated organics analogous to BrMA exists (Noszticzius and Bódiss 1979, Adamciková and Sevcik 1982, Sevcik and Adamciková 1985). Second, as it was shown by Varga, Györgyi and Kőrös (1985) even in the case of the classical BZ reaction (substrate: malonic acid), the main source of bromide is *not* the bromomalonic acid.

The other candidate for a bromide producing reaction in a BZ system starts with hypobromous acid:

$$HOBr \longrightarrow Br^-$$

$$RH \qquad ROH+H^+$$

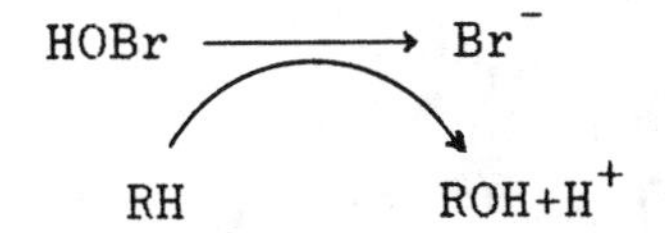

(E3)

where RH is some reducing organic material. For example, in the case of oxalic acid substrate (E3) is the well-studied reaction

$$HOBr \longrightarrow Br^-$$

$$(COOH)_2 \qquad 2CO_2 + H_2O + H^+$$

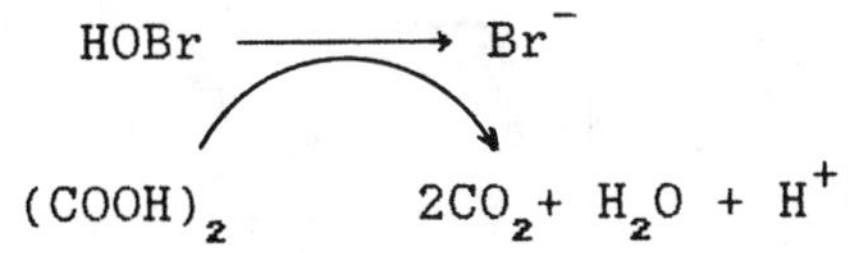

Naturally, there are important chemical consequences of the two different pathways (O5) and (E3). Nevertheless, what we want to emphasize here is not the difference in the chemistry but the difference of the resulting two separate oscillatory networks. Applying (O5) leads to the classical Oregonator and to a network which we shall call "parallel". Involving (E3) instead of (O5) leads to an Explodator-type model with a characteristic "serial" structure.

Explodator and Oregonator:

Serial and parallel oscillatory networks

The two different oscillatory networks are depicted in Fig.2. and 3. To complete those networks an additional recovery reaction was added to both of them:

$$Y \longrightarrow products$$

(R)

(R) is necessary to remove the control intermediate from the system. Otherwise, the continuously present control intermediate would prevent the recurrence of the autocatalytic process (T) and no sustained oscillations would occur. The main difference between the two networks is the way of forming Z. In the Explodator type network Z is formed *from* the autocatalytic intermediate X (that is in a "serial" way). In the Oregonator-type network Z is formed *simultaneously* with X (that is in a "parallel" way).

As most of the theoretical analysis in the recent years focussed on the Oregonator type models (Showalter, Noyes, and Bar-Eli 1978, Tyson 1985, Barkley, Ringland, and Turner 1987, Gáspár and Showalter 1988, Barkley 1988), that is on parallel networks, our theoretical efforts focussed on serial networks, that is Explodator-type models. Finally, let us remark that according to the latest experimental results (Noszticzius, McCormick,Swinney, and Schelly 1987, Försterling and Noszticzius 1989, Försterling, Murányi, Noszticzius 1990) both serial and parallel oscillatory networks play an important role in the mechanism of the BZ reaction.

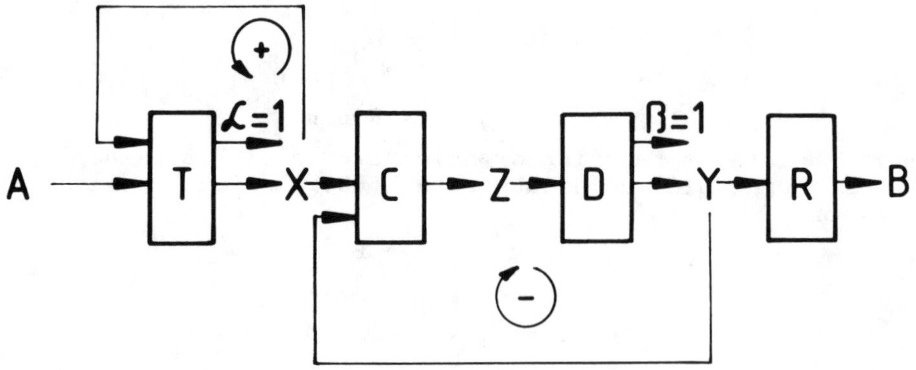

Figure 2. Serial oscillatory network (Explodator type).

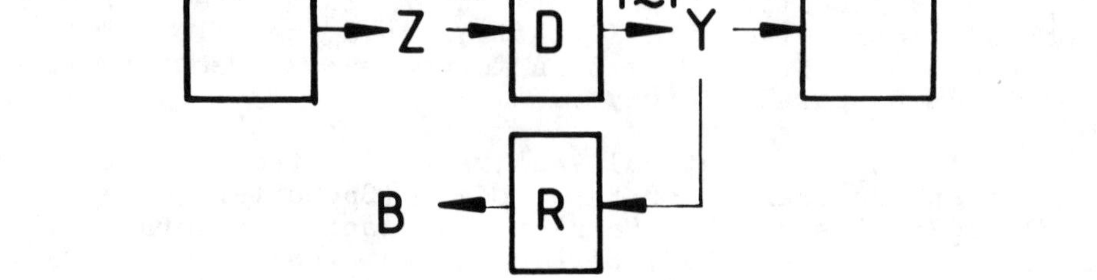

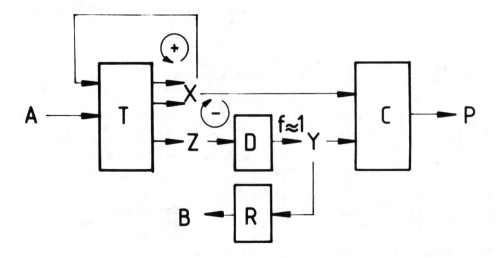

Figure 3. Parallel oscillatory network (Oregonator type).

3. Explodator models and the Lotka-Volterra scheme

The chemical reactions of the Explodator model as depicted in Fig.2. are the following ones:

$$A + X \longrightarrow (1+\alpha)X \quad (E1)$$

$$X + Y \longrightarrow Z \quad (E2)$$

$$Z \longrightarrow (1+\beta)Y \quad (E3)$$

$$Y \longrightarrow B \quad (E4)$$

Assuming that $\alpha=1$, $\beta=1$, and (E2) together with (E3) are merged into a single reaction we get the famous Lotka-Volterra scheme:

$$A + X \longrightarrow 2X \quad (LV1)$$

$$X + Y \longrightarrow 2Y \quad (LV2)$$

$$Y \longrightarrow B \quad (LV3)$$

Thus it is clear that the Explodator model of (E1) - (E4) and the Lotka-Volterra model (LV1) - (LV3) are closely related.

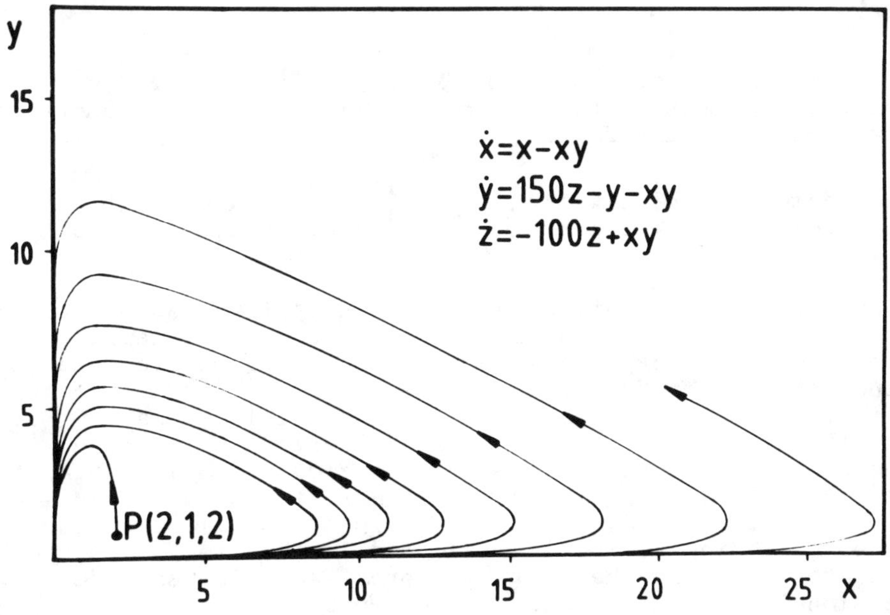

Figure 4. Spiralling explosion of an Explodator core.

In the followings (E1)-(E4) will be referred as a three dimensional explodator "core". The Lotka-Volterra mechanism is often criticized because of its conservative oscillations and its structural instability. The aforementioned three dimensional explodator core is structurally stable; and it always displays an explosive behavior (Kertész 1984). See Fig. 4. as an illustrative example. There is no upper limit for the amplitude of the evergrowing oscillations in Fig. 4. The name "Explodator" refers to this spiralling explosion of the model. Naturally, real systems are not explosive ones. That difficulty can be overcome, however, and explosion can be easily avoided by adding one more reaction - a so-called "limiting reaction" - to the explodator core. This way we obtain limited explodator models which can produce limit cycle oscillations. A simple example for such a limiting reaction is reaction (L). That reaction represents a constant inflow of the autocatalytic intermediate X :

$$\xrightarrow{\mu} X \qquad \text{(L)}$$

Here μ is the rate of production of X in reaction (L). μ is an important parameter of the system: if μ is increased gradually starting from zero, at a critical value μ_0 Hopf bifurcation occurs and the stationary point of the system becomes stable for all μ values above μ_0. Below that critical value the stationary point is unstable, but, as it was proven by Tang (1989), there is a stable limit cycle in the parameter region $0<\mu<\mu_0$.

Other limiting reactions can work in a similar way. Thus, the essence of our method to construct limit-cycle oscillators can be summarized in the following scheme:

[**explodator** core] + [**limiting** reaction] ⟶ [Hopf bifurcation and **limit-cycle** oscillation]

Explodator cores containing two consecutive autocatalytic process can be created in different ways. For example, the three-dimensional explodator core can be derived from the two dimensional Lotka-Volterra scheme by adding a third intermediate Z to the system. Another possibility is to stay with the two dimensional Lotka-Volterra-like scheme but changing the mechanism of the consecutive autocatalytic reactions. This way we can construct two dimensional explodator cores. Again, limit cycle oscillators can be created by adding a limiting reaction to the two dimensional explodator core. In the next chapter we discuss such two dimensional oscillatory chemical networks.

4. The generalized LV-model. Two dimensional Explodators

As we have seen, the original Lotka-Volterra model is a conservative one; its oscillations are not of the limit cycle type. There are two possibilities to obtain limit cycle oscillation by modifying the LV-model, namely:

-adding new reactions and/or new intermediates to the model,
-using more complicated (higher-order) rate laws.

The three dimensional Explodator (Noszticzius, Farkas and Schelly 1984a, 1984b) is an example for the former approach, while the latter approach will be illustrated in this section.

Let us consider the following mechanism (Farkas and Noszticzius 1985b):

$$\hat{p}X \longrightarrow (\hat{p}+1)X \qquad \text{(GLV'1)}$$

$$pX+qY \longrightarrow (p+q)Y \qquad \text{(GLV'2)}$$

$$\hat{q}Y \longrightarrow \qquad \text{(GLV'3)}$$

This mechanism with mass-action kinetics yields the kinetic equations:

$$\dot{x} = k_1'x^{\hat{p}} - pk_2'x^p y^q \qquad (1.a)$$

$$\dot{y} = pk_2'x^p y^q - k_3'y^{\hat{q}} \qquad (1.b)$$

Here k_i' (i=1,2,3) are rate constants, x and y are the concentrations of the intermediates X and Y respectively. We did not indicated the components what are in great abundance, or more precisely, whose concentrations can be considered constant in time (pool of chemicals, Gray and Scott 1986). Obviously, such components must be involved in the mechanism (GLV').

The exponents $\hat{p}$, p, q, $\hat{q}$ are assumed to be not lesser than one, for mathematical convenience. Exponents lesser than one lead to the violation of the Lifshitz condition and, consequently, the uniqueness of the solution would not be guaranteed in this case. To overcome this difficulty, see details in (Farkas and Noszticzius 1985b). Remark that this difficulty does not occur for the case of zero exponent.

In the reaction (GLV'2) one molecule of X transforms into exactly one molecule of Y. Abandoning this requirement, the positive term in the rhs of (1.b) will not be the same as the negative term on the rhs of (1.a). Instead of (1), we consider a system whose rhs involves four different terms, and in this section we will use the term "Generalized Lotka-Volterra model" (GLV in short) for the following chemical mechanism:

autocatalytic production of X $\quad \xrightarrow{X} X \quad$ (GLV1)

autocatalytic production of Y from X $\quad X \xrightarrow{Y} Y \quad$ (GLV2)

decay of Y $\quad Y \longrightarrow \quad$ (GLV3)

with the kinetic equations:

$$\dot{x} = k_1 x^{\hat{p}} - k_2 x^p y^q \qquad (2.a)$$

$$\dot{y} = k_3 x^p y^q - k_4 y^{\hat{q}} \qquad (2.b)$$

Here the rate constants k_i (i=1,2,3,4) are positive quantities, and the exponents are assumed to be not lesser than one. Dot denotes differentiation with respect to time t, and the variables x and y denotes concentrations, and hence, they cannot be negative.

If all the exponents are equal to 1, and $k_2 = k_3$, we get the original LV model.

The exceptional case, when

$$\hat{p}q + p\hat{q} - \hat{p}\hat{q} = 0 \qquad (3)$$

is investigated in detail by Dancsó, H. Farkaş, M. Farkas, and Szabó (1990). For this case, the nullclines $\dot{x}=0$ and $\dot{y}=0$ are coincide if a further condition is fulfilled; in this case all the points in the nullcline are fixed points. The point which separates the stable and unstable fixed points moves as a parameter changes, that is a "zip" bifurcation occurs (M. Farkas 1984, Dancsó, H. Farkas, M. Farkas, and Szabó 1990).

Now we consider the generic case when

$$\hat{p}q + p\hat{q} - \hat{p}\hat{q} \neq 0 \qquad (4)$$

In this generic case there exists a unique fixed point of the system (2) in the positive quadrant (X>0,Y>0). By an appropriate scaling (that is by a transformation $X=\alpha x$, $Y=\beta z$, $Z=\gamma y$, $T=\delta t$ with properly chosen values $\alpha,\beta,\gamma,\delta$), this unique positive fixed point can be moved into (1,1), and we obtain the transformed equations:

$$\dot{X} = X^{\hat{p}} - X^p Y^q \qquad (5.a)$$

$$\dot{Y} = C(X^p Y^q - Y^{\hat{q}}) \qquad (5.b)$$

where C is a positive constant (Dancsó, H. Farkas, M. Farkas, and Szabó 1990). Now the dot denotes differentiation with respect to the transformed time T.

A candidate for Liapunov function.

To this system let us introduce the following function:

$$V(X,Y) = \Phi_p(X) + \frac{1}{C}\Phi_q(Y) \qquad (6)$$

where

$$\Phi_n(u) \equiv \begin{cases} u - \dfrac{1}{1-n} u^{1-n} & \text{if } n>1 \\ u - \ln u & \text{if } n=1 \end{cases}$$

The level curves $V(X,Y)=E$ (constant) form the same structure as the orbits of the original LV model do: they are closed curves in the positive quadrant, nested around the fixed point (1,1) (Farkas and Noszticzius 1985b, Dancsó, H. Farkas, M. Farkas, and Szabó 1990).

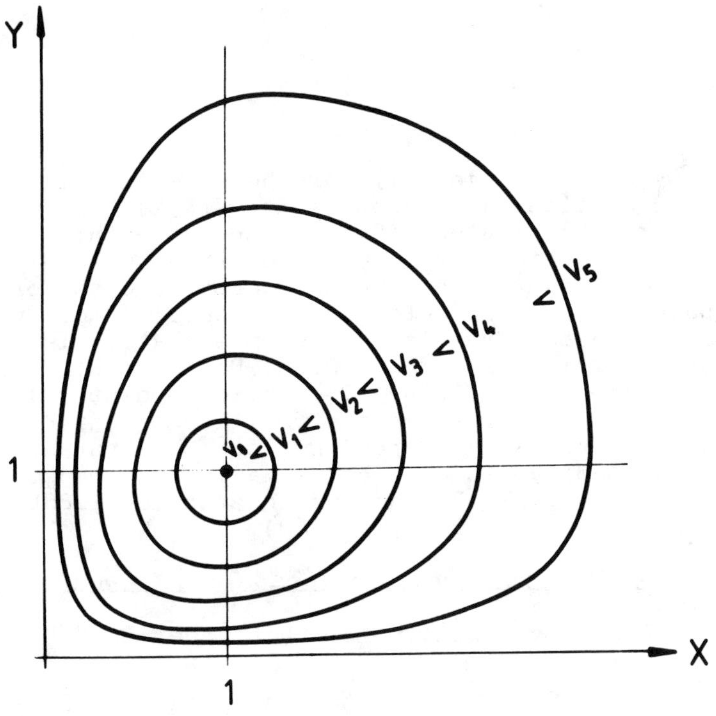

Figure 5. The level curves of V(X,Y) in (6)

The Eulerian derivative $\dot{V}$ with respect to the system (5) is:

$$\dot{V} = \frac{\partial V}{\partial X}\dot{X} + \frac{\partial V}{\partial Y}\dot{Y} = (1-X^{-p})(X^{\hat{p}}-X^{p}Y^{q}) + (1-Y^{-q})(X^{p}Y^{q}-Y^{\hat{q}}) \qquad (7)$$

The function V is a useful tool to reveal the global qualitative behavior of the solutions of (5) for different regions of the parameters.

Conservative case: $\hat{p}=p$, $\hat{q}=q$

The intermediates X and Y are produced in one of the reactions of GLV, and are consumed in the other. If the reaction orders are equal to each other, the system is conservative. Indeed, the function V(X,Y) defined in (6) is a first integral of (5), because

$$\dot{V} = 0 \tag{8}$$

follows from (7) for the present case.

The qualitative behavior of the conservative system is the same as that of the original LV model (p=1=q).

Explosive case: $\hat{p}\geq p$, $q\geq\hat{q}$, and $\hat{p}+q>p+\hat{q}$

In this case it follows from (7) that

$$\dot{V} \geq 0 \tag{9}$$

The equality may be valid only for the exceptional values X=1 or Y=1. The relation (9) means that V(X,Y) is a Liapunov function for the system (5) in the present case. The trajectories cross the level curves of V in outward direction. Since the axes X and Y are invariant sets, and the origin is a saddle point, the trajectory X=0 comes into, while the trajectory Y=0 comes from it. Hence, the system is explosive (Farkas and Noszticzius 1985a, 1985b, Dancsó, H. Farkas, M. Farkas, and Szabó 1989, 1990). The fixed point (1,1) is a source of the trajectories in the positive quadrant and the sink is at the infinity. Depending on the parameters, the process of tending to infinity may be monotonous or spiralling for large T (Farkas and Noszticzius 1985b, Farkas, Noszticzius, Savage and Schelly 1989).

Stable (dissipative) case: $\hat{p}\leq p$, $q\leq\hat{q}$, and $\hat{p}+q<p+\hat{q}$

V(X,Y) is a Liapunov function for this case, too, but now

$$\dot{V} \leq 0 \tag{10}$$

follows from (7). So the trajectories cross the level curves of V always "inward". The fixed point is globally asymptotically stable. It is an attractor and its basin is the whole open positive quadrant. Starting from any initial concentrations, the final state will be the stationary state (1,1).

Critical Hopf bifurcation

We can summarize the above results in a figure (Fig. 6).

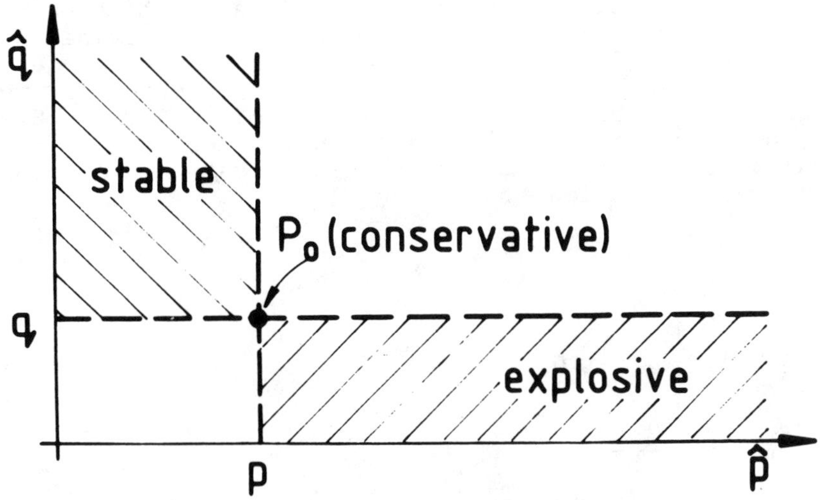

Figure 6. Stable, conservative and explosive regions of the GLV system.

Here the values p and q are considered as fixed ones, and Fig. 6. illustrates the dependence on the parameters $\hat{p}$, $\hat{q}$. The point P_o represents the conservative case. The stable (dissipative) and explosive regions are also indicated in the figure. For the other ("white") regions of the figure the function V is not a Liapunov function; its time derivative is not definite.

Let an oriented curve be given in the parameter plane $(\hat{p},\hat{q})$, and let it go from the stable region into the explosive region via the conservative point P_o. To this curve g we can introduce a bifurcation parameter μ. The fixed point (1,1) is stable for $\mu<\mu_o$, and unstable for $\mu>\mu_o$. The value μ_o corresponds to the point P_o. At the value $\mu=\mu_o$ a Hopf bifurcation occurs.This "critical" Hopf bifurcation is not a typical one.The typical kinds of Hopf bifurcations are the subcritical and the supercritical ones. In case of subcritical Hopf bifurcation an unstable limit cycle coexists with a stable fixed point, while in case of supercritical Hopf bifurcation a stable limit cycle coexists with an unstable fixed point. The critical Hopf bifurcation provides us no limit cycle, but infinite number of periodic orbits existing only at the bifurcation point. Those periodic orbits are orbitally stable in Liapunov sense (Nemitskii and Stepanov 1960), but the term "stable" is not used for such cases in natural sciences. E.g. in case of mechanical equilibrium the similar case is called *neutral* equilibrium.

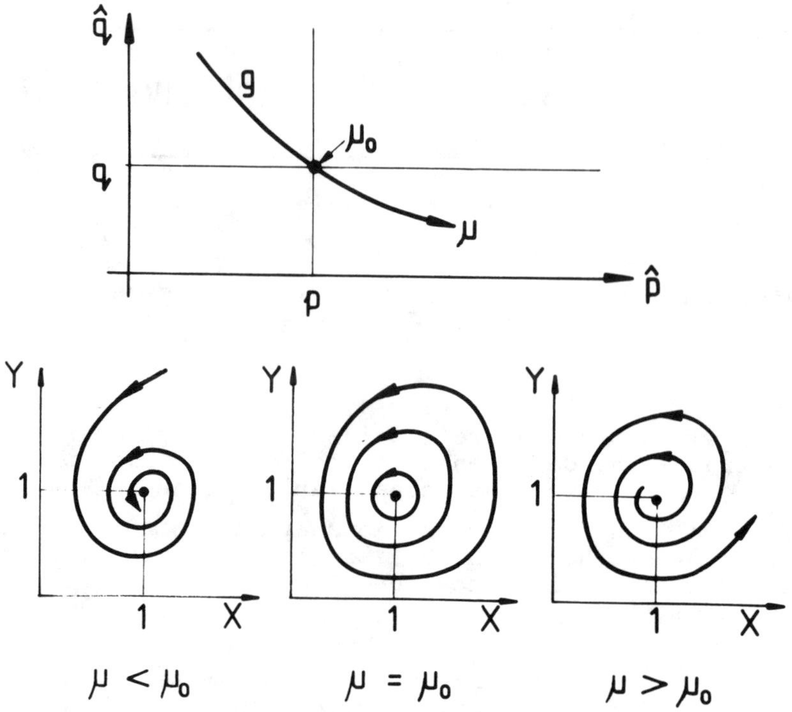

Figure 7. Critical Hopf bifurcation

Local analysis of the GLV model

We summarize here the results of the complete local investigation of the GLV model (Dancsó, H. Farkas, M. Farkas, and Szabó 1989, 1990). The local investigation provides us a local bifurcation diagram concerning the character of the stationary point as well as that of the Hopf bifurcation.

Three kinds of local bifurcations can be observed in Fig.8.

a) Saddle-node transition. This occurs when crossing the hyperbola

$$r \equiv \hat{p}q + p\hat{q} - \hat{p}\hat{q} = 0$$

The fixed point is a saddle if $r<0$. The hyperbola has another branch, which is not indicated in Fig. 8., lying outside the positive quadrant.

Recall that in the case r=0 the number of fixed points in the open positive quadrant is zero or infinity, depending on the parameters.

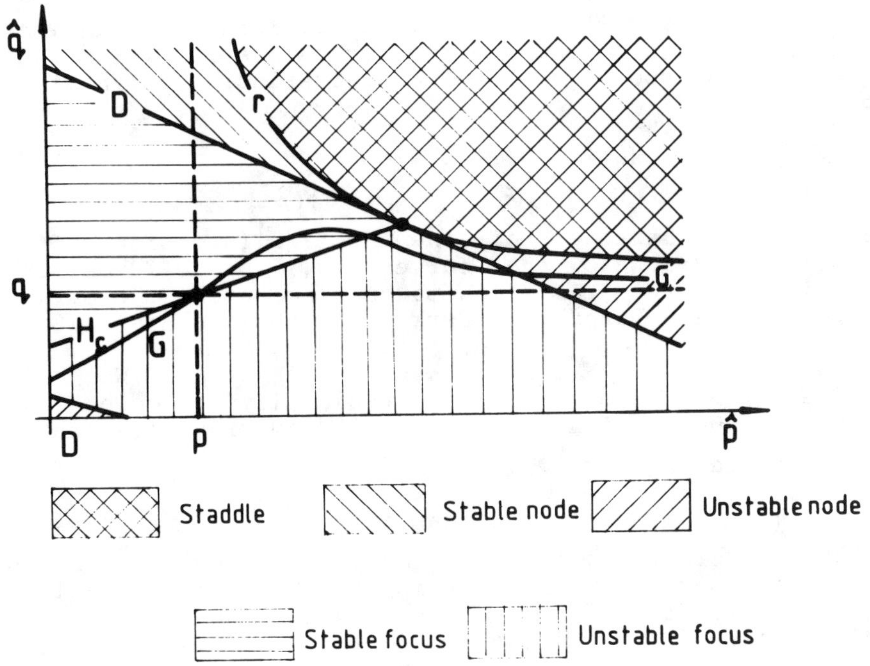

Figure 8. Local bifurcation diagram of the GLV system.

<u>b) Node-focus transition.</u> The discriminant D of the linearized variational equations at the fixed point (1,1) is:

$$D = (\hat{p}-p+C(\hat{q}-q))^2 - 4Cpq$$

The equation D=0 represents a pair of parallel straight lines which are tangential to the hyperbola r=0. The fixed point is a focus if D<0 (i.e. between the straight lines), and it is node or saddle if D>0. Along the straight lines D=0 the fixed point is a degenerate node (the characteristic equation has a real double root).

<u>c) Hopf bifurcation.</u> Hopf bifurcation takes place when crossing the straight line $H_C=0$, where

$$H_C = \hat{p}-p + C(q-\hat{q})$$

This line goes through the point $P_o=(p,q)$. The fixed point is a stable focus below the transition line (i.e. if $H_C>0$), and an unstable focus above that line. The slope of the line $H_C=0$ depends on the parameter valued C: this fact is reflected in the index C.

Notice that the listed transition lines have a unique common point at which the line D=0 is tangent to the hyperbola r=0. The straight line $H_C=0$ can be continued in the saddle region, but in that region it does not determine any qualitative change: it belongs to those parameter values for which the sum of the two real roots of the characteristic polynomial is zero.

The sign of the quantity

$$G=p(\hat{p}-p)(\hat{q}-1)(\hat{q}+q) - q(\hat{q}-q)(\hat{p}-1)(\hat{p}+p)$$

determines the character of the Hopf bifurcation. If the transition occurs below the curve G=0, the bifurcation is subcritical, if the transition takes place above the curve, it is supercritical (Dancsó, H. Farkas, M. Farkas, Szabó 1990).

The Hopf theorem (Hassard, Kazarinoff, and Wan 1981) guarantees the existence of a stable limit cycle in the "unstable focus" region near the line $H_C=0$ provided that the bifurcation is supercritical.

Discussion

The GLV model includes some special cases which have particular importance. The most important special case is the original LV model, of course ($\hat{p}=p=\hat{q}=q=1$). Among the rather simple models we treated two explodator-type models:

-LVB model (Lotka-Volterra-Brusselator, Farkas, Noszticzius, Savage, and Schelly 1989). In this model the second autocatalytic reaction was borrowed from the Brusselator model: $\hat{p}=p=\hat{q}=1$ and $q=2$.

-LVA model (Lotka-Volterra-Autocatalator, Farkas and Noszticzius 1985b). In this case, the first autocatalytic reaction was borrowed from the Autocatalator model (Gray and Scott 1988): $\hat{p}=2$ and $p=\hat{q}=q=1$.

The Sel'kov oscillator is also a special case of the GLV model with $\hat{p}=0$, $p=1$, $q=g$, $\hat{q}=1$ (Sel'kov 1968).

Poland also dealt with some models which are included in the scheme GLV (Poland 1989). Concerning other special cases of the GLV model, see further references in (Farkas and Noszticzius 1985b).

The Liapunov technique used above can be applied without any difficulty even for non-polynomial cases. The essential point in this respect is: we may find a Liapunov function if the rate velocity of the second autocatalytic reaction (transformation of X into Y) is a product of the type f(X)g(Y) (Farkas and Noszticzius 1985b).

5. THE "SIMPLEST" OSCILLATING MODEL

The Lotka-Volterra model is a good starting point in modeling of chemical oscillations. It is very simple from a chemical point of view and rather simple from a mathematical point of view. However, there were successful attempts on different bases to construct simple chemical models showing limit cycle oscillation. The most famous and widely studied model is the Brusselator (Prigogine and Lefever 1968, Gray, Scott, and Merkin 1988).Recently Gray and Scott (1986) constructed an other model, the "Autocatalator" which shows oscillation and other exotic phenomena. Escher (1980) gave a receipt how to construct chemical models for prescribed ellipses limit cycles. These models have a common feature: they include third-order (non-elementary) chemical reactions.

From a mathematical point of view, there is no doubt that the simplest oscillating model is this:

$$\dot{x} = -y \qquad \dot{y} = x \tag{11}$$

The integral curves of this system are circles centered at the origin. In a chemical system the variables x and y are concentrations, and consequently, hey must not be negative. To fit this condition, we have to move the center into the positive quadrant. The general linear system of this type is:

$$\dot{x} = K_1 - K_2 y \qquad \dot{y} = K_3 x - K_4 \tag{12}$$

Now the integral curves are ellipses nested around the point $(K_4/K_3, K_1/K_2)$. Even this model is not reasonable in chemistry: it involves "negative cross-effect" terms ($-K_2y$ and $-K_4$), which are not allowed in chemistry (Hárs and Tóth 1979). These terms are responsible for an unpleasant property of the system(12) : the positive quadrant is not positively invariant. (A set *B* is is said positively invariant under a system of differential equations if any solution remains in *B* for $t>t_0$, provided that the initial state at t_0 were in *B*.)

However, the undesirable terms can easily be interpreted as "resultants" of chemically reasonable reactions. E.g., the term $-K_2y$ can be considered as a consequence of two reactions:

2a) production of a third intermediate Z in a reaction catalyzed by Y: $\xrightarrow{Y} Z$

2b) a reaction between X and Z yielding irrelevant products:

$$X + Z \longrightarrow$$

The balance equation for Z is:

$$\dot{z} = k_{2a}y - k_{2b}xz \tag{13}$$

If the reaction 2b) is very fast, then the quasi-stationary approximation is justified for Z, and consequently, the equation (13) reduces to

$$0 = k_{2a}y - k_{2b}xz \qquad (14)$$

This trick had already been used by Korzukhin (1967), who introduced a six-dimensional oscillating model. We have constructed a four-dimensional model which reduces to (12) in the quasi-stationary approximation. Also, it was possible to find models consisting of biochemical reactions which follows this mathematical structure (Dancsó and Farkas 1989, Dancsó, H. Farkas, M. Farkas, and Szabó 1989).

Local study of the model.

After a proper scaling, the differential equations of our model takes the form (Wittmann 1990):

$$\dot{X} = a(1-XZ) \qquad \dot{Y} = b(X-YW)$$
$$\dot{Z} = c(Y-XZ) \qquad \dot{W} = d(1-YW) \qquad (15)$$

This model does not imply "negative cross-effect" terms. On the other hand, in the limit case $c \longrightarrow \infty$ and $d \longrightarrow \infty$ we get the approximation:

$$\dot{X} = a(1-Y) \qquad \dot{Y} = b(X-1)$$
$$0 = Y - XZ \qquad 0 = 1 - YW \qquad (16)$$

which is identical to (12) apart from a scale transformation.

Hence, the solutions of (15) are approximately conservative oscillations for large values c and d. The orbits are located on the two-dimensional surface $Z=Y/X$, $W = 1/Y$ of the four-dimensional state space. The approximation (16) fails when a state variable tends to zero: the positive orthant is positively invariant under the system (15) but not under (16).

The system (15) has a unique positive stationary state: (1,1,1,1). The characteristic polynomial of the linearized system is:

$$f(\lambda) = \lambda^4 + (a+b+c+d)\lambda^3 + (a+c)(b+d)\lambda^2 + abc\lambda + abcd \qquad (17)$$

To draw the local bifurcation diagram, we use the method of parametric representation (Gilmore 1981, Farkas, Gyökér, and Wittmann 1989, 1990). First, we introduce new parameters by the transformations:

$$U=a+c \qquad V=b+d \qquad R=abc \qquad S=abcd \qquad (18)$$

We will consider U and V constant, and investigate the character of the fixed point (1,1,1,1) as a function of R and S. The results will be expressed in terms of the new parameters (U,V,R,S), but hey can be readily interpreted in the terms of the original parameters (a,b,c,d) by use of the formulas:

$$d=S/R \qquad b=V-S/R \qquad a+c=U \qquad ac=R/(V-S/R) \qquad (19)$$

Given positive values U,V,R,S, we can find the corresponding positive values for a,b,c,d if and only if

$$S < RV\left(1-\frac{4R}{U^2V}\right) \qquad (20)$$

that is we should restrict ourselves to the shadowed parabolic region of Fig. 9.

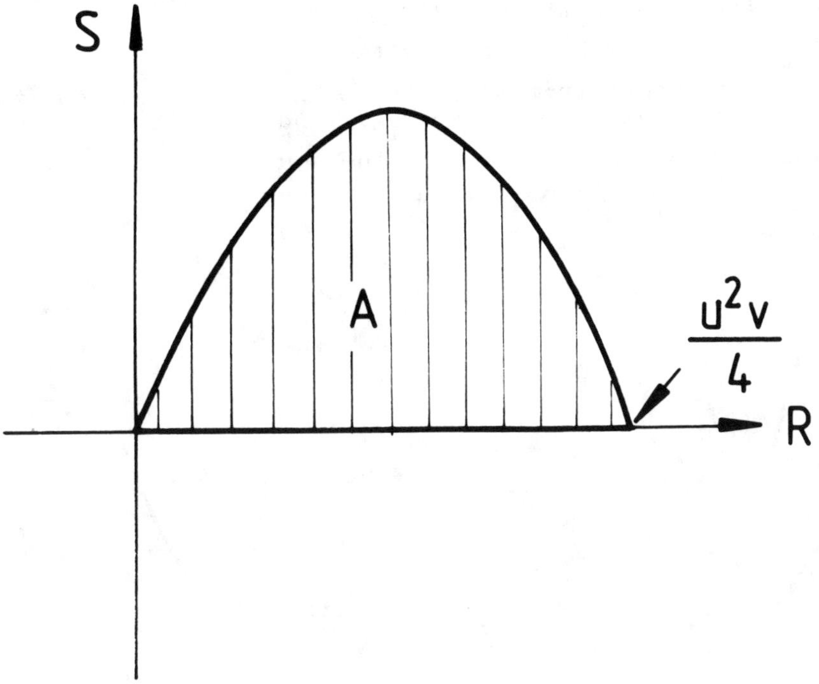

Figure 9.
The admissible region of parameters: $0<S<RV(1-4R/(U^2V))$.

The characteristic polynomial (17) can be expressed in terms of the new parameters:

$$f(\lambda) = \lambda^2(\lambda+U)(\lambda+V) + R\lambda + S \qquad (21)$$

The discriminant of the algebraic equation $f(\lambda)=0$ plays a very important role in the character of the fixed point. Recall that the discriminant D is equal to zero if and only if the polynomial has a multiple root (Korn and Korn 1961), in other words if

$$f(\lambda)=f'(\lambda)=0 \qquad (22)$$

Using the method of parametric representation, the curve D=0 can be given comfortably in a parametric form:

$$R(\lambda) = -4\lambda^3 - 3(U+V)\lambda^2 - 2UV\lambda$$
$$S(\lambda) = 3\lambda^4 + 2(U+V)\lambda^3 + UV\lambda^2 \qquad (23)$$

where the multiple root λ is considered the parameter of the curve.

The qualitative forms of the functions $R(\lambda)$ and $S(\lambda)$ are such as depicted in Fig.10 (Wittmann 1990). It can be proved (Farkas, Gyökér and Wittmann 1989) that

$$S'(\lambda) = -\lambda \; R'(\lambda) \qquad (24)$$

and therefore, apart from $\lambda=0$, these functions takes their extremes at the same values of λ (λ_1 and λ_2 in Fig. 10.). These values are the solutions of the equation

$$f''(\lambda) = 12\lambda^2 + 6(U+V)\lambda + 2UV = 0 \qquad (25)$$

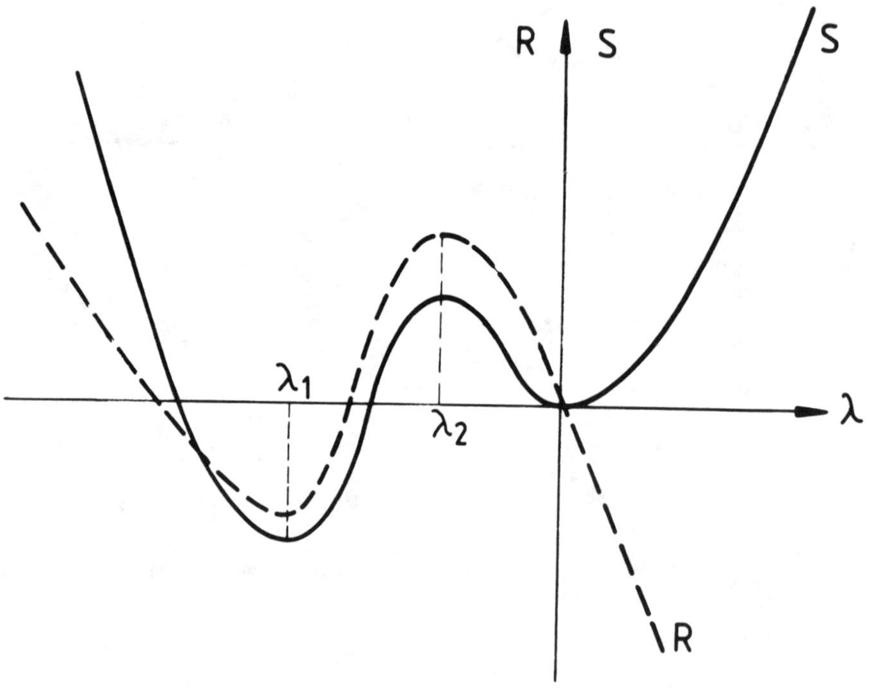

Figure 10. The graph of $R(\lambda)$ and $S(\lambda)$ in (23)

Figure 11. shows the curve D=0, the arrows indicate the increasing direction of the curve parameter λ. This curve separates parameter regions with respect to the number of real roots of the characteristic polynomial $f(\lambda)$. Even we can determine the actual values of these roots: given a point in the parameter plane (R,S), we draw all the straight lines tangential to the curve D=0, and the parameter values λ at the tangent points are just the roots (Farkas, Gyökér and Wittmann 1989). Hence, there are three different regions, B_0, B_2, B_4, in which the polynomial $f(\lambda)$ has 0, 2 and 4 roots respectively.

The cusp points belongs to the values λ_1 and λ_2 determined by the equation (25). There is another peculiar point C_x, where the curve intersects itself. At this point the polynomial has two double roots. Hence, the coordinates of this point are:

$$R_x=-(U+V)((U-V)^2)/8 \qquad S_x=(U-V)^4/64$$

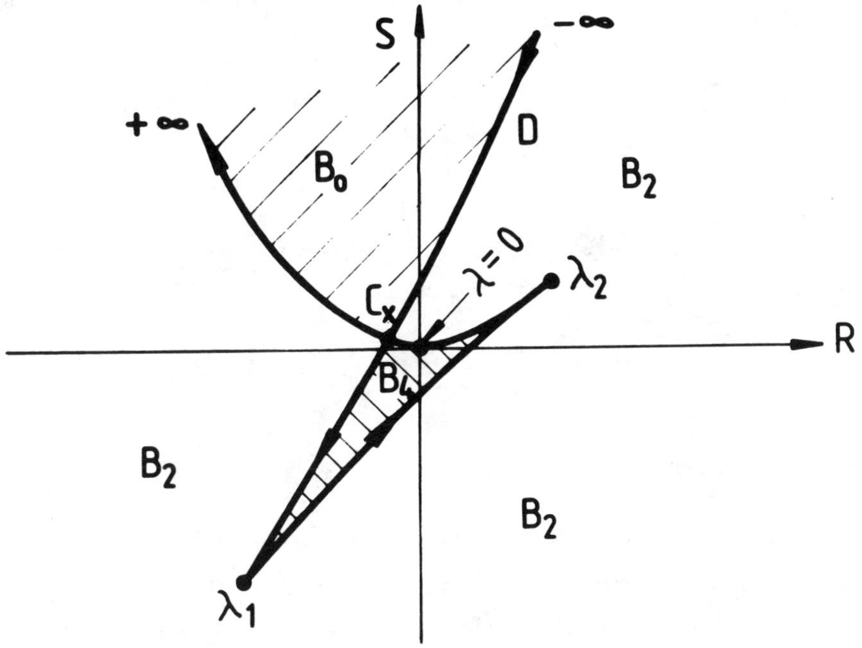

Figure 11. The curve D=0.

Now we are interested only in the admissible region A. Here the number of the real roots is 2 or 4. Indeed

$$f(-\infty)=+\infty \qquad f(-V)=-RV+S=-ab^2c<0 \qquad f(0)>0 \tag{26}$$

and consequently, there are at least two negative real roots. In the set $A \cap B_4$ there exist four negative real roots.

Positive real roots are ruled out by Descartes's rule.

Let us turn to the problem of stability. There are two typical ways in which a system can lose its stability:

I) A single real root becomes zero (changes its sign from - to +).
II) The real part of a pair of conjugate complex roots becomes zero (changes its sign from - to +).

Bifurcation of the type I cannot take place in our case: it could occur only at $S=0$, but we assumed $S>0$.

Bifurcation of the type II is the Hopf bifurcation. To find the Hopf bifurcation diagram we use the parametric representation method again.

Reformulating the Hopf bifurcation problem, we search for points in the plane (R,S) for which the sum of two (real or complex) roots is zero. In other words, together with a root λ its negative $-\lambda$ is also a root:

$$f(\lambda) = 0 \qquad f(-\lambda) = 0 \tag{27}$$

By addition and subtraction of these equations we get two new equations for the even and odd parts of the polynomial $f(\lambda)$:

$$f(\lambda)+f(-\lambda)=0 \qquad f(\lambda)-f(-\lambda)=0 \tag{28}$$

In our case:

$$S+UV\lambda^2+\lambda^4=0 \qquad R+(U+V)\lambda^2=0 \tag{29}$$

Let H^* denote the locus of the wanted points. H^* is a curve given in parametric form by (29):

$$R = -(U+V)\lambda^2 \qquad S = -UV\lambda^2-\lambda^4 \tag{30}$$

H^* is a parabola, and its equation in explicit form is:

$$S = \frac{R}{(U+V)^2}[UV(U+V)-R] \tag{31}$$

From the parametric form (30) one can see that only a part of H^* corresponds to Hopf bifurcation, namely, the part belonging to $\lambda^2<0$ (purely imaginary pair of roots). The other part of H^* (dashed line in Fig. 12) corresponds to the case when the sum of two real roots is zero.

<u>Discussion.</u>

Using the above results, Fig. 13. represents the complete local bifurcation diagram of our model. The diagram indicates three subregions of the admissible region A:

A_I : four negative real roots,

A_{II} : two negative real roots and two complex roots with negative real part,

A_{III} : two negative real roots and two complex roots with positive real part.

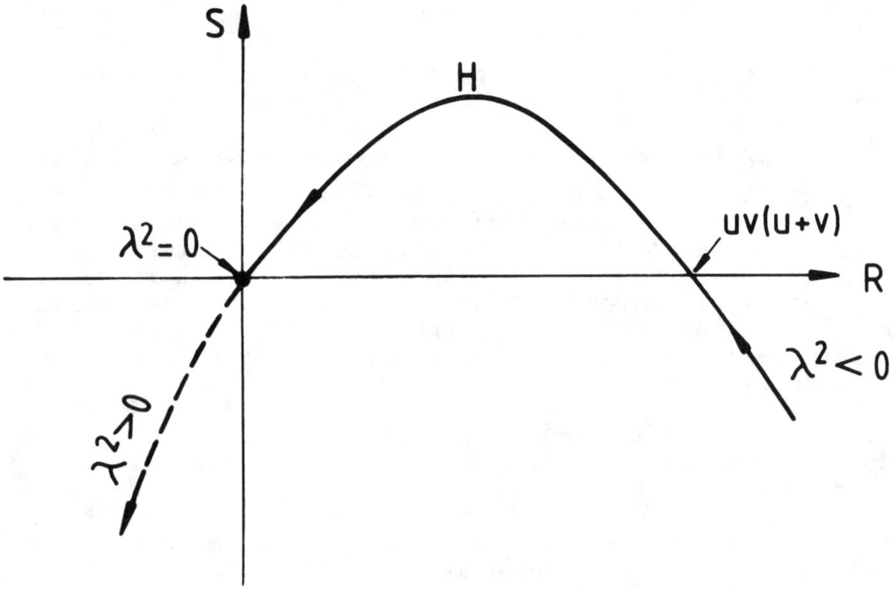

Figure 12. The curve H^*.

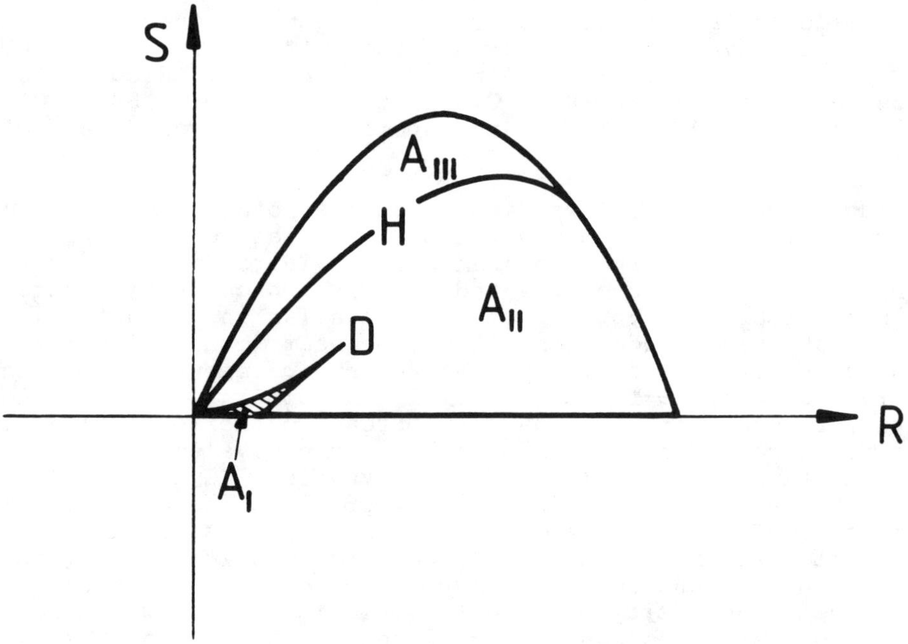

Figure 13. The complete local bifurcation diagram.

Hopf bifurcation takes place when crossing the parabola arch H. Numerical calculations show that in the unstable region A_{III} there exists a stable limit cycle (Dancsó and Farkas 1989). The oscillations are approximately sinusoidal as it was expected.

6. Thermodynamics and chemical dynamics

Thermodynamics claims universal validity: its laws should apply to every process and every phenomenon in natural sciences. On the other hand, most branches of natural sciences have their own terminology (so does chemical dynamics, too), independently of thermodynamics.In chemistry one can say: "Of course, our equations are only approximations. More exactly, there must exist reactions -allowedly, with very small rates- which were neglected in our treatment."

In the models detailed above we neglected the reverse reactions. Gray , Scott, and Merkin (1988) investigated the effects caused by the reverse reactions for the Brusselator model. Simon (1990) found that in case of the LVA model already one reverse reaction makes the system stable in a sense: there exists a bounded attractor for this system. At the same time, due to higher-order nonlinearities, more complex behavior (e.g. multistationarity) can be observed. The unique positive stationary state of LVA does not lie on the *thermodynamic branch* (Glansdorff and Prigogine 1970), that is it is not a continuation of the equilibrium state.

For some other aspects of the relation between thermodynamics and chemical dynamics we refer to the literature (Truesdell 1969, Gyarmati 1970, Nicolis and Prigogine 1977,Nicolis and Baras 1983, Babloyantz 1986, Gyarmati and Lengyel 1987, Oláh 1988, Oláh and Bódiss 1989, and especially Lengyel's article in this volume.).

What do thermodynamic laws mean to the other topics of natural sciences? In our opinion, the answer is not so straightforward as expected in general. Thermodynamics works with concepts such as energy and entropy, which -in general- are not primary concepts in other fields. As an illustration, let us take heat conduction. Theory of heat conduction as a dynamic theory uses only temperature as a primary concept: the theory must account for the spatial and temporal distribution of temperature. In the usual approach, thermodynamics is involved through the fundamental laws or equivalently, through the variational principles of thermodynamics (Gyarmati 1970, Farkas 1975a, 1975b).But if we consider the phenomenon of pure heat conduction *phenomenologically*, the straightforward treatment takes the initial and boundary conditions together with eventual sources as *input* and the temperature distribution as *output* of a deterministic blackbox. General plausible postulates can be formulated in these terms, and some secondary concept such as heat current can be derived theoretically (Farkas 1980).

Similar, purely dynamic approach was applied to a nonlinear electric model (Farkas 1978). In this model elementary work, voltage and charge are the primary concepts, and using a very general thermodynamic reasoning, it was possible to separate the work into a "conservative" and a "dissipative" part. The former one leads us to the concept of energy in this case.

It is generally accepted that for a closed system the second law of thermodynamics implies the existence of a globally stable unique equilibrium. Hence, a possible dynamic (entropy-free) formulation of the second law is:

Proposition 1. Any closed system has a globally stable point attractor.

The classical thermodynamic reasonings are based on the impossibility of *perpetuum mobile*'s. The corresponding dynamic formulation is:

Proposition 2. No closed system has a nontrivial periodic solution.

At first sight, the relations between the second law of thermodynamics and the above propositions are:

Second law → *Proposition 1* → *Proposition 2*

However, we think, that -using some additional plausible postulates- the implications may go into equivalences at least in some special cases.

The second law of thermodynamics determines the direction of the processes. By reversing the direction of time ($t \rightarrow -t$), we get a dual model in which the processes go in the reverse direction: this way we get a dissipative model from an explosive one and vice versa. By the second law, any real closed system must be dissipative.

The second law of thermodynamics and chemical oscillations

It is well known that Belousov's original manuscript on the first BZ type oscillating reaction (substrate: citric acid) was rejected by his reviewers, and he was not able to publish his discovery for many years (Winfree 1984). His reviewers were of the opinion that Belousov's observations were against the second law of thermodynamics. Really, sustained oscillations in a closed system around the chemical equilibrium would be against the second law. Of course, this is not the case with oscillating chemical systems. First, sustained oscillations can occur in open systems only. Second,these oscillations take place around an unstable steady state far from the chemical equilibrium. In other words, chemical oscillations require open systems which are far from equilibrium, and the second law in its usual form does not apply here. While this explanation is completely satisfactory, some further questions arise inevitably. The most important among these questions is the following one. Is there any evolutionary[1] law, similar to the second law of thermodynamics, a general law predicting the evolution of open dissipative systems far from equilibrium? Before answering that question, however, the idea of open systems far from equilibrium should be discussed first.

Open systems maintained far from equilibrium

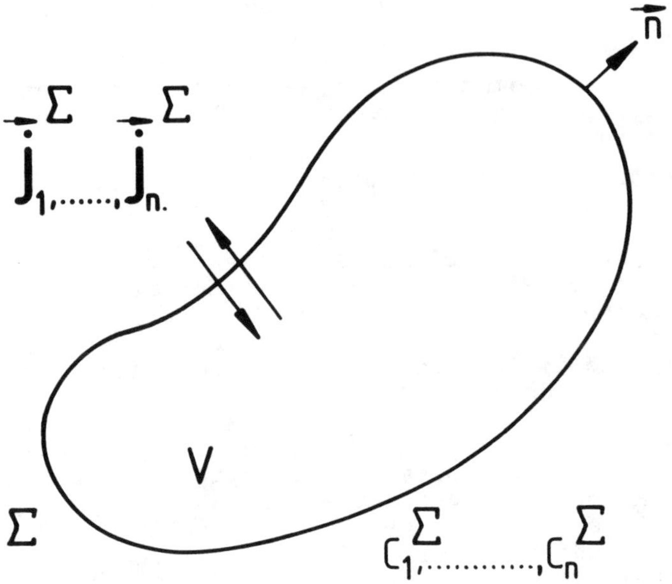

Figure 14. Schematic drawing of an open system (Nicolis and Prigogine 1977). V denotes the volume of the system and Σ stands for its surface, $\vec{n}$ is the outer normal of the surface. Current densities on the surface $\vec{j}_i^{\Sigma}$, or concentrations on the surface c_i^{Σ} are given by the boundary conditions.

In classical thermodynamics two different types of systems are regarded; thermodynamic systems with and without an environment. In the latter case the system is isolated.

[1]Here we use the term *evolutionary* in the same sense as it was used by Prigogine's school (see e.g. Babloyantz 1986). That is *evolutionary criteria* determines the direction of processes but not the details of the evolution.

According to the second law an isolated system evolves toward its equilibrium which state is characterized by the entropy maximum. In the former case there is an environment surrounding the system. That environment is in equilibrium within itself, its intensive variables (temperature, pressure, chemical potential etc.) are uniform in space and do not change in time. Moreover, the environment is big enough compared to the system. This way any change in the state of the environment caused by the system can be neglected. The environment and the system are separated by the surface of the system. That interface acts as a "thermodynamic wall". Even such systems surrounded by a steady, uniform equilibrium environment evolve toward a uniquely determined equilibrium state. Thus both types of classical thermodynamic systems (i.e. systems with and without an environment) evolve toward a final equilibrium state, and they cannot be maintained indefinitely in a state far from equilibrium.

Open systems are different: they can be fixed permanently in a state far from equilibrium. It is necessary to emphasize that the above definition is somewhat different from the usual definition of an open system. According to most authors any system exchanging substances with an external world is an open system. For example, a conventional drawing of an open system is depicted in Fig. 14. However, if there is a homogeneous steady equilibrium environment surrounding the system in Fig. 14. an evolution of the system toward a final equilibrium cannot be avoided. In other words, a far from equilibrium situation cannot be maintained here in spite of the possibility for an exchange of different chemical components through the surface. We have to realize that such a system is not a permanently open one; mass transport will cease gradually in this case.

On the other hand, if the system has more than one -e.g. two - environments like in Fig. 15., then a situation far from equilibrium can be sustained indefinitely. In the following we shall call such systems as open ones. Naturally, Fig. 14. can depict a real open system as well provided that there is a non-uniform or any other type of non-equilibrium environment surrounding the system.

The actual means to maintain a non-equilibrium situation can be rather different. For example, the system can be fed from two different reservoirs by diffusion. If there is no stirring within the system then chemical reaction and diffusion can generate chemical patterns or waves[1] like the "chemical pinwheels" in an annular gel reactor (Noszticzius, Horsthemke, McCormick, Swinney and Tam 1987). Another type of system is the continuously fed stirred tank reactor (CSTR) where the different environments are reservoirs, and pumps generate fixed flows on the boundaries of the system. Here different

---

[1]For a recent review see Ross, Müller, and Vidal (1988).

steady states, simple and complex oscillations and chemical chaos can be observed (Noszticzius, McCormick and Swinney 1987, 1989). Another way to prevent a system from evolving toward equilibrium is to maintain the initial concentration of some reactive starting materials. This is the so-called "pool of chemicals" approximation used extensively in the different Lotka-Volterra schemes of this paper. If the consumption of the starting materials is relatively slow then the pool of chemicals approximation can work for a while even for closed systems as well. For example, several hundred oscillations can be observed in certain BZ systems even in a batch reactor.

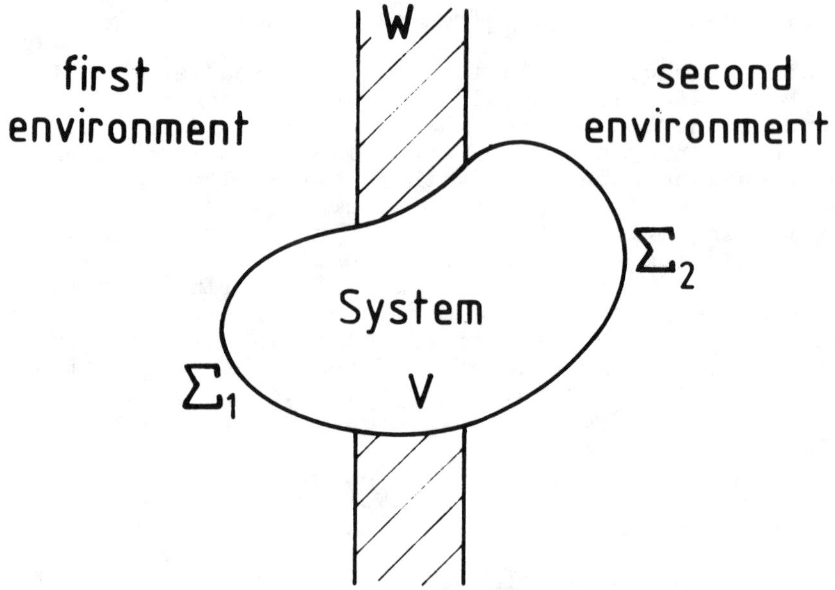

Figure 15. Schematic drawing of an open system with two environments. "W" denotes a wall isolating the two environments from each other. Σ_1 and Σ_2 are permeable surfaces of the system separating it from the first and the second environment respectively.

Evolutionary criterion for open systems

Now we can return to our original question concerning the existence of some evolutionary law -similar to the second law of thermodynamics- for open systems maintained far from equilibrium. Dynamic system theory provides such an evolutionary criterion for dissipative systems (see for example Thompson and Stewart 1986). It is known that dissipative dynamic systems exhibit a start up transient after which the motion settles down towards some form of long term recurrent behavior or asymptotic state. That recurrent behavior can be a steady state, or a periodic, or even a chaotic motion. Using the terms of nonlinear dynamics the system, after a transient, settles down on an *attractor* in the multidimensional phase space. Some well known examples for such attractors are the point, the periodic, the quasi-periodic and the strange attractor. Now, can we regard the existence of an attractor for a dissipative system as a "generalization" of the second law of thermodynamics? The answer is yes in a certain sense. To support that view let us regard the open system depicted in Fig. 15. In addition, let us assume that there is a parameter μ which expresses somehow the difference between the intensive variables of the two environments (e.g. temperature or concentration difference between the two environments). The most important point is that whenever $\mu=0$ the two environments should be identical. This is a case of classical thermodynamics; we have, in fact, only one environment if $\mu=0$.(Naturally there are cases where more than one parameter has to be adjusted. In that case μ is a vector and the two environments are equal only if $\vec{\mu}=0$. Nevertheless, the number of parameters does not affect our further considerations.) If there is only one environment the system evolves toward an equilibrium. With the terminology of nonlinear dynamics the systems evolves toward a point attractor in this case. Now, if we increase μ gradually in the close neighborhood of $\mu=0$, we still have a point attractor which is a stable steady state. The set of these steady states form the so-called "thermodynamic branch" of the system (Glansdorff and Prigogine 1970). It is important that linear systems do not leave that branch. If the system is nonlinear, however, then at a certain μ value other attractors can appear via different bifurcations (Noszticzius, Wittmann and Stirling 1985, 1987), and the system can leave the thermodynamic branch. This is the case when the nonlinear system is regarded to be "far from equilibrium" (i.e. when it is not on the thermodynamic branch any more). Nevertheless, even in such a case, a dissipative dynamic system evolves toward one of its attractors independently of the quality or the quantity of these attractors. The case $\mu=0$ can be regarded as a special one, when there is only one point attractor (second law of thermodynamics).

Thus the existence of attractors can be regarded as a general formulation of the second law of thermodynamics for dissipative dynamic systems.

Acknowledgment

The authors thank Mary Noszticzius, Éva Kovács and István Noszticzius for their help in preparing the manuscript.

6. References

Adamciková, L. and P. Sevcik. 1982. A completely inorganic oscillating system of the Belousov-Zhabotinskii type. Int.J.Chem.Kinetics 14: 735-738.

Babloyantz, A. 1986. Molecules, dynamics & life. New York: Wiley..

Barkley, D., J. Ringland, and J.S. Turner. 1987. Observation of a torus in a model of the Belousov-Zhabotinskii reaction. J.Chem.Phys. 87: 3812-3823.

Barkley, D. 1988. Slow manifolds and mixed-mode oscillations in the Belousov-Zhabotinskii reaction. J.Chem.Phys. 89: 5547-5559.

Belousov, B.P. 1951. A periodic reaction and its mechasnism. A Russian manuscript translated to English. In Oscillations and traveling waves in chemical systems. Eds. R.J. Field and M. Burger. 605-613. New York: Wiley (1985).

Bonhoffer, K.F. 1948. Über periodische chemische Reaktionen I. Z.Elektrochem. 51: 24-29.

Clarke, B.L. 1976. Stability of the bromate-cerium malonic acid network I. Theoretical formulation. J.Chem.Phys 64: 4165-4178.

Dancsó, A. and H. Farkas 1989. On the "simplest" oscillating chemical system. Periodica Polytechnica CE (in press).

Dancsó, A., H. Farkas, M. Farkas,. Gy. Szabó. 1989. Hopf bifurcation in some chemical models. React.Kin.Catal.Lett. (in press).

Dancsó, A., H. Farkas, M. Farkas, and Gy. Szabó 1990. Investigations on a class of generalized Lotka-Volterra schemes. (submitted).

De Kepper, P. and K. Bar-Eli 1983. Dynamical properties of the Belousov-Zhabotinskii reaction in a flow system. Theoretical and experimental analysis. J.Phys.Chem. 87: 480-488.

Epstein, I.R. 1984. Complex dynamical behavior in "simple" chemical systems. J.Phys.Chem. 88: 187-198.

Erdi, P. and J. Toth. 1989. Mathematical models of chemical reactions. Theory and application of deterministic and stochastic models. Manchester: University Press.

Escher, C. 1980. Models of chemical reaction systems with exactly evaluable limit cycle oscillations and their bifurcation behaviour. Ber.Bunsenges.Phys.Chem. 84: 387-391.

Farkas, H. 1968. The reformulation of the Gyarmati principle in a generalized "Γ" picture. Z.Phys.Chem. 239: 124-132.

Farkas, H. 1975a. Error estimation for approximations of the solution of the one-dimensional stationary heat conduction eq. on the basis of the Governing Principle of Dissipative Processes. Int.J.Engng.Sci. 13: 1029-1033.

Farkas, H. 1975b. On the phenomenological theory of heat conduction. Int.J.Engng.Sci. 13: 1034-1053.

Farkas, H. 1978. Thermodynamic concepts for a class of one-ports. Acta Phys.Hung. 45: 317-326.

Farkas, H. 1980. Generalization of the Fourier law. Periodica Polytechnica ME 24: 291-308.

Farkas, H. 1982. A new proof and generalization of the Maximum principlle of heat conduction. J.Non-Equilibrium Thermodyn. 7: 355-362.
Farkas, H. and Z. Noszticzius. 1985a. Use of Liapunov functions in dissipative and explosive models. Ber.Bunsenges.Phys. Chem. 89: 604-605.
Farkas, H. and Z. Noszticzius. 1985b. Generalized Lotka-Volterra shcemes. Construction of two-dimensional explodator cores and their Liapunov functions via "critical" Hopf bifurcations. J.Chem.Soc.Faraday Trans. 2. 81: 1487-1505.
Farkas, H., V. Kertész, and Z. Noszticzius 1986. Explodator and bistability. React.Kin.Catal.Lett. 32: 301-306.
Farkas, H. and Z. Noszticzius 1987. Mathematical problems in modelling of the Belousov-Zhabotinsky systems in "Proceedings of the Eleventh Internationa Conference on Nonlinear Oscillations", Eds. M. Farkas, V. Kertész, and G. Stépan, J. Bolyai Math.Soc., Budapest.
Farkas, H. , S. Gyökér, and M. Wittmann, 1989. Investigation of global equilibrium bifurcations by the method of parametric representation (in Hungarian). Alk.Mat. Lapok: (in press).
Farkas, H., Z. Noszticzius, C.R.Savage, and Z.A. Schelly 1989. Two-dimensional explodators 2. Global analysis of the Lotka-Volterra-Brusselator (LVB) model. Acta Phys.Hung. (in press).
Farkas, H., S. Gyökér, and M. Wittmann. 1990. (Work in progress.)
Farkas, M. 1984. Zip bifurcation in a competition model. Nonlin.Anal.TMA 8: 1245-1309.
Field, R.J., E. Kőrös, and R.M. Noyes 1972. Oscillations in chemical systems, Part 2. Thorough analysis of temporal oscillations in the Ce-BrO_3-malonic acid system. J.Am.Chem.Soc. 94: 8649-8664.
Field, R.J. and R.M. Noyes 1974. Oscillations in chemical systems. Part 4. Limit cycle behavior in a model of a real chemical reaction. J.Chem.Phys. 60: 1877-1884.
Field, R.J. 1985. Experimental and mechanistic characterization of bromate-ion-driven chemical oscillations and traveling waves in closed systems. In: Oscillations and Traveling waves in chemical systems. Eds. R.J. Field and M. Burger. 55-92. New York: Wiley.
Field, R.J. and P.M. Boyd.1985. Bromine-hydrolysis control in the cerium ion - oxalic acid - acetone Belousov-Zhabotinskii oscillator. J.Phys.Chem. 89: 3707-3714.
Försterling, H.D., and Z. Noszticzius. 1989. An additional negative feedback loop in the classical Belousov-Zhabotinsky reaction: malonyl radical as a second control intermediate. J.Phys.Chem. 93: 2740-2748.
Försterling, H.D., S. Murányi, and Z. Noszticzius 1990. Evidence of malonyl radical controlled oscillations in the Belousov-Zhabotinsky reaction (malonic acid - bromate - cerium - system). J.Phys.Chem. 94: (in press).
Franck, U.F. 1978. Chemical oscillators. Angew.Chem.Int.Ed. 17: 1-15.

Ganapathisubramanian, N. and R.M. Noyes. 1982. Bromate-driven oscillators in the presence of excess silver ion. J.Phys.Chem. 86: 5155-5157.
Gilmore, R. 1981. Catastrohe theory for scientists and engineers. New York: Wiley.
Glansdorff, P. and I. Prigogine 1970. Non-equilibrium stability theory. Physica 46: 344-366.
Gáspár, V. and K. Showalter 1988. Period lenthening and associated bifurcations in a two-variable, flow Oregonator. J.Chem.Phys. 88: 778-791.
Gray, P. and S.K. Scott 1985. Sustained oscillations and other exotic patterns of behavior in isothermal reactions. J.Phys.Chem. 89: 22-32.
Gray, P. and S.K. Scott 1986. A new model for oscillatory behaviour in closed systems:The Autocatalator. Ber.Bunsenges.Phys.Chem. 90: 985-996.
Gray, P., S.K. Scott, and J.H. Merkin 1988. The Brussalator model of oscillatory reactions. J.Chem.Soc.Faraday Trans. I. 84: 993-1012.
Gyarmati, I. 1970. Non-equilibrium thermodynamics. Field Theory and variational principles. Berlin: Springer.
Hárs, V. and J. Tóth. 1979.On the inverse problem of reaction kinetics. Colloq.Math.Soc.Janos Bolyai 30: 363-379. (Qualitative Theory of Differential equations, Szeged).
Hassard, B., N.D. Kazarinoff, and Y.H. Wan. 1981. Theory and applications of Hopf bifurcations. Cambridge: University Press.
Hayes, D.,K.H. Schmidt, and D. Meisel. 1989. Growth mechanism of silver halide clusters from the molecule to the colloidal particle. J.Phys.Chem. 93: 6100-6109.
Kertész, V. 1984. Global mathematical analysis of the "Explodator". Nonlin. Anal. TMA 8: 941-961.
Kertész, V. and H. Farkas. 1989. Local investigation of bistability problems in physico-chemical systems. Acta Chim.Hung. (in press).
Korn, G.A. and T.M. Korn. 1961. Mathematical Handbook for scientists and Engineers. McGraw-Hill, New York.
Korzukhin, M.D. 1967. KPBPS. Moscow. Cited in A.M. Zhabothinky. 1974. Concentration autooscillations (in Russian). Moscow: Nauka.
Lengyel, S. and I. Gyarmati. 1981. On the thermodynamics of elementary chemical reactions in homogeneous systems. J.Chem.Phys. 75: 2384-2389.
Lengyel, S. 1990. Consistency of chemical kinetics with thermodynamics.(In this volume).
Li Yue-Xian 1985. Hierarchical coupling and coupled behavior of dynamic subunits in living systems. Thesis for M.Sci., Institute of Biophysics, Chinese Acad.Sci., Beijing.
Lotka, A.J. 1920. Undamped oscillations derived from the law of mass-action. J.Am.Chem.Soc. 42: 1595-1599.
Luo, Y. and I.R. Epstein 1990. Feedback analysis of mechanisms for chemical oscillators. In Advances in Chemical Physics (in press).
Maselko, J. 1982. Determination of bifurcation in chemical systems. An experimental method. Chem.Phys. 67: 17-26.

Maselko, J. and H.L. Swinney 1986. Complex periodic oscillations and Farey arithmetic in the Belousov-Zhabotinskii reaction. J.Chem.Phys. 85: 6430-6441.
Nicolis, G. and I. Prigogine. 1977. Self-organization in Nonequilibrium Systems. New York: Wiley.
Nicolis, G. and F. Baras. 1984. (eds) Chemical instabilities. NATO ASI Series. D. Reidel, Dordrecht.
Nemitskii, V.V. and V.V. Stepanov 1960. Qualitative Theory of differential equations. Princeton: University Press.
Noszticzius, Z. 1977. Periodic carbon monoxide evolution in an oscillating reaction. J.Phys.Chem. 81: 185-186.
Noszticzius, Z. and J. Bódiss. 1979. A heterogeneous chemical oscillator. The Belousov-Zhabotinsky-type reaction of oxalic-acid. J.Am.Chem.Soc. 101: 3177-3182.
Noszticzius, Z. 1979a. Oscillating reaction of the Belousov-Zhabotinsky type with oxalic acid - acetone substrate (in Hungarian). Magyar Kemiai Folyoirat 85: 330-331.
Noszticzius, Z. 1979b. Non-bromide controlled oscillations in the Belousov-Zhabotinsky reaction of malonic acid. J.Am. Chem.Soc. 101: 3660-3663.
Noszticzius, Z. and J. Bódiss. 1980. Contribution to the chemistry of the Belousov-Zhabotinsky type reactions. Ber. Bunsenges.Phys.Chem. 84: 366-369.
Noszticzius, Z. and Farkas H. 1981. An old model as a new idea in the modelling of the oscillatiing BZ reaction systems. in Modelling of Chemical reaction systems. Eds. E.H. Ebert, P. Deuflhard and W. Jaeger , 275-281. Heidelberg: Springer.
Noszticzius, Z., H.Farkas, and Z.A. Schelly. 1984a. Explodator: a new skeleton mechanism for the halate driven chemical oscillators. J.Chem.Phys. 80: 6062-6070.
Noszticzius, Z., H. Farkas, and Z.A. Schelly 1984b. Explodator and Oregonator: parallel and serial oscillatory networks. A comparison. React.Kinet.Catal.Lett. 25: 305-311.
Noszticzius, Z., P. Stirling and M. Wittmann. 1985. Measurement of bromine removal rate in the oscillatory BZ reaction of oxalic acid. Transition from limit cycle oscillations to excitability via saddle-node infinite period bifurcation J.Phys.Chem. 89: 4914-4942.
Noszticzius, Z. and W.D. McCormick. 1987. Comment on "A quantitative and comparative study on silver ion perturbed Belousov-Zhabotinsky systems". J.Phys.Chem. 91: 4430-4431.
Noszticzius, Z., M. Wittmann, and P. Stirling. 1987. Bifurcation from excitability to limit cycle oscillations at the end of the induction period in the classical Belousov-Zhabotinsky reaction. J.Chem.Phys. 86: 1922-1926.
Noszticzius, Z., W.D. McCormick, H.L. Swinney, and Z.A. Schelly. 1987. Parallel and serial networks in the mechanism of the oscillating Belousov-Zhabotinsky reaction. The Tandem oscillator. Acta Pol.Scand. Ch. 178: 57-77.
Noszticzius, Z., W.D. McCormick, and H.L. Swinney. Effect of trace impurities on a bifurcation structure in the Belousov-Zhabotinskii reaction and preparation of high-purity malonic acid J.Phys.Chem. 91: 5129-5134.
Noszticzius, Z., W. Horsthemke, W.D. McCormick, H.L. Swinney and W.J. Tam. 1987. Sustained chemical waves in an annular gel reactor: a chemical pinwheel. Nature 329:619-620.

Noszticzius, Z. and W.D. McCormick. 1988. Estimation of the rate constant of the $Ag^+ + Br^- \rightarrow AgBr$ reaction. The possibility of non-bromide-controlled oscillations in the Belousov-Zhabotinsky reaction. J.Phys.Chem. 92: 374-376.
Noszticzius, Z., W.D. McCormick and H.L. Swinney. 1989. Use of bifurcation diagrams as fingerprints of chemical mechanisms. J.Phys.Chem. 93: 2796-2800.
Noyes, R.M., R.J. Field, and R.C. Thompson 1971. Mechanism of reaction of bromine(V) with weak one-electron reducing agents. J.Am.Chem.Soc. 93: 7315-7316.
Noyes, R.M. 1984. An alternative to the stoichiometric factor in the Oregonator model. J.Chem.Phys. 80: 6071-6078.
Noyes, R.M., R.J. Field, H.D. Försterling, E.Kőrös, and P. Ruoff. 1989. Controversial interpretetions of Ag^+ perturbation of the Belousov-Zhabotinsky reaction. J.Phys.Chem. 93: 270-174.
Oláh, K. 1988. Thermostatics, thermodynamics and thermokinetics. Acta Chimica Hung. 125:117-130.
Oláh, K. and J.Bódiss. 1989. Stoichiometric coefficients as chemical charges. React.Kin.Catal.Lett. 39: 163-167.
Poland, D. 1989. The effect of clustering on the Lotka-Volterra model. Physica D 35: 148-166.
Prigogine, I. and R. Lefever. 1968. Symmetry breaking instabilities II. J.Chem.Phys. 48: 1695-1700.
Rábai, Gy., Gy. Bazsa, M.T. Beck. 1979. Design of reaction systems exhibiting overshoot-undershoot kinetics. J.Am. Chem.Soc. 101: 6746-6748.
Ross, J., S.C. Müller, and C. Vidal. 1988. Chemical waves. Science 240: 460-465.
Ruoff, P. and B. Schwitters 1984. Theoretical study of Ag^+-induced oscillations and excitations in the classical homogeneous Belousov-Zhabotinsky reaction using the Oregonator model. J.Phys.Chem. 88: 6421-6429.
Ruoff, P. and R.M. Noyes 1986. An amplified Oregonator model simulating alternative excitabilities, transitions in types of oscillations, and temporary bistability in a closed system. J.Chem.Phys. 84: 1413-1423.
Ruoff, P. and J. Vestvik. 1989. Potentiometric and spectrophotometric studies of the silver bromide reaction in 1M sulfurice acid and its relevance to silver ion perturbed bromete-driven oscillators. J.Phys.Chem. 93: 7798-7801.
Russo, T. 1990. Modelling of a silver-ion perturbed Belousov-Zhabotinskii oscillator. J.Phys.Chem. 94: (in press).
Schwitters, B. and P. Ruoff. 1986. Simulation of bromate-driven oscillations in the presence of excess silver ions using the Oregonator model. J.Phys.Chem. 90: 2497-2501.
Sel'kov, E.E. 1968. Self-oscillations in glycolysis. Europ.J.Biochem. 4: 79-86.
Sevcik, P. and L. Adamciková. 1985. The oscillating Belousov-Zhabotinsky type reactions with saccharides. J.Phys.Chem. 89: 5178-5179.
Showalter, K., R.M. Noyes, and K. Bar-Eli. 1978. A modified Oregonator model exhibiting complicated limit cycle behavior in a flow system. J.Chem.Phys. 69: 2514-2524.
Simon, P. 1990. (Work in progress.)

Swinney, H.L. and J.C. Roux. 1984. Chemical chaos. in Non-equilibrium dynamics in chemical systems. Ed. C. Vidal and A. Pacault. 124-140. Berlin: Springer.
Tang, B. 1989. On the existence of periodic solutions in the Limited Explodator model for the Belousov-Zhabotinskii reaction. Nonlin.Anal.TMA. 13: 1359-1374.
Thompson, J.M.T. and H.B. Stewart. 1986. Nonlinear dynamics and chaos. New York: Wiley.
Thompson, R.C. 1971. Reduction of bromine(V) by cerium (III) manganese(II) and neptunium(V) in aqueous sulfuric acid. J.Am.Chem.Soc. 93: 7315.
Truesdell, C. 1969. Rational Thermodynamics. New York: McGraw-Hill.
Tyson, J.J. 1984. Relaxation oscillations in the revised Oregonator. J.Chem.Phys. 80: 6079-6082.
Tyson, J.J. 1985. A quantitative account of oscillations, bistability and traveling waves in the Belousov-Zhabotinskii reaction. In: Oscillations and traveling waves in chemical systems. Eds. R.J. Field and M. Burger. 93-144. New York: Wiley.
Varga, M., L. Györgyi, and E. Kőrös. 1985. Bromate oscillators: elucidation of the source of bromide ion and modification of the chemical mechanism. J.Am.Chem.Soc. 107: 4780-4781.
Varga, M. and E. Kőrös 1986. Thorough study of bromide control in bromate oscillators. 4. A quantitative and comparative study on silver ion perturbed Belousov-Zhabotinsky systems. J.Phys.Chem. 90: 4373-4376.
Vavilin, V.A. and A.N. Zhabotinskii. 1969. Autocatalytic oxidation of Ce(III) by bromate I.(in Russian). Kinet.Katal. 10: 83-88.
Volterra, V. 1931. In Animal ecology. Ed. R.W. Chapman. 409-448. New York: McGraw-Hill.
Wittmann, M. 1990. (Work in progress.)
Zhabotinsky, A.N. 1964. Periodic processes of the oxidation of malonic acid in solution (in Russian). Biofizika 9: 306-311.
Zhabotinsky, A.N. 1985. The early period of systematic studies of oscillations and waves in chemical systems. In Oscillations and traveling waves in chemical systems. Eds. R.J. Field and M. Burger. 1-6. New York: Wiley.

Transport Processes in Electrolytes and Membranes

Katrine Seip Førland
Divisions of Inorganic Chemistry
The Norwegian Institute of Technology
The University of Trondheim N-7034 Trondheim, Norway

Tormod Førland and Signe Kjelstrup Ratkje
Physical Chemistry the Norwegian Institute of Technology
The University of Trondheim N-7034 Trondheim, Norway

ABSTRACT
A new approach to irreversible thermodynamics is reviewed (Section 2). The main feature is that thermodynamic components and operationally defined quantities are used consistently. Equations for transport processes are developed emphasizing the physical principles, using a minimum of mathematics. Relations are revealed that reduce the number of parameters needed to describe a process.

To demonstrate the applicability of irreversible thermodynamics to a variety of practical problems, some examples are discussed. Section 3 deals with the glass electrode and cells with liquid junctions. Section 4 reviews transport of two cations and water in a cation exchange membrane. Section 5 describes transport combined with chemical reaction in a battery and in muscular contraction. Section 6 reviews some non-isothermal systems. Common to all examples is that a better understanding can be achieved, and in some cases, new information on the systems is obtained through the application of the new approach.

1. INTRODUCTION

The main objective of irreversible thermodynamics is to describe completely and quantitatively interacting transport processes. Simultaneous and interacting transport processes, e.g., diffusion and electric current, or transport of heat coupled to transport of mass or electric charge are common phenomena in different fields such as electrochemistry, chemical engineering, biochemistry and biophysics, metallurgy, and geology.

The basis for dealing with transport phenomena in irreversible thermodynamics is the entropy production by the process. The transport is caused by forces given by variations in thermodynamic functions.

The equilibrium state is a dynamic state where the rates in opposite directions are equal. To obtain simple equations for transport the system must not be far from equilibrium, so that transport rates in opposite directions are near equal. On a molecular scale the transport may be pictured as a movement of a particle over an energy barrier. In the direction of the force the rate of transfer is proportional to

$\exp(-E_a/kT)$ where E_a is the activation energy and k is the Bolzmann constant. The rate in the opposite direction is $\exp((-E_a+\Delta E)/kT)$ where $+E_a-\Delta E$ is the activation energy for transport in the opposite direction. When ΔE is negative, the net rate of transport in the forward direction is positive. For a symmetrical barrier it is proportional to

$$\exp(-E_a/kT) - \exp((-E_a+\Delta E)/kT)$$

$$= \exp(-E_a/kT)(1-\exp(-\Delta E/kT)) = (1-1+\Delta E/kT+....)\exp(-E_a/kT)$$

If the ratio $\Delta E/kT$ has a low numerical value, we can neglect all higher terms in the expansion of the exponential function. Then the rate of transport will be proportional to ΔE and we obtain a linear relation between the transport rate and the force. For some transports across a biological membrane the force, $-\Delta E$, is large compared to kT and the transport rate is given by an exponential function (Garlid 1989).

Furthermore we assume that the local thermodynamic functions for a system not in equilibrium, are well defined and are the same functions of the local state variables as for a system in equilibrium. Local refers to a volume, small compared to the volume of the whole system, but large compared to the size of a molecule. The requirement of local equilibrium is closely connected to the definition of components in a system where transport takes place.

The choice of components to describe a system is a central point in our treatment of transport processes. The following definition will be used:

> The number of thermodynamic components in a system is the minimum number of neutral chemical species necessary to describe any part of the system (Lewis 1961)(Compare also Gibbs 1875).

For a system of electrolytes our treatment is the same as for systems of fused salts (Førland 1964), we do not use ions as components. We shall use thermodynamic components to describe local composition in systems where transport takes place. We assume local equilibrium. This implies that the rate of chemical reaction is fast compared to the transport rate.

From this follows that we do not use affinity $A = -\Sigma\nu_i\mu_i = -\Delta G$ as a force. Here ν_i is a stoichiometric coefficient for a chemical reaction with a Gibbs energy change ΔG. Locally we assume $A = 0$. Even when A has a nonzero value, it does not contribute directly to a transport process in an isotropic system according to the Curie principle. Over a larger distance in a system containing membranes (anisotropic system) the scalar A can contribute to a vectorial transport. For this kind of system, however, we shall see how the scalar force can be replaced by vectorial forces (Section 5).

2. PRINCIPLES

2.1. Entropy Production.

To simplify the derivation we consider a unidirectional transport in a system as shown in Figure 1. One of the components is a membrane which makes a pressure gradient possible. The cross section of the membrane is of unit area. In order to describe the movement of components we need a frame of reference, and it is convenient to choose the membrane as frame of reference. We assume no laminar flow and no acceleration in any part of the system.

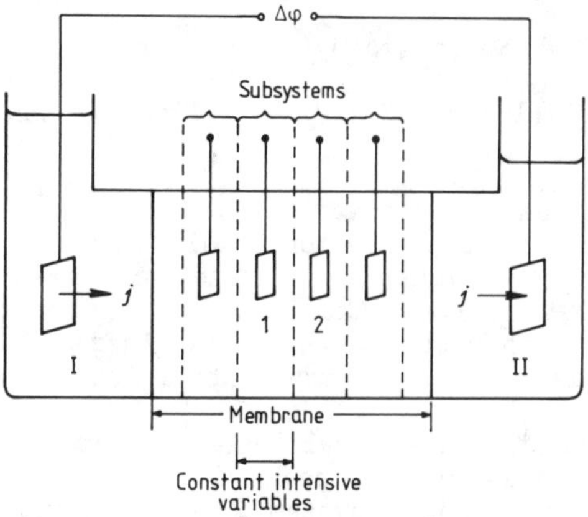

Figure 1. Transfer of heat, matter and charge in an adiabatic system composed of a series of subsystems (Førland 1980). Reprinted with permission from Pergamon Journals Ltd.

We shall deal with systems where we can assume local electroneutrality. A transport process may cause a small change in charge separation in an electrolyte, but the contribution to the change in electrostatic energy of the system is negligible (Førland 1981).

To derive an equation for entropy production we divide the system into small subsystems as shown in Figure 1. They are small compared to the total system, but sufficiently large to allow thermodynamic functions to be well defined. The continuous change in the intensive variables (such as p, T, μ_i) is described by functions varying in steps between subsystems. Intensive variables are constant within a subsystem.

The entropy of each subsystem is a function of internal energy, U, volume, V, and number of moles of each component, n_i.

$$S = f(U,V,n_i) \tag{2.1}$$

We shall first consider transport between only two open subsystems 1 and 2 which together form a closed and adiabatic system (compare Førland 1982). Using the first law for the total system we have

$$dU_1 + dU_2 = - p_1 dV_1 - p_2 dV_2 \tag{2.2}$$

From eqs (2.1) and (2.2) we can calculate the change in entropy for each subsystem and obtain the entropy production of the pair of subsystems (Førland 1988a, Section 2.1.2)

$$dS = dS_1 + dS_2 = -\frac{dq_1}{T_1} - \sum_i S_{i,1}dn_i + \frac{dq_2}{T_2} + \sum_i S_{i,2}dn_i$$

$$= \frac{dq_2}{T_2} - \frac{dq_1}{T_1} + \sum_i (S_{i,2} - S_{i,1})dn_i \qquad (2.3)$$

where dq_1 is the heat leaving subsystem 1, at a temperature T_1, dq_2 is the heat received by subsystem 2 at a temperature T_2, S_i is the partial molar entropy of a component and dn_i is number of moles transferred. It can be shown that when $|T_2 - T_1| = |\Delta T| << T_1$, eq. (2.3) can be transformed to

$$dS = \Delta(\frac{1}{T})dq_1 - \frac{1}{T}\sum_i \Delta\mu_{i,T}dn_i \qquad (2.4)$$

where $\Delta\mu_{i,T}$ is the change in chemical potential with changes in composition and pressure. Similar exchange of heat and mass takes place between all other pairs of subsystems, and the entropy production per unit length can be obtained.

To obtain a well defined electric potential, $\Delta\phi$, (or a gradient, $\nabla\phi$), we introduce reversible test electrodes in each subsystem. There will be no net electric current on the test electrodes by charge transfer in the total system.

The entropy is a state function. Therefore any convenient path from the initial to the final state may be chosen for calculating the change in entropy of a subsystem and thus for calculating entropy production in a pair of subsystems. The change in composition for each subsystem can be described by transfer of neutral components from one subsystem to the next one even though the mechanism of transfer was ion migration.

When an electric work dw_{el} is carried out on the pair of subsystems in addition to pressure volume work, the first law gives

$$dU_1 + dU_2 = -p_1dV_1 - p_2dV_2 + dw_{el} \qquad (2.5)$$

The electric work can be expressed as $-\Delta\phi \cdot dQ$ where dQ is the electric charge transferred and $\Delta\phi$ is the electric potential of the cell consisting of two subsystems. Together with eq. (2.1) this gives the entropy production for a pair of subsystems

$$dS = \Delta(1/T)dq - \frac{1}{T}\sum_i \Delta\mu_{i,T}dn_i - (\Delta\phi/T)dQ \qquad (2.6)$$

In this derivation of entropy production the measurable electric potential entered the equation via external electric work as expressed in the first law.

Suppose all transfers between pairs of subsystems of unit cross section take place during the time $\mathrm{d}t$. From eq. (2.6) we obtain the entropy production per unit time and unit volume

$$\theta = \frac{\mathrm{d}}{\mathrm{dx}}\left(\frac{\mathrm{d}S}{\mathrm{d}t}\right) = \frac{\mathrm{d}(1/T)}{\mathrm{dx}}(\mathrm{d}q/\mathrm{d}t) - \frac{1}{T}\sum_{i}\frac{\mathrm{d}\mu_{i,T}}{\mathrm{dx}}(\mathrm{d}n_i/\mathrm{d}t) - \frac{1}{T}\frac{\mathrm{d}\phi}{\mathrm{dx}}(\mathrm{d}Q/\mathrm{d}t) \tag{2.7}$$

where x is the length coordinate. We define fluxes

$$J_q = \mathrm{d}q/\mathrm{d}t,\ J_i = \mathrm{d}n_i/\mathrm{d}t,\ j = \mathrm{d}Q/\mathrm{d}t \tag{2.8}$$

Here j is the electric current density in faraday $m^{-2}s^{-1}$. Multiplying eq. (2.7) by T we obtain the dissipation function

$$T\theta = -(\mathrm{dln}T/\mathrm{dx})J_q - \sum_{i=1}^{n-1}(\mathrm{d}\mu_{i,T}/\mathrm{dx})J_i - (\mathrm{d}\phi/\mathrm{dx})j \tag{2.9}$$

where n is the number of components. This equation is often written in the abbreviated form, where k = n+1 and where X_j are forces.

$$T\theta = \sum_{j=1}^{k} X_j J_j \tag{2.10}$$

When one component in a mixture is chosen as frame of reference, its flux is equal to zero, and the force - conjugate flux product does not contribute to $T\theta$. All changes in chemical potentials, $\mathrm{d}\mu_{i,T}$ are connected by the Gibbs-Duhem equation. This means that all chemical potentials but one can be varied independently. Thus we can conclude that the forces contributing to $T\theta$ are independent.

The above derivation of the expression for the entropy production is different from the one commonly used. Local entropy production in an electrolyte is usually obtained by introducing the electric force on ions in the expression for the first law of thermodynamics (deGroot 1962). In a slightly different procedure the electrostatic potential is introduced in the Gibbs equation (Katchalsky 1965). In both cases the unmeasurable electrostatic potential is eliminated by introducing the electrochemical potential (compare eq. (2.36)). The usual approach distinguishes between two kinds of entropy change, d_iS entropy production and d_eS entropy exchange with surroundings. We do not need this distinction since we obtain the entropy production directly as a sum of changes in two state functions.

2.2. The Flux Equations

It is postulated in irreversible thermodynamics that a flux, J_i can be expressed as a linear homogeneous, function of the forces

$$J_i = \sum_{j=1}^{k} L_{ij}X_j \quad ; \quad (i=1,\ \ldots,\ k) \tag{2.11}$$

The phenomenological coefficients, L_{ij}, are independent of the forces. The cross coefficients, L_{ij} with $i \neq j$, are related by the Onsager reciprocal relations (Onsager 1931a,b),

$$L_{ij} = L_{ji} \tag{2.12}$$

A complete and consistent description of all transport processes in a system requires all the equations (2.10), (2.11) and (2.12) (Coleman 1960), (Førland 1988a, Section 3.2).

Many irreversible processes approach stationary state with time. For systems in a stationary state with transfers over a temperature difference of only a few degrees, forces can be expressed by differences across the system instead of gradients. The transports may be pictured as transports between two reservoirs. Both reservoirs are so large that intensive variables undergo only negligible changes by the transfers between them. In the stationary state the composition and gradients in the transport region do not change with time. There is a change in entropy in the two reservoirs, and they form a pair of sub-systems. The entropy production for the total system is given by eq. (2.6):

The cross section may vary through the system, and it is convenient to define fluxes as flows of heat, matter and charge per unit time over the whole cross section of the system between the two reservoirs (not over unit area as in eq. (2.7)). The fluxes of heat and matter over the whole cross section will be given the same symbols as before, J_q and J_i, while the electric current is given the symbol I to distinguish it from current density. The corresponding forces are $-\Delta \ln T$, $-\Delta\mu_{i,T}$ and $-\Delta\phi$. The dissipated energy per unit time for the total system is

$$T(\mathrm{d}S/\mathrm{d}t) = -\Delta \ln T J_q - \Sigma \Delta\mu_{i,T} J_i - \Delta\phi I \tag{2.13}$$

The corresponding flux equations are

$$J_j = -\bar{L}_{j1}\Delta \ln T - \sum_{i=2}^{k-1} \bar{L}_{ji}\Delta\mu_{i,T} - \bar{L}_{jk}\Delta\phi; \quad (j=1,\ldots,k) \tag{2.14}$$

In eq. (2.14) J_1 is the heat flux J_q, and $J_2,\ldots, J_{k-1}$ are the material fluxes, over the whole cross section, while J_k is the current, I. The coefficients, $\bar{L}_{ij}$ are given for the whole cross section. They have average values over the path of transport. The Onsager reciprocal relations are assumed valid for the average coefficients:

$$\bar{L}_{ij} = \bar{L}_{ji} \tag{2.15}$$

The coefficients in eq. (2.14) will depend on temperature, composition and pressure. When changing composition both $\Delta\mu_{i,T}$ and coefficients will change. The use of eq. (2.14) is therefore limited to systems where the forces $-\Delta \ln T$ and $-\Delta\mu_{i,T}$ have small numerical values. For large values of these forces we need the local values of the coefficients L_{ij}

to calculate fluxes. A method of obtaining local values of transport coefficients is discussed in section 4.2.

2.3. Relations between Phenomenological Coeffients.

Since entropy production by a process cannot be negative, and the coefficients are independent of forces, the direct coefficients, L_{ii}, cannot be negative. Furthermore there is a restriction on the numerical value of a cross coefficient in relation to the direct coefficients. In eq. (2.10) we can make all forces except X_1 and X_2 equal to zero.

For a system of two forces and fluxes:

$$J_1 = L_{11}X_1 + L_{12}X_2 \tag{2.16a}$$

$$J_2 = L_{21}X_1 + L_{22}X_2 \tag{2.16b}$$

one of the forces, e.g. X_2, may be eliminated from the equations. Then

$$J_1 = (L_{11} - L_{12}L_{21}/L_{22})X_1 + (L_{12}/L_{22})J_2$$

or

$$J_1 = l_{11}X_1 + (L_{12}/L_{22})J_2 \tag{2.17}$$

where

$$l_{11} = L_{11} - L_{12}L_{21}/L_{22} \tag{2.18}$$

The coefficient l_{11} is a direct coefficient, coupling the flux J_1 to its main driving force, X_1. With independent forces we can choose a value of X_2 that gives $J_2 = 0$. This reduces transport in the system to one flux, $J_1 = l_{11}X_1$, and the direct coefficient l_{11} cannot be negative. If the fluxes are linearly dependent ($J_1 = \alpha J_2$), we have $l_{11} = 0$ and the coupling coefficient $\alpha = L_{12}/L_{22}$. With independent forces and fluxes

$$L_{11}L_{22} > L_{12}^2 \tag{2.19}$$

A consequence of our method, where the composition is described by electrically neutral components, is that some of the coefficients L_{ij} depend on the kind of electrodes used to measure $\Delta\phi$. Consider e.g. the isothermal cell

$$\mathrm{Ag(s)|AgCl(s)|HCl(aq,}c_{\mathrm{I}})||\mathrm{HCl(aq,}c_{\mathrm{II}}|\mathrm{AgCl(s)|Ag(s)} \tag{a}$$

With water as frame of reference the flux equations are

$$J_{HCl} = - L_{11}\nabla\mu_{HCl} - L_{12}\nabla\phi \quad (2.20a)$$

$$j = - L_{21}\nabla\mu_{HCl} - L_{22}\nabla\phi \quad (2.20b)$$

When $\nabla\mu_{HCl}$ = 0, the following is obtained from eqs (2.21)

$$J_{HCl}/j = = L_{12}/L_{22} = t_{HCl} \quad (2.21)$$

Here J_{HCl}/j is the number of moles HCl transferred with one faraday electric charge. It may be named the *transference coefficient* of HCl, t_{HCl} (Førland 1988a, Section 4.1.1). With AgCl/Ag electrodes it is equal to the transference number of H^+, $t_{HCl} = t_{H^+}$. If we change the electrodes to H^+/H_2 electrodes, we have a cell

$$(Pt)H_2(g,1\ atm)|HCl(aq,c_I)||HCl(aq,c_{II})|H_2(g,1\ atm)\ (Pt) \quad (b)$$

with the flux equations

$$J'_{HCl} = - L'_{11}\nabla\mu_{HCl} - L'_{12}\nabla\phi' \quad (2.22a)$$

$$j' = - L'_{21}\nabla\mu_{HCl} - L'_{22}\nabla\phi' \quad (2.22b)$$

For this system we obtain the transference coefficient of HCl, t'_{HCl}, when $\nabla\mu_{HCl}$ = 0.

$$J'_{HCl}/j' = L'_{12}/L'_{22} = t'_{HCl} = - t_{Cl^-} \quad (2.23)$$

There are simple exact relations between the two sets of coefficients L_{ij} and L'_{ij} (Førland 1988a, Section 4.1.2).

There are some other relations between coefficients that reduce the number of parameters needed to describe completely the transport in a system. Consider the cell (a). It can be shown (Førland 1988a, Section 4.1.1) that the coefficients in eqs (2.20) are connected by

$$L_{11} = L_{21} \quad (2.24)$$

in dilute solutions, where we can assume that the ions H^+ and Cl^- do not migrate as neutral HCl molecules. For cell (b) we obtain in a similar way

$$L'_{11} = - L'_{21} \quad (2.25)$$

A similar kind of relation is obtained for a cell with an electrolyte containing two salts e.g. HCl and NaCl (Førland 1988a, Section 4.1.4)

$$\mathrm{Ag(s)|AgCl(s)|HCl}(\mathrm{aq},c_{\mathrm{HCl,I}}),\mathrm{NaCl}(\mathrm{aq},c_{\mathrm{NaCl,I}})|| \qquad \text{(c)}$$

$$\mathrm{HCl}(\mathrm{aq},c_{\mathrm{HCl,II}}),\mathrm{NaCl}(\mathrm{aq},c_{\mathrm{NaCl,II}})|\mathrm{AgCl(s)|Ag(s)}$$

with the flux equations

$$J_{\mathrm{HCl}} = - L_{11}\nabla\mu_{\mathrm{HCl}} - L_{12}\nabla\mu_{\mathrm{NaCl}} - L_{13}\nabla\phi \qquad (2.26a)$$

$$J_{\mathrm{NaCl}} = - L_{21}\nabla\mu_{\mathrm{HCl}} - L_{22}\nabla\mu_{\mathrm{NaCl}} - L_{23}\nabla\phi \qquad (2.26b)$$

$$j = - L_{31}\nabla\mu_{\mathrm{HCl}} - L_{32}\nabla\mu_{\mathrm{NaCl}} - L_{33}\nabla\phi \qquad (2.26c)$$

Assuming again that the ions do not migrate as neutral molecules, we obtain the relations

$$L_{11} + L_{21} = L_{31} \qquad (2.27a)$$

$$L_{12} + L_{22} = L_{32} \qquad (2.27b)$$

If we also assume that there is no significant interaction between transport of H^+ and Na^+ apart from that due to the requirement of local electroneutrality, $(J_{\mathrm{HCl}})_{\nabla\phi=0} = - L_{11}\nabla\mu_{\mathrm{HCl}}$, we have

$$L_{12} = 0 \qquad (2.28)$$

With the Onsager reciprocal relations and eqs (2.27a), (2.27b) and (2.28) the number of independent coefficients in eqs (2.26) is reduced from nine to three. These coefficients could be the three direct coefficients L_{11}, L_{22} and L_{33}. As these coefficients can be expressed by mobilities, u, we can alternatively choose u_{H^+}, u_{Na^+} and u_{Cl^-} as the set of parameters to describe the transport. The latter set has the advantage of being fairly independent of composition as long as the solution is dilute. The connection between the two sets is

$$L_{11} = (1/F)c_{\mathrm{H}^+}u_{\mathrm{H}^+} \qquad (2.29)$$

$$L_{22} = (1/F)c_{\mathrm{Na}^+}u_{\mathrm{Na}^+} \qquad (2.30)$$

$$L_{33} = (1/F)(c_{\mathrm{H}^+}u_{\mathrm{H}^+} + c_{\mathrm{Na}^+}u_{\mathrm{Na}^+} + c_{\mathrm{Cl}^-}u_{\mathrm{Cl}^-}) \qquad (2.31)$$

If the electrolytes in cell (c) were separated by a cation conducting membrane, we would obtain similar relations between coefficients. We now have sixteen coefficients and six equations from the Onsager reciprocal relations. If the membrane is completely impermeable to anions, we have an exact coupling between fluxes

$$J_{HCl} + J_{NaCl} - j = 0 \qquad (2.32)$$

We then obtain a set of four exact equations similar to eqs (2.27). In general for any kind of coupling between fluxes

$$\sum_i \alpha_i J_i = 0 \qquad (2.33)$$

where α_i are force-independent coupling coefficients. When the forces are independent, we obtain the following relations between the phenomenological coefficients (deGroot 1969), (Førland 1988a, Section 5.5.2).

$$\sum_i \alpha_i L_{ij} = 0 \quad ; \quad (j = 1,2, ...) \qquad (2.34)$$

2.4. Nernst-Planck Flux Equations.

The Nernst-Planck equation for the flux of the H^+ ion in an electrolyte solution can be written in the form (Nernst 1888), (Planck 1890)

$$J_{H^+} = (1/F)\ c_{H^+} u_{H^+}\ (\nabla\mu_{H^+} + \nabla\psi) \qquad (2.35)$$

where u_{H^+} is the mobility of the H^+ ion, $\nabla\mu_{H^+}$ is the gradient in the single ion chemical potential of the ion, and $\nabla\psi$ is the gradient in electrostatic potential. Introducing the electrochemical potential gradient of an ion (Guggenheim 1929)

$$\nabla\tilde{\mu}_i = \nabla\mu_i + z_i\nabla\psi \qquad (2.36)$$

where z_i is the charge of the ion we have

$$J_{H^+} = (1/F)\ c_{H^+} u_{H^+}\ \nabla\tilde{\mu}_{H^+} \qquad (2.37)$$

The electrochemical potential gradient is defined by eq. (2.36) and it is a measurable quantity. The two terms on the right hand side of eq. (2.36) are not measurable. It can be shown that $\Delta\tilde{\mu}_{H^+}$ from the left hand side to the right hand side of an electrochemical cell with two H^+/H_2 electrodes is equal to the electric potential, $\Delta\phi'$, of the cell (Katchalsky 1965), (Førland 1988 a, Section 4.1.2). In a similar way one finds that $\Delta\tilde{\mu}_{Cl^-}$ is equal to minus the electric potential, $\Delta\phi$, of a cell with two Cl^- - reversible electrodes, cells (a) and (c).

$$-\Delta\phi = \Delta\mu_{Cl^-} = \Delta\mu_{Cl^-} - \Delta\psi \tag{2.38}$$

Equation (2.37) can be compared with eq. (2.26a) when we introduce eqs (2.27) and (2.28). We obtain

$$J_{H^+} = J_{HCl} = - L_{11}(\nabla\mu_{HCl} + \nabla\phi) = - L_{11}(\nabla\mu_{HCl} - \nabla\mu_{Cl^-} + \nabla\psi)$$

$$= (c_{H^+}u_{H^+}/F)\ \nabla\mu_{H^+} \tag{2.39}$$

Thus the Nernst - Planck equation is not valid unless eqs (2.27) and (2.28) are valid.

For high concentrations of the salts deviations from eqs (2.27) and (2.28) are significant (Miller 1967).

3. THE emf FOR ISOTHERMAL CELLS WITH IRREVERSIBLE TRANSPORT

Measurable thermodynamic properties of a system or of a section of the system are independent of the choice of components. When using thermodynamic components in the description of an irreversible process taking place in, e.g., an electrochemical cell, one can divide the cell into sections. For each section one can calculate the change in Gibbs energy separately. The results can be added up to give the change in Gibbs energy by the process for the whole cell (Førland 1989a).

Let us consider an isothermal electrochemical cell where diffusion takes place and electric current passes through the cell. From flux equations using only operationally defined quantities it was shown that the change in Gibbs energy by the process is separable into two parts (Førland 1960) (Førland 1988a, Section 4.2.2).

$$dG = dG(t) + dG(Q) \tag{3.1}$$

where $dG(t)$ is time dependent and charge independent, while $dG(Q)$ is charge dependent and time independent. This corresponds to superposition of charge transport and diffusion (Ekman 1978). For small currents where the Joule heat can be neglected, we have

$$dG(Q) + \Delta\phi dQ = 0 \tag{3.2}$$

This equation is consistent with the Onsager reciprocal relations (Førland 1988a, Appendix 2.5). The change in Gibbs energy per faraday transferred is given by the equation,

$$\Delta G(Q) + \Delta\phi = 0 \tag{3.3}$$

This means that one can obtain the electromotive force, $E = \Delta\phi/F$, for an electrochemical cell from the change in Gibbs energy by charge transfer. This method for determination of emf is an alternative to the integration of the equation for $\nabla\phi$ when $j = 0$, compare eq. (2.26c).

3.1. An Electrochemical Cell with Liquid Junction and Glass Membrane.

As an example we shall examine the expression for the emf of the following cell (Førland 1989a):

$$\mathrm{Ag(s)|AgCl(s)|KCl(aq},c_3)\overset{\text{l.j.}}{||}\mathrm{HCl(aq},c_1),\mathrm{NaCl(aq},c_2)\overset{\text{g.m.}}{||}$$

$$\mathrm{HCl(aq},c_0)|\mathrm{AgCl(s)|Ag(s)}$$

A schematic illustration of the cell is given in Figure 2.

A small charge, ΔQ, is allowed to pass through the cell. The corresponding change in Gibbs energy per faraday is calculated for each section of the cell. Thereafter they are added up to give $\Delta G(Q)$ for the cell, and finally an expression for E is found.

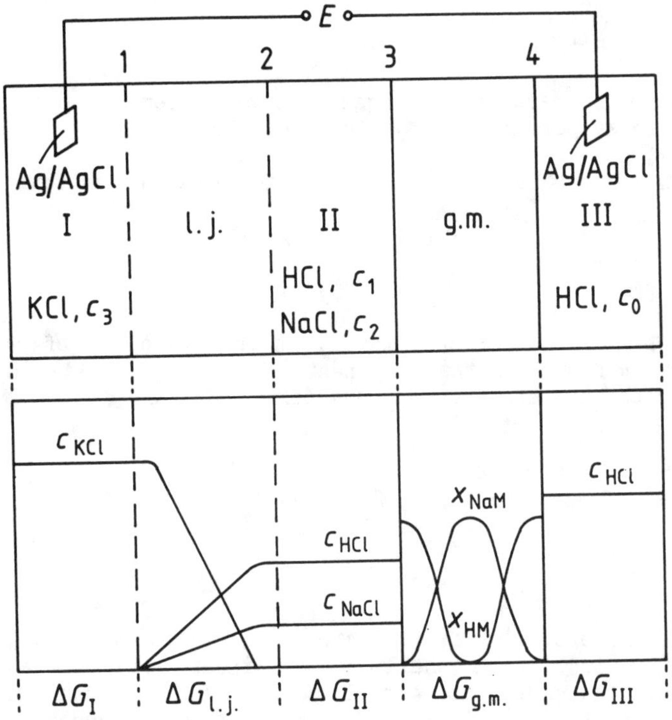

Figure 2. An electrochemical cell with Ag/AgCl electrodes, a liquid junction, l.j., and a glass membrane, g.m., separating the solutions. The cell is divided into five sections at the boundaries 1, 2, 3 and 4. The lower part of the figure illustrates concentrations in the solutions and the mole fractions in the glass membrane. The symbol for the Gibbs energy change for each section is given at the bottom.

When an electric current passes through the cell, chemical reactions take place at the electrodes, and transport of matter takes place in the electrolytes. Figure 3 illustrates the changes taking place when one faraday of positive charge passes from left to right through the cell.

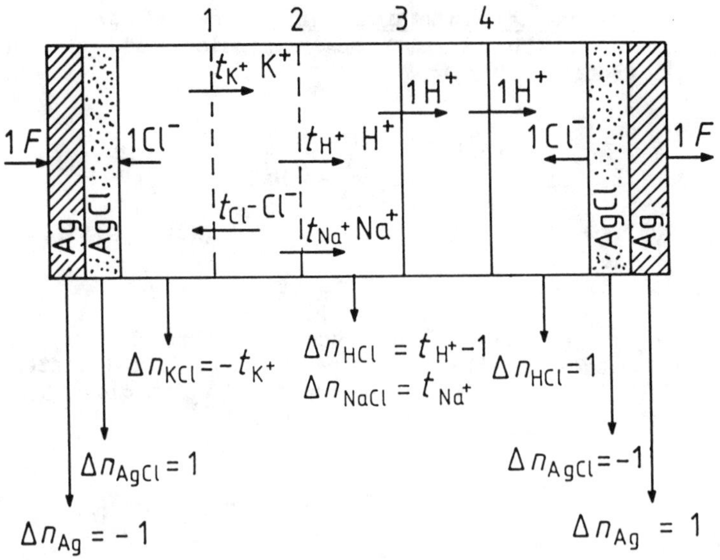

Figure 3. Material balance. Changes resulting from the passage of one faraday positive charge from left to right.

In *compartment* I there is the following change in Gibbs energy:

$$\Delta G_I = \mu_{AgCl} - \mu_{Ag} - \mu_{KCl,I} t_{K^+} \tag{3.4}$$

In *the liquid junction* the solvent, water, is the frame of reference. The change in the contents of the components HCl, NaCl and KCl is given by the change in the transference numbers for H^+, Na^+ and K^+.

$$\Delta G_{l.j.} = -\int_{(1)}^{(2)} \Sigma \mu_i dt_i = -\Big|_{(1)}^{(2)} \Sigma \mu_i t_i + \int_{(1)}^{(2)} \Sigma t_i d\mu_i \tag{3.5}$$

The summation is over the three components. Hertz developed a similar equation (Hertz 1980). We shall assume linear gradients and constant mobilities of the ions in the liquid junction. Furthermore, for simplicity we shall assume $t_{Cl^-} = ½$, and use an extended Debye-Hückel equation to express the mean activity coefficient, $y_\pm$. (When the solutions are not dilute, this will not give an accurate result for the activity term). We obtain

$$\Delta G_{l.j.} = -\Sigma(\mu_i t_i)_{II} + \mu_{KCl,I} t_{K^+} + RT \ln \frac{c_1 + c_2}{c_3} + RT \ln \frac{y_{\pm,II}}{y_{\pm,I}} + \Delta G_{corr} \tag{3.6}$$

The correction term, ΔG_{corr}, originates from the difference in mobilities for the different ions. A more detailed investigation of the activity term and the correction term has been carried out (Johnsen 1989).

In *compartment* II we shall assume a perfect proton selective glass membrane with $t_{H^+} = 1$ across the interface. Then we obtain

$$\Delta G_{II} = - \int_{(2)}^{(3)} \Sigma \mu_i t_i = \Sigma(\mu_i t_i)_{II} - \mu_{HCl,II} \tag{3.7}$$

For a perfect *glass membrane* where $t_{H^+} = 1$ at both interfaces and the nearest layers, $dt_{H^+} = 0$ and there is no contribution to $\Delta G(Q)$. In deeper layers both H^+ and Na^+ take part in charge transfer. The corresponding change in Gibbs energy is:

$$\Delta G_{g.m} = - \int_{(3)}^{(4)} (\mu_{HM} dt_{H^+} + \mu_{NaM} dt_{Na^+}) \tag{3.8}$$

where μ_{HM} and μ_{NaM} are the chemical potentials of the glass membrane in hydrogen form and sodium form respectively.

The integration from (3) to (4) may be divided into two steps. A limit between the two steps is chosen in the core of the membrane where all cation sites are occupied by Na^+. The two integrals will have opposite signs and, for an ideal membrane, the same numerical value, hence we have

$$\Delta G_{g.m.} = 0 \tag{3.9}$$

In *compartment* III we have

$$\Delta G_{III} = - \mu_{AgCl} + \mu_{Ag} + \mu_{HCl,III} \tag{3.10}$$

The overall change in Gibbs energy per faraday for the cell is the sum of the changes in all five sections,

$$\Delta G(Q) = \Delta G_I + \Delta G_{l.j.} + \Delta G_{II} + \Delta G_{g.m.} + \Delta G_{III} \tag{3.11}$$

or

$$\Delta G(Q) = RT \ln \frac{c_1 + c_2}{c_3} + RT \ln \frac{y_{\pm,II}}{y_{\pm,I}} + \Delta G_{corr} - \mu_{HCl,II} + \mu_{HCl,III} \tag{3.12}$$

Some of the terms in eq. (3.12) are constant. If we have an Ag/AgCl reference electrode on the left hand side of the cell, the solution in compartment I is a saturated solution of KCl, i.e., c_3 and $y_{\pm,I}$ are constant. When ΔQ is small, $\mu_{HCl,III}$ is essentially constant. This is particularly so when the concentration of HCl in compartment III is high such as in the glass electrode used for pH measurements. Compartment II may contain solutions of different compositions, and $\mu_{HCl,II}$ is a variable term:

$$\mu_{HCl,II} = \mu^{o}_{HCl} + RT \ln (c_{H^+}c_{Cl^-}y^2_{\pm})_{II} \tag{3.13}$$

Here μ^{o}_{HCl} is a constant, $c_{H^+} = c_1$ and $c_{Cl^-} = c_1 + c_2$. Introducing the expression for $\mu_{HCl,II}$ in eq. (3.12), collecting all the constant terms in one term, and replacing $\Delta G(Q)$ with $-FE$ and ΔG_{corr} with $-FE_{corr}$, we obtain

$$E = \frac{RT}{F} \ln c_1 + \frac{RT}{F} \ln y_{\pm,II} + E_{corr} + \text{const} \tag{3.14}$$

where

$$E_{corr} = \frac{RT}{F} \frac{(u_{K^+} - u_{Cl^-})c_3 - (u_{H^+} - u_{Cl^-})c_1 - (u_{Na^+} - u_{Cl^-})c_2}{(u_{K^+} + u_{Cl^-})c_3 - (u_{H^+} + u_{Cl^-})c_1 - (u_{Na^+} + u_{Cl^-})c_2} \times$$

$$\ln \frac{(u_{K^+} + u_{Cl^-})c_3}{(u_{H^+} + u_{Cl^-})c_1 + (u_{Na^+} + u_{Cl^-})c_2} \tag{3.15}$$

Here u_i are ionic mobilities in the electrolyte. The same expression for the emf is obtained by considering the electrochemical cell as composed of three simpler cells in series (Førland 1988a, Section 8.2.1). The potentials of these cells are calculated using only operationally defined quantities.

The conventional method uses single ion activities to calculate the single electrode potential of the ion selective electrode and the liquid junction potential separately. Data are not available for an exact calculation, therefore approximations are introduced. When non-operational quantities such as single ion activities are used in calculations, one cannot test the validity of the different approximations separately.

Since we use operationally defined quantities, the validity of any approximation introduced can in principle be checked by experiment.

Our method gives an insight into the problems of ion selective electrodes from a new angle. This insight may be useful in the search for new and better electrodes (Johnsen, 1989).

4. TRANSPORT PROPERTIES OF A CATION EXCHANGE MEMBRANE

In this section we shall review experimental data for transport of two salts, water and electric current across a cation-exchange membrane (Førland 1988a, Chapter 9). With four independent fluxes and forces there are six independent transport coefficients (Førland 1988a, Section 5.5.2). Although the reduction in the number of coefficients from the original 16 is substantial, the experimental effort to obtain the remaining six coefficients is still great. It is therefore useful to know which ones of the coefficients are the larger ones and how they can be estimated.

Transport experiments have been carried out using a cation exchange membrane CR61 AZL 386 (CR). In diffusion measurements the solutions on the two sides of the membrane differ in composition. An important question is then: How do we relate the measured average transport coefficients to local membrane composition? In order to answer this question we need to know membrane - solution equilibrium compositions.

4.1. Exchange Equilibrium Membrane - Solution.

The composition and thermodynamic properties are usually known for the solutions in contact with the membrane when measuring transports in membranes. A cation exchange membrane in contact with an electrolyte has cation sites M^- filled to electroneutrality with the cations of the electrolyte. The exchange equilibrium membrane-solution for an aqueous solution of KCl and $SrCl_2$ can be written as

$$SrCl_2(aq) + 2KM = 2KCl(aq) + SrM_2 \tag{4.1}$$

The membrane components are KM and SrM_2. Figure 4 relates membrane compositions to solution compositions in the membrane. Equivalent fractions, $x_{SrCl_2} = 2n_{SrCl_2}/(2n_{SrCl_2} + n_{KCl})$ in the solution and $x_{SrM_2} = 2n_{SrM_2}/(2n_{SrM_2} + n_{KM})$ in the membrane, are given on the abscissa and the ordinate respectively. The membrane, CR, contains 40% by weight of water. The thermodynamic equilibrium constant for eq. (4.1) is $K = 5.6$ at 25°C (Holt 1985).

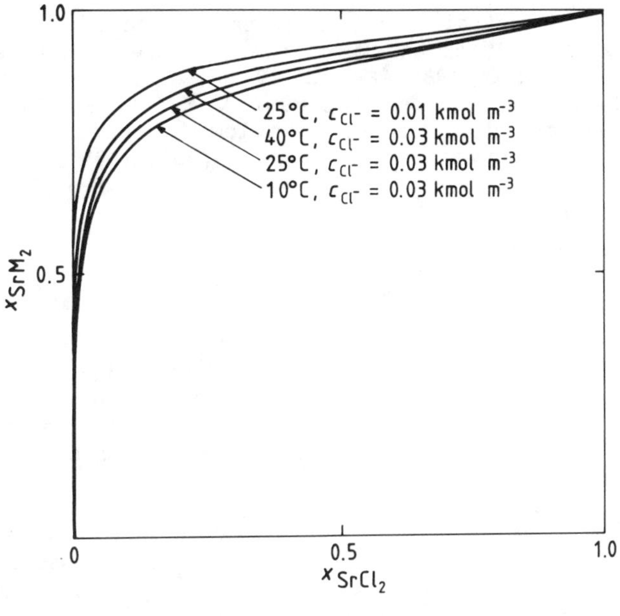

Figure 4. Equilibrium isotherms for the cation exchange membrane CR. Equivalent fraction of Sr in membrane, x_{SrM_2}, as function of equivalent fraction of Sr in solution, x_{SrCl_2}, for different temperatures and two different total concentrations.

Theories developed for simple ionic mixtures (Førland 1962) can be applied to equilibria similar to eq. (4.1). Two models may be used to derive equations for the entropy of mixing in the membrane. (1) Assuming a random distribution of K^+, Sr^{2+} and cation vacancies, CV, we obtain

$$\Delta S_{\text{mix}} = -R(2n_{\text{SrM}_2} + n_{\text{KM}})(x_{\text{SrM}_2}\ln x_{\text{SrM}_2} + x_{\text{KM}}\ln x_{\text{KM}}) \tag{4.2}$$

(2) If Sr^{2+} and CV are associated, the approximate entropy of mixing is given by

$$\Delta S_{\text{mix}} \approx -R(n_{\text{SrM}_2}\ln x_{\text{SrM}_2} + n_{\text{KM}}\ln x_{\text{KM}}) \tag{4.3}$$

In both cases the energy of mixing is given by

$$\Delta U_{\text{mix}} = b x_{\text{SrM}_2} x_{\text{KM}} \tag{4.4}$$

where b is a constant. The volume change by mixing SrM_2 and KM is small and $\Delta U_{\text{mix}} \approx \Delta H_{\text{mix}}$ at atmospheric pressure. It follows that $\Delta G_{\text{mix}} = b x_{\text{SrM}_2} x_{\text{KM}} - T\Delta S_{\text{mix}}$.

We have shown that the distribution of the ion pairs K^+/Sr^{2+}, Na^+/H^+ and Na^+/K^+ across the polymer matrix of the membrane CR can be described by such equations. (Holt 1985, Skrede 1987). Thermodynamic data for the mixture of K^+ and Sr^{2+} are shown in Figure 5. Model (2) gives the best fit to the experimental results. At low concentrations of Sr^{2+}, the vacancy probably dissociates from the Sr^{2+} (Holt 1985).

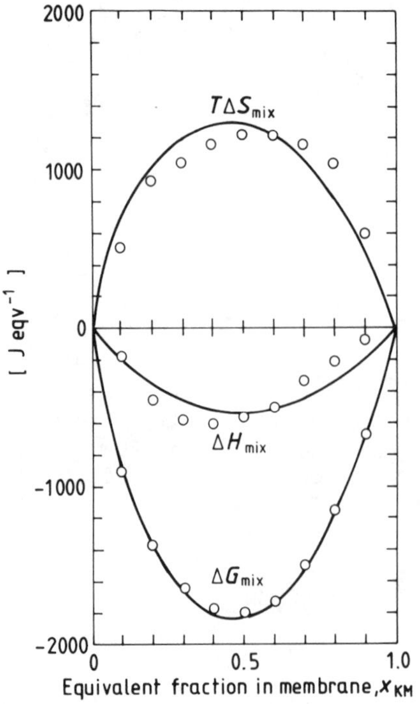

Figure 5.
Thermodynamic properties of mixtures as functions of equivalent fraction of K^+, x_{KM}, for mixing of KM and SrM_2.

Curves represent values calculated from a lattice model assuming association between Sr^{2+} and one cation vacancy, (Sr^{2+}CV), and a random distribution of K^+ and (Sr^{2+}CV). The circles represent values calculated from experimental results.

4.2. Local Diffusion Coefficients from Integral Measurements.

We shall consider diffusion through a perfect and homogeneous cation exchange membrane separating two electrolytes of different composition. For simplicity we shall choose an example where the left hand side electrolyte is a mixture of HCl(aq) and NaCl(aq), while the right hand side electrolyte is pure HCl(aq). The total concentration of chloride is the same on both sides. Each electrolyte is of constant properties, gradients occur only in the membrane. We shall assume the situation of stationary state.

In the thermodynamic sense the system contains the four components, HCl, NaCl, H_2O and HM, where HM denotes the membrane in the hydrogen form. The membrane in the sodium form, NaM, is not a component in the thermodynamic sense. Assuming local equilibrium inside the membrane, and that the composition anywhere in the membrane corresponds to an equilibrium composition of HCl and NaCl in aqueous solution, NaM can be expressed by the other components in the equilibrium equation, NaM + HCl(aq) $\rightleftarrows$ NaCl(aq) + HM. When diffusion takes place through the membrane, the changes taking place are described by fluxes of the neutral components, J_{HCl}, J_{NaCl} and J_{H_2O}. The membrane in the hydrogen form is chosen as the frame of reference. This means that the flux of HM, J_{HM}, is chosen as equal to zero, and the other fluxes are given relative to J_{HM}. The reference, HM, is constant for all membrane compositions (Førland 1988a, Section 5.5).

Assuming a constant chloride concentration and a constant pressure for a dilute solution in equilibrium with the membrane of any composition, the chemical potential of H_2O is the same everywhere in the membrane, and the fluxes of HCl and NaCl may be written

$$J_{HCl} = - l_{11}\nabla\mu_{HCl} - l_{12}\nabla\mu_{NaCl} \quad (4.5a)$$

$$J_{NaCl} = - l_{21}\nabla\mu_{HCl} - l_{22}\nabla\mu_{NaCl} \quad (4.5b)$$

The gradients in chemical potentials refer to an electrolyte solution in equilibrium with the membrane. In the experiment J_{NaCl} was measured. It can be shown that $l_{11} + l_{12} = 0$, $l_{21} + l_{22} = 0$ and $l_{11} = l_{22}$. (Førland 1988a, Section 5.5.2). When $\nabla\mu_{H_2O} = 0$, we obtain from the Gibbs-Duhem equation, $\nabla\mu_{HCl} = - (x_{NaCl}/x_{HCl})\nabla\mu_{NaCl}$. Thus eq. (4.5b) can be replaced by the expression

$$J_{NaCl} = - (l_{22}/x_{HCl})\nabla\mu_{NaCl} \quad (4.6)$$

or, for gradients only in the x-direction,

$$J_{NaCl} = - (l_{22}/x_{HCl})d\mu_{NaCl}/dx \quad (4.7)$$

From a diffusion experiment we can record J_{NaCl} for a known difference in chemical potential between the two sides of the membrane, $\Delta\mu_{NaCl}$.

This is an integral measurement giving an average coefficient. We shall see how a local coefficient, l_{ij}, can be obtained (compare Holt 1981).

A limiting case, the biionic cell with the flux of NaCl, J'_{NaCl}, will be used as an auxiliary in developing the formula for l_{22}. In Figure 6 two diffusion cells are compared. Cell (a) is a biionic cell with pure NaCl(aq) on the left hand side of the membrane and pure HCl(aq) on the right hand side. In cell (b) the left hand side solution is a mixed solution. The concentration of Cl^- is the same in all electrolyte solutions. The concentration profiles drawn in the membranes represent concentrations in the electrolyte in equilibrium with the membrane. The mixed solution on the left hand side of cell (b) is in equilibrium with a membrane composition identical to the membrane composition at distance x in cell (a). Since coefficients of the flux equations are determined by the local composition, and are independent of gradients, l_{ij} must be the same at x in cell (a) and at 0 in cell (b).

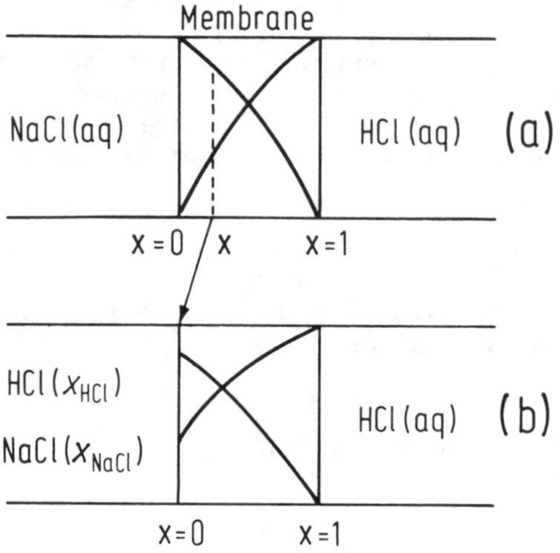

Figure 6. Diffusion in a membrane. Cell (a): biionic cell, cell (b): mixed electrolyte on left hand side. Hypothetical concentration profiles for solutions in equilibrium with membrane are shown. The membrane is of unit thickness, and x is the distance from the left hand side of the membrane.

The ratio between the stationary state fluxes in the two cells is

$$J_{NaCl}/J'_{NaCl} = r \tag{4.8}$$

The range of membrane compositions in cell b from 0 to 1 is the same as the range from x to 1 in cell (a). The flux J'_{NaCl} can be expressed by the diffusion coefficient l_{22}, and the chemical potential gradient for NaCl, $d\mu'_{NaCl}/dx$ at x

$$J'_{NaCl} = -(l_{22}/x_{HCl})d\mu'_{NaCl}/dx \tag{4.9}$$

we also have

$$J_{NaCl} = -(l_{22}/x_{HCl})d\mu_{NaCl}/dx = rJ'_{NaCl} = -(l_{22}/x_{HCl})rd\mu'_{NaCl}/dx \tag{4.10}$$

Thus the relation between the gradients in the two cells is

$$d\mu_{NaCl}/dx = rd\mu'_{NaCl}/dx \tag{4.11}$$

This is valid for any value of x with the corresponding value of x_{NaCl}. There is a similar relation for the slope of the secants to the concentration profiles:

$$\frac{\Delta\mu_{NaCl}}{1} = r\frac{\Delta\mu'_{NaCl}}{1 - x} \tag{4.12}$$

Since $\Delta\mu_{NaCl} = \Delta\mu'_{NaCl}$, we must have

$$r = 1 - x \tag{4.13}$$

We may differentiate eq. (4.8):

$$dJ_{NaCl} = J'_{NaCl}dr = - J'_{NaCl}dx \tag{4.14}$$

With eq. (4.9) we obtain

$$J'_{NaCl} = (l_{22}/x_{HCl})(d\mu'_{NaCl}/dJ_{NaCl})J'_{NaCl} \tag{4.15}$$

Assuming ideal mixture in the electrolyte we obtain

$$l_{22} = \frac{1}{RT}\, x_{NaCl}x_{HCl}(dJ_{NaCl}/dx_{NaCl}) \tag{4.16}$$

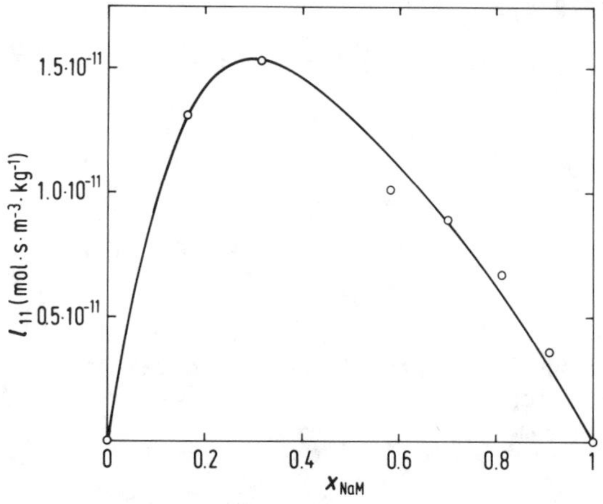

Figure 7. The diffusion coefficient l_{11} as a function of membrane composition for the system HM-NaM.

Thus we can obtain local values of $l_{22} = l_{11}$ from J_{NaCl} as a function of x_{NaCl}. Combining this with data on membrane - solution equilibrium we obtain the local diffusion coefficient as function of membrane composition.

4.3. Conductivity of the Membrane.

The flux equations for transport of two salts and water, and electric current density may be written:

$$J_i = - \Sigma L_{ij} \nabla \mu_j - L_{i4} \nabla \phi \quad ; \qquad (i=1,2,3,4) \tag{4.17}$$

Fluxes of the two salts and water are numbered 1,2 and 3 respectively, while the electric current density, $j = J_4$.

The coefficient L_{44} represents the conductivity of the membrane. It can be expressed by the mobilities of the ions in the membrane, u_i, and the equivalent fractions of the ions in the membrane, x_i. For a membrane that is a mixture of NaM and HM we have

$$L_{44} = \frac{c}{F}(x_{NaM} u_{Na^+} + x_{HM} u_{H^+}) \tag{4.18}$$

where c is the concentration of cation sites in the membrane. The conductivity, $\kappa = L_{44} F^2$. Experimental results for κ as a function of x_{HM} are presented in Figure 8. The curved line indicates that mobilities vary with the composition of the membrane. We used the following empirical equations for the mobilities as functions of composition:

$$u_{H^+} = u^{o}_{H^+}(1-kx_{NaM}) \tag{4.19a}$$

$$u_{Na^+} = u^{o}_{Na^+}(1-kx_{HM}) \tag{4.19b}$$

where $u^{o}_{H^+}$ and $u^{o}_{Na^+}$ are the mobilities of H^+ in a pure HM membrane and of Na^+ in a pure NaM membrane, respectively. The parameter k is an interaction constant. Values for such constants have been obtained using curve fitting procedures (Ratkje 1988).

Ionic transference numbers were calculated using the relations

$$t_{H^+} = x_{HM} u_{H^+}/(x_{HM} u_{H^+} + x_{NaM} u_{Na^+}) \tag{4.20a}$$

$$t_{Na^+} = x_{NaM} u_{Na^+}/(x_{HM} u_{H^+} + x_{NaM} u_{Na^+}) \tag{4.20b}$$

Calculated values agree well with experimental values obtained by a Hittorf method (Førland 1988a, Section 9.2.1).

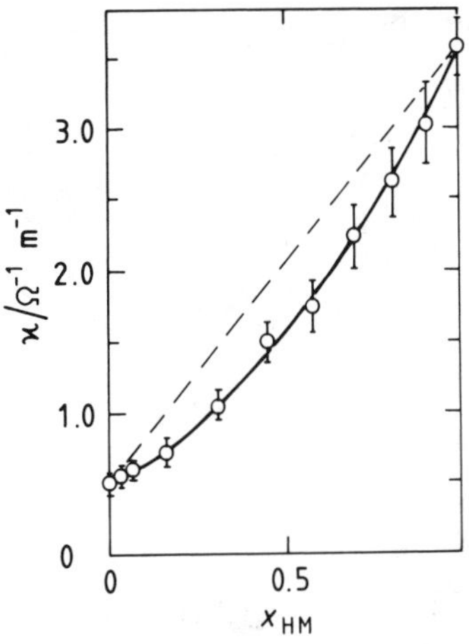

Figure 8.
Membrane conductivity as a function of equivalent fraction of HM in a membrane consisting of HM and NaM. The $c_{Cl^-} = 0.03$ kmol/m^3 in the external solutions. Circles with bars indicate experimental points with uncertainities. See text for curved line.

4.4. Validity of Nernst-Planck Flux Equations.

For a perfect cation exchange membrane the current density is given by the fluxes of cations through the membrane. For the NaM - HM membrane we thus have $J_1 + J_2 = j$, which implies the following relation between coefficients (Førland 1988a, Section 5.4.2):

$$L_{11} + L_{12} = L_{14} \tag{4.21}$$

When assuming $L_{12} = 0$ and $\Delta\mu_3 = 0$, the flux equations, eqs (4.17) become identical to the Nernst-Planck flux equations;

$$J_1 = - L_{14}(\nabla\mu_1 + \nabla\phi); \quad J_2 = - J_{24}\ (\nabla\mu_2 + \nabla\phi) \tag{4.22}$$

The coefficient L_{14} (and L_{24}) can be determined experimentally by a Hittorf method when the conductivity of the membrane is known. The coefficient L_{11} (and L_{22}) can be determined from a diffusion experiment since the diffusion coefficients are related to the phenomenological coefficients of eqs (4.17) by the following equation (Førland 1988a, Section 5.5.2):

$$l_{ij} = L_{ij} - L_{i4}L_{4j}/L_{44} \tag{4.23}$$

Values for L_{11} and L_{14} obtained in this manner, are shown in Figure 9. Values obtained for L_{12} from eq. (4.21) are also indicated. The results shown that the Nernst-Planck equations may be a good first approximation to a description of transport of two salts in the membrane. The gradient in the chemical potential of water is usually insignificant.

When Nernst-Planck equations are valid, one can obtain diffusion coefficients from electric mobilities. Since $t_1 = L_{14}/L_{44}$ and $t_2 =$

L_{24}/L_{44} (Førland 1988a, Section 5.5.2). We obtain from eq. (4.23)

$$l_{11} = t_1 t_2 \kappa / F^2 \tag{4.24}$$

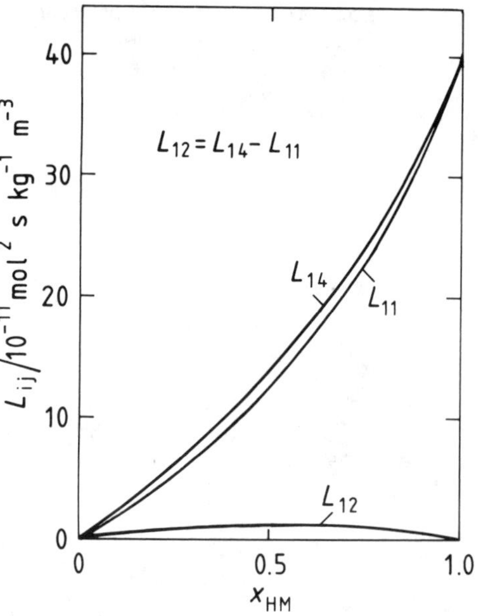

Figure 9.
Variations in coefficients L_{12}, L_{11} and L_{14} with composition of membrane for the system HM - NaM.

4.5. Transport of Water.

The transference coefficient for water, t_{H_2O}, represents the flux of water with the flux of electric charge. It is usually determined by measurements of streaming potential (Brun 1967).

There is only a small variation in t_{H_2O} with concentration in the aqueous solution in equilibrium with the membrane, when the solution is dilute and contains one salt of a monovalent or divalent cation (Trivijitkasem 1980). At higher concentrations of salt in the solution, or with a trivalent cation present, the membrane does not behave as a perfect cation exchange membrane, and anions may enter the membrane. This leads to a decrease in the value of t_{H_2O}.

When there are two different cations present in a membrane, t_{H_2O} varies with composition. There is not a linear relationship between t_{H_2O} and t_{K^+} or between t_{H_2O} and t_{H^+}, see Figure 10. This means that we cannot interprete the transport of water as a constant number of water molecules transported with each kind of ion.

Osmotic and hydraulic water fluxes are due to the driving forces $\Delta\mu_{H_2O}(c)$ and $V_{H_2O}\Delta p$ respectively. For an isothermal system we have $\Delta\mu_i = \Delta\mu_i(c) + V_i\Delta p$. We reported different values for the flux of water as a function of $\Delta\mu_{H_2O}(c)$ when $\Delta p = 0$ and the flux of water as a function of $V_{H_2O}\Delta p$ when $\Delta\mu_{H_2O}(c) = 0$ (Førland 1988a, Section 9.3.2). This difference is probably caused by extra transport of bulk solution when pressure difference is applied. More experiments are needed.

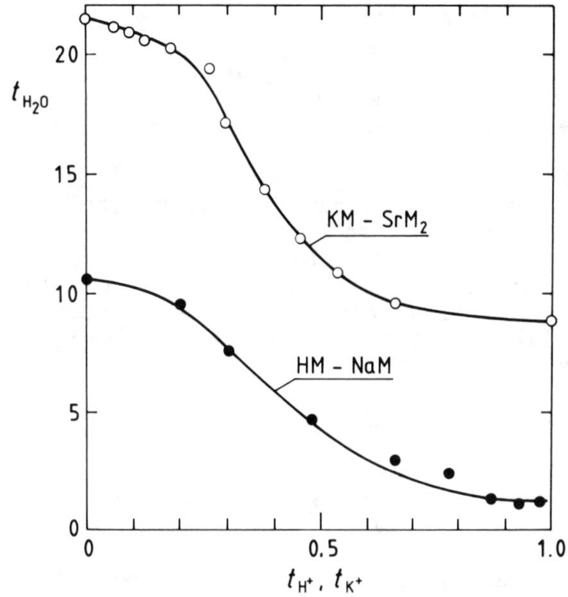

Figure 10.
The transference coefficient for water, t_{H_2O} as a function of the ionic transference number t_{H^+} for the system HM-NaM and t_{K^+} for the system KM-SrM$_2$.

The exchange equilibrium membrane - solution is of central importance to the behaviour of a membrane in an aqueous solution. The theory of mixtures (Førland 1962 and 1964) ought to be tested systematically for ion-exchange membranes in equilibrium with different salt solutions.

When gradients in the chemical potential of water are insignificant, the Nernst-Planck equation is a useful approximation for the description of transport of two cations in the membrane. An estimate of the transport properties of a membrane can be obtained by the following procedure: The membrane conductivity as a function of electrolyte composition is recorded. Using empirical equations of the type given in eqs (4.19), mobilities of the ions in the membrane may be estimated. Ionic transference numbers in the membrane are obtained from eqs (4.20), and diffusion coefficients are obtained from eq. (4.23). More experimental data are needed to check the validity of eqs (4.19).

5. CHEMICAL REACTIONS IN ANISOTROPIC SYSTEMS A TREATMENT WITHOUT USING AFFINITY

In an isotropic system there may not be any coupling between scalar and vectorial quantities (Prigogine 1955). This is often called the Curie principle. In the case where J_1 and X_1 are scalars while J_2 and X_2 are vectors, the cross coefficients must be equal to zero, $L_{12} = L_{21} = 0$.

An important consequence of the Curie principle is that a spontaneous chemical reaction in an isotropic system cannot be utilized for producing mechanical work.

In an electrochemical cell with a combination of ion exchange membranes, however, the energy of a chemical reaction (in addition to the electrode reactions) can be converted into electric energy and *vice versa*. In many anisotropic biological systems a chemical reaction can cause a flux of components across a membrane (active transport). Also, a chemical reaction can produce mechanical work in a muscle.

For two examples we shall see how one can avoid using affinity, A, as a force. We shall use only vectorial forces and fluxes. This description permits a more detailed analysis of the processes.

5.1. An Electrochemical Cell Utilizing Acid-Base Neutralization. Electrochemical cells with a combination of membranes can be utilized for energy-storage batteries. The battery is discharged by the mixing of electrolytes in solution. When this mixing leads to a chemical reaction, the Gibbs energy of the reaction is converted to electric energy. One may recharge the batteries with a direct current forcing the reaction in the opposite direction and demixing of the solutions - or simply by replacing the solutions with new unmixed solutions.

We shall consider the following electrochemical cell:

$$\mathrm{Ag(s)|AgCl(s)|\underset{I}{KCl(aq)}|^{A}|\underset{II}{HCl(aq)}|^{C}|\underset{III}{H_2O}|^{A}|\underset{IV}{KOH(aq)}|^{C}|\underset{V}{KCl(aq)}|AgCl(s)|Ag(s)}$$

The cell has a number of compartments, I, II, III, IV and V, separated by cation exchange membranes $|^C|$ and anion exchange membranes $|^A|$. The electrodes are AgCl|Ag electrodes, and the compartments are filled with electrolytes as shown above (Makange 1980), (Førland 1983).

The cell is an electrochemical energy producer, and Figure 11 shows the directions for the movement of current carriers when an electric current passes from left to right through the cell. The ions H^+ and OH^- react to form water in compartment III and the cell reaction is

$$\mathrm{HCl(II) + KOH(IV) \rightarrow H_2O(III) + KCl(V)} \tag{5.1}$$

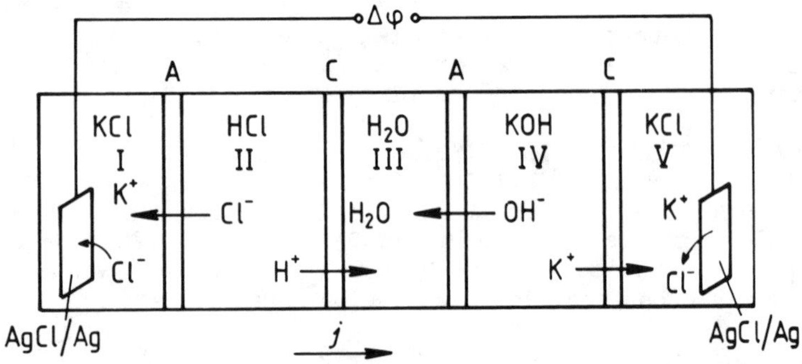

Figure 11. A schematic picture showing spontaneous transfers in the cell with a closed circuit. Reproduced by permission of The Electrochemical Society, Inc. (Førland 1983).

The theoretical emf of the cell. We may calculate the change in Gibbs energy for each compartment of the cell when one faraday of positive charge passes from left to right through the cell. The sum of these changes is equal to $\Delta G(Q) = -\Delta\phi$ (compare section 3).

$$\Delta G(Q) = \Delta G_{I} + \Delta G_{II} + \Delta G_{III} + \Delta G_{IV} + \Delta G_{V}$$

$$= 0 - \mu_{HCl,II} + \mu_{H_2O,III} - \mu_{KOH,IV} + \mu_{KCl,V} \tag{5.2}$$

In order to operate with independent forces only thermodynamic components are used. Local equilibrium is assumed for the cell reaction HCl + KOH = H_2O + KCl anywhere in the cell. (In order to have local equilibrium in compartments II, III and IV, there must be a minute amount of KCl present.) With local equilibrium only three of the four species can be considered as components. We may choose KCl, HCl and H_2O as our set of components.

The chemical potential of KOH in compartment IV is expressed by the components

$$\mu_{KOH,IV} = \mu_{H_2O,IV} + \mu_{KCl,IV} - \mu_{HCl,IV} \tag{5.3}$$

With dilute solutions in all compartments the activity of water is close to unity, and the chemical potential of water is approximately the same in all compartments. With perfect selective membranes and no concentration variation within any compartment, we obtain

$$-\Delta G(Q) = \Delta\phi = -(\mu_{HCl,IV} - \mu_{HCl,II} + \mu_{KCl,V} - \mu_{KCl,IV}) \tag{5.4}$$

Assuming ideal solutions we can replace chemical potentials with concentration products. When introducing the relation $\mu - \mu^{\circ} = RT\ln c_{+}c_{-}$ in eq. (5.4), the chemical potentials in the standard state cancel. Furthermore, when the concentration of solute is the same in compartments II, IV and V, we obtain

$$\Delta\phi = RT(\ln c_{H^+,II} - \ln c_{H^+,IV}) \tag{5.5}$$

We may subtract and add the same quantity, $RT\ln c_{H^+,III}$, to the right hand side of eq. (5.5), and we obtain

$$\Delta\phi = RT\ [\ln (c_{H^+,II}/c_{H^+,III}) - \ln (c_{H^+,IV}/c_{H^+,III})] \tag{5.6}$$

The emf of the cell at 25°C is

$$E = -\ 0.059\ [(pH_{II}-pH_{III}) + (pOH_{IV}-pOH_{III})] \tag{5.7}$$

For a cell with pure water in compartment III (pH = pOH = 7), pH = 1 in compartment II, and pH = 13 in compartment IV, we obtain E = 0.71 V. In order to increase the voltage one can repeat a unit extending from the middle of compartment I to the middle of compartment V one or several times.

The conductivity of the cell with pure water in compartment III is very low. In order to improve the conductivity an electrolyte must be added. The addition of simple salts dissociating into cations and anions lead to substantial losses in emf, because this will lead to uneven concentrations and variations in pH in the compartment.

Some organic substances form *zwitterions*, $^{+}HY^{-}$, which can accept a proton to form H_2Y^{+}, and donate a proton to form Y^{-} (e.g. amino acids). The ions H_2Y^{+} and Y^{-} will be current carriers. An aqueous solution of a zwitterion will keep a constant pH, the pH of the *isoelectric point*. We may add a zwitterion to the water in compartment III to improve the conductivity of the cell.

Experimental studies show that the addition of aminonaphtalenedisulphonic acid is beneficial. The addition has two effects. First, it increases the conductivity of the water and thus the ohmic loss in electric potential is smaller. Secondly, the solution of the zwitterion has a high buffer capacity, and thus there will be less polarization in compartment III. (Førland 1988a, Section 8.3.4)

The present cell converts chemical energy into electric energy. The chemical energy is stored as acid and base separated by membranes.

The force of a chemical reaction is a scalar, while transport of electric charge is a vector. In the present cell the membranes make the system anisotropic, and scalar - vector coupling may take place. The coupling of a vectorial flux and a scalar force implies a vectorial average phenomenological coefficient.

We used an alternative treatment. Assuming local equilibrium in the cell, local affinity is equal to zero. Using thermodynamic components, only three species out of the four are components. We chose HCl, KCl and H_2O as components, while the chemical potential of the species KOH was expressed by the chemical potentials of the other three species. In this way we avoided affinity as a force and obtained only vectorial forces. The overall driving force in the cell is the difference in chemical potential for HCl, between compartments II and IV. It is balanced by $\Delta\phi$. Local forces are gradients in chemical potentials and local phenomenological coefficients are scalars. This treatment permits a detailed analysis of the transport process.

5.2. Muscular Contraction.

The muscle converts chemical energy into mechanical work. Efficiencies (mechanical work/energy input) of almost 0.45 have been reported (Hill 1964), hence the muscle competes favourably with the best steam engines. To achieve such a high efficiency, the energy conversion must operate close to equilibrium and the linear equations of irreversible thermodynamics may be applied to muscular contraction.

When a muscle contracts two kinds of filaments, myosin and actin, move relative to each other whereby chemical energy in the form of ATP is consumed. A mechanism for the creation of the force between the myosin and the actin filaments is suggested (Førland 1985, Førland 1988a).

Model. The force between the myosin and the actin filaments is created by the cross bridges protruding from the myosin filament, see Figure 12. A cross bridge consists of a flexible section and a myosin head, which can form a bond to actin.

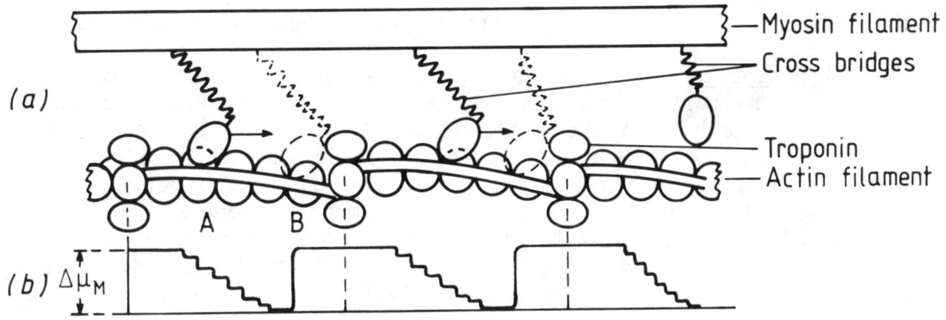

Figure 12. (a) Force generation by moving myosin heads along an actin filament in a sarcomere. (b) Chemical potential difference of myosin heads during movement.

After attachment the myosin head moves along an interval between two troponin sites on the actin filament at a constant angle between the myosin head and the actin filament. Figure 12 pictures a movement of a myosin head from left to right along the actin filament. A myosin head attaches itself to actin at a position A near the left hand side of the interval. By stretching and bending the elastic part of the cross bridge the myosin head moves towards the right forming stepwise stronger bonds to actin. This exerts a force, which moves the myosin filament relative to the actin filament. At position B the myosin head has arrived at the point of maximum bond strength. Here the bond between myosin and actin must be broken in order that the myosin head proceeds to the next interval. The myosin-actin bond is broken by the formation of a strong myosin - ATP bond. ATP is subsequently hydrolysed, and a new myosin - actin bond can be formed with release of ADP and phosphate. The movements of different myosin heads are out of phase, and the result is a smooth release of energy in very small steps. This is a new description of muscular contraction. It is in agreement with the observation that myosin heads do not change orientation while exerting force (Cooke 1982). The mechanism commonly referred to presumes that the myosin heads make a 45° rotation (Huxley 1969).

The source of energy. The direct source of energy in muscular contraction is ATP. When a muscle performs work and ATP is converted to ADP, the concentration of ATP is still close to constant (Carlson 1974). The consumed ATP is rapidly restored by a reaction between ADP and PCr (phosphocreatine). The concentration of PCr in a muscle is much higher than the concentration of ATP, and PCr represents a way of storing energy ready for use. In energy calculations it is common to consider PCr as the source of energy. Changes in internal energy, ΔU, can be obtained from observed values of heat, q, and work, w, when a muscle contracts. Since pressure-volume work by muscle contraction is negligible, changes in internal energy can be replaced by enthalpy changes, $\Delta U_{ATP} \approx \Delta H_{ATP}$ and $\Delta U_{PCr} \approx \Delta H_{PCr}$.

The more accurate determinations of q and w were obtained with PCr as the source of energy (Carlson 1974), and the result is

$$\Delta U_{PCr} = \Delta H_{PCr} = -\ 46.4\ \text{kJ mol}^{-1} \tag{5.8}$$

Within the limits of error ΔU_{PCr} and ΔU_{ATP} are equal (Bendall 1969).

Application of irreversible thermodynamics. The derivation will be based on the mechanism involving moving myosin heads described above. A myosin head attached to the actin filament moves in a stepwise manner along an interval on the actin filament because the strength of the bond to actin increases for each step. The change in bond strength over the interval can be expressed as a change in chemical potential of the myosin head. In this first part of the cyclic process for myosin heads, work is performed. Some of the energy in the first part of the cycle is dissipated as heat. In the second part of the cycle, where the actin-myosin bond is broken, all the energy is dissipated as heat.

We shall apply irreversible thermodynamics to the first part of the cyclic process for the myosin heads, the part where work is performed. Our system, pictured schematically in Figure 12, consists of the right hand side halves of a sarcomere. We have chosen the actin filament as the frame of reference for transport. The myosin heads (and the myosin filament) move to the right when work is performed. The change in chemical potential per mole attached myosin heads, $\Delta\mu_M$, over an interval of length ℓ, gives an average force $X_M = -\Delta\mu_M/\ell$. The conjugate flux, J_M, is equal to $c_M v$, where c_M is the concentration of myosin heads attached to the actin filament and v is their velocity $v = d\ell/dt$. This velocity is equal to the velocity of the myosin filament relative to the neighbouring actin filament. When a muscle carries out work ($-w$) a mechanical force pulls the myosin heads in the opposite direction of X_M. The average mechanical force over the interval ℓ is w/ℓ. This is a force operating on one mole attached myosin heads, and the conjugate flux is $c_M v$.

The process can be considered as quasi-stationary (Caplan 1983). The dissipation function (dissipated energy per unit volume per unit time) is

$$T\Theta = J_M(-(\Delta\mu_M/\ell)) + c_M v(w/\ell) = c_M v(w-\Delta\mu_M)/\ell \tag{5.9}$$

For comparison of theoretical relations and experimental results it is convenient to calculate the dissipation function per mole attached myosin heads (dissipated energy per mole per unit time).

$$T\Theta/c_M = v(w - \Delta\mu_M)/\ell \tag{5.10}$$

The corresponding flux (or velocity) equation is

$$v = L(w - \Delta\mu_M)/\ell = \frac{1}{R}(w - \Delta\mu_M) \tag{5.11}$$

where $R = \ell/L$ is resistance per mole attached myosin heads over the interval ℓ.

The linear relation in eq. (5.11) is expected to be valid when the change in $(w - \Delta\mu_M)/N_A$ per step is small compared to kT (see Section

1.). Here N_A is the Avogadro constant, the number of myosin heads in one mole.

Energy balance. The energy change for the complete cycle is the sum of the energy changes for the first part of the cycle ΔU_I, and for the second part, ΔU_{II}.

$$\Delta U_{PCr} = \Delta U_I + \Delta U_{II} \tag{5.12}$$

For the first part of the cycle the dissipated energy is equal to $(T\Theta/c_M)\Delta t$ where $\Delta t = \ell/v$ is the time needed for the myosin heads to travel the distance ℓ. The dissipated energy according to eq (5.10) is equal to $w - \Delta\mu_M$, and according to eq. (5.11) it is equal to Rv. Thus the corresponding absorbed heat, q_1 is a linear function of contraction velocity:

$$q_1 = - Rv \tag{5.13}$$

The Gibbs energy change for the first part of the cycle ΔG_I is given by $\Delta\mu_M = \Delta G_I \approx \Delta U_I - T\Delta S_M$.

The term $T\Delta S_M$ involves the change in partial molar entropy for myosin heads when moving from position A on the actin to position B (see Figure 12). The heat absorbed is $q' = T\Delta S_M$. This heat is independent of velocity.

In the second part of the cycle no work is performed, and the whole energy ΔU_{II} is converted to heat, which is also independent of velocity.

The total heat absorbed by the cyclic process is

$$q = q_1 + q' + \Delta U_{II} = q_1 + q_2 = - Rv + q_2 \tag{5.14}$$

The velocity-independent parts of the total heat absorbed, $q' + \Delta U_{II}$, are collected in the one constant term, q_2. For the highest velocity, where $w = 0$, we have

$$\Delta U_{PCr} = - Rv_{max} + q_2 \tag{5.15}$$

A muscle consumes ATP (and thereby PCr) during isometric tetanus, when no work is performed. This may be interpreted as a breaking and re-forming of an actin - myosin bond at the same position B on the actin filament consuming one ATP per cycle. We may assume that this process will also occur for low values of v, involving gradually fewer myosin heads as v increases. The dissipated energy due to reattachment at B is expected to decrease rapidly with increasing v.

The mole fractions of myosin heads going through the first and second process are x_1 and x_2 respectively. The total heat absorbed per mole PCr consumed is

$$q = x_1(-Rv+q_2) + x_2\Delta U_{PCr}$$

or

$$q = x_1(-Rv+q_2) + x_2(-Rv_{max}) + x_2q_2 \qquad (5.16)$$

The term $x_2(-Rv_{max}) = q_3$ is the additional heat absorption caused by the second process. It represents a loss in energy since it cannot be converted to work. We have

$$q_3 = x_2(-Rv_{max}) = q - q_2 + x_1Rv \qquad (5.17)$$

With increasing contraction velocity, x_2 decreases and thus $q_3 \rightarrow 0$. In the limit eq. (5.16) becomes identical to eq. (5.14).

The energy balance for the whole process is

$$\Delta U_{PCr} = q_1 + q_2 + q_3 + w = -\ 46.4 \text{ kJ mol}^{-1} \qquad (5.18)$$

The three contributions to the dissipation of energy are separable because of their different dependence on v. Figure 13 shows how they add up to the total heat absorption for varying contraction velocity in the frog sartorius muscle.

From Figure 13 the following values can be read. The slope of the slanting line gives $q_1 = -1.3\ v$ kJ mol^{-1} where the units for v is nanometre per millisecond, i.e. 10^{-6} m s^{-1}. The intercept with the ordinate for $v = 0$ gives $q_2 = -14$ kJ mol^{-1}. The heat q_3 is the difference between the total heat, q, and $q_1 + q_2$. The $\Delta\mu_M$ is a state function and thus of constant value. For the highest velocity where $w = 0$, we obtain from eq. (5.11) $\Delta\mu_M = -Rv_{max} = -32.4$ kJ mol^{-1}.

The value of $\Delta\mu_M$ agrees well with values obtained from different experiments, - 34 kJ mol^{-1} from equilibrium constants (Taylor 1979) and -36 kJmol^{-1} from maximum isometric tension (Førland 1985).

A specific mechanism for muscle contraction was used as the basis for formulating equations of irreversible thermodynamics. The change in chemical potential of myosin heads is used as the force in the present derivation, not the affinity of the energy supplying reaction. The forces and flux used are directly related to the mechanism proposed.

Irreversible thermodynamics may reveal relations between properties. In the present case we found the value for $\Delta\mu_M$ from the maximum velocity and we were able to confirm this value by means of results from independent experiments.

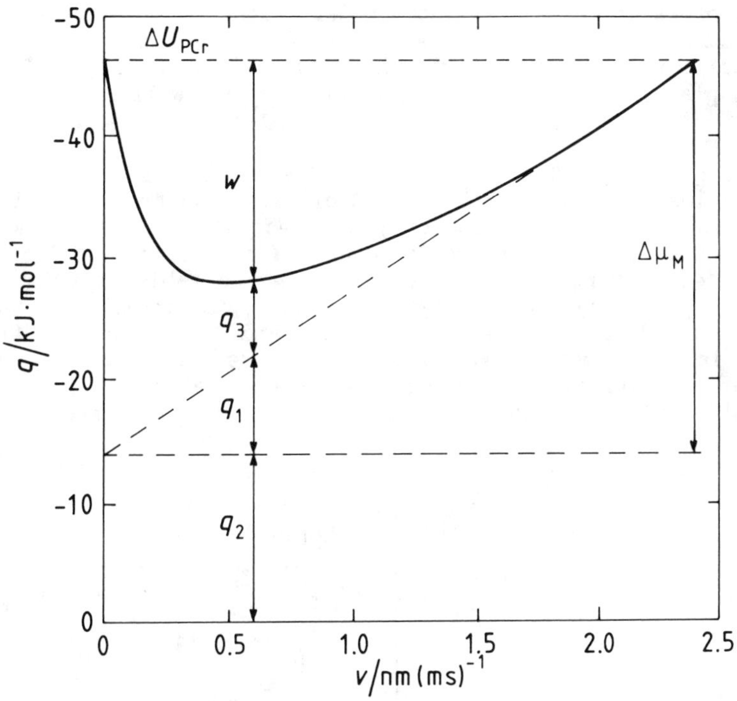

Figure 13. Absorbed heat per mole PCr as a function of contraction velocity - experimental curve (Bendall 1969) (Førland 1985) Reproduced by permission of The Biophysical Society. See text for explanations of symbols and numerical values.

Irreversible thermodynamics helps to explore the consequences of a proposed mechanism and it can be a useful tool for developing and checking theories about mechanisms in biological processes. In principle the irreversible thermodynamics is independent of mechanism. It cannot prove any specific mechanism, but it can exclude some mechanisms as impossible.

6. SYSTEMS WITH TEMPERATURE GRADIENTS

In a system with a temperature gradient the conjugate force - flux pair, $-\nabla \ln T$ - J_q contributes to the dissipation function. There may be an interdependence between a heat flux and a mass or charge flux. The Dufour effect, thermal diffusion (Soret effect), thermal osmosis, Peltier effect, thermoelectric effect (Seebeck effect) are examples of such interdependence. They can all be treated by the methods of irreversible thermodynamics (Førland 1988a, Chapter 6). We shall review one example of thermal osmosis, namely frost heave (Førland 1988a, Sections 6.2 and 11.1), (Førland 1988b), and one example of the Peltier effect in a thermocell (Førland 1988a, Sections 6.6.1, 6.6.2. and 11.2).

6.1. Frost Heave, an Example of Thermal Osmosis.

When a soil containing water freezes, its surface heaves, often unevenly; this is called frost heave. In frost heave the soil expands more than the expansion due to the freezing of the bound water in the unfrozen soil.

Excess of water forms ice lenses under the surface. This excess of water is transported to the ice lenses during freezing. Figure 14 gives a schematic picture of a vertical column of soil where frost heave is taking place. The transport of water and heat will be considered in two steps.

<u>First step.</u> The pores of the fine-grain soil act as a set of capillaries. These are filled by capillary rise from the table of water. As we shall see when discussing the second step, water is transported away from level 2. If there is no access to air, the void must be filled with water flowing through the soil from the table of water. The force for this flux is caused by a pressure reduction at level 2. The temperature difference between levels 1 and 2 will cause a flux of heat. There is experimental evidence that there is little or no coupling between the flux of water and the measurable heat flux in the first step. (Takashi 1980), (Førland 1988a, Section 11.1.3), (Førland 1988b).

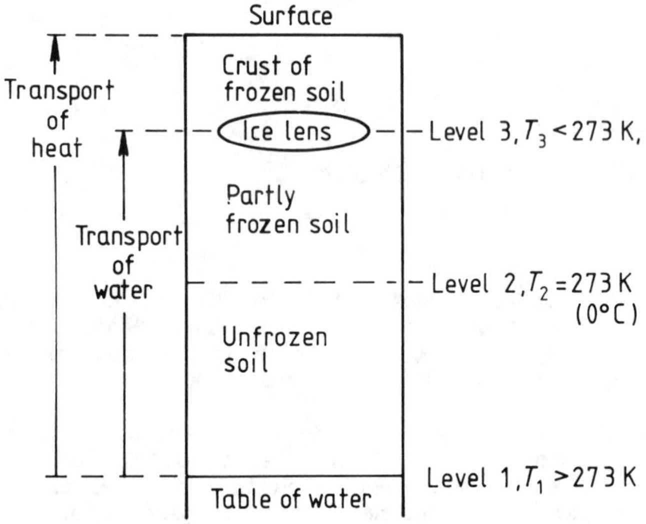

Figure 14. A section of freezing soil. Schematic picture of frost heave.

At level 2 there is an equilibrium between ice in the pores and liquid water. The relative amounts of the two phases can change without a change in Gibbs energy. We include in the first step the freezing of all the transported water at level 2. The water is transported in liquid form in the second step, and this means that the second step will involve the remelting of the ice at level 2.

<u>Second step.</u> The ice melts at level 2 and heat is absorbed. Liquid water is transported through pores of the partly frozen soil in the direction of decreasing temperature. The temperature is below 0°C. Liquid water below 0°C is a well-known phenomenon in capillaries. In the smallest pores the water will remain liquid at temperatures several degrees below 0°C, while ice can form at 0°C and atmospheric pressure in the largest pores. At a sufficiently low temperature even the water in the smallest pores will freeze. Below this temperature there is a solid crust of ice which does not permit any transport of water. The transport process from level 2 to level 3 can be treated as thermal osmosis.

Thermal osmosis may occur over a membrane when the transfer of mass across the phase boundaries to the membrane leads to absorption or release of heat. In frost heave ice melts at level 2 absorbing heat. The partly frozen soil between levels 2 and 3 serves as a membrane. At level 3 the water freezes and heat is released. Under a solid crust of ice the thermal osmosis leads to an increased pressure at level 3. The thermal osmosis is a slow process.

In thermal osmosis the fluxes of heat and material are both dependent on the two forces, given by temperature difference and pressure difference. For slow processes we may assume local equilibrium and the flux equations may be written,

$$J_q = -\bar{l}_{11}\Delta\ln T - \bar{l}_{12}\Delta\mu_{\mathrm{ice},T} \tag{6.1a}$$

$$J_{\mathrm{w}} = -\bar{l}_{21}\Delta\ln T - \bar{l}_{22}\Delta\mu_{\mathrm{ice},T} \tag{6.1b}$$

We have $\Delta\ln T = \ln(T_3/T_2)$, and with pure ice at both levels we have

$$\Delta\mu_{\mathrm{ice},T} = V_{\mathrm{ice}}\Delta p \tag{6.2}$$

where V_{ice} is the molar volume of ice and Δp is the difference between the pressure acting on the ice lens at level 3 and on an ice crystal at level 2. Ice and liquid water are in equilibrium at level 2 and the pressure on the ice is the same as the pressure on the water in the pores.

The maximum possible frost heave pressure is the pressure that completely suppresses the flux of water, $J_{\mathrm{w}} = 0$. This pressure is found from eqs (6.1b) and (6.2)

$$\Delta p_{J_{\mathrm{w}}=0} = -(1/V_{\mathrm{ice}})(\bar{l}_{21}/\bar{l}_{22})\Delta\ln T \tag{6.3}$$

Further, since $\bar{l}_{12} = \bar{l}_{21}$. we have

$$(J_q/J_{\mathrm{w}})_{\Delta T=0} = \bar{l}_{12}/\bar{l}_{22} = q_{\mathrm{w}}^* \tag{6.4}$$

where q_{w}^* is the heat of transfer. The transfer of liquid water between levels 2 and 3 gives only negligible contribution to the transfer of heat. This is in agreement with the lack of coupling between J_q and J_{w} in the first step. The main contribution to the heat of transfer is given by the melting of ice at level 2 with absorption of heat and the freezing of the water at level 3 with release of heat. Thus the heat of transfer is equal to the enthalpy of melting:

$$q_{\mathrm{w}}^* = \Delta H_{\mathrm{m}} \tag{6.5}$$

The maximum frost heave pressure can be expressed in terms of the enthalpy of melting. From eqs (6.3-6.5) we have

$$\Delta p_{J_W=0} = -(\Delta H_m/V_{ice})\Delta \ln T \tag{6.6}$$

The enthalpy of melting varies with pressure and temperature. The variation with pressure is negligible within the pressure ranges of frost heave, while the variation with temperature may be of importance when there is a large temperature difference between levels 2 and 3. The variation of V_{ice} with pressure and temperature is negligible.

When the numerical value of ΔT is much smaller than T, $\Delta \ln T \simeq \Delta T/T$. Since $\Delta H_m/T_m = \Delta S_m$, we obtain.

$$\Delta p_{J_W=0} = -(\Delta S_m/V_{ice})\Delta T \tag{6.7}$$

At 0^oC and 1 atm pressure $\Delta S_m = 22.0$ J K^{-1} mol^{-1} and $V_{ice} = 1.96 \times 10^{-5}$ m^3 mol^{-1}. When the temperature of the ice is not too low, we thus have

$$\Delta p_{J_W=0} = -11.2 \times 10^5 \Delta T \text{ Pa} \tag{6.8}$$

If a soil permits transport of water down to a temperature of -5^oC, an ice lens will grow at this temperature. The flux of water will not be suppressed until a pressure of more than 50 atm has been built up.

In eq. (6.1b) $\bar{l}_{22}\Delta H_m$ may be substituted for $\bar{l}_{21}$ (see eqs (6.4) and (6.5), and further $\Delta S_m \Delta T$ for $\Delta H_m \Delta \ln T$. We may also substitute $V_{ice}\Delta p$ for $\Delta \mu_{ice}$ (see eq. (6.2)):

$$J_W = -\bar{l}_{22}(\Delta S_m \Delta T + V_{ice}\Delta p) \tag{6.9}$$

The flux of water, J_W, is equal to the rate of growth for the ice lens, and $\bar{l}_{22}$ is the average hydraulic permeability for the total transport path for water. The value of $\bar{l}_{22}$ may chiefly be determined by low local values of l_{22} close to the ice lens.

Frost heave was investigated experimentally under controlled conditions (Takashi 1980). Their result show that

$$\Delta p_{J_W=0}/\Delta T = -11.4 \text{ kg cm}^{-2} = -11.2 \times 10^5 \text{ Pa} \tag{6.10}$$

when $|\Delta T|$ is not too large, in good agreement with eq. (6.8). Small deviations for larger temperature differences may be explained by the change in ΔS_m with temperature. Larger deviations may be caused by experimental imperfections (Førland 1988a, Section 11.1.3.) (Førland 1988b).

Frequently one can observe several distinct ice lenses at different levels. On the basis of a nucleation theory, an explanation of this phenomenon was suggested (Førland 1989b).

In the industrial production of aluminium metal, cathode heaving is a problem (Sørlie 1989). It may be useful to consider the phenomenon as similar to frost heave, and apply the theories of thermal osmosis.

In geology some cases of magmatic differentiation and fractionization may be better understood by applying the theories of thermal osmosis.

6.2. The Peltier Effect in a Thermocell.

The heat balances are important factors in the industrial use of electrochemical cells. In order to obtain a detailed picture of temperatures in the cell, it is not sufficient to know the total heat of cell reaction and the heat supplied to the cell (or removed). We must also have knowledge about e.g., heat absorbed and heat evolved at the individual electrodes, the Peltier effects at the electrodes.

A thermocell is a non-isothermal electrochemical cell. The two electrodes are at different temperatures. One can observe a Seebeck effect; the temperature difference causes an electromotice force. A current through the cell causes Peltier effects.

In an electrochemical cell ions are current carriers in the electrolyte and there are chemical reactions at the electrodes, where electrons are taking over as current carriers through the electronic conductors.

The first cell that we shall consider consists of two electrodes of the metal A and a solid or liquid electrolyte AX:

$$(T)\mathrm{A}|\mathrm{AX}|\mathrm{A}(T + \Delta T) \qquad \text{(a)}$$

The left hand side electrode is kept at a constant temperature T by means of the heat reservoir I, while the right hand side electrode is kept at a constant temperature $T + \Delta T$ by means of the heat reservoir II. Both junctions between electrode and outer circuit are kept at the same temperature T_0. The cell is shown in Figure 15.

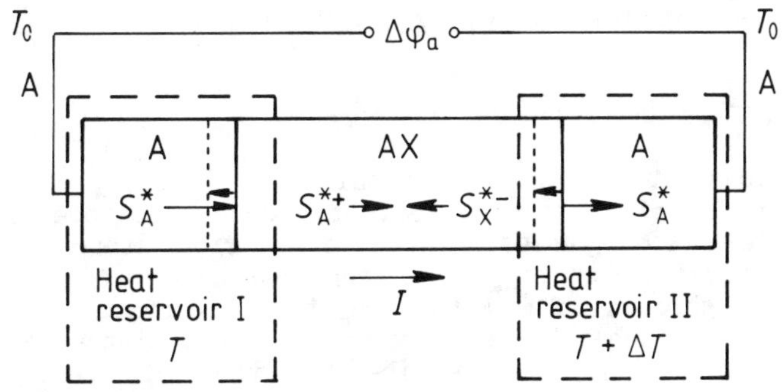

Figure 15. Schematic arrangement of the non-isothermal cell A|AX|A. See text for explanation of the symbols .

The following changes take place upon the passage of one faraday of positive charges from left to right through the inner circuit of the cell. At the left hand side electrode one mole A is removed and t_X- mole AX is supplied, while one mole A is supplied at the right hand side electrode and t_X- mole AX is removed. When A is shifted from the left hand side electrode to the right hand side one in this way, the barycentre of the electrolyte, AX, moves relative to the walls of the container.

When the system is isobaric, there are only two forces, $-\Delta \ln T$ and $-\Delta\phi_a$. The dissipated energy per unit time is

$$\frac{T\mathrm{d}S}{\mathrm{d}t} = -\Delta \ln T J_q - \Delta\phi_a I \qquad (6.11)$$

The corresponding flux equations are:

$$J_q = -\bar{L}_{11}\Delta \ln T - \bar{L}_{12}\Delta\phi_a \qquad (6.12a)$$

$$I = -\bar{L}_{21}\Delta \ln T - \bar{L}_{22}\Delta\phi_a \qquad (6.12b)$$

The Peltier heat is defined:

$$(J_q/I)_{\Delta T=0} = \bar{L}_{12}/\bar{L}_{22} = \pi_{T,a} \qquad (6.13)$$

The electric potential for $I = 0$ is obtained from eq. (6.12b);

$$\Delta\phi_a = -(\bar{L}_{21}/\bar{L}_{22})\Delta \ln T \qquad (6.14)$$

By the Onsager reciprocal relations we have $\bar{L}_{12} = \bar{L}_{21}$, and for small $|\Delta T|$ we can write $\Delta \ln T \approx (1/T)\Delta T$. Hence we have

$$\Delta\phi_a = -\pi_{T,a}\Delta \ln T = -(\pi_{T,a}/T)\Delta T \qquad (6.15)$$

where $\pi_{T,a}/T$ is the entropy transferred.

We shall consider the reversible balance of entropy across the metal - electrolyte interface for the left hand side electrode when one faraday of positive charge passes from left to right through the cell. The interface receives entropy from the heat reservoir, $\pi_{T,a}/T$, entropy transported through the metal to the interface, S^*_A, and entropy transported through the electrolyte to the interface, $t_{X^-}S^*_{X^-}$. The disappearance of one mole A liberates the molar entropy, S_A. This balances the entropy consumed by the formation of t_X- mole AX with a molar entropy, S_{AX}, and the entropy transported through the electrolyte away from the interface, $t_{A^+}S^*_{A^+}$;

$$\pi_{T,a}/T + S_A^* + S_A + t_{X^-}S_{X^-}^* = t_{X^-}S_{AX} + t_{A^+}S_{A^+}^* \quad (6.16)$$

Rearranging the equation we obtain

$$\pi_{T,a}/T = - S_A - S_A^* + (t_{X^-}S_{AX} + t_{A^+}S_{A^+}^* - t_{X^-}S_{X^-}^*) \quad (6.17)$$

The values of t_{A^+} and t_{X^-} depend on the chosen frame of reference, while we always have $t_{A^+} + t_{X^-} = 1$. The entropy transferred is independent of the chosen frame of reference, and the term inside the bracket in the equation above has the same value independent of frame of reference. From this we can infer that

$$S_{AX} = S_{A^+}^* + S_{X^-}^* \quad (6.18)$$

From eqs. (6.15), (6.17) and (6.18) we obtain

$$\Delta\phi_a/\Delta T = - \pi_{T,a}/T = (S_A^* + S_{X^-}^*) - (S_{AX} - S_A) = S_A + S_A^* - S_{A^+}^* \quad (6.19)$$

The term $(S_A^* + S_{X^-}^*)$ contains transport quantities. The entropy transported through the metal, S_A^*, is small, showing only small variations with the nature of the metal. The entropy transported through the electrolyte, $S_{A^+}^*$ and $S_{X^-}^*$, may contribute substantially to the Peltier heat. The term $(S_{AX} - S_A)$ contains thermodynamic entropies and originates from the chemical reactions. This term may give the major contribution to the Peltier heat.

We shall consider a cell similar to the one studied above, except for the electrodes. The electrodes of solid conductor A, reversible to the cation, are replaced by electrodes reversible to the anion, X_2 gas in contact with platinum metal;

$$(T)X_2|AX|X_2\ (T + \Delta T) \quad (b)$$

For the cell we find

$$\Delta\phi_b/\Delta T = - \pi_{T,b}/T = (S_A^* - S_{A^+}^*) + (S_{AX} - \tfrac{1}{2}S_{X_2}) \quad (6.20)$$

From eqs (6.19) and (6.20) we obtain

$$\Delta\phi_a/\Delta T - \Delta\phi_b/\Delta T = S_A + \tfrac{1}{2}S_{X_2} - S_{AX} = - \Delta S \quad (6.21)$$

All entropies transported with the current cancel, and $\Delta S = S_{AX} - S_A - \tfrac{1}{2}S_{X_2}$ is the entropy change for the cell reaction of the isothermal cell $A|AX|X_2(Pt)$.

Substantial Peltier effects at the electrodes was found in the electrolysis of water in molten sodium hydroxide with platinum electrodes (Ito 1981), (Ito 1982), (Ito 1984), (Ito 1985).

The cell studied was

$$(Pt)O_2(g),\ H_2O(g)|NaOH(l)|H_2O(g),\ H_2(g)\ (Pt) \qquad (c)$$

When one faraday of positive charge passes from left to right through the inner circuit of the cell, one half mole $H_2(g)$ is gained at the cathode, while one mole $H_2O(g)$ is lost. At the anode one fourth mole $O_2(g)$ and one half mole $H_2O(g)$ is gained. The total reaction is

$$\tfrac{1}{2}H_2O(g) = \tfrac{1}{2}H_2(g) + \tfrac{1}{4}O_2(g) \qquad (6.22)$$

The OH^- produced at the cathode migrates through the molten sodium hydroxide to the anode where it is consumed.

In order to find the Peltier heats in cell (c) we may study the following two cells:

$$(T)(Pt)O_2(g),H_2O(g)|NaOH(l)|H_2O(g),O_2(g)(Pt)(T + \Delta T) \qquad (d)$$

and

$$(T)(Pt)H_2(g),H_2O(g)|NaOH(l)|H_2O(g),H_2(g)(Pt)(T + \Delta T) \qquad (e)$$

For cell (d) we find

$$\pi_{T,\mathrm{d}}/T = -\ (S^*_{Pt} + S^*_{OH^-}) + \tfrac{1}{4}S_{O_2} + \tfrac{1}{2}S_{H_2O} \qquad (6.23)$$

and for cell (e) find

$$\pi_{T,\mathrm{e}}/T = -\ (S^*_{Pt} + S^*_{OH}) - \tfrac{1}{2}S_{H_2} + S_{H_2O} \qquad (6.24)$$

In the experimental determination of the Peltier heats for cell (c) a reference electrode was combined with each half cell at different temperatures (Ito 1984). Corresponding values for $\Delta\phi_d$ and $\Delta T = T_2 - T_1$ were obtained by the following measurements:

(T_1) Ref. electrode$|NaOH(l)|H_2O(g),O_2(g)(Pt)(T_1)$ $\quad\Delta\phi_1$

(T_1) Ref. electrode$|NaOH(l)|H_2O(g),O_2(g)(Pt)(T_2)$ $\quad\Delta\phi_2$

For cell (d) with identical electrodes, $\Delta\phi_d = \Delta\phi_2 - \Delta\phi_1$ when $T = T_1$ and

$T + \Delta T = T_2$. In a similar way corresponding values for $\Delta\phi_e$ and ΔT were obtained.

The Peltier heat for each electrode in cell (c) can be obtained from the experimental values of $\Delta\phi/\Delta T$. At 350°C (623 K) we have

$$\pi_d = -(\Delta\phi_d/\Delta T)T = 72 \text{ kJ faraday}^{-1}$$

when $p_{H_2O} = 0.2$ atm and $p_{O_2} = 0.8$ atm

and

$$-\pi_e = +(\Delta\phi_e/\Delta T)T = -63 \text{ kJ faraday}^{-1}$$

when $p_{H_2O} = 0.2$ atm and $p_{H_2} = 0.8$ atm

For each faraday of charge passing through the cell, 72 kJ are absorbed at the anode, while 63 kJ are released at the cathode. Unless compensated for, this leads to cooling of the anode and heating of the cathode.

The entropy transported with one mole OH^- ions in molten sodium hydroxide is $S^*_{OH^-}$. Its value can be obtained from eq. (6.23) or eq. (6.24) when the value is known for all the other terms in the equation.

The entropies for H_2O, H_2 and O_2 under the given conditions are calculated from data given in tables of thermodynamic data (Barin 1973). The transported entropy for positive charge in platinum (Moore 1973) is $S^*_{Pt} = -1 \text{ JK}^{-1} \text{ faraday}^{-1}$.

The average value of $S^*_{OH^-}$ obtained from eqs (6.23) and (6.24) is

$$S^*_{OH^-} = 54 \text{ JK}^{-1} \text{ faraday}^{-1} \text{ at } 350^{\circ}\text{C} \tag{6.25}$$

According to eq. (6.18) we have

$$S_{NaOH(l)} = S^*_{Na^+} + S^*_{OH^-} \tag{6.26}$$

The value of $S_{NaOH(l)}$ at 350°C is 138 $\text{JK}^{-1} \text{ mol}^{-1}$ (Barin 1973), and thus we have

$$S^*_{Na^+} = 84 \text{ JK}^{-1} \text{ faraday}^{-1} \text{ at } 350^{\circ}\text{C} \tag{6.27}$$

The example treated here shows that the Peltier heats of the electrodes may be substantial, and much larger than the enthalpy of reaction for the cell. A large contribution to the Peltier heat of an electrode comes from the reactants disappearing and products being formed at the electrodes. For known half-reactions one can easily estimate this contribution from known values of the entropies of the species. The transported entropy may also give a considerable contribution to the Peltier heat. Information about transported entropies, however, is scarce. Research is needed in order to obtain more values.

The Peltier heats may be of importance for the heat balance in many electrochemical cells. In the aluminium electrolysis the Peltier effect causes a cooling of the anode and a heating of the cathode, probably exceeding the contributions to the heat balance from Joule heat and overpotential under normal conditions (Ratkje 1990a).

An investigation of thermocells with AgCl/Ag electrodes and an aqueous solution of a monovalent chloride as the electrolyte, show large heat effects at the electrodes, several orders of magnitude larger than the electric work obtained from the cell (Ratkje 1990b).

ACKNOWLEDGEMENT

Several parts of this work have been reproduced by permission 1990 of John Wiley and Sons, Ltd. (Førland 1988a).

REFERENCES

Barin,I. and O.Knacke eds.1973.Thermochemical Properties of Inorganic Substances. Berlin: Springer

Bendall,J.R.1969.Muscles, Molecules and Movement. London: Heinemann.

Brun,T.S. and D.Vaula.1967.Correlation of measurements of electroosmosis and streaming potentials in ion exchange membranes. Ber. Bunsenges. Physik. Chem. 71: 824-829.

Caplan,S.R. and A.Essig.1983.Bioenergetics and Linear Nonequilibrium Thermodynamics. Harvard: University Press

Carlson,F.D. and D.R.Wilkie.1974.Muscle Physiology. New York: Prentice-Hall.

Coleman,B.D. and C.Truesdell.1960.On the resiprocal relations of Onsager. J. Chem. Phys. 33: 28-31.

Cooke,R., M.S.Cowder, and D.D.Thomas.1982.Orientation of spin labels attached to cross bridges in contracting muscle fibres. Nature (London) 300: 776-778.

de Groot,S.R. and P.Mazur.1962.Non-Equilibrium Thermodynamics. Amsterdam: North-Holland

Ekman,A., S.Liukkonen, and K.Kontturi.1978.Diffusion and electric conduction in multicomponent electrolyte systems. Electrochim. Acta 23: 243-250.

Førland,T.1960.The diffusion process and the diffusion potential in relation to the E.M.F. of concentration cells. Acta Chem. Scand. 14: 1381-1388.

Førland,T.1962.Thermodynamic properties of simple ionic mixtures. Disc. Faraday. Soc. 32: 122-127.

Førland,T.1964.Thermodynamic properties of fused salt systems. In Fused Salts, ed. B.R. Sundheim. New York: McGraw-Hill.

Førland,T. and S.K.Ratkje.1980.Entropy production by heat, mass charge transfer and spesific chemical reactions. Electrochim. Acta 25: 157-164.

Førland,T. and S.K.Ratkje.1981.Small contributions to emf from changes in electrostatic energy. Electrochim. Acta 26: 49-52.

Førland,K.S. and T.Førland.1982.Colligative properties - a unified thermodynamic treatment. Education in Chemistry 19: 12-14.

Førland,K.S., T.Førland,A.A.Makange and S.K.Ratkje.1983.The coupling between transport of electric charge and chemical reaction. J. Electrochem. Soc. 130: 2376-2380.
Førland,T.1985.On the mechanism of muscular contraction. Biophys. J.47: 665-671.
Førland,K.S.,T.Førland, and S.K.Ratkje.1988a.Irreversible Thermodynamics. Theory and Applications. Chichester: Wiley
Førland,K.S.,T.Førland, and S.K.Ratkje.1988b.Frost heave. Proc Fifth Internat Conf. on Permafrost,Norway: 344-348.
Førland,K.S. and T.Førland.1989a.The calculation of emf for cells with irreversible transport. Proc. Fifth Scientific Session Ion-Selective Electrodes, Hungary.
Førland,K.S. and T.Førland.1989b.Dynamics of ice lens formation during frost heave. Lecture at Fifth Meeting on Ground Freezing., Sapporo, Japanese Society of Snow and Ice.
Garlid,K.D.,A.D.Beavis, and S.K.Ratkje.1989.On the nature of leaks in energy-transducing membranes. Biochim. Biophys. Acta. 976: 109-120.
Gibbs,J.W.1875.On the equilibrium of heterogeneous substances. In The Scientific Papers of J.W. Gibbs. Vol 1, Thermodynamics, III, 1961 New York: Dover Publications.
Guggenheim,E.A.1929.The conceptions of electrical potential difference between two phases and the individual activities of ions. J. Phys. Chem. 33: 842-849.
Hertz,H.G.1980.Electrochemistry. Berlin: Springer
Hill,A.V.1964.The efficiency of mechanical power development during muscular shortening and its relation to load. Proc. R. Soc. (B)159: 319-324
Holt,T.,S.K.Ratkje,K.S.Førland, and T.Østvold.1981. Hydrostatic pressure gradients in ion exchange membranes during mass and charge transfer. J. Membr. Sci. 9: 69-82.
Holt,T.,T.Førland, and S.K.Ratkje.1985.Cation exchange membranes as solid solutions. J. Membr. Sci. 25: 133-151.
Huxley,H.E.1969.The mechanism of muscular contraction. Science 164: 1356-1366.
Ito,Y.,H.Kaiya, and S.Yoshizawa.1981.Water vapor electrolysis using molten sodium hydroxide electrolyte. Energy Dev. Jpn. 3: 153-164.
Ito,Y.,F.R.Foulkes, and S.Yoshizawa.1982.Energy analysis of a steady-state electrochemical reaction. J. Electrochem. Soc. 129: 1936-1943.
Ito,Y.,H.Kaiya,S.Yoshizawa,S.K Ratkje, and T.Førland.1984. Electrode heat balances of electrochemical cells. J. Electrochem. Soc. 131: 2504-2509.
Ito,Y.,H.Hayashi,N.Hayafuji, and S.Yoshizawa.1985.Energy flow through ß-alumina solid electrolyte. Electrochimica Acta 30: 701-703.
Johnsen,E.E.1989.On Liquid Junction Potentials and a Solid State pH Electrode, Dr.ing.thesis No.46.Division of Physical Chemistry, Norw.Inst.Technology, University of Trondheim.
Katchalsky,A. and P.F.Curran.1965.Non-equilibrium Thermodynamics in Biophysics: Harvard University Press.
Lewis,G.N. and M.Randall.1961.Thermodynamics. (revised by Pitzer, K.S. and L. Brewer) 2nd ed. New York: McGraw-Hill.
Makange,A.A.1980.Elektrolytisk celle, Patent 143774 No (appl. 772235, F. 24 June 77)29 Dec.

Miller,D.G.1967.Application of irreversible thermodynamics to electrolyte solutions II. J. Phys. Chem. 71: 616-32.
Moore,J.P. and R.S.Graves.1973.Absolute Seebeck coefficient of platinum from 80 to 340 K and the thermal and electrical conductivities of lead from 80 to 400 K. J. Appl. Phys. 44: 1174-1178.
Nernst,W.1988.Zur Kinetik der in Lösung befindlichen Körper. Z. Phys. Chem. 2: 613-637.
Onsager,L.1931a.Resiprocal relations in irreversible processes I. Phys. Rev. 37: 405-426.
Onsager,L.1931b.Resiprocal relations in irreversible processes II. Phys. Rev. 38: 2265-2279.
Planck,M.1890.Über die Potentialdifferenz zwischen zwei verdünten Lösungen binäre Electrolyte. Am. Phys. Chem. 40: 561-576.
Prigogine,I.1955.Thermodynamics of Irreversible Processes. Springfield: Thomas.
Ratkje,S.K.,T.Holt, and M.G.Skrede.1988.Cation membrane transport: Evidence for local validity of Nernst-Planck equations. Ber. Bunsenges. Physik. Chem. 92: 825-832.
Ratkje,S.K.1990a.Local heat changes during aluminium electrolysis submitted to J. Electrochem. Soc.
Ratkje,S.K.,T.Ikeshoji, and K.Syverud.1990b.Heat- and internal energy changes at electrodes and junctions in thermocells submitted to J. Electrochem. Soc.
Skrede,M.G. and S.K.Ratkje.1987.Cation exchange membranes as solid solutions with Na^+/H^+ and Na^+/K^+. Zeitschr. Phys. Chem. Neue Folge 155: 211-222.
Sørlie,M. and H.A.Øye.1989.Cathodes in Aluminium Electrolysis. Düsseldorf: Aluminium-Verlag.
Takashi,T.,T.Ohrai,H.Yamamoto, and J.Okamoto.1980.Upper limit of heaving pressure derived by pore-water pressure measurements of partially frozen soil. The 2nd International Symposium on Ground Freezing, Norw. Inst. Technology, Trondheim, 713-725.
Taylor,E.W.1979.Mechanism of actomyosin ATPase and the problem of muscle contraction. CRC Crit. Rev. Biochem. 6: 103-164.
Trivijitkasem,P.and T.Østvold.1980.Water transport in ion-exchange membranes. Electrochim. Acta 25: 171-178.

SYMBOLS

|A|	Anion exchange membrane
A	Affinity of chemical reaction, J
|C|	Cation exchange membrane
c	Concentration, mol m^{-3}
E	Electromotive force, emf, V
E_a	Activation energy
F	Faraday constant
G	Gibbs energy, J
dG(Q)	Charge-dependent change in Gibbs energy, J
dG(t)	Time-dependent change in Gibbs energy
I	Electric current, flux of charge, C s^{-1}, faraday s^{-1}
J_i	Flux of component i, mol $m^{-2}s^{-1}$
J_q	Measurable heat flux, J $m^{-2}s^{-1}$
j	Current density, flux of charge through unit, A m^{-2}, faraday m^{-2} s^{-1}
K	Thermodynamic equilibrium constant
k	Boltzmann constant, J K^{-1}
k	Parameter
L, L_{ij}	Phenomenological coefficient
L_{ii}	Direct coefficient
L_{ij}	Cross coefficient, $j \neq i$
l, l_{ij}	Diffusion coefficient
ℓ	Length (of cell, system, interval etc.)
||	Liquid junction
M^-	Cation site in membrane
n	Number of components
n_i	Amount of component i, mol
p	Pressure, Pa
Q	Quantity of electricity, electric charge, C, faraday
q	Heat, J
q_i^*	Heat of transfer, component i, J mol^{-1}
R	Gas constant, J K^{-1} mol^{-1}
R	Resistance coefficient
S	Entropy, J K^{-1}
S_A^*	Entropy transported through conductor A, J K^{-1} $faraday^{-1}$
$S_{X^-}^*$	Entropy transported by ion X^-, J K^{-1} mol^{-1}
T	Temperature, K
$T\theta$	Dissipation function J $m^{-3}s^{-1}$
t	Time, s
t_i	Transference number of the ion i or transference coefficient of neutral component i, mol $faraday^{-1}$
U	Internal energy, J
u_i	Electric mobility of the ion i, m^2 s^{-1} V^{-1}
v	Rate of transport, velocity, m s^{-1}, nm $(ms)^{-1}$
V	Volume, m^3
X	Force
X_i	Driving force for flux of component i, J mol^{-1} m^{-1}
x	Distance in the x-direction, m
x_i	Mole fraction of component i or equivalent fraction of ion i
y	Activity coefficient, concentration basis
z_i	Charge of the ion
w	Mechanical work, muscles J mol^{-1}
w_{el}	Electric work, J
α	Coupling coefficient

Θ	Entropy production per unit volume and unit time J m^{-3} s^{-1} K^{-1}
κ	Electric conductivity, A $V^{-1}m^{-1}$
μ_i	Chemical potential of component (species, ion) i, J mol^{-1}
$\tilde{\mu}_i$	Electrochemical potential of single ion, J mol^{-1}
$\nabla\mu_i(c)$	Gradient in chemical potential caused by gradient in concentration only, J mol^{-1} m^{-1}
$\Delta\mu_{i,T}$	Change in chemical potential at constant temperature, J mol^{-1}
ν_i	Stoichiometric coefficient for species i
π	Peltier heat, J $faraday^{-1}$
$\Delta\phi$	Electric potential, J $faraday^{-1}$
$\nabla\phi$	Gradient in electric potential, J $faraday^{-1}$ m^{-1}
ψ	Electrostatic potential, J $faraday^{-1}$

Thermodynamics of Ionized Systems

Stefan Wiśniewski
Institute of Heat and Refrigeration Engineering
Technical University of Lódź

The application of thermodynamics in describing of ionized systems is given. Formation of plasma and calculation of its composition is discussed. As examples of application of nonequilibrium thermodynamics to ionized systems, the phenomena in the thermionic diode and the magnetohydrodynamic generator are considered.

PRINCIPAL NOTATION

A	Richardson - Dushmann constant
B	magnetic field strength
c_p	specific heat at constant pressure
e	specific internal energy
e_e	electric charge per unit number of electrons
E	electric field strength
$\tilde{h}_i$	partial specific enthalpy of component i
j_e	electron flux
J_e	electric current density
J_E	internal energy flux
j_i	diffusion flux of componet i
J''_q	heat flux in the entropy balance equation
J_s	entropy flux
k	number of components
K_p	equilibrium constant
L_{ab}	phenomenological coefficient
M	molar mass
p	pressure
Q''^{*}_e	heat of transport of electrons
R	gas constant
s	specific entropy
$\tilde{s}_i$	partial specific entropy of component i
T	temperature

v velocity
V_i ionization potential
x_i mass fraction of component i
z_i molar fraction of component i
β degree of ionization
λ thermal conductivity
μ_i chemical potential of component i
$\tilde{\mu}_i$ electrochemical potential of component i
ρ density of substance
σ entropy source strength
φ electric potential
ϕ work function
ψ electrostatic potential
Ψ dissipation function

THE FORMATION OF PLASMA

With a rise in temperature gas molecules dissociate into simpler molecules or even individual atoms. At even higher temperatures, there occurs a process called ionization in which electrons are detached from atoms. The number of electrons leaving the electronic shell of an atom determines the ionization multiplicity (single ionization, double ionization, multiple ionization). The ionization of a gas results in the production of positive ions and electrons, forming an electronic gas.

A solution of gases containig a considerable number of ionized atoms is called plasma. The ions and the electrons occur in plasma in such relations that their resulting electrical charge is equal to zero, i.e. the plasma is electrically quasineutral. The number of ions and electrons is so high that it decides of macroscopic thermodynamic properties of the gas.

The degree of ionization of the gas is defined as the ratio of the number of kilomoles of ionized atoms n_i to the number of kilomoles of atoms before the ionization n_b, that is

$$\beta = \frac{n_i}{n_b} \tag{1}$$

The energy required to remove an outer-shell electron from its normal position in the un-ionized atom to outside the nuclear interaction range determines the ionization potential of a gas (Table 1). Large amounts of energy are required to ionize gases and this energy is in turn evolved during recombination in which electrons combine with positive ions.

The energy needed to change the energy level or to detach an outer-shell electron from an atom may come from the collision of the atom with an electron or positive ion, from absorption of radiation energy, and from the collision of atoms which remain un-ionized at high temperatures. All the molecules of a gas at high temperatures have high velocities and, hence, also have sufficient kinetic energy to detach electrons in collisions with atoms.

Table 1. Potentials of single (double) ionization

Substance	Ionization potential V_i in volts	Substance	Ionization potential V_i in volts
Ar	15.75 (27.8)	Li	5.39 (75.62)
Ba	5.19	Mg	7.65
Ca	6.09	N	14.5 (29.5)
Cd	8.99	Na	5.138 (47.29)
Cs	3.893 (25.1)	Ne	21.56 (40.9)
H	13.595	O	13.6 (34.9)
He	24.58 (54.2)	Rb	4.176 (27.5)
Hg	10.4	Sr	5.67
K	4.339 (31.81)	Xe	12.10
Kr	14.0 (24.7)	Zn	9.39

In the equilibrium state the kinetic energy of atoms satisfies the Maxwell distribution and the mean kinetic energy of atom is equal to

$$E_m = \frac{3}{2} kT. \qquad (2)$$

Thus the temperature may be expressed in the units of energy.

If 1 eV = $1.60206 \cdot 10^{-19}$ J and the Boltzmann's constant k = $1.38041 \cdot 10^{-23}$ J/K, we have

$$1 \text{ eV} = 11605 \text{ K} \cong 11600 \text{ K}.$$

During ionization atom A (subscript a) decomposes into positive ion A^+ (subscript k) and negative electron E^- (subscript e) in accordance with the reaction

$$A \rightleftharpoons A^+ + E^-.$$

In the equlibrium state the sum of all molar[1] electrochemical potentials $\tilde{\mu}_i$, each multiplied by the relevant stoichiometric coefficient, is zero:

$$\sum_i \nu_i \tilde{\mu}_i = 0, \tag{3}$$

$$\tilde{\mu}_a - \tilde{\mu}_k - \tilde{\mu}_e = 0. \tag{4}$$

The electrochemical potentials of electrons and ions $\tilde{\mu}_e$ and $\tilde{\mu}_k$ can be expressed in terms of the chemical potentials μ_e and μ_k the electrostatic potential ψ in vacuum, and the charge per kilomole of electrons e_e or the charge per kilomole of positive ions e_k = = $- e_e$, respectively (by simple ionization). Hence, the electrochemical potential per kilomole of electrons is

$$\tilde{\mu}_e = e_e \psi + \mu_e. \tag{5}$$

The electrochemical potential per kilomole of positive ions is

$$\tilde{\mu}_k = - e_e \psi + \mu_k. \tag{6}$$

Combination of equations (4), (5), and (6), with $\tilde{\mu}_a = \mu_a$, leads to

$$\mu_a - \mu_k - \mu_e = 0. \tag{7}$$

As is seen, the equilibrium for ionzed gas has two similar forms which differ only in that electrochemical potentials appear in equation (4) whereas chemical potentials do in equation (7).

This is an alternative form of equation (4) which corresponds to

---

[1] In equations from (3) to (17) and from (42) to (106) all specific quantities are per kilomole as the unit of the amount of substance.

the well-known equilibrium condition of a multicomponent fluid.

Although ions and electrons interact through the intermediary of Coulomb forces, in the case of low pressures and high temperatures the gas under consideration may be assumed to obey Dalton's law.

The chemical potential corresponds to the partial free specific enthalpy

$$\mu_i = \tilde{h}_i - T\tilde{s}_i, \qquad (8)$$

where: $\tilde{h}_i$ - partial specific enthalpy, $\tilde{s}_i$ - partial specific entropy of the component i.

For perfect monoatomic gases, the chemical potential of the component i has the form

$$\mu_i(T,p) = c_{pi}(T - T_o) + c_{pi}T \ln \frac{T}{T'_o} + RT \ln \frac{p_i}{p_o} + \mu_{io} =$$

$$= c_{pi}T - c_{pi}T \ln T + RT \ln p_i + h_{io}(T_o) - T\tilde{s}_{io}(T'_o, p_o), \qquad (9)$$

where: $h_{io}(T_o)$ and $\tilde{s}_{io}(T'_o,\ p_o)$ are the specific enthalpy and the specific entropy of the component i, respectively, in the reference state at the temperature T_o or T'_o and the pressure p_o.

Thus the equilibrium condition of the ionized gas considered has the form

$$T \sum_{i=1}^{3} \nu_i c_{pi} - T (\ln T) \sum_{i=1}^{3} \nu_i c_{pi} + RT \ln \frac{p_k p_e}{p_a} +$$

$$+ \sum_{i=1}^{3} \nu_i h_{io} - T \sum_{i=1}^{3} \nu_i \tilde{s}_{io} = 0. \qquad (10)$$

In this equation the change in heat capacity during the ionization reaction occurs at constant pressure and at the reference temperature $T_o = 0$ K, called the ionization potential of substance considered (Table 1)

$$V_i = \sum_{i=1}^{3} \nu_i h_{io} = h_{ko} + h_{eo} - h_{ao} \qquad (11)$$

and the specific heat at constant pressure for plasma

$$c_p = \sum_{i=1}^{3} \nu_i c_{pi}. \tag{12}$$

Equation (10) may now be written in the form of the Saha equation

$$K_p(T) = \frac{p_k p_e}{p_a} = BT^{c_p/R} \exp\left(-\frac{V_i}{RT}\right), \tag{13}$$

where the constant B is defined as

$$B = \exp\left[\frac{1}{R}\sum_{i=1}^{3} \nu_i\left(\tilde{s}_{io} - c_{pi}\right)\right], \tag{14}$$

while $K_p(T)$ is the equilibrium constant for the reaction of single ionization of the monoatomic gas under consideration and its meaning is similar to that of the equilibrium constant for the dissotiation reaction.

All the particles considered are single. For an perfect monoatomic gas the specific heat at constant pressure is

$$c_p = \frac{5}{2} R. \tag{15}$$

For an ionized monoatomic gas under conditions such that the solution of un-ionized gas along with the products of its ionization may be treated as an perfect gas the Saha equation becomes

$$K_p(T) = \frac{p_k p_e}{p_a} = BT^{5/2} \exp\left(-\frac{V_i}{RT}\right). \tag{16}$$

As is seen from Table 1, the alkali metals, especially caesium, have the lowest ionization potentials, and hence the greatest ionizability.

If the plasma has such a large density, that the intermolecular forces must be taken into account, then in the Saha equation the component pressures are replaced by the fugacities of components defined as

$$\ln f_i = \ln z_i p - \frac{1}{RT}\int_0^p \left[v_{id} - \tilde{v}_i \right] dp. \quad (17)$$

where: v_{id} - specific volume of perfect gas, $\tilde{v}_i$ - partial specific volume of the component i, z_i - molar fraction.

The dissociation of diatomic hydrogen gas H_2 begins at a temperature of about 2000 K, and above 5000 K hydrogen has been almost completely dissociated, that is to say, consists of individual atoms H. At temperatures in the interval from 2000 to 5000 K hydrogen is partially dissociated and, knowing its dissociation constant, one can determine its composition in the equilibrium state on the basis of relations given in the classical thermodynamics.

Ionization of hydrogen begins above 8000 K. Hydrogen atoms lose negative electrons so that only positive protons remain. Since the hydrogen atom consists of only one proton and one electron, hydrogen can be only singly ionized.

Hydrogen ionization proceeds according to the scheme

$$H \rightleftarrows H^+ + E^-,$$

that is, monatomic hydrogen H (subscript H) gives rise to protons H^+ (subscript p) and electrons E^- (subscript e), forming, respectively, proton gas and electron gas which are treated as perfect gases. The molar fractions of these components will be labelled, respectively, as z_H - the molar fraction of monatomic hydrogen, z_p - the molar fraction of protons, and z_e - the molar fraction of electrons, whereby

$$z_H + z_p + z_e = 1. \quad (18)$$

In electrically neutral hydrogen plasma the number of protons and electrons is the same, and hence

$$z_p = z_e, \quad (19)$$

$$z_H = 1 - 2z_e = 1 - 2z_p. \quad (20)$$

The total pressure of hydrogen plasma is

$$p = p_H + p_p + p_e, \quad (21)$$

and, since

$$p_e = p_p, \tag{22}$$

for electrically neutral plasma it may also be written as

$$p = p_H + 2p_e = p_H + 2p_p. \tag{23}$$

The component pressures of monatomic hydrogen, protons, and electrons, respectively, are

$$p_H = z_H p, \quad p_p = z_p p, \quad p_e = z_e p. \tag{24}$$

These component pressures of hydrogen plasma are related by the ionization equilibrium constant which is a function of only the temperature, this constant being determined by means of the Saha equation (16)

$$K_p = \frac{p_p p_e}{p_H} = \frac{z_p z_e}{z_H} p. \tag{25}$$

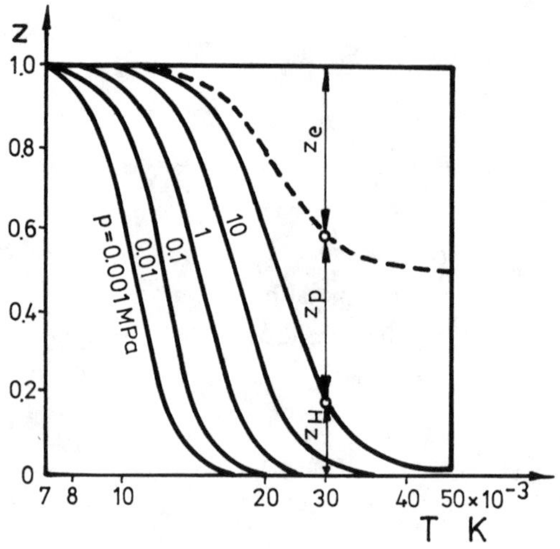

Figure 1. The molar composition of hydrogen plasma.

The molar composition of hydrogen plasma is given in Figure 1.

The equilibrium constant for hydrogen ionization can be associated with the degree of ionization β by means of the relations below. By the time equilibrium is reached, $n_b \beta$ kilomoles out of n_b kilomoles of hydrogen will have been ionized; the amount of un-ionized hydrogen remaining, in kilomoles, is

$$n_H = n_b (1 - \beta). \tag{26}$$

At the same time, ionization products will have evolved in the form of protons and electrons with the same number of kilomoles:

$$n_p = n_e = n_b\beta. \tag{27}$$

The total number of kilomoles of plasma components in the equilibrium state is

$$n = n_H + n_p + n_e = n_b (1 + \beta). \tag{28}$$

If the total pressure of the plasma is p, the respective component pressures are:

$$p_H = \frac{1 - \beta}{1 + \beta} p, \qquad p_p = p_e = \frac{\beta}{1 + \beta} p. \tag{29}$$

The equilibrium constant for hydrogen ionization may be written in terms of the degree of ionization as

$$K_p = \frac{\beta^2}{1 - \beta^2} p. \tag{30}$$

The degree of ionization of hydrogen can be calculated from the equilibrium constant by the relation

$$\beta = \sqrt{\frac{K_p}{p + K_p}} = \left[1 + \frac{p}{K_p} \right]^{-0.5}. \tag{31}$$

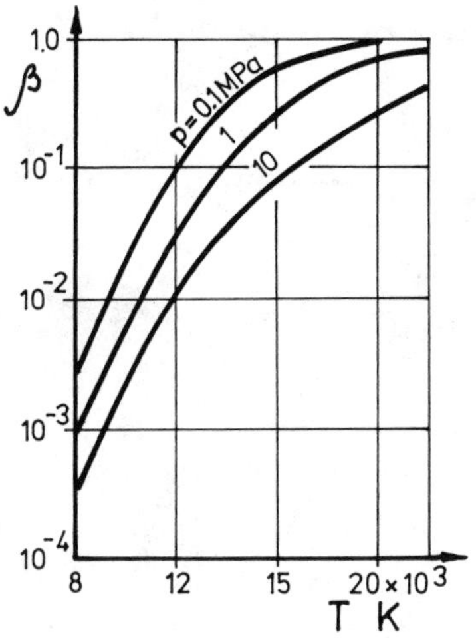

Figure 2. The degree of ionization of hydrogen plasma vs. the temperature and the pressure.

The degree of hydrogen ionization is plotted in Figure 2 against the temperature and the pressure.

At ionization of some substances e.g. lithium, helium only two neighbouring degrees of ionization occur at the same time. The equilibrium constant for single ionization of lithium can be calculated by analogy with equation (30), while the degree of single ionization, by analogy with equation (31).

For double ionization of lithium or some other gas subject to multiple ionization (e.g. helium) with only two adjacent degrees of ionization the reasoning proceeds as follows. By the time equilibrium is attained $n_{(1)b}\beta_{(2)}$ out of $n_{(1)b}$ kilomoles singly-ionized lithium has become doubly-ionized, leaving

$$n_{(1)} = n_{(1)b}(1 - \beta_{(2)}) \tag{32}$$

kilomoles not doubly-ionized. At the same time ionization products have appeared in the form of doubly-ionized lithium, the amount being

$$n_{(2)} = n_{(1)b}\,\beta_{(2)} \tag{33}$$

and $n_{(1)b}$ and $n_{(1)b}\beta_{(2)}$ kilomoles of electrons from single and double ionization, respectively, the total being

$$n_e = n_{(1)b}\,(1 + \beta_{(2)}). \tag{34}$$

The overall number of kilomoles of components of lithium plasma in which there is double ionization is

$$n = n_{(1)} + n_{(2)} + n_e = n_{(1)b}\,(2 + \beta_{(2)}) \tag{35}$$

When the total pressure of the plasma is p, the component pressures are

$$p_{(1)} = \frac{1 - \beta_{(2)}}{2 + \beta_{(2)}}\,p,\quad p_{(2)} = \frac{\beta_{(2)}}{2 + \beta_{(2)}}\,p,\quad p_e = \frac{1 + \beta_{(2)}}{2 + \beta_{(2)}}\,p. \tag{36}$$

The equilibrium constant for the double ionization lithium, calculated on the basis of the degree of ionization, is

$$K_{p(2)} = \frac{p_{(2)}p_e}{p_{(1)}} = \frac{(1 + \beta_{(2)})\,\beta_{(2)}}{(2 + \beta_{(2)})\,(1 - \beta_{(2)})}\,p. \tag{37}$$

The equation allows the degree of double ionization to be calculated:

$$\beta_{(2)} = -\frac{1}{2} + \sqrt{\frac{1}{4} + \frac{2}{1 + \dfrac{p}{K_{p(2)}}}}\,. \tag{38}$$

Analogous reasoning for triple ionization shows that the component pressures are

$$p_{(2)} = \frac{1 - \beta_{(3)}}{3 + \beta_{(3)}}\, p, \quad p_{(3)} = \frac{\beta_{(3)}}{3 + \beta_{(3)}}\, p, \quad p_e = \frac{2 + \beta_{(3)}}{3 + \beta_{(3)}}\, p, \qquad (39)$$

and the equilibrium constant is

$$K_{p(3)} = \frac{p_{(3)} p_e}{p_{(2)}} = \frac{(2 + \beta_{(3)})\, \beta_{(3)}}{(3 + \beta_{(3)})\,(1 - \beta_{(3)})}\, p, \qquad (40)$$

whereas the degree of triple ionization is

$$\beta_{(3)} = -1 + \sqrt{1 + \frac{3}{1 + \frac{p}{K_{p(3)}}}}\,. \qquad (41)$$

THERMODYNAMICS OF THE THERMIONIC DIODE

Thermionic generators were applied in the 1950s as electric energy sources for communication satellites. Other applications are also possible but are not wide-spread up till now. Thermionic generators belong to the group of devices for direct conversion of internal energy to electric energy.

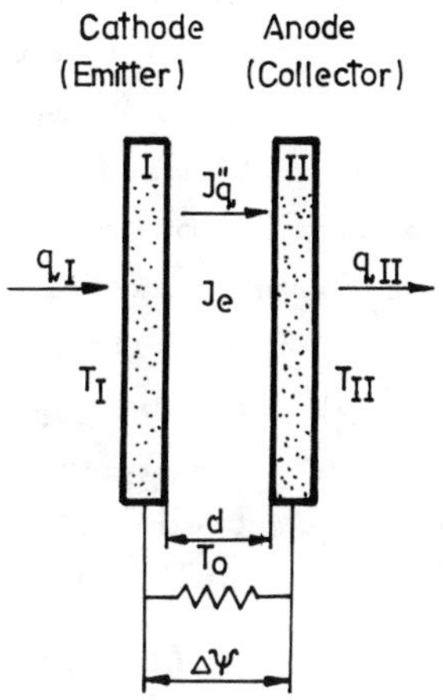

Figure 3. Diagram of a thermionic generator.

The element of the thermionic generator (Figure 3) contains a hot cathode (emitter) which emitts electrons, and a cold anode (collector) to which electrons tend. Great electron fluxes are obtained at temperatures in the interval 1800÷2800 K. The thermal efficiency of thermionic generators is in the range of 10÷20%. The advantage of thermonic generators is the lack of moving elements, while the disadvantage is generation of direct current of a low voltage. The low voltage (0,5÷1 V) requires series connection of thermionic elements. An important problem is a suitable choice of materials for electrodes. Typical meterials used for emitters are high melting metals, e.g.: tungsten, molybdenum, rhenium, tantalum. Typical collector materials are : caesium on a tungsten oxide or silver oxide matrix and on a tungsten or nickel matrix.

The thermal loading of the electrode surfaces is usually 10÷20 W/cm^2. This requires good heat input and good heat abstraction directly from the source. One solution of this problem is the application of heat pipes, mainly for the high temperature emitter.

As a result of the reciprocal repulsion of electrons having the electric charge of the same sign, a space charge is produced in the gap between the electrodes. For this reason, part of electrons emitted from the emitter do not reach the collector and return to the emitter. To reduce the disadvantageous action of the space charge on the generator power, the electrodes must be brought closer to each other, which smooths the nonlinearity of the potential distribution or the space charge must be neutralized by the introduction of positive ions. This is attained by the introduction of an easily ionized gas at low pressure, e.g. caesium at a pressure of $10 \div 10^3$ Pa.

Phenomena occuring in the thermionic diode can be considered by means of thermodynamic methods. The characteristic features of these phenomena are: low temperatures of the plasma (due to the resistance of the electrode materials to high temperatures) low pressures of the plasma (to avoid the recombination of ions and electrons to atoms), small distances between the electrodes (phenomena are considered as one-dimensional ones with a linear distribution of the electrostatic potential in the gap), low velocities of the substance molecules (their kinetic energy and the dissipation of the kinetic energy may be omitted), the plasma is transparent for thermal radiation.

In the first place, we shall consider the equilibrium of a partially ionized vapour with the solid phase of the same substance which is a conductor of electricity. The vapour pressure is so small that the vapour may be viewed as an perfect gas. The solid phase will be indicated by a superscript s, and the gas by a superscript g.

In addition to equality of temperatures, the condition for equilibrium is that the electrochemical potentials of the atoms, positive ions, and electrons in the gas and solid phases be equal:

$$\tilde{\mu}_a^g = \tilde{\mu}_a^s = \mu_a^g = \mu_a^s, \qquad \tilde{\mu}_k^g = \tilde{\mu}_k^s, \qquad \tilde{\mu}_e^g = \tilde{\mu}_e^s. \tag{42}$$

For un-ionized atoms, by relation (9) is

$$\mu_a^g = c_{pa}^g T - c_{pa}^g T \ln T + RT \ln p_a + h_{ao}^g - T\tilde{s}_{ao}^g, \tag{43}$$

where the component pressure of the vapour of atoms is

$$p_a = C_a T^{c_{pa}^g/R} \exp\left[\frac{\mu_a^g - h_{ao}^g}{RT}\right], \tag{44}$$

where the value of the constant for the given substance is

$$C_a = \exp\left[\frac{\tilde{s}_{ao}^g - c_{pa}^g}{R}\right]. \tag{45}$$

For a monatomic gas, since $c_{pa}^g/R = 5/2$, the component pressure is

$$p_a = C_a T^{5/2} \exp\left[\frac{\mu_a^g - h_{ao}^g}{RT}\right]. \tag{46}$$

By relations (5), (6), and (42), the chemical potentials of ions and electrons of the solid phase can be related to their electrochemical potential in the gas phase:

$$\tilde{\mu}_k^s = \mu_k^g - e_e\psi, \qquad \tilde{\mu}_e^s = \mu_e^g + e_e\psi. \tag{47}$$

The electrochemical potential of ions or electrons in the solid phase can also be made dependent on ϕ_k or ϕ_e, the work function of ions or electrons in the solid phase, and on ψ^o, the limiting value of the electrostatic potential external to the solid phase but right at the phase interface:

$$\tilde{\mu}_k^s = e_e(\phi_k - \psi^o), \qquad \tilde{\mu}_e^s = e_e(\phi_e + \psi^o). \tag{48}$$

The physical interpretation of the work function is based on the following reasoning. The specific energy of component i, which possesses electric charge, is

$$e_{Ei} = e_i(s,v) + e_i\psi,$$

where $e_i = e_e$ for electrons and $e_i = -e_e$ for positive ions. In the state with zero specific internal energy ($e = 0$) and zero specific entropy ($s = 0$) external to the solid phase the specific energy of component i is

$$e_{Eio} = e_i\psi^o.$$

Through the free enthalpy the chemical potential is related to the work of a reversible isobaric-isothermal process by

$$\left[\frac{dW_{p,T}}{dn_i}\right]_{rev} = -\left[\frac{\partial G}{\partial n_i}\right]_{p,T} = -\left(\tilde{\mu}_i - e_{Eio}\right) = -\left(\tilde{\mu}_i - e_i\psi^o\right),$$

and, when equation (48) is taken into account, also by

$$\phi_i = \pm\left[\frac{dW_{p,T}}{d(e_i n_i)}\right]_{rev}.$$

The work funkction ϕ_i for the surface of a solid phase is equal to the work done per unit charge if an infinitesimal amount of component i is removed reversibly to the solid phase from a point external to the surface and from a state of zero entropy and zero internal energy.

Comparison of relations (47) and (48) for the solid-phase surface reveals that

$$\mu_i^g = e_i\phi_i \quad \text{(at surface).} \tag{49}$$

By relation (10) the component pressure of ions or electrons is obtained in a form analogous to that of equation (46) derived above for atoms:

$$p_i = C_i T^{c^g_{pi}/R} \exp\left[\frac{\mu_i^g - h_{io}^g}{RT}\right], \tag{50}$$

where the constant C_i is

$$C_i = \exp\left[\frac{\tilde{s}^g_{io} - c^g_{pi}}{R}\right]. \tag{51}$$

The reference level for the molar enthalpy of ions and electrons in the gas phase is chosen so that h^g_{io} at the reference temperature $T_o = 0$ K. In this case the molar enthalpy of atoms in the reference state is, by relation (11), equal to the ionization potential taken with a minus sign, that is

$$h^g_{ao} = -V_i, \qquad \text{when} \qquad T_o = 0 \text{ K}. \tag{52}$$

If these reference values are used for the enthalpies of atoms, ions, and electrons, then in the case of equilibrium with the solid-phase surface equations (44) and (52) give the component pressures for un-ionized atoms,

$$p_a = C_a T^{c^g_{pa}/R} \exp\left[\frac{\mu^s_a + V_i}{RT}\right], \tag{53}$$

and the component pressures of ions or electrons at the solid-phase surface,

$$p_{is} = C_i T^{c^g_{pi}/R} \exp\left[\pm \frac{e_i \phi_i}{RT}\right], \tag{54}$$

where the minus sign in front of the fraction applies to ions, and the plus sign to electrons.

Electrons have a monatomic structure, hence $c^g_{pe}/R = 5/2$ and the component pressure of electrons in equilibrium with the surface of the solid phase is

$$p_{es} = C_e T^{5/2} \exp\left[\frac{e_e \phi_e}{RT}\right]. \tag{55}$$

Phenomena occurring in the thermionic diode can be considered in an approximate manner by means of the methods employed by nonequilibrium thermodynamics. This reasoning is based on the assumption of local thermodynamic equilibrium and linear phenomenological equations. Thus, it is strictly valid only for small temperature and potential differences but it also does allow a qualitative appraisal of the phenomena when these conditions are not met. In the case now under discussion (Figure 4).

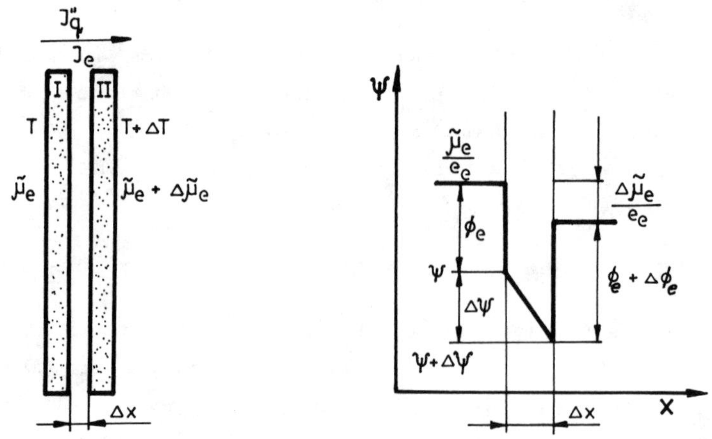

Figure 4. Linear distribution of the electrostatic potential in a very narrow gap between parallel surface of conductors I and II with only small differences between the temperatures, potentials, and work functions.

the gap Δx between the surfaces of conductors I and II is very narrow, the electrochemical potentials for electrons in the conductors are $\tilde{\mu}_e$ and $\tilde{\mu}_e + \Delta\tilde{\mu}_e$, respectively, whereas the temperatures are T and $T+\Delta T$. An electric current of electron flux J_e and a heat flux J_q'' flow between the surfaces of conductors I and II.

While considerating the thermionic diode it is most convenient to take the velocity of positive ions as the reference velocity, and then the diffusion flux of positive ions will be equal to zero. In the equations, only the electron diffusion flux j_e will occur.

The balance equation of the internal energy e for the plasma contained in the gap has the form

$$\rho \frac{de}{dt} = - \frac{\Delta J_E}{\Delta x} + e_e j_e E, \tag{56}$$

where: ρ - density of substance, t - time, J_E - internal energy flux, E - electric field strength.

The balance equation of the entropy s is

$$\rho \frac{ds}{dt} = \frac{\rho}{T}\frac{de}{dt} - \frac{\rho}{T}\tilde{\mu}_e \frac{dz_e}{dt} =$$

$$= - \frac{\Delta}{\Delta x}\left[\frac{J_E - \tilde{\mu}_e j_e}{T}\right] - \frac{1}{T^2} J_E \frac{\Delta T}{\Delta x} - \frac{j_e}{T}\left[T \frac{\Delta}{\Delta x}\left(\frac{\tilde{\mu}_e}{T}\right) - e_e E\right] =$$

$$= - \frac{\Delta J_s}{\Delta x} + \sigma , \qquad (57)$$

where: T – temperature, $\tilde{\mu}_e$ – electrochemical potential of electrons, z_e – molar fraction of electrons, J_s –entropy flux, σ – entropy source strength.

As is seen the entropy source strength has now the form

$$\sigma = J_E \frac{\Delta}{\Delta x}\left(\frac{1}{T}\right) - \frac{1}{T} j_e \left[T \frac{\Delta}{\Delta x}\left(\frac{\tilde{\mu}_e}{T}\right) - e_e E\right], \qquad (58)$$

and the dissipation function

$$\Psi_s = T\sigma = - J_E \frac{\Delta T}{T\Delta x} - j_e\left[T \frac{\Delta}{\Delta x}\left(\frac{\tilde{\mu}_e}{T}\right) - e_e E\right] =$$

$$= - J''_q \frac{\Delta T}{T\Delta x} - j_e \left[\frac{\Delta\tilde{\mu}_e}{\Delta x}\right]_T , \qquad (59)$$

where the heat flux found in the entropy balance equation has been introduced

$$J''_q = J_E - \tilde{h}_e j_e \qquad (60)$$

and the isothermal gradient of the electrochemical potential of electrons

$$\left[\frac{\Delta\tilde{\mu}_e}{\Delta x}\right]_T = \left[\frac{\Delta\mu_e}{\Delta x}\right]_T + e_e \frac{\Delta\varphi}{\Delta x} = \left[\frac{\Delta\mu_e}{\Delta x}\right]_T - e_e E, \qquad (61)$$

$$\left[\frac{\Delta\mu_e}{\Delta x}\right]_T = \frac{\Delta\mu_e}{\Delta x} + \tilde{s}_e \frac{\Delta T}{\Delta x}. \qquad (62)$$

Considerations concerning the foregoing effect in the thermionic diode are analogous to the analysis of a flow of electric current and heat through a homogeneous conductor. By equation (59) the dissipation function times the separation between the conductor surfaces gives

$$\Psi \, \Delta x = - J''_q \frac{\Delta T}{T} - j_e \left(\Delta \tilde{\mu}_e \right)_T . \tag{63}$$

On assuming that conductors I and II are made of the same substance in the chemical respect and have the same pressure acting on them, it can be found by equation (49) that the electron work function ϕ_e is a function only of the temperature:

$$\left(\Delta \phi_e \right)_{\Delta T=0} = 0. \tag{64}$$

The definition (48) of the electron work function

$$\phi_e = \frac{\tilde{\mu}_e}{e_e} - \psi \tag{65}$$

implies that the increment in electrostatic potential in the gap is

$$\Delta \psi = \frac{\left(\Delta \tilde{\mu}_e \right)_{\Delta T=0}}{e_e}, \tag{66}$$

and since the electric current density is

$$J_e = j_e e_e, \tag{67}$$

the dissipation function can be written as

$$\Psi \, \Delta x = - J''_q \frac{\Delta T}{T} - J_e \Delta \psi . \tag{68}$$

The differences $- \Delta T / T$ and $- \Delta \psi$ are taken for the thermodynamic forces, whereas the quantities taken for the fluxes are:
the heat flux in the entropy balance equation

$$J''_q = - L_{qq} \frac{\Delta T}{T} - L_{qe} \Delta \psi, \tag{69}$$

the electric current density

$$J_e = - L_{eq} \frac{\Delta T}{T} - L_{ee} \Delta \psi . \tag{70}$$

By the Onsager reciprocal relations the matrix of phenomenological coefficients is symmetrical,

$$L_{eq} = L_{qe}. \tag{71}$$

The condition that the dissipation function be positive implies that

$$L_{qq} > 0, \qquad L_{ee} > 0, \qquad L_{qq}L_{ee} - L_{qe}^2 > 0 \tag{72}$$

In special cases the two conductors may be at the same temperature ($\Delta T = 0$), have the same electrostatic potentials ($\Delta\psi = 0$), or there may be no flow of electric current ($J_e = 0$).

For isothermal conditions equations (69) and (70) yield the heat of transport of electrons

$$Q''^{*}_{e} = \left[\frac{J''_q}{j_e} \right]_{\Delta T=0} = e_e \left[\frac{J''_q}{J_e} \right]_{\Delta T=0} = e_e \frac{L_{qe}}{L_{ee}}. \tag{73}$$

Elimination of the electrostatic potential difference from equations (69) and (70) yields the heat flux in the form

$$J''_q = \frac{L_{qe}}{L_{ee}} J_e - \left[L_{qq} - \frac{L_{qe}^2}{L_{ee}} \right] \frac{\Delta T}{T} = \frac{Q''^{*}_{e}}{e_e} J_e - L_k \Delta T, \tag{74}$$

where we have introduced a new phenomenological coefficient defined as

$$L_k = \left[L_{qq} - \frac{L_{qe}^2}{L_{ee}} \right] \frac{1}{T}. \tag{75}$$

The physical meaning of this coefficient can easily be determined by considering the heat flux when no electric current is flowing:

$$\left[J''_q \right]_{J_e=0} = - L_k \Delta T. \tag{76}$$

The phenomenological coefficient L_k thus is the heat flux set up by a unit temperature difference in the absence of an electric current.

The density of electric current flowing through the isothermal diode can be determined from equation (70) as

$$\left(J_e \right)_{\Delta T=0} = - L_{ee}\Delta\psi \ . \tag{77}$$

The electron gas in the gap and the electron gas emitted by the surface of zero reflectivity and in equilibrium with the first one are treated as an perfect gas having an electron velocity distribution in accordance with the Maxwell distribution. As it results from the statistical physics considerations (Hatsopoulos et al. 1965), the fraction f of electrons each of the electric charge emitted from surface I, and having a kinetic energy component greater than $\epsilon\Delta\psi$ in the direction perpendicular to the surface is given by the relation

$$f = \exp\left(- \frac{\epsilon\Delta\psi}{kT} \right) = \exp\left(- \frac{e_e\Delta\psi}{RT} \right). \tag{78}$$

The density J_e of the current flowing from surface I to II under the condition in question, that is, with a zero electrostatic potential gradient at surface I and zero re-emission from surface II, is called the saturation current density

$$J_{es} = AT^2 \exp\left(\frac{e_e\phi_e}{RT} \right), \tag{79}$$

where the quantity A is called the Richardson-Dushman constant. As follows from theoretical considerations, the Richardson-Dushman constant for pure metals is

$$A = 120{,}4 \ \mathrm{A\ cm^{-2}K^{-2}} \tag{80}$$

and does not depend on the material of the conductor on whose surface the saturation current density is determined. In actual fact, the magnitude measured of the Richardson-Dushman constant does depend on the kind of conductor material (table 2).

The density J_e of the current flowing from surface I to II can be calculated from the saturation current densities J_{esI} and J_{esII} at surfaces I and II as the difference between the density of the current flowing from surface I to II and that flowing in the opposite direction

$$J_e = fJ_{esI} - J_{esII} = J_{esI} \exp\left(- \frac{e_e\Delta\psi}{RT_I} \right) - J_{esII}. \tag{81}$$

Table 2. Work Function ϕ_e and Measured Value of the Richardson-Dushman Constant A_m for Some Conductors

Conductor	Work Function ϕ_e in volts	Richardson Constant A_m in A $cm^{-2}K^{-2}$
Ag	4.08	60.2
Al_2O_3	3.77	1.4
BaO	3.44÷1.66	2.5
BaO-SrO	1÷1.2	0.1÷1
Cr	4.6	48
Cs	1.81	160
K	1.9÷246	
LaB_2	2.7	14
Mo	4.2	51
Na	2.5	
Nb	4	
Ni	2.77	26.8
Pt	5.32	32
Re	4.7	720
Ta	4.2	55
ThC_2	3.2	200
ThO_2	2.6	5
W	4.5	75
W-Ba	1.6÷2	1÷2
W-Cs	1.4	3
W-Th	2.6	3
ZrC	3.8	134
ZrO_2	3.4	0.35

The isothermal energy of transport of electrons with the potential energy ψ and kinetic energy greater than $e_e\ \Delta\psi$ is equal to

$$E_e^* = e_e\left(\frac{J_E}{J_e}\right)_{\Delta T=0} = e_e\left(\psi + \Delta\psi + \frac{2RT}{e_e}\right) . \qquad (82)$$

in accordance with equation (81) is

$$(J_e)_{\Delta T=0} = \left[J_{es} \exp\left(- \frac{e_e \Delta\psi}{RT} \right) - \left(J_{es} + \Delta J_{es} \right) \right]_{\Delta T=0} . \tag{83}$$

The saturation current density is a function of the temperature, pressure, and chemical composition of the conductor, but does not depend on the value of the electrostatic potential at the conductor surface. Hence

$$(\Delta J_{es})_{\Delta T=0} = 0 \tag{84}$$

and consequently

$$(J_e)_{\Delta T=0} = J_{es} \left[\exp\left(- \frac{e_e \Delta\psi}{RT} \right) - 1 \right] . \tag{85}$$

For small potential differences, to which linear nonequilibrium thermodynamics is confined, equation (85) takes on the form

$$(J_e)_{\Delta T=0} = - J_s \frac{e_e \Delta\psi}{RT} . \tag{86}$$

Comparison of relations (86) and (77) enables the phenomenological coefficient concerning current conduction to be defined as

$$L_{ee} = J_{es} \frac{e_e}{RT} . \tag{87}$$

The heat of transport of electrons may be calculated from the isothermal energy of transport of electrons (82)

$$Q_e^{''*} = E_e^{*} - h_e = e_e(\psi + \Delta\psi) - \frac{1}{2} RT . \tag{88}$$

Now that the phenomenological coefficient L_{ee} and the heat of transport have been determined, equation (70) expressing the electric current density can be rearranged by discarding the second-order term $\Delta\psi\Delta T$ and then

$$J_e = - L_{ee} \left(\frac{L_{eq}}{L_{ee}} \frac{\Delta T}{T} + \Delta\psi \right) =$$

$$= - \frac{J_{es}}{R} \left[\left(2RT + e_e\psi - h_e \right) \frac{\Delta T}{T^2} + \frac{e_e \Delta\psi}{T} \right] . \tag{89}$$

The potential drop across the interelectrode gap can be written as

$$\Delta\psi = \psi_{II} - \psi_I = \phi_{eI} - \phi_{eII} + \frac{1}{e_e}\Delta\tilde{\mu}_e \ . \tag{90}$$

The electric current density can thus be put in the form

$$J_e = AT_I^2 \exp\left[\frac{e_e\phi_{eII} - \Delta\tilde{\mu}_e}{RT_I}\right] - AT_{II}^2 \exp\left[\frac{e_e\phi_{eII}}{RT_{II}}\right] , \tag{91}$$

hence the density J_e of the current flowing through the diode does not depend on the work function ϕ_{eI} for removal of electrons from the emitter.

Equation ,defining the electric current density in a homogeneous conductor, can be cast into the form

$$J_e dx = -\frac{1}{\rho e_e}\left[S^*_{eB}\, dT + d\tilde{\mu}_e\right] , \tag{92}$$

where the subscript B refers to the material of electrode B.

This equation can be used to integrate the differential of the electrochemical potential of electrons in conductor B, of cross-sectional area B_o , connecting the diode with the resistive load; in this operation the electrostatic potential difference $e_e\varphi$ between the terminals of the resistive load,

$$\tilde{\mu}_{eII} - \tilde{\mu}_{eI} = \Delta\tilde{\mu}_e = -e_e\int_{T_I}^{T_{II}} \varepsilon_B dT - e_e J_e \int_I^{II} \frac{\rho_B}{B} dx + e_e\varphi , \tag{93}$$

must be added to the right-hand side. Hence, the potential difference across the electrodes is

$$\varphi = \frac{\Delta\tilde{\mu}_e}{e_e} + \int_{T_I}^{T_{II}} \varepsilon_B dT + J_e \int_I^{II} \frac{\rho_\beta}{B_0} dx \ . \tag{94}$$

where ρ_B is the resistivity of conductor B and ε_B is the absolute Seebeck coefficient of conductor B, defined by the equation

$$\varepsilon_B = -\frac{\Delta\tilde{\mu}_{eB}}{e_e \Delta T} = \frac{1}{e_e} S^*_{eB} \ . \tag{95}$$

The energy flux $J_{EI,II}$ flowing through the thermionic diode is equal to the difference between the energy flux associated with the electrons emitted by the emitter I and the energy flux associated with the electrons emitted by the collector II, with due account for the energy transport by radiation between the emitter and collector surfaces:

$$J_{EI,II} = J_{EI} - J_{EII} + J_r , \tag{96}$$

where the radiant energy flux is calculated as

$$J_r = \frac{\sigma}{\dfrac{1}{\varepsilon_I(T_I)} + \dfrac{1}{\varepsilon_{II}(\overline{T})} - 1}\left(T_I^4 - T_{II}^4\right) = F\left(T_I^4 - T_{II}^4\right) . \tag{97}$$

In this formula the relative emissivity ε_{II} of the collector is calculated for the geometric mean of the emitter and collector temperatures,

$$\overline{T} = \sqrt{T_I\ T_{II}} . \tag{98}$$

The symbol σ = 5.729 × 10^{-8} W m^{-2} K^{-4} is the radiation constant for a blackbody surface.

The internal energy flux can be determined by taking account of the electric current density, as given by expression (91)

$$J_{EI,II} = AT_I^2\left(\psi_{II} + \frac{2RT_I}{e_e}\right)\exp\left(\frac{e_e\phi_{eII} - \Delta\tilde{\mu}_e}{RT_I}\right) +$$

$$- AT_{II}^2\left(\psi_{II} + \frac{2RT_{II}}{e_e}\right)\exp\left(\frac{e_e\phi_{eII}}{RT_{II}}\right) + J_r . \tag{99}$$

or

$$J_{EI,II} = J_e\psi_{II} + \frac{2RA}{e_e}\left[T_I^3\exp\left(\frac{e_e\phi_{eII} - \Delta\tilde{\mu}_e}{RT_I}\right) +\right.$$

$$\left. - T_{II}^3\exp\left(\frac{e_e\phi_{eII}}{RT_{II}}\right)\right] + J_r . \tag{100}$$

For the limiting case $T_{II} - T_I \to dT$ and $\tilde{\mu}_{eII} - \tilde{\mu}_{eI} \to d\tilde{\mu}_e$, taking relation (91) into consideration, equation (100) yields the infinitesimal energy flux between the electrodes:

$$\delta J_{EI,II} = \psi_{II}\delta J_e - \frac{2J_{es}}{e_e}\left[\left(3RT - e_e\phi_{eII}\right)\frac{dT}{T} + d\tilde{\mu}_e\right] + dJ_r =$$

$$= \left[\psi_{II} + \frac{2RT}{e_e}\right]\delta J_e - \frac{2J_{es}}{e_e} R\mathrm{d}T + \mathrm{d}J_r \, . \qquad (101)$$

The heat flux q_I required to maintain the emitter at the constant temperature T_I is the sum of energy fluxes flowing between emitter and collector ($J_{EI,II}$), given by equation (100) and extracted from the emitter by conductor B (J_{EIB}):

$$q_I = J_{EI,II} + J_{EIB} \, . \qquad (102)$$

The energy flux carried off by the conductor can be determined by means of the equation

$$J_{EIB} = -\frac{J_e}{e_e}E_e^* - B_o\lambda_{\infty B}\left[\frac{\mathrm{d}T_B}{\mathrm{d}x}\right]_{x=0} = -(TS_{eB}^* + \mu_e)\frac{J_e}{e_e} +$$

$$- B_o\lambda_{\infty B}\left[\frac{\mathrm{d}T_B}{\mathrm{d}x}\right]_{x=o} \, . \qquad (103)$$

A minus sign has been put in front of the first term on the right-hand side of the equation since J_e is the electric current density flowing from emitter to conductor B, that is, opposite to the direction taken for the x-axis. If the conductor B of length l is adiabatically insulated and if its resistivity ρ_B , thermal conductivity $\lambda_{\infty B}$, and Seebeck coefficient ε_B do not depend on the temperature, then the gradient temperature at the junction of conductor B with the emitter is

$$\left[\frac{\mathrm{d}T_B}{\mathrm{d}x}\right]_{x=0} = \frac{T_{II} - T_I}{l} + \frac{J_e^2 \rho_B l}{2B_o^2 \lambda_{\infty B}} \, . \qquad (104)$$

Finally, when the values of the fluxes as given by equations (101) and (103) are inserted into equation (102) and when relation (104) is taken into account, the heat flux q_I carried off from the emitter is obtained for the limiting case $T_{II} - T_I \to \mathrm{d}T$:

$$q_I = \frac{\delta J_e}{e_e}\left[e_e\psi - \mu_e - TS_{eB}^* + 2RT\right] - \left[\frac{2J_{es}R}{e_e} + 4FT^3 + \frac{B_o\lambda_{\infty B}}{l}\right]\mathrm{d}T =$$

$$= \delta J_e\left[\frac{2RT}{e_e} - \phi_e - T\varepsilon_B\right] - \left[\frac{2J_{es}R}{e_e} + 4FT^3 + \frac{B_o\lambda_{\infty B}}{l}\right]\mathrm{d}T \, . \qquad (105)$$

The thermal efficiency of the thermionic generator treated as a heat engine is the ratio of the power $P = \varphi J_e$ to the heat flux delivered to the emitter:

$$\eta = \frac{P}{q_I} = \frac{\varphi J_e}{q_I} . \qquad (106)$$

EQUATIONS FOR PLASMA IN AN ELECTROMAGNETIC FIELD

For the direct enthalpy conversion in electric energy, a magnetohydrodynamic generator (MHD) may also be applied. The operating principle of the magnetohydrodynamic generator consists in obtaining electric current directly from the kinetic energy of partially ionized gas. This current is generated by the motion of electrical charges in the external magnetic field. Owing to the action of the Lorentz forces, particles with opposite charges move towards different electrodes incorporated in the electrical circuit (Figure 5).

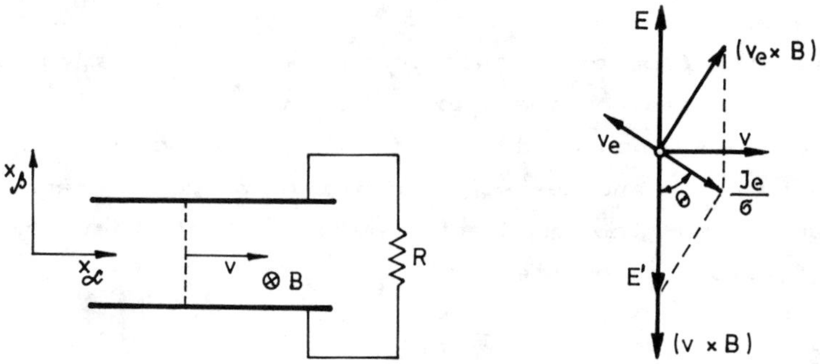

Figure 5. Continuous-electrode MHD generator.

Electric current flows in the MHD generator perpendicular to the direction of both the plasma flow and the magnetic induction. Under the effect of the interaction between the external magnetic field and the ionized particles of the plasma, the plasma velocity decreases as the electric current is drawn from the electrodes.

Plasma occurring in the magnetohydrodynamic generator can be considered as a k - component medium consisting of atoms (subscript a), positive ions (subscript k), and electrons (subscript e). Each of the components is characterized by temperature, component pressure, mass or molar fraction, and three velocity components. The state of the plasma under study also depends on radiation temperature and on the electric and magnetic field strength E and B which, being vectors, have three components each. Generally, in order to determine the local state of such a plasma, one must know the values of $6k + 7$ variables related to each other by $6k + 7$ equations of a thermodynamic ($6k + 1$ equations) and electro-magnetic (6 equations) nature. The number of unknowns is reduced considerably if one temperature is assumed for all the components and for radiation.

Low density plasma can in general be treated as an perfect gas and can have the equations of state like those for perfect gases applied to it, that is

$$p = \rho RT \ , \tag{107}$$

$$\mathrm{d}e = c_v \mathrm{d}T \ , \qquad \mathrm{d}h = c_p \mathrm{d}T \ . \tag{108}$$

The fundamental equations for an nonviscous and unpolarized medium are applicable to plasma. These are equations of:

the conservation of charge

$$\rho \frac{\mathrm{d}e}{\mathrm{d}t} = - \nabla \cdot \mathrm{J}_e \ , \tag{109}$$

the conservation of the amount of substance

$$\frac{\mathrm{d}\rho}{\mathrm{d}t} = - \rho(\nabla \cdot \mathrm{v}) \ , \tag{110}$$

the momentum balance

$$\rho \frac{\mathrm{d}\mathrm{v}}{\mathrm{d}t} = - p_{,\alpha} + \sum_{i=1}^{k} \rho_i e_i \Big[\mathrm{E} + (\mathrm{v} \times \mathrm{B}) \Big] \ , \tag{111}$$

the internal energy balance

$$\rho \frac{\mathrm{d}e}{\mathrm{d}t} = - \nabla \cdot \mathrm{J}_{\mathrm{E}} - \rho p \frac{\mathrm{d}\mathrm{v}}{\mathrm{d}t} + \mathrm{J}_e \cdot \Big[\mathrm{E} + (\mathrm{v} \times \mathrm{B}) \Big] \ , \tag{112}$$

the entropy balance

$$\rho \frac{\mathrm{d}s}{\mathrm{d}t} = - \nabla \cdot \mathrm{J}_s - \frac{1}{T^2} \mathrm{J}_{\mathrm{E}} \cdot \nabla T - \frac{1}{T} \sum_{i=1}^{k} \mathrm{j}_i \cdot \Bigg\{ T \nabla \left[\frac{\mu_i}{T} \right] + - e_i \Big[\mathrm{E} + (\mathrm{v} \times \mathrm{B}) \Big] \Bigg\} \ , \tag{113}$$

from which the entropy source strength results

$$\sigma = J_E\cdot\nabla\left(\frac{1}{T}\right) - \frac{1}{T}\sum_{i=1}^{k} j_i\cdot\left\{T\nabla\left(\frac{\mu_i}{T}\right) - e_i\left[E + (v\times B)\right]\right\} =$$

$$= J''_q\cdot\nabla\left(\frac{1}{T}\right) - \frac{1}{T}\sum_{i=1}^{k-1} j_i\cdot\left\{\nabla_T(\mu_i - \mu_k) + \right.$$

$$\left. - (e_i - e_k)\left[E + (v\times B)\right]\right\} \geq 0 \ . \tag{114}$$

and the dissipation function

$$\Psi = T\sigma = - J''_q\cdot\nabla \ln T - \frac{1}{T}\sum_{i=1}^{k-1} j_i\cdot\left\{\nabla_T(\mu_i - \mu_k) + \right.$$

$$\left. - (e_i - e_k)\left[E + (v\times B)\right]\right\} \geq 0 \ . \tag{115}$$

For perfect gases, the isothermal gradient of the chemical potential can be writen as

$$\nabla_T\mu_i = - T\ \nabla_T\ \tilde{s}_i = R_i T\ \nabla\ \ln p_i = \frac{1}{\rho_i}\nabla p_i \ . \tag{116}$$

Thus for ideal gases, the expressions

$$X_q = - \nabla\ \ln T \ , \tag{117}$$

$$X_i = - \frac{1}{\rho_i}\nabla p_i + \frac{1}{\rho_k}\nabla p_k + (e_i - e_k)\left[E + (v\times B)\right] = - (\tilde{v}_i - \tilde{v}_k)\nabla p +$$

$$- p\left[\frac{1}{\rho_i}\nabla z_i - \frac{1}{\rho_k}\nabla z_k\right] + (e_i - e_k)\left[E + (v\times B)\right]$$

$$(i = 1,2,3,\ldots,k-1) \tag{118}$$

can be taken as the thermodynamic forces, conjugated with the generalized flows in the form of the heat flux J''_q in the entropy balance equation and the diffusion flux j_i .

Now the phenomenological equations can be written for the heat fluxes which occur in the entropy balance equation,

$$J''_q = L_{qq} \cdot X_q + \sum_{i=1}^{k-1} L_{qi} \cdot X_i \ , \qquad (119)$$

and for the diffusion flux of component m,

$$j_m = L_{mq} \cdot X_q + \sum_{i=1}^{k-1} L_{mi} \cdot X_i \ . \qquad (120)$$

In the presence of a magnetic field, plasma becomes nonisotropic and the phenomenological coefficients are tensors of rank two, associated by Onsager-Casimir relations of the form

$$L_{qi}(B) = \tilde{L}_{iq}(-B) \ , \qquad L_{mi}(B) = \tilde{L}_{im}(-B) \ , \qquad (121)$$

The above equations may be adapted to particular cases. If totally-ionized plasma is formed from a chemically pure gas, plasma is a binary fluid consisting exclusively of positive ions (subscript k) and electrons (subscript e).

For neutral plasma the electrical charge of the positive ions is equal to minus the charge of the same number of electrons, and hence per kilomole we have

$$M_k e_k + M_e e_e = 0 \ . \qquad (122)$$

The sum of the diffusion fluxes of all components is zero and hence the ion diffusion flux is thus equal to the electron diffusion flux in the opposite direction

$$j_k = - j_e \ . \qquad (123)$$

The density of the electric current arising as a result of diffusion of electrically charged particles is

$$J_e = e_k j_k + e_e j_e = e_e \left[\frac{M_e}{M_k} + 1 \right] j_e = e_e M_e \left[\frac{1}{M_k} + \frac{1}{M_e} \right] j_e \ . \qquad (124)$$

In many cases the molar mass of ions is much greater than that of electrons, $M_k \gg M_e$, and then equation (124) simplifies to

$$J_e = e_e j_e \ . \qquad (125)$$

The electron diffusion flux is defined by means of the phenomenological equation

$$j_e = -j_k = -L_{eq}\cdot\nabla\ln T - L_{ee}\cdot\left\{(\tilde{v}_e - \tilde{v}_k)\nabla p +\right.$$

$$\left. + \frac{R_eR_k}{R}\frac{T}{x_ex_k}\nabla x_e - e_e\left(1 + \frac{M_e}{\overline{M}_k}\right)\left[E + (v\times B)\right]\right\} . \qquad (126)$$

When the heat of transport of electrons $Q_e^{''*}$, defined by the relation

$$Q_e^{''*} L_{ee} = L_{eq} , \qquad (127)$$

is introduced, the electron diffusion flux becomes

$$j_e = -L_{ee}\cdot\left\{Q_e^{''*}\nabla\ln T + (\tilde{v}_e - \tilde{v}_k)\nabla p + \frac{R_eR_k}{R}\frac{T}{x_ex_k}\nabla x_e +\right.$$

$$\left. - e_e\left(1 + \frac{M_e}{\overline{M}_k}\right)\left[E + (v\times B)\right]\right\} . \qquad (128)$$

If only an electromagnetic field acts on the plasma, the electric current density due to the diffusion of charged particles is

$$J_e = e_e^2\left(1 + \frac{M_e}{\overline{M}_k}\right)^2 L_{ee}\cdot\left[E + (v\times B)\right] = \sigma\cdot\left[E + (v\times B)\right] , \qquad (129)$$

where electrical conductivity has been defined as

$$\sigma = e_e^2\left(1 + \frac{M_e}{\overline{M}_k}\right)^2 L_{ee} \approx e_e^2 L_{ee} . \qquad (130)$$

In view of the above, the phenomenological coefficient is

$$L_{ee} \approx \frac{\sigma}{e_e^2} . \qquad (131)$$

The electron flux can be defined in terms of the electrical conductivity as

$$j_e = -\frac{\sigma}{e_e^2}\cdot\left\{Q_e^{''*}\nabla\ln T + (\tilde{v}_e - \tilde{v}_k)\nabla p + \frac{R_eR_k}{R}\frac{T}{x_ex_k}\nabla x_e +\right.$$

$$\left. - e_e\left(1 + \frac{M_e}{\overline{M}_k}\right)\left[E + (v\times B)\right]\right\} . \qquad (132)$$

Elimination of the thermodynamic force X_e from the system of phenomenological equations

$$J''_q = \mathbf{L}_{qq}\cdot X_q + \mathbf{L}_{qe}\cdot X_e \ , \tag{133}$$

$$j_e = \mathbf{L}_{eq}\cdot X_q + \mathbf{L}_{ee}\cdot X_e \ , \tag{134}$$

leads to

$$J''_q = \mathbf{L}_{qe}\cdot\mathbf{L}_{ee}^{-1}\cdot j_e + (\mathbf{L}_{qq} - \mathbf{L}_{qe}\cdot\mathbf{L}_{eq}\cdot\mathbf{L}_{ee}^{-1})\cdot X_q =$$

$$= \mathbf{L}_{qe}\cdot\mathbf{L}_{ee}^{-1}\cdot j_e - \boldsymbol{\lambda}_{\infty}\cdot\nabla T \ , \tag{135}$$

where

$$\boldsymbol{\lambda}_{\infty} = \frac{1}{T}(\mathbf{L}_{qq} - \mathbf{L}_{qe}\cdot\mathbf{L}_{eq}\cdot\mathbf{L}_{ee}^{-1}) \tag{136}$$

is the thermal conductivity in the absence of any flow of electric current.

In the case of incompletely ionized plasma, three components are considered: un-ionized atoms (subscript a), positive ions (subscript k), and electrons (subscript e).

The phenomenological equations (119) and (120) in this case take the form:
heat flux in the entropy balance equation

$$J''_q = \mathbf{L}_{qq}\cdot X_q + \mathbf{L}_{qk}\cdot X_k + \mathbf{L}_{qe}\cdot X_e \ , \tag{137}$$

diffusion flux of positive ions

$$j_k = \mathbf{L}_{kq}\cdot X_q + \mathbf{L}_{kk}\cdot X_k + \mathbf{L}_{ke}\cdot X_e \ , \tag{138}$$

diffusion flux of electrons

$$j_e = \mathbf{L}_{eq}\cdot X_q + \mathbf{L}_{ek}\cdot X_k + \mathbf{L}_{ee}\cdot X_e \ . \tag{139}$$

The diffusion flux of un-ionized atoms is

$$j_a = - j_k - j_e \ , \tag{140}$$

and hence the diffusion flux of charged particles is

$$j_{ch} = j_e + j_k = - j_a \ . \tag{141}$$

The electric current density for neutral plasma (122) is

$$J_e = e_k j_k + e_e j_e = e_e\left[j_e - \frac{M_e}{M_k}\,j_k\right] = e_e M_e\left[\frac{j_e}{M_e} - \frac{j_k}{M_k}\right] \ . \tag{142}$$

In the case of incomplete ionization of plasma the thermodynamic forces are

$$X_q = -\nabla \ln T\ , \tag{143}$$

$$X_k = -(\tilde{v}_k - \tilde{v}_a)\nabla p - p\left[\frac{1}{\rho_k}\nabla z_k - \frac{1}{\rho_a}\nabla z_a\right] +$$

$$+\ e_k\left[E + (v\times B)\right]\ , \tag{144}$$

$$X_e = -(\tilde{v}_e - \tilde{v}_a)\nabla p - p\left[\frac{1}{\rho_e}\nabla z_e - \frac{1}{\rho_a}\nabla z_a\right] +$$

$$+\ e_e\left[E + (v\times B)\right]\ . \tag{145}$$

The phenomenological equations (135) to (139) can be transformed by introducing the heats of electron and ion transport, $Q_e^{''*}$ and $Q_k^{''*}$, respectively, which for the given ternary system are related by

$$L_{qe} = \tilde{L}_{eq} = L_{ee}\,Q_e^{''*} + L_{ke}\,Q_k^{''*}\ , \tag{146}$$

$$L_{qk} = \tilde{L}_{kq} = L_{ek}\,Q_e^{''*} + L_{kk}\,Q_k^{''*}\ . \tag{147}$$

Consequently, the heat flux in the entropy balance equation is

$$J_q'' = Q_e^{''*}\,j_e + Q_k^{''*}\,j_k + (L_{qq} - L_{eq}\,Q_e^{''*} - L_{kq}\,Q_k^{''*})\cdot X_q\ , \tag{148}$$

the diffusion flux of positive ions is

$$j_k = L_{ke}\cdot(X_e + Q_e^{''*}\,X_q) + L_{kk}\cdot(X_k + Q_k^{''*}\,X_q)\ , \tag{149}$$

and the diffusion flux of electrons is

$$j_e = L_{ee}\cdot(X_e + Q_e^{''*}\,X_q) + L_{ek}\cdot(X_k + Q_k^{''*}\,X_q)\ . \tag{150}$$

In the stationary state, that is, in the absence of diffusion flows, equations (148) yields a relation of the form of Fourier's law

$$J_q'' = -\boldsymbol{\lambda}_\infty\cdot\nabla T\ , \qquad j_a = j_k = j_e = 0\ , \tag{151}$$

with the thermal conductivity in the stationary state

$$\lambda_{\infty} = \frac{1}{T}(L_{qq} - L_{eq} Q_e^{"*} - L_{kq} Q_k^{"*}) \ . \qquad (152)$$

THE HALL EFFECT IN PLASMA

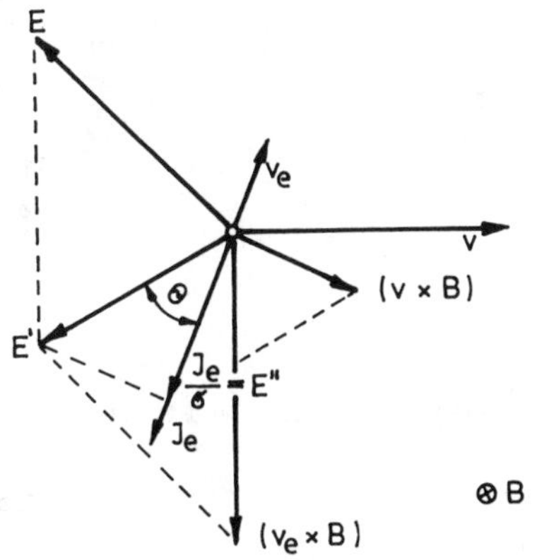

Figure 6. Arrangement of vectors explaining the Hall effect in plasma.

An electromagnetic field acting on plasma causes the Hall effect to appear. This is a cross-effect, the essence of which is that a flow of electric current perpendicular to the magnetic displacement sets up a potential difference in the direction perpendicular to both previous vectors (Figure 6). The magnetic field strength B is perpendicular to the paper and directed out of it. Plasma flowing in the direction of the x_{α}-axis with barycentric velocity v is subjected to an induced force determined by the vector product (v×B). Accordingly, the strength E of a motionless external electric field can be replaced by an electric field strength in a coordinate system moving together with the fluid and specified by the resultant of the vectors E and (v×B), that is,

$$E' = E + (v \times B) \ . \qquad (153)$$

The current flowing in the direction of vector E' suffers a change under the influence of the current produced by the Hall effect. In actual fact the electrons move with velocity v_e and the electric field strength relative to the moving electrons is given by

$$E" = E' + (v_e \times B) \ . \qquad (154)$$

If the scalar electrical conductivity σ of the plasma is introduced and the current transported by the ions is neglected this yields for the electric current density the expression

$$J_e = -v_e\rho_e = \sigma\left[E' + (v_e \times B)\right] = \sigma E'' . \tag{155}$$

where ρ_e is the electronic charge density. The last term in this equation, $\sigma(v_e \times B)$, specifies the Hall current density.

The electrical conductivity σ of the plasma can be expressed in terms of the electron mobility ν_e and electronic charge density ρ_e as

$$\sigma = \nu_e\rho_e , \tag{156}$$

whence

$$\sigma v_e = \nu_e\rho_e v_e = -\nu_e J_e , \tag{157}$$

and hence the electric current density may also be written as

$$J_e = \sigma E' - \nu_e(J_e \times B) , \tag{158}$$

where the Hall current density is $-\nu_e(J_e \times B)$.

The Hall angle θ between vectors E' and $E'' = J_e/\sigma$ is determined by

$$\tan\theta = \frac{v_e B\sigma}{J_e} = \nu_e B . \tag{159}$$

With the choice of coordinate system as in Figure 6 the respective components of the electric current density are:

$$J_{e\alpha} = \sigma E'_\alpha - \nu_e J_{e\beta} B , \tag{160}$$

$$J_{e\beta} = \sigma E'_\beta + \nu_e J_{e\alpha} B , \tag{161}$$

$$J_{e\gamma} = \sigma E'_\gamma . \tag{162}$$

If the coordinate system is chosen so that the electric vector is in the $x_\alpha x_\beta$-plane and if the Hall effect coefficient

$$\beta_e = \nu_e B \tag{163}$$

is introduced, equations (160) and (161) transform into

$$J_{e\alpha} = \frac{\sigma(E'_\alpha - \beta_e E'_\beta)}{1 + \beta_e^2} , \tag{164}$$

$$J_{e\beta} = \frac{\sigma(E'_{\beta} + \beta_e E'_{\alpha})}{1 + \beta_e^2} . \qquad (165)$$

The conductivity tensor $\boldsymbol{\sigma}$ of ionized gas in a electromagnetic field then is of the form

$$\boldsymbol{\sigma} = \begin{vmatrix} \frac{\sigma}{1 + \beta_e^2} & -\frac{\sigma\beta_e}{1 + \beta_e^2} & 0 \\ \frac{\sigma\beta_e}{1 + \beta_e^2} & \frac{\sigma}{1 + \beta_e^2} & 0 \\ 0 & 0 & 0 \end{vmatrix} . \qquad (166)$$

MAGNETOHYDRODYNAMIC GENERATORS

Three main types of MHD generators are considered below, with the Hall effect being taken into account.

Continuous-electrode MHD Generator (Figure 5). As a result of the Hall effect the flow of electrons is deflected from the x_{β}-direction perpendicular to the electrode surface. In this arrangement the components of the electric field are $E_{\alpha} = 0$, $E'_{\alpha} = 0$, and $E'_{\beta} = E_{\beta} - vB$. If

$$\alpha = \frac{E_{\beta}}{vB} \qquad (167)$$

denotes the ratio of the strength of the acting electric field to the magnetic induction for the open circuit, the components of the electric current density can be written as [equations (164) and (165)]

$$J_{e\alpha} = \frac{\sigma vB\beta_e(1 - \alpha)}{1 + \beta_e^2} , \qquad (168)$$

$$J_{e\beta} = -\frac{\sigma vB(1 - \alpha)}{1 + \beta_e^2} . \qquad (169)$$

The power generated per unit volume of plasma in this case amounts to

$$P = -E_\beta J_{e_\beta} = \frac{\sigma v^2 B^2 \alpha(1 - \alpha)}{1 + \beta_e^2} . \quad (170)$$

The ratio of the power conveyed to the electrodes per unit volume of plasma to the power used to ensure plasma flow is the electrical efficiency of the generator. In the case in question the power supplied is

$$-J_{e\beta} vB = \frac{\sigma v^2 B^2 (1 - \alpha)}{1 + \beta_e^2} . \quad (171)$$

Hence the electrical efficiency of the continuous-electrode generator is

$$\eta_e = \frac{E_\beta J_{e\beta}}{J_{e\beta} vB} = \alpha . \quad (172)$$

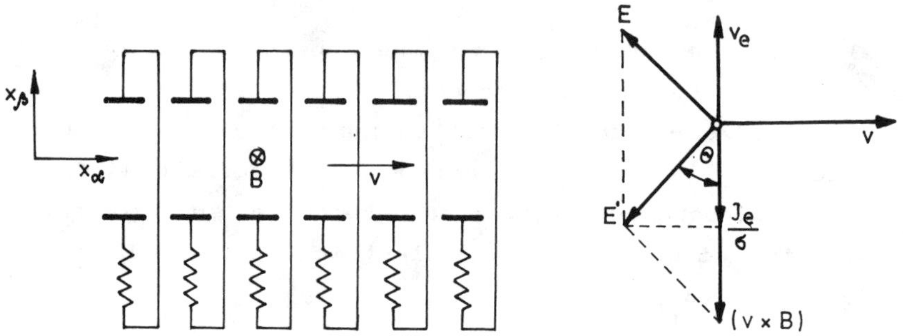

Figure 7. Segmented-electrode MHD generator.

Segmented-electrode MHD Generator (Figure 7). The undesirable influence that the Hall effect has on the generation of electricity can be eliminated by segmenting the electrodes, in which case the components of the electric field strength are

$$E'_\alpha = \beta_e E'_\beta , \qquad E_\alpha = \beta_e \left[E_\beta - vB\right] , \quad (173)$$

and the components of the electric current density are

$$J_{e\alpha} = 0 , \qquad J_{e\beta} = -\sigma\left[E_\beta - vB\right] = \sigma vB(1 - \alpha) \quad (174)$$

The power generated per unit volume of plasma in this case amounts to

$$P = \sigma v^2 B^2 \alpha(1 - \alpha) \ , \tag{175}$$

which corresponds fully to the previous case when the Hall effect is neglected [equation (171) with Hall coefficient $\beta_e = 0$]. The electrical efficiency of the generator is the same as in the previous case, that is to say, is given by equation (172).

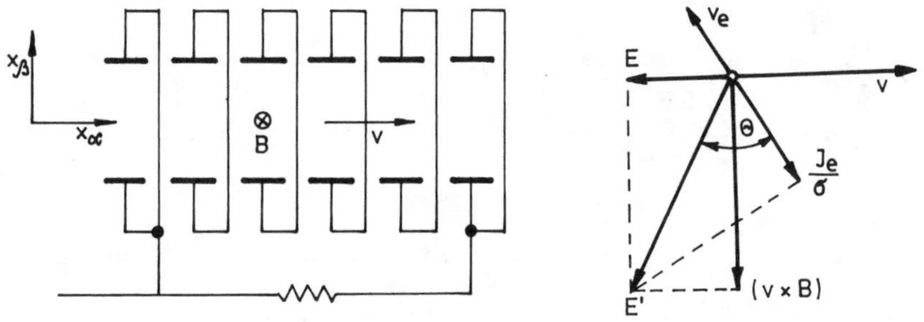

Figure 8. Hall generator.

Hall Generator (Figure 8). Instead of eliminating its influence on the generation of electricity, one can use the Hall effect to perform work in an electrical circuit in the manner shown in the drawing. In this case the respective components of the electric field strength are

$$E_\beta = 0 \ , \qquad E'_\alpha = E_\alpha \ , \qquad E'_\beta = - vB \ , \tag{176}$$

whereas the components of the electric current density are

$$J_{e\alpha} = \frac{\sigma(E_\alpha + \beta_e vB)}{1 + \beta_e^2} \ , \tag{177}$$

$$J_{e\beta} = \frac{\sigma(\beta_e E_\alpha - vB)}{1 + \beta_e^2} \ . \tag{178}$$

For an open electrical circuit, since $J_{e\alpha} = 0$,

$$E_{\alpha o} = -\beta_e vB \ , \qquad J_{e\beta o} = -\sigma vB \ , \tag{179}$$

and hence the parameter α specifying the ratio of the electric field strength in the x_α-direction with the circuit closed to that with the circuit open is

$$\alpha = \frac{E_\alpha}{E_{\alpha o}} = -\frac{E_\alpha}{\beta_e vB} \ . \tag{180}$$

The power generated per unit volume of plasma is now equal to

$$P = \frac{\sigma v^2 B^2 \alpha \beta_e^2 (1 - \alpha)}{1 + \beta_e^2} \ . \tag{181}$$

Now, the power consumed is

$$-J_{e_\beta} vB = \frac{\sigma v^2 B^2 (1 + \alpha\beta_e^2)}{1 + \beta_e^2} \ , \tag{182}$$

whereby the electrical efficiency of the Hall generator is

$$\eta_e = \frac{\alpha\beta_e^2 (1 - \alpha)}{1 + \alpha\beta_e^2} \ . \tag{183}$$

REFERENCES

Chen F.F., 1984. Introduction to Plasma Physics and Controlled Fusion. Plenum Press, New York, London.

Finkelnburg W. and Maecker H., 1965. Elektrische Bögen und Termisches Plasma, in: Hanbuch der Physik. Springer-Verlag, Berlin.

Hatsopoulos G.N. and Keenan J.H., 1965. Principles of General Thermodynamics. John Wiley and Sons, Inc., New York.

Soo S.L., 1968. Direct Energy Conversion. Prentice-Hall, Inc., Englewood Cliffs, New York.

Thermodynamic Analysis of Thermo-electric Effects in a Metal–Metal Couple

A. G. Guy
University of Florida
Gainesville, FL 32611, USA

ABSTRACT
The phenomenological equations of nonequilibrium thermodynamics are employed to treat the thermo-electric coefficients. The correct form of these equations is discussed, and the Seebeck coefficient (absolute thermopower) of a single isolated conductor is shown to be the logical choice for the basic coefficient because it is determined by the enthalpy of the electronic carriers. A thermodynamic analysis also furnishes a quantitative description of the equilibrium condition in a metal-metal couple, such as a gold-silver couple. The excess electrons that enter the gold have their maximum concentration at the interface. The concentration decreases exponentially with distance from the interface, dropping to 1/e of the maximum value at a depth of 1 nm. The internal electric field behaves similarly. No appreciable energy barrier exists at a metal-metal interface since the electrochemical potentials in the two metals are equal. The contact potential has no influence on the thermo-electric behavior of a metal-metal couple; consequently, the Seebeck and Peltier coefficients for a couple are related in a simple way to the corresponding coefficients for the single metals.

1. INTRODUCTION

The concept of thermodynamic electrochemical potential is widely used, both by chemists and by physicists. Nevertheless, in several important instances this useful concept has not yet been effectively utilized. An example is the junction zone between two metals, where a contact potential V_c occurs in the form of an extremely steep gradient in electrical potential. Physicists have pictured this potential as a barrrier to the motion of electrons, and they have explained the absence of any unusual effect on electrical conductivity as being the result of quantum-mechanical tunneling (Pollack and Seitchik 1969). The concept of electrochemical potential can be used to show, however, that no unusual barrier exists at a junction, and therefore no special mechanism is needed to explain electrical conduction across the junction.

Misconceptions concerning metal-metal junctions often have their origins in the erroneous use of the chemical potential of electrons, μ, instead of the electrochemical potential, $\bar{\mu}$,

$$\bar{\mu} = \mu + qV \tag{1.1}$$

where q is the charge on an electron and V is the electrical potential. For example, Dugdale (1977) states that the condition for equilibrium of electrons when metals 1 and 2 are in contact is $\mu_1 = \mu_2$: the actual condition is $\bar{\mu}_1 = \bar{\mu}_2$. Similarly, Kireev (1978) argues that if prior to contact $\mu_1' > \mu_2'$, then upon contact some electrons transfer from metal 1 to metal 2 until $\mu_1 = \bar{\mu}_2$. This argument omits the paramount effect, the potential V generated in the process of electron transfer.

The electrochemical potential is useful for describing thermo-electric effects in a single metal as well as in a couple composed of two different metals. For this reason, Section 2 briefly treats the behavior of a single metal in a temperature gradient, dT/dx. Section 3 then describes the equilibrium condition in a metal-metal couple ($dT/dx = 0$), and Section 4 considers the effect of a temperature gradient. For simplicity the analyses are limited to the usual case of chemically homogeneous metals or alloys.

2. THERMO-ELECTRIC EFFECTS IN A SINGLE METAL

Nonequilibrium thermodynamics (de Groot and Mazur 1962) shows that thermo-electric effects can be treated conveniently using the following phenomenological equations for the electric current density, I, and the reduced heat flux, J',

$$I = \frac{eL_{ee}}{T}\left(\frac{d\bar{\mu}}{dx}\right)_T + \frac{eL_{eq}}{T^2}\frac{dT}{dx} \tag{2.1}$$

$$J' = J + \frac{yIh}{e} = -\frac{L_{qe}}{T}\left(\frac{d\bar{\mu}}{dx}\right)_T - \frac{L_{qq}}{T^2}\frac{dT}{dx} \tag{2.2}$$

Here e is the absolute value of the electronic charge, T is temperature in kelvin, J is the flux of heat determined experimentally, y is the concentration of electrons expressed as the ratio of conduction electrons to atoms, h is the partial enthalpy of an electron, and L_{ee}, L_{qe}, L_{eq} and L_{qq} are phenomenological coefficients. Expressions for L_{ee} and L_{qq} are easily derived (Callen 1985) and are found to be,

$$\frac{L_{ee}}{T} = \frac{\sigma}{e^2} \qquad\qquad \frac{L_{qq}}{T^2} = \lambda \tag{2.3}$$

where σ is the electrical conductivity and λ is the thermal conductivity. The other two phenomenological coefficients are connected by an Onsager relation,

$$L_{qe} = L_{eq} \tag{2.4}$$

Many previous treatments of thermo-electric phenomena, including those of Callen (1985) and of Domenicali (1954), have employed the total gradient $d\bar{\mu}/dx$ rather than the gradient at constant temperature, $(d\bar{\mu}/dx)_T$. This procedure fails to identify the enthalpy as the source of thermopower, a fact demonstrated by Eq. (2.13) below.

For an analysis of thermo-electricity, Eqs. (2.1) and (2.2) have a simpler form because the electric current, I, is zero;

$$0 = \frac{L_{ee}}{T}\left[\left(\frac{d\mu}{dx}\right)_T - e\frac{dV_g}{dx}\right] + \frac{L_{eq}}{T^2}\frac{dT}{dx} \tag{2.5}$$

$$J_q = -\frac{L_{qe}}{T}\left[\left(\frac{d\mu}{dx}\right)_T - e\frac{dV_g}{dx}\right] - \frac{L_{qq}}{T^2}\frac{dT}{dx} \tag{2.6}$$

Two sources of the internal electrical potential, V, are usefully distinguished; $V = V_g + V_a$. V_g is a portion generated by displacement of electrons; see Eq. (2.8) below, and V_a is the usual internal potential attributable to an externally applied electrical source. Since no applied voltage is involved in the simplest case of thermo-electricity, only V_g appears in Eqs. (2.5) and (2.6).

If the Onsager relation, Eq. (2.4), is used to eliminate L_{eq} and L_{qe} from Eqs. (2.5) and (2.6), the result is,

$$J_q = \frac{L_{ee}\left[\left(\frac{d\mu}{dx}\right)_T - e\frac{dV_g}{dx}\right]^2}{\frac{dx}{dT}} - \frac{L_{qq}}{T^2}\frac{dT}{dx} \tag{2.7}$$

The appropriate expression for $(d\mu/dx)_T - e(dV_g/dx)$ is obtained as follows from Eq. (2.5). The present case of a single conductor differs significantly from that of two joined conductors (a thermocouple). Whereas an electric current can be produced in a thermocouple circuit, a temperature gradient cannot cause an electric current at steady-state in a single uniform conductor even if the conductor is in the form of a continuous loop. At each point in the conductor the local temperature gradient, dT/dx, results in two influences that tend to displace the electrons: (1) a change $d\mu^o/dT$ of the chemical potential; see Eq. (3.1): and (2) the "dragging" action of the heat flux, J_q, on the electrons. The displaced electrons represent "excess charge" in the sense of Poisson's equation and generate an electric field, $E_g = -dV_g/dx$, that prevents any net flow of electric current at the steady state. In a metal $(d\mu/dx)_T$ is negligibly small; therefore,

$$\left(\frac{d\mu}{dx}\right)_T - e\frac{dV_g}{dx} \simeq - e\frac{dV_g}{dx} = - e\frac{dV_g}{dT}\frac{dT}{dx} = - e\varepsilon\frac{dT}{dx} \tag{2.8}$$

where $\varepsilon = dV_g/dT$ is the absolute thermopower. Substitution of Eq. (2.8) and the two relations of Eq. (2.3) in Eq. (2.7) gives the following useful expression,

$$J_q = - (\lambda - \sigma T\varepsilon^2)\frac{dT}{dx} \tag{2.9}$$

Thus, a straight-forward elimination of the cross coefficients yields the accepted phenomenological equation needed for an adequate analysis of thermal conductivity in a chemically uniform conductor (Guy 1977). Comparison of Eqs. (2.6) and (2.9) shows that,

$$L_{qe} = \frac{\sigma T^2\varepsilon}{e} \tag{2.10}$$

A further step in the treatment of thermo-electricity is the prediction of L_{qe} in terms of a thermodynamic property of the electrons (Guy 1983a); namely, their partial enthalpy, h. When Eq. (2.2) is evaluated for ordinary (isothermal) electrical conduction it takes the form,

$$J'_q = 0 + \frac{yIh}{e} = e\frac{L_{qe}}{T}\frac{dV_a}{dx} \tag{2.11}$$

When the expression for I from Eq. (2.1) for these conditions is substituted in Eq. (2.11) the result is,

$$L_{qe} = L_{eq} = - yhL_{ee} = - \frac{\sigma Th}{e^2} \tag{2.12}$$

Elimination of L_{qe} between Eqs. (2.10) and (2.12) gives,

$$h = - \frac{eT\varepsilon}{y} \tag{2.13}$$

This relation evaluates h from experimental data on absolute thermopower, ε, and therefore is a first step toward a phenomenological thermodynamic calculation of the behavior of electronic devices (Guy 1983b). The relation of Eq. (2.13) can also be used to demonstrate, by means of Eqs. (2.1) and (2.2), that Eq. (2.9) remains valid even if an electric current, I, flows during a measurement of thermal conductivity (Guy 1983c).

Recognition of the relation of ε to h, Eq. (2.13), makes possible the following simple derivations of the Peltier coefficient, π, and of the Thomson coefficient, τ, in terms of the Seebeck coefficient, ε. The Peltier coefficient is defined in terms of the flux of heat, J', accompanying an isothermal flow of electric current,

$$J' = -\pi I \qquad (2.14)$$

The source of J' is the enthalpy, h, of the electrons that constitute the electric current. This fact can be shown analytically by use of Eq. (2.2) since J is zero for an isothermal process;

$$J' = \frac{yIh}{e} = -\pi I = -IT\varepsilon \qquad (2.15)$$

The final expression is obtained by the use of Eq. (2.13). Thus, the relation between the Peltier and Seebeck coefficients is,

$$\pi = T\varepsilon \qquad (2.16)$$

This is the accepted relation (Callen 1985), but the present simple derivation is new.

The Thomson coefficient τ is defined in terms of the heat absorbed, $\delta q'$, when an electronic unit of positive charge, e, flows from a region in which the temperature is T to a region in which the temperature is T + dT,

$$\delta q' = \tau\, dT \qquad (2.17)$$

The following analysis of the flow of electric current is independent of the flow of heat caused by the temperature gradient. The explanation is similar to that given above following Eq. (2.13). Consider a steady flow of a very small electric current (composed of electrons) in a temperature gradient. Since the electrons are now entering the thermodynamic system with energy -eV and leaving with energy -e(V+dV), the first law of thermodynamics must now be in the form for use with electrical systems and take account of the resulting change in energy, edV. No pressure work is done on the system; therefore the first law of thermodynamics requires that an addition of heat δq furnish the total additional energy dE,

$$\delta q = dE = \frac{\partial h}{\partial T}\, dT + e\,\frac{dV}{dT}\, dT \qquad (2.18)$$

The first term on the right is the change in enthalpy per electron, and the second term is the change in electrical energy.

For a monovalent metal such as gold or silver, $y=1$ and therefore h in Eq. (2.13) becomes $h=-eT\varepsilon$. Consequently,

$$\frac{\partial h}{\partial T} = -eT\,\frac{d\varepsilon}{dT} - e\varepsilon \qquad (2.19)$$

Since the second term on the right in Eq. (2.18) can be written in terms of $\varepsilon = dV/dT$, the final form of Eq. (2.18) is,

$$\delta q = -eT\,\frac{d\varepsilon}{dT}\, dT = -\tau\, dT \qquad (2.20)$$

The minus sign, compared to Eq. (2.17), occurs because $\delta q'$ is defined for <u>positive</u> charge. Equation (2.20) leads to the accepted relation,

$$\tau = eT\,\frac{d\varepsilon}{dT} \qquad (2.21)$$

between the Thomson and Seebeck coefficients, but again the method of derivation is new.

Although it has been recognized that each of the three thermo-electric coefficients describes a behavior of a single metal, previous derivations (Callen 1985) have required the use of a two-metal couple to obtain the relations of Eqs. (2.14) and (2.21). The present derivation, employing only the metal in question, is possible because of the quantitative expression for the enthalpy of a conduction electron, Eq. (2.13).

3. EQUILIBRIUM IN AN ISOTHERMAL COUPLE

The major applications of thermo-electric effects employ two conductors (metals, semiconductors, etc.) joined to form a couple and subjected to a temperature gradient. The relatively simple case of a metal-metal thermocouple is considered below in Section 4. A useful preliminary, however, is the present consideration of the condition of equilibrium of the electrons in a metal-metal couple in the absence of a temperature gradient.

The chemical potential, μ, that enters in Eq. (1.1) has the general definition,

$$\mu = \mu^{o} + kT \ln (fy) \tag{3.1}$$

where μ^{o} is a convenient reference value, k is Boltzmann's constant, T is temperature in kelvin, f is an activity coefficient and y is the fractional concentration of electrons. A thermodynamic analysis of equilibrium in a planar metal-metal couple can be made with the use of Eqs. (1.1) and (3.1) and the standard methods of electrostatics. The details have been presented elsewhere (Guy 1983d), but the quantitative values will be given here for the case of a metal-metal couple composed of a sheet of gold joined to a similar sheet of silver to form a specimen 2 mm in thickness. The variations in properties across the thickness direction (x-direction) will be considered.

As shown by data on the work functions of silver and of gold, the chemical potential of the electrons is higher in silver than in gold. Consequently, when the two metals are first joined, electrons transfer from silver to gold. The quantity of charge transferred, σ, per m^2 of sheet area is about $\sigma = 10^{-7}$ C/m^2, corresponding to the fraction 10^{-14} of the conduction electrons in the sheet of silver. The accompanying excess charges (positive in silver, negative in gold) lead to electrical fields and eventually to an abrupt change in electrical potential (the contact potential, V_c) at the interface between the two metals.

The following quantitative expression is obtained for the variation of excess charge density, ρ, in C/m^3 as a function of distance, x, across a member of the couple. In the case of the gold sheet,

$$\rho = - \frac{a\sigma[\exp (ax) + \exp (-ax)]}{[\exp (ad) - \exp (-ad)]} \tag{3.2}$$

where $x = 0$ at the free surface and $x = d$ at the interface. The numerical value of the constant for typical metals is about $a = 10^{9}$ m^{-1}. The value of ρ is a maximum at the interface, 10^{2} C/m^3. This value decreases exponentially with distance into the gold and falls to 1/e of the maximum at a distance $1/a = 1$ nm below the interface. A derivation based on Eq. (1.1) shows that the internal electric field, E, is proportional to $d\rho/dx$. The maximum value, $E = 0.33$ V/m, exists at the interface. Like the charge density, the electric field decreases extremely rapidly with distance from the interface. The quantitative distributions of excess charge and electric field given by the thermodynamic analysis are roughly compatible with the current ideas of "a surface layer of charge" and "zero field in the interior". The electrical potential is essentially constant in each metal; the contact potential $V_c = 0.69$V is divided equally between the two metals, giving values of + 0.69/2 in the silver and - 0.69/2 in the gold.

4. THERMO-ELECTRIC EFFECTS IN A METAL-METAL COUPLE

The usual treatment of thermo-electric effects considers, not a single metal as was done in Section 2, but a pair of metals joined to form a couple. In view of the results presented in Section 3, a metal that is part of a couple differs from the same isolated metal in two respects: (1) the average fractional concentration of conduction electrons has changed by the negligibly small amount, 10^{-14}; and (2) the electrical potential is different by a fraction of a volt. Neither of these differences is significant for analyses based on Eqs. (2.1) and (2.2) because only changes in $\bar{\mu}$ are considered, not the absolute value of $\bar{\mu}$. A feature of a metal-metal couple that would appear to have a bearing on thermo-electric phenomena is the jump in electrical potential, V_C, that occurs at the interface. In fact, the contact potential, V_C, depends only on the chemical potentials of the two metals in contact and is essentially independent of the temperature gradient or of the existence of an electric current. Rather then producing a flow of electrons, V_c equalizes $\bar{\mu}_1$ and $\bar{\mu}_2$ and thus prevents the occurrence of an electric current.

The nature of the Seebeck effect in a metal-metal couple can be illustrated by a comparison with the corresponding behavior of a single metal. Consider two equal lengths of metal 1 joined at each end to produce two junctions, one to be heated to temperature T_H and the other held at a lower temperature, T_C. The total difference in potential between the two junctions is given by,

$$\Delta V_1 = \int_{T_C}^{T_H} \varepsilon_1 \, dT \tag{4.1}$$

where ε_1 is the Seebeck coefficient of metal 1. Because the same difference of potential exists in both wires, no flow of current occurs. If, however, the two wires are different metals, 1 and 2, the net difference of potential measured (by potentiometer) at the colder juntion is,

$$\Delta V_{12} = - (\Delta V_1 - \Delta V_2) = \int_{T_C}^{T_H} (\varepsilon_2 - \varepsilon_1) \, dT \tag{4.2}$$

The quantity $(\varepsilon_2 - \varepsilon_1)$ is often written ε_{12} and referred to as the Seebeck coefficient of the pair of metals 1-2.

Similarly, there is no generation of "Peltier heat" at a junction between two wires of metal 1 because the same flux of heat, J_1' given by Eq. (2.14), flows into the junction and out of the junction. At a junction between metals 1 and 2, however, the rate of absorption of heat, dQ/dt, per unit volume is,

$$- (J_1' - J_2') = (\pi_2 - \pi_1) \, I \tag{4.3}$$

The quantity $(\pi_2 - \pi_1)$ is often written π_{12} and referred to as the Peltier coefficient of the junction between metals 1 and 2.

Thus, the analysis in this section demonstrates that the thermo-electric phenomena in a metal-metal couple (thermocouple) are explained completely by the thermo-electric behavior of an isolated metal. Effects associated with the junction, such as contact potential, do not influence thermo-electric behavior. The basic thermo-electric effect remains the Seebeck coefficient, ε, since the Thomson coefficient π is given by $\pi = T\varepsilon$, Eq. (2.16).

5. CONCLUSIONS

In view of the fact that Lord Kelvin used thermodynamics to elucidate thermo-electric effects 150 years ago, is there anything new to be reported on this subject? The answer is definitely "yes", as shown by the following summary of significant advances.

1. An analysis based on nonequilibrium thermodynamics has shown that the Seebeck coefficient, ε, is determined by the partial enthalpy, h, of the conduction electrons.

2. The Seebeck coefficient should, for this reason, be treated as the basic one of the three thermo-electric coefficients.

3. Because of the relation in (1), the Kelvin relations among the thermo-electric coefficients can be derived for a single conductor.

4. In an isothermal metal-metal couple the distributions of excess charge and internal electric field can be described quantitatively.

5. The results in (4) permit a detailed analysis of the behavior of a metal-metal couple in a temperature gradient (thermocouple).

6. The internal electric field described in (4) gives a new insight into the nature of the contact potential in a metal-metal couple.

REFERENCES

Callen, H.B. 1985. Thermodynamics and an Introduction to Thermostatistics, 2nd ed. 323. Wiley, New York.

de Groot, S.R. and Mazur, P. 1962. Nonequilibrium Thermodynamics, 33. North Holland, Amsterdam.

Domenicalli, C.A. 1954. Reviews of Modern Physics 26: 237.

Dugdale, J.S. 1977. The Electrical Properties of Metals and Alloys, 20. Edward Arnold, London.

Guy, A.G. 1977. Proceedings of Conference on Thermophysical Properties 7: 105-108.

Guy, A.G. 1983a. Journal of Nonequilibrium Thermodynamics 8: 119-126.

Guy, A.G. 1983b. Solid-State Electronics 26: 433-436.

Guy, A.G. 1983c. Effect of Simultaneous Electric Current on Thermal Conductivity. In Thermal Conductivity 16, ed. D.C. Larson, 405-408. Plenum, New York.

Guy, A.G. 1983d. Applied Physics Communications 2; 131-142.

Kireev, P.S. 1978. Semiconductor Physics, 516. Mir, Moscow.

Pollack, S.R. and Seitchik, J.A. 1969. Applied Solid-State Science 1: 375.

APPENDIX

This appendix has been added because a referee has raised the following interesting point in connection with Eq. (2.11). In general a partial molar quantity (such as h) is not an absolute quantity but depends on several factors, including the reference state. A discussion of this topic can be found for example on page 102 in K. Denbigh's "Principles of Chemical Equilibrium", 3rd. ed., Cambridge, 1971. In the present case of pure metals there is only one evident choice of reference state. Consequently, the value of h is unambiguous.

A related topic is the manner in which the partial enthalpy, h, enters into the "reduced heat flux", J', Eq. (2.2). The explanation is given on page 26 of the reference by de Groot and Mazur listed above.

Nonequilibrium Thermodynamics of Sufaces

D. Bedeaux
Department of Physical and Macromolecular Chemistry
Gorlaeus Laboratories, University of Leiden
P.O. Box 9502, 2300 RA Leiden, The Netherlands

ABSTRACT

The theory of nonequilibrium thermodynamics is formulated for a multicomponent system consisting of two bulk phases which are separated by an interfacial region. The excesses of the mass, momentum, energy and entropy densities and of their respective currents are described as singular contributions on the dividing surface to the density and current distributions. Both the position and the local curvature of the interface may depend on the time. Chemical reactions and mass transport through and into the interfacial area are possible. An explicit expression for the excess entropy production is derived. This makes it possible to give a complete list of all the forces and fluxes which may play a role in the description of the dynamical processes near and in the interface. Onsager relations are formulated.

1. INTRODUCTION

In this paper we will discuss the application of the theory of nonequilibrium thermodynamics, containing both the Onsager relations and an explicit expression for the entropy production, to boundary layers between two bulk phases. For a discussion of the general field of nonequilibrium thermodynamics and in particular its application to bulk phases we refer to the book on this subject by De Groot and Mazur [De Groot 1962]. The equilibrium thermodynamics of a socalled surface of discontinuity were already extensively discussed by Gibbs [Gibbs 1906]. A general method for the application of nonequilibrium thermodynamics to surfaces of discontinuity consistent with the equilibrium theory for surface thermodynamics formulated by Gibbs was given by Bedeaux, Albano and Mazur [Bedeaux 1976].

A major difficulty in such an analysis is that the surface of discontinuity may not only move through space but also has a time–dependent curvature. This makes it necessary to use time–dependent orthogonal curvilinear coordinates for the more difficult parts of the analysis. In section 2 we will discuss the mathematical description and the use of time–dependent curvilinear coordinates in this context in more detail.

The theory gives an explicit expression for the entropy production at the surface. Using this expression one may identify the appropriate forces and fluxes. Onsager relations for the linear constitutive coefficients relating these forces and fluxes may then be given. The expressions for the fluxes through the surface of discontinuity from one bulk phase to the other lead to what are usually called boundary conditions. Examples of the linear constitutive coefficients contained in these boundary conditions are for instance the slip coefficient and the temperature jump coefficient. Some fluxes give the flow from the bulk phases into the surface of discontinuity. An example is the flow of surface active material from the bulk to the interface. Finally some fluxes describe the flow along the interface. They are most similar to the usual fluxes in the bulk except for being two dimensional in the comoving reference frame. As we shall discuss, the entropy production at the interface contains the value of the fluxes into and through the interface in the comoving frame of reference. The fact that these fluxes should be calculated in a frame of reference in which the surface is at rest gave rise to considerable discussion among workers in the field of membrane transport [Mikulecky 1966].

An important property of the surface is that as a consequence of its existence the translational symmetry of the system is broken. As a consequence the tensorial nature of the fluxes and forces at the interface appearing in the interfacial entropy production differs from the corresponding behaviour in the bulk phases. At the interface there is still translational symmetry along the surface and rotational symmetry around the normal on the surface. As a consequence the force–flux pairs contributing to the interfacial entropy production are 2×2 tensors, 2–dimensional vectors and scalars. In the bulk regions one has 3×3 tensors, 3–dimensional vectors and scalars. At the interface the normal component of a 3–dimensional current, as for instance the asymptotic values of the bulk heat currents at the dividing surface, appear as a scalar fluxes in the interfacial entropy production while the parallel parts appear as 2–dimensional vectors. Similarly a 3–dimensional tensor should be split up in a 2×2 tensor, two 2–dimensional vectors and a scalar. As a result the number of force–flux pairs in the interfacial entropy production is considerably larger than the corresponding number in the entropy production in the bulk. In view of Curie's symmetry principle, fluxes only depend on forces of the same tensorial nature. Even though this reduces the number of constitutive coefficients considerably the number of linear constitutive coefficients describing the behaviour of the system at the interface is still considerably larger than the number needed in the bulk regions. At the interface many processes may couple which do not couple to one another in the bulk. This is the reason why phenomena like active transport for instance may occur at surfaces of discontinuity.

The original paper by Bedeaux, Albano and Mazur [Bedeaux 1976] considered the special case of an interface between two immiscible one–component fluids. A crucial element in their analysis was the description of the excess contributions to the densities and currents using δ–functions in the corresponding density and flow fields at the socalled dividing surface. Why the use of such δ–function singularities is mathematically sound and the proper definition of the excess densities and currents as a function of the position along the interface is discussed by Albano, Bedeaux and Vlieger [Albano 1979]. Subsequent work extended the original analysis to multi–component systems with mass transport through the interface[Kovac 1977; 1981] and to systems in which electromagnetic effects play a role [Wolf 1979; Albano 1980; 1987]. A review of much of this work and on some of the work on the description of the fluctuations of the interface has been given by Bedeaux [Bedeaux 1986].

In this paper we will restict ourselves to the present state of the nonequilibrium thermodynamic description of interfaces. We will consider multicomponent systems in which chemical reactions may take place in the bulk phases as well as on the surface of discontinuity. For a discussion of the influence of electromagnetic fields we refer to [Albano 1987]. The emphasis will be on interfaces between two fluid phases but much of the analysis is also applicable if one or both bulk phases are solids. After a discusion of mathematical tools for the analysis in section 2 we will proceed to discuss conservation laws in section 3 – 5, the entropy balance in section 6 and the resulting linear laws in section 7. While the formalism discussed in these sections describes all these properties both in the two bulk phases as well as at the dividing surface, the emphasis in the presentation will be on these properties for the interface. Onsager relations for the interfacial constitutive coefficients will be given in section 7. In the last section we shortly discuss what to do with the very large number of constitutive coefficients which appear in the linear laws. In particular we indicate how one may simplify the description in various cases. We will not elaborate on the molecular foundation of the application of the method of nonequilibrium thermodynamics to interfaces discussed in this paper. As usual such a molecular foundation is easy to discuss in general terms but difficult to make mathematically precise [Bedeaux 1986; Ronis 1978].

2. THE MATHEMATICAL DESCRIPTION OF INTERFACES

In order to give the time–dependent location of the dividing surface, it is convenient to introduce a set of time dependent orthogonal curvilinear coordinates [Morse 1953]: $\xi_i(\vec{r},t)$, $i = 1, 2, 3$, where $\vec{r} = (x, y, z)$ are the Cartesian coordinates and t the time. These curvilinear coordinates are chosen in such a way that the time–dependent location of the dividing surface is given by

$$\xi_1(\vec{r},t) = 0 \tag{2.1}$$

The dynamical properties of the system are described using balance equations; for instance, balance equations for the mass densities of the various components. Consider as an example the balance equation for the density per unit of volume of an as yet unspecified variable $d(\vec{r},t)$

$$\frac{\partial}{\partial t} d(\vec{r},t) + \text{div}\, \vec{J}_d(\vec{r},t) = \sigma_d(\vec{r},t) \tag{2.2}$$

Here $\vec{J}_d(\vec{r},t)$ is the current of the variable d and $\sigma_d(\vec{r},t)$ the production of d in the system, both at the position $\vec{r}$ and the time t. In our description of interfaces d, $\vec{J}_d$ and σ_d vary continuously in the regions occupied by the bulk phases. At the dividing surface, however, these fields contain singular contributions equal to the properly defined [Albano 1979] excess of these fields near the surface of discontinuity. Thus these fields have the following form

$$d(\vec{r},t) = d^-(\vec{r},t)\, \theta^-(\vec{r},t) + d^s(\vec{r},t)\, \delta^s(\vec{r},t) + d^+(\vec{r},t)\, \theta^+(\vec{r},t) \tag{2.3}$$

$$\vec{J}_d(\vec{r},t) = \vec{J}_d^-(\vec{r},t)\, \theta^-(\vec{r},t) + \vec{J}_d^s(\vec{r},t)\, \delta^s(\vec{r},t) + \vec{J}_d^+(\vec{r},t)\, \theta^+(\vec{r},t) \tag{2.4}$$

$$\sigma_d(\vec{r},t) = \sigma_d^-(\vec{r},t)\, \theta^-(\vec{r},t) + \sigma_d^s(\vec{r},t)\, \delta^s(\vec{r},t) + \sigma_d^+(\vec{r},t)\, \theta^+(\vec{r},t) \tag{2.5}$$

In these expressions θ^- and θ^+ are the time–dependent characteristic functions of the two bulk phases; 1 in one phase and 0 in the other. Using the time–dependent curvilinear coordinates these characteristic functions may be written as

$$\theta^{\pm}(\vec{r},t) \equiv \theta(\pm\xi_1(\vec{r},t)) \tag{2.6}$$

Here $\theta(s)$ is the Heavyside function; 1 if s is positive and 0 if s is negative. Furthermore, $\delta^s(\vec{r},t)$ is a δ–function on the dividing surface which is defined in terms of the curvilinear coordinates as

$$\delta^s(\vec{r},t) \equiv |\text{grad}\, \xi_1(\vec{r},t)|\, \delta(\xi_1(\vec{r},t)) \tag{2.7}$$

The characteristic functions for the bulk phases restrict an integration over the volume to the regions occupied by the corresponding phase. Similarly $\delta^s(\vec{r},t)$ restricts an integration over the volume to a two–dimensional integral over the dividing surface.

It is clear from the above expressions for the fields that d^s, $\vec{J}^s_d$ and σ^s_d depend on the position along the dividing surface alone; thus one has for instance

$$d^s(\vec{r},t) = d^s(\xi_2(\vec{r},t),\xi_3(\vec{r},t),t) \tag{2.8}$$

and similarly for $\vec{J}^s_d$ and σ^s_d. An important consequence of this is that the spatial derivatives of d^s, $\vec{J}^s_d$ and σ^s_d normal to the dividing surface are zero.

When one substitutes the expressions (2.3) – (2.5) into the balance equation (2.1), one obtains spatial and temporal derivatives of the characteristic functions $\theta^{\pm}$ and of δ^s. For the spatial derivative of the characteristic functions one has

$$\text{grad }\theta^{\pm}(\vec{r},t) = \vec{n}(\vec{r},t)\ \delta^s(\vec{r},t) \tag{2.9}$$

Here $\vec{n}$ is the normal on the dividing surface defined by

$$\vec{n}(\xi_2,\xi_3,t) \equiv \vec{a}_1(\xi_1 = 0,\xi_2,\xi_3,t) \tag{2.10}$$

where $\vec{a}_i$ is the unit vector in the direction of increasing ξ_i given by

$$\vec{a}_i(\vec{r},t) \equiv h_i(\vec{r},t)\ \text{grad }\xi_i(\vec{r},t) \quad \text{with} \quad h_i(\vec{r},t) \equiv |\text{grad }\xi_i(\vec{r},t)|^{-1} \tag{2.11}$$

These unit vectors which play an important role when one uses curvilinear coordinates are defined in each point in space and not only on the dividing surface like the normal $\vec{n}$. They form an orthonormal set for all $\vec{r}$ and t:

$$\vec{a}_i(\vec{r},t).\vec{a}_j(\vec{r},t) = \delta_{ij} \tag{2.12}$$

For the gradient of the surface δ–function one may show that [Bedeaux 1976]

$$\text{grad }\delta^s(\vec{r},t) = \vec{n}\ \vec{n}.\vec{\nabla}\ \delta^s(\vec{r},t) \tag{2.13}$$

where $\vec{\nabla} \equiv (\partial/\partial x, \partial/\partial y, \partial/\partial z)$ is the Cartesian gradient. An important aspect of both eq.(2.9) and eq.(2.13) is the fact that the gradient only has a component in the direction normal to the dividing surface. While this property is intuitively obvious, it nevertheless plays a crucial role.

Even though this has not been explicitly indicated in the above equation, the direction of the normal depends on the position along the interface and the time. In the equations below we will similarly not always explicitly indicate the dependence on position and time as this dependence is usually clear from the context.

In order to give expressions for the time derivatives we introduce the velocity field of the curvilinear coordinate system relative to the fixed Cartesian coordinates

$$\vec{w}(\xi_1, \xi_2, \xi_3, t) \equiv \frac{\partial}{\partial t}\vec{r}(\xi_1, \xi_2, \xi_3, t) \tag{2.14}$$

The more physical velocity field of the dividing surface is given in terms of the velocity field of the curvilinear coordinate system by

$$\vec{w}^s(\xi_2, \xi_3, t) = \vec{w}(\xi_1 = 0, \xi_2, \xi_3, t) \tag{2.15}$$

Using the velocity of the dividing surface, one may show that [Bedeaux 1976] the time derivative of the characteristic functions for the bulk phases is given by

$$\frac{\partial}{\partial t}\theta^{\pm}(\vec{r},t) = w_n^s(\vec{r},t)\,\delta^s(\vec{r},t) \tag{2.16}$$

Here the subscript n indicates the normal component. Similarly one may show that [Bedeaux 1976] the time derivative of the surface δ–function is given by

$$\frac{\partial}{\partial t}\delta^s(\vec{r},t) = -\,w_n^s(\vec{r},t)\,\vec{n}.\vec{\nabla}\,\delta^s(\vec{r},t) \tag{2.17}$$

It follows from the above equations that the time derivative of the characteristic functions and of the surface δ–function only contain the normal component of the surface velocity. The motion of the geometrical surface is in fact given in terms of the normal component of this velocity. The parallel component of the interfacial velocity should be related to material flow along the interface. This will become clear in its definition below.

The above formulae for the gradient and the time derivative of the characteristic functions and of the surface δ–function make it possible to substitute the explicit form of d, $\vec{J}_d$ and σ_d given in eqs.(2.3)–(2.4) into the balance equation (2.2) and one obtains, after some rearranging, the following more detailed formula for the balance equation

$$[\frac{\partial}{\partial t}d^-(\vec{r},t) + \mathrm{div}\,\vec{J}_d^-(\vec{r},t) - \sigma_d^-(\vec{r},t)]\,\theta^-(\vec{r},t)$$

$$+ [\frac{\partial}{\partial t} d^+(\vec{r},t) + \text{div}\, \vec{J}_d^+(\vec{r},t) - \sigma_d^+(\vec{r},t)]\, \theta^+(\vec{r},t)$$

$$+ [\frac{\partial}{\partial t} d^s(\vec{r},t) + \text{div}\, \vec{J}_d^s(\vec{r},t) - \sigma_d^s(\vec{r},t)$$

$$+ J_{d,n}^+(\vec{r},t) - J_{d,n}^-(\vec{r},t) - w_n^s(\vec{r},t)(d^+(\vec{r},t) - d^-(\vec{r},t))]\, \delta^s(\vec{r},t)$$

$$+ [J_{d,n}^s(\vec{r},t) - w_n^s(\vec{r},t) d^s(\vec{r},t)]\, \vec{n}.\vec{\nabla}\, \delta^s(\vec{r},t) = 0 \qquad (2.18)$$

This formula contains four different contributions, all of which are equal to zero. Thus, it follows from the first term that

$$\frac{\partial}{\partial t} d^-(\vec{r},t) + \text{div}\, \vec{J}_d^-(\vec{r},t) = \sigma_d^-(\vec{r},t) \qquad \text{for } \xi_1(\vec{r},t) < 0 \qquad (2.19)$$

and it follows from the second term that

$$\frac{\partial}{\partial t} d^+(\vec{r},t) + \text{div}\, \vec{J}_d^+(\vec{r},t) = \sigma_d^+(\vec{r},t) \qquad \text{for } \xi_1(\vec{r},t) > 0 \qquad (2.20)$$

The above two equations are the balance equations in the bulk regions and these equations have there usual form. As the above analysis shows these equations together with the restriction of the area where they are valid are contained in the general form of the balance equation (2.2) together with the explicit form of the fields given in eq.(2.3)–(2.5). The third term gives the balance equation for the interface

$$\frac{\partial}{\partial t} d^s(\vec{r},t) + \text{div}\, \vec{J}_d^s(\vec{r},t) + J_{d,n}^+(\vec{r},t) - J_{d,n}^-(\vec{r},t) - w_n^s(\vec{r},t)(d^+(\vec{r},t) - d^-(\vec{r},t)) = \sigma_d^s(\vec{r},t)$$

$$\text{for } \xi_1(\vec{r},t) > 0 \qquad (2.21)$$

In this equation the first two terms on the left hand side and the term on the right hand side are similar to the terms found in the bulk regions. The third term on the left hand side gives the flow of d leaving the surface of discontinuity into the + phase in the comoving frame. Similarly the fourth term gives the flow from the – phase into the surface of discontinuity in the comoving frame. Finally the fifth term in the above equation accounts for the rate of increase of d^s due to the motion of the dividing surface. The last term in eq.(2.18) gives

$$J_{d,n}^s(\vec{r},t) - w_n^s(\vec{r},t) d^s(\vec{r},t) = 0 \qquad (2.22)$$

This condition expresses the important fact that in a reference frame moving along with the interface the excess current $\vec{J}_d^s$ is pointed along the interface. While of course it is difficult to see how it could be any other way, it is nevertheless important to realize that this property is contained in the general form of the balance equation (2.2) together with the explicit form of d, $\vec{J}_d$ and σ_d given in eqs.(2.3)–(2.5).

If one would therefore contemplate describing a system in which the excess current is not along the interface in the comoving frame of reference one must necessarily modify eqs.(2.2)–(2.5). In this context one may in fact wonder why it is not necessary to add terms to d, $\vec{J}_d$ and σ_d which are proportional to normal derivatives of arbitrary order of the surface δ–function. In [Albano 1979] we discuss the underlying motivation for our choice in more detail. We will restrict ourselves here to the observation that the proper choice of the variables is crucial in this context. This aspect is most apparent for a charge double layer. As the charge double layer has no excess charge the charge distribution has no contribution proportional to δ^s in this system. It does have a contribution proportional to $\vec{n}\vec{n}.\vec{\nabla}\, \delta^s$, however. As there is no excess charge one may, however, use the polarization density as a variable rather than the charge density. This polarization density only has a contribution proportional to δ^s which illustrates our point. If one would use the charge density anyway this is of course perfectly alright as long as one realises that the above equations should be accordingly modified.

An important quantity of a surface of discontinuity is its curvature which is defined in the following way

$$C(\xi_2, \xi_3, t) \equiv -[\, h_1^{-1} \frac{\partial}{\partial \xi_1} \ln(h_2 h_2)]_s = \frac{1}{R_1} + \frac{1}{R_2} \tag{2.23}$$

The subindex s indicates that the value of the corresponding quantity is taken for $\xi_1 = 0$ [Morse 1953]. R_1 and R_2 are the so called radii of curvature. One may show [Bedeaux 1976] that the curvature is minus the divergence of the normal on the surface

$$C = -(\vec{\nabla}.\vec{n})_s \tag{2.24}$$

In order to make the dynamic description of the system complete one must also have an equation of motion for the surface. It may be shown that the normal on the surface satifies the following equation

$$\frac{\partial}{\partial t}\vec{n}(\vec{r},t) = -(\vec{1} - \vec{n}\vec{n}) \cdot (\vec{\nabla} w^s_n)_s \qquad (2.25)$$

where $\vec{1}$ is a unit tensor. The time derivative of the normal is thus given in terms of the gradient of the normal component of the velocity of the surface.

As a matter of convenience we introduce the average of the extrapolated bulk values at the dividing surface:

$$d_+(r,t) \equiv \frac{1}{2}[d^+(\xi_1 = 0, \xi_2, \xi_3, t) + d^-(\xi_1 = 0, \xi_2, \xi_3, t)] \qquad (2.26)$$

Another convenient combination is the jump of these extrapolated values accross the surface

$$d\_(r,t) \equiv d^+(\xi_1 = 0, \xi_2, \xi_3, t) - d^-(\xi_1 = 0, \xi_2, \xi_3, t) \qquad (2.27)$$

Similar definitions of the average and the jump of the extrapolated bulk values will be used for all bulk fields. Using the above definition of the jump one may write the balance equation for the interface (2.21) in the following form

$$\frac{\partial}{\partial t} d^s(\vec{r},t) + \text{div}\, \vec{J}^s_d(\vec{r},t) + J^+_{d,n,-}(\vec{r},t) - w^s_n(\vec{r},t) d\_(\vec{r},t) = \sigma^s_d(\vec{r},t)$$

$$\text{for } \xi_1(\vec{r},t) > 0 \qquad (2.28)$$

This is a little more compact then the original form.

3. CONSERVATION OF MASS

Consider a system consisting of n components among which r chemical reactions are possible. The balance equation for the mass density ρ_ℓ of component ℓ is written in the form

$$\frac{\partial}{\partial t}\rho_\ell + \text{div}\, \rho_\ell \vec{v}_\ell = \sum_{j=1}^{r} \nu_{\ell j} J_j \qquad (3.1)$$

where $\vec{v}_\ell$ is the velocity of component ℓ and $\nu_{\ell j} J_j$ is the production of component ℓ in the j–th chemical reaction. All these quantities have the form given in eqs.(2.3)–(2.5). In order to clarify the use of this notation we now give this form again explicitly

$$\rho_\ell(\vec{r},t) = \rho_\ell^-(\vec{r},t)\,\theta^-(\vec{r},t) + \rho_\ell^s(\vec{r},t)\,\delta^s(\vec{r},t) + \rho_\ell^+(\vec{r},t)\,\theta^+(\vec{r},t) \tag{3.2}$$

$$\rho_\ell(\vec{r},t)\vec{v}_\ell(\vec{r},t) = \rho_\ell^-(\vec{r},t)\,\vec{v}_\ell^-(\vec{r},t)\,\theta^-(\vec{r},t) + \rho_\ell^s(\vec{r},t)\,\vec{v}_\ell^s(\vec{r},t)\,\delta^s(\vec{r},t) + \rho_\ell^+(\vec{r},t)\,\vec{v}_\ell^+(\vec{r},t)\,\theta^+(\vec{r},t) \tag{3.3}$$

$$J_j(\vec{r},t) = J_j^-(\vec{r},t)\,\theta^-(\vec{r},t) + J_j^s(\vec{r},t)\,\delta^s(\vec{r},t) + J_j^+(\vec{r},t)\,\theta^+(\vec{r},t) \tag{3.4}$$

The similarity of eq.(3.2) and eq.(2.3) are clear and needs no further discussion. We only note that while $\rho_\ell^\pm$ are mass densities per unit of volume that ρ_ℓ^s is a mass density per unit of surface area. For the mass current of component ℓ given in eq.(3.3) we have as usual written this current as the product of the mass density and a velocity. This procedure in fact defines the velocity field as the quotient of the mass current and the mass density. In the bulk regions this will be clear. The surface velocity field $\vec{v}^s$ is defined as the ratio of the mass current per unit of surface area and the mass density per unit of surface area both for component ℓ. Notice the fact that the surface velocity $\vec{v}_\ell^s$ and the bulk velocities $\vec{v}_\ell^\pm$ have the same dimensionality. This clearly is not the case for the surface and the bulk mass currents. The advantage of the use of specific quantities like the the velocity is that one may consider for instance the difference of the velocity of a component in the bulk with the velocity of the same or in fact other components on the surface. This may not be done for the bulk and surface mass currents as there dimensionality is in fact different. It is important to realize, however, that an equation like eq.(2.4) may only be written down for the mass current and not for the velocity field. Finally, $\nu_{\ell j}J_j^\pm$ is the production of component ℓ in the j–th chemical reaction per unit of volume in the bulk regions, while $\nu_{\ell j}J_j^s$ is the production of component ℓ in the j–th chemical reaction per unit of surface area on the surface.

The quantity $\nu_{\ell j}$ divided by the molecular mass of component ℓ is proportional to the stoichiometric coefficient with which ℓ appears in the chemical reaction j. Since mass is conserved in each chemical reaction we have

$$\sum_{\ell=1}^{n} \nu_{\ell j} = 0 \tag{3.5}$$

As a consequence the total mass

$$\rho \equiv \sum_{\ell=1}^{n} \rho_\ell \quad \Longleftrightarrow \quad \begin{cases} \rho^{\pm} \equiv \sum_{\ell=1}^{n} \rho_\ell^{\pm} & \text{for the bulk regions} \\ \rho^{s} \equiv \sum_{\ell=1}^{n} \rho_\ell^{s} & \text{for the interface} \end{cases} \tag{3.6}$$

is a conserved quantity. This may easily be verified if one sums eq.(3.1) over ℓ and uses eq.(3.5). This in fact gives

$$\frac{\partial}{\partial t} \rho + \operatorname{div} \rho\vec{v} = 0 \tag{3.7}$$

where the barycentric velocity field is defined by

$$\rho\vec{v} \equiv \sum_{\ell=1}^{n} \rho_\ell \vec{v}_\ell \quad \Longleftrightarrow \quad \begin{cases} \rho^{\pm} \vec{v}^{\pm} \equiv \sum_{\ell=1}^{n} \rho_\ell^{\pm} \vec{v}_\ell^{\pm} & \text{for the bulk regions} \\ \rho^{s} \vec{v}^{s} \equiv \sum_{\ell=1}^{n} \rho_\ell^{s} \vec{v}_\ell^{s} & \text{for the interface} \end{cases} \tag{3.8}$$

The condition for the normal component of the mass current, cf. eq.(2.22), gives for the normal components of the various interfacial velocity fields

$$v_{\ell,n}^{s} = v_n^{s} = w_n^{s} \tag{3.9}$$

Thus the excess mass of the different components and as a consequence also the barycentric motion of the interface have the same normal velocity as the dividing surface. It is clear that a proper choice of the time–dependent position of the dividing surface is in fact only acceptible if $v_n^s = w_n^s$. The fact that the excesses of all components have the same velocity in the normal direction is necessary to assure that the interface does not split up in two or more different interfaces. This is not a phenomenon we want to describe. We now use the freedom in the choice of the curvilinear coordinate system to choose

$$\vec{v}^{s} = \vec{w}^{s} \tag{3.10}$$

The advantage of this is that also the velocity of the curvilinear coordinate system at the dividing surface in the direction parallel to the dividing surface has a direct physical meaning.

Using the general balance equation (2.21) for the excess mass density of component ℓ one obtains using also eq.(3.9)

$$\frac{\partial}{\partial t}\rho_\ell^s + \text{div}\,\rho_\ell^s\vec{v}_\ell^s + [\,\rho_\ell(\,v_{\ell,n} - v_n^s\,)]_- = \sum_{j=1}^{r} \nu_{\ell j} J_j^s \qquad (3.11)$$

As is clear from this equation the rate of change of the excess mass of component ℓ is given in terms of the divergence of the interfacial mass current of component ℓ, the flow of mass of component ℓ from the bulk regions into the surface of discontinuity and the rate of change due to the excess of the chemical reaction rates in the surface of discontinuity.

Similarly, one finds for the total excess mass

$$\frac{\partial}{\partial t}\rho^s + \text{div}\,\rho^s\,\vec{v}^s - [\,\rho\,(\,v_n - v_n^s\,)]_- = 0 \qquad (3.12)$$

It is convenient to define a barycentric time derivative for the dividing surface

$$\frac{d^s}{dt} \equiv \frac{\partial}{\partial t} + \vec{v}^s.\,\text{grad} \qquad (3.13)$$

This barycentric time derivative gives the time rate of change of a variable in a reference frame which moves along with the excess surface mass. Using this barycentric time derivative in eq.(3.12) one finds

$$\frac{d^s}{dt}\rho^s + \rho^s\,\text{div}\,\vec{v}^s - [\,\rho\,(\,v_n - v_n^s\,)]_- = 0 \qquad (3.14)$$

for the mass balance at the interface

The interfacial and the bulk diffusion fluxes of component ℓ are defined by

$$\vec{J}_\ell^s \equiv \rho_\ell^s\,(\vec{v}_\ell^s - \vec{v}^s) \qquad \text{and} \qquad \vec{J}_\ell^\pm \equiv \rho_\ell^\pm\,(\vec{v}_\ell^\pm - \vec{v}^\pm) \qquad (3.15)$$

It follows from the definition of the barycentric velocities that

$$\sum_{\ell=1}^{n} \vec{J}_\ell^s = 0 \qquad \text{and} \qquad \sum_{\ell=1}^{n} \vec{J}_\ell^\pm = 0 \qquad (3.16)$$

which implies that as usual only n – 1 diffusion currents are independent. In view of the fact that the velocities of all the components in the direction normal to the interface are identical to v_n^s it follows that

$$J^s_{\ell,n} = 0 \qquad (3.17)$$

The interfacial diffusion currents are therefore along the dividing surface. If we define the interfacial and bulk mass fractions by

$$c^s_\ell \equiv \rho^s_\ell / \rho^s \qquad \text{and} \qquad c^\pm_\ell \equiv \rho^\pm_\ell / \rho^\pm \qquad (3.18)$$

one finds upon substitution in the interfacial balance equation (3.11) for component ℓ

$$\rho^s \frac{d^s}{dt} c^s_\ell + \text{div}\, \vec{J}^s_\ell + [(v_n - v^s_n)\rho(c_\ell - c^s_\ell) + J_{\ell,n}]_- = \sum_{j=1}^{r} \nu_{\ell j} J^s_j \qquad (3.19)$$

Of course eq.(3.11) or eq. (3.19) are completely equivalent and one should simply use the one that is most convenient for any particular application.

As an intermesso we now return to the general balance equation discussed in section 2. At the interface this balance equation is given by

$$\frac{\partial}{\partial t} d^s + \text{div}\, \vec{J}^s_d + [J_{d,n} - v^s_n\, d]_- = \sigma^s_d \qquad (3.20)$$

Furthermore the normal component of the current satisfies

$$J^s_{d,n} - v^s_n\, d^s = 0 \qquad (3.21)$$

The purpose of the intermesso is to see how the introduction of specific densities

$$d \equiv \rho\, a \quad \Longleftrightarrow \quad \begin{cases} a^\pm \equiv d^\pm / \rho^\pm & \text{for the bulk regions} \\ a^s \equiv d^s / \rho^s & \text{for the interface} \end{cases} \qquad (3.22)$$

gives rise to alternative equations for the balance at the interface. For this purpose it is furthermore convenient to define currents in the comoving reference frame

$$\vec{J}_d \equiv \rho\, a\, \vec{v} + \vec{J}_a \quad \Longleftrightarrow \quad \begin{cases} J^\pm_d \equiv \rho^\pm a^\pm + \vec{J}^\pm_a & \text{for the bulk regions} \\ J^s_d \equiv d^s \rho^s + \vec{J}^s_a & \text{for the interface} \end{cases} \qquad (3.23)$$

Here $\rho\, a\, \vec{v} = d\, \vec{v}$ is the so-called convective contribution to the current. Upon substitution of these definitions into eq.(3.20) one obtains, using also the balance equation for the excess mass,

$$\rho^s \frac{d^s}{dt} a^s + \text{div}\, \vec{J}_a^s + [(v_n - v_n^s)\rho(a - a^s) + J_{a,n}]_- = \sigma_d^s \tag{3.24}$$

Furthermore eq.(3.21) gives

$$J_{a,n}^s = 0 \tag{3.25}$$

These alternative equations, which are of course equivalent to the original ones, will be often used in the analysis below. It must be emphasised that this is entirely a matter of personal choice. Using the original form of these equations yields exactly the same results. This in particular also when the dividing surface can be chosen such that $\rho^s = 0$. We shall come back to this point below.

4. CONSERVATION OF MOMENTUM

The equation of motion of the system is given by

$$\frac{\partial}{\partial t} \rho\vec{v} + \vec{\nabla}.\, [\rho\vec{v}\vec{v} + \vec{\vec{P}}] = \sum_{\ell=1}^{n} \rho_\ell \vec{F}_\ell \tag{4.1}$$

where the convective contribution to the momentum flow is equal to

$$\rho\, \vec{v}\, \vec{v} = \rho^-\, \vec{v}^-\, \vec{v}^-\, \theta^- + \rho^s\, \vec{v}^s\, \vec{v}^s\, \delta^s + \rho^+\, \vec{v}^+\, \vec{v}^+\, \theta^+ \tag{4.2}$$

Furthermore $\vec{\vec{P}}$ is the pressure tensor that gives the rest of the momentum flow and $\vec{F}_\ell$ is an external force acting on component ℓ. The pressure tensor can be written in terms of the hydrostatic pressures p^- and p^+ in the bulk regions, the surface tension and the viscous pressure $\vec{\vec{\Pi}}$ in the following way

$$\vec{\vec{P}} = p^- \vec{\vec{1}}\, \theta^- - \gamma\, (\vec{\vec{1}} - \vec{n}\vec{n})\, \delta^s + p^+ \vec{\vec{1}}\, \theta^+ + \vec{\vec{\Pi}} \tag{4.3}$$

The viscous pressure tensor may also have an excess in the dividing surface. As is usual in the discussion of the bulk regions we assume that the system possesses no intrinsic internal angular momentum, this neither in the bulk nor at the interface. As a consequence the viscous contributions are symmetric [De Groot 1962; Waldmann 1967]. Eq.(3.25) gives for this case

$$\vec{n}.\overset{\rightarrow}{\overset{\rightarrow}{\Pi}} = \overset{\rightarrow}{\overset{\rightarrow}{\Pi}}.\vec{n} = 0 \qquad \text{and} \qquad \vec{n}.\overset{\rightarrow}{\vec{P}} = \overset{\rightarrow}{\vec{P}}.\vec{n} = 0 \tag{4.4}$$

Consequently the excess of the viscous pressure tensor and of the total pressure tensor are symmetric 2x2 tensors. It is at this point that the choice of the pressure tensor has limited the discussion to fluid interfaces. It is not difficult to give the alternative expressions for the case that one or both phases are solids.

We restrict the discussion to conservative forces which can therefore be written in terms of time–independent potentials:

$$\vec{F}_\ell = -\,\text{grad}\,\Psi_\ell \tag{4.5}$$

Because of these external forces the momentum is not conserved. The balance equation for the excess interfacial momentum (4.1) can now be written in a form analoguous to eq.(3.23) which gives

$$\rho^s \frac{d^s}{dt}\vec{v}^s + \vec{\nabla}.\,[-\gamma(\overset{\rightarrow}{\overset{\rightarrow}{1}} - \vec{n}\vec{n}) + \overset{\rightarrow}{\overset{\rightarrow}{\Pi}}{}^s] + \vec{n}.\,[(\vec{v} - \vec{v}^s)\rho(\vec{v} - \vec{v}^s) + \overset{\rightarrow}{\vec{P}}\,]_- = \sum_{\ell=1}^{n} \rho_\ell^s \vec{F}_{\ell,s} \tag{4.6}$$

Here the subscript s indicates as usual the value of the corresponding quantity at the dividing surface. Notice in this context the fact that the external force is continuous at the dividing surface. Another point which should be realized is that after taking the gradient or the divergence of a quantity which is only defined on the surface, which when one differentiates merely implies that this quantity does not depend on ξ_1, one should take the result for $\xi_1 = 0$. The gradient operator may otherwise introduce spurious ξ_1 dependence. We will not indicate this fact explicitly in the formulae, by setting for instance $(\text{grad}\,\gamma)_s$ instead of grad γ, for ease of notation.

In order to study the force resulting from the surface tension we write

$$\vec{\nabla}.\,[\,\gamma\,(\overset{\rightarrow}{\overset{\rightarrow}{1}} - \vec{n}\vec{n})] = \vec{\nabla}.\,\gamma - \gamma\,\vec{n}\,\text{div}\,\vec{n} = \vec{\nabla}.\,\gamma - \gamma\,\vec{n}\,C \tag{4.7}$$

where we made use of the fact that the gradient of a quantity like the surface tension which is only defined on the surface is always directed along the dividing surface. As is clear from this equation the gradient of the surface tension gives rise to a force along the interface while in the normal direction one obtaines a force equal to the curvature times the surface tension. For an interface in an equilibrium situation, ie when $\vec{v} = 0$ and $\overset{\rightarrow}{\overset{\rightarrow}{\Pi}} = 0$, eq.(4.6) reduces to

$$\vec{\nabla}.\gamma - \gamma\vec{n}\,C + \vec{n}\,p\_ = \sum_{\ell=1}^{n} \rho_\ell^s \vec{F}_{\ell,s} \tag{4.8}$$

The balance of forces for the equilibrium situation normal to the surface is thus given by

$$-\gamma C + p\_ = -\gamma C + p_s^+ - p_s^- = \sum_{\ell=1}^{n} \rho_\ell^s F_{\ell,s,n} \tag{4.9}$$

This is a direct generalization of the Laplace equation for the hydrostatic pressure difference in terms of the surface tension and the curvature. The balance of forces for the equilibrium situation parallel to the surface is given by

$$-\vec{\nabla}\gamma = \sum_{\ell=1}^{n} \rho_\ell^s \vec{F}_{\ell,s}.(\vec{1} - \vec{n}\vec{n}) \tag{4.10}$$

In the non–equilibrium case both the Laplace equation and the above equation for the forces parallel to the surface contain additional terms due to the inertial term, the viscous pressure and the convective contribution to the momentum current. The contributions may easily be found using eq.(4.6).

In the analysis of the conservation of energy in the next section we need an equation for the time rate of change of the kinetic energy density. As we are in particular considering the behaviour of the interface we will restrict ourselves to the excess kinetic energy. In general we will refer to [De Groot 1962] for a discussion of bulk contributions and of the relevant equations for the bulk regions. For the excess kinetic energy one finds using eq.(4.6)

$$\rho^s \frac{d^s}{dt}\frac{1}{2}|\vec{v}^s|^2 = \rho^s \vec{v}^s . \frac{d^s}{dt}\vec{v}^s$$

$$= -\vec{\nabla}.(\vec{\vec{P}}^s.\vec{v}^s) + \vec{\vec{P}}^s : \vec{\nabla}\,\vec{v}^s - \vec{n}.[(\vec{v} - \vec{v}^s)\rho(\vec{v} - \vec{v}^s) + \vec{\vec{P}}\,]\_ . \vec{v}^s + \sum_{\ell=1}^{n} \rho_\ell^s \vec{F}_{\ell,s}.\vec{v}^s \tag{4.11}$$

Similarly we need an expression for the time rate of change of the potential energy of the excess mass:

$$\rho^s \Psi^s \equiv \sum_{\ell=1}^{n} \rho_\ell^s \Psi_{\ell,s} \tag{4.12}$$

Using eq.(3.19) one finds

$$\rho^s \frac{d^s}{dt} \Psi^s = \rho^s \frac{d^s}{dt} \sum_{\ell=1}^{n} \Psi_{\ell,s} c_\ell^s = \sum_{\ell=1}^{n} \Psi_{\ell,s} \rho^s \frac{d^s}{dt} c_\ell^s + \rho^s \sum_{\ell=1}^{n} c_\ell^s \vec{v}^s . \vec{\nabla} \Psi_{\ell,s}$$

$$= - \sum_{\ell=1}^{n} \Psi_{\ell,s} \vec{\nabla} . \vec{J}_\ell^s - [(v_n - v_n^s)\rho(\Psi - \Psi^s) + \sum_{\ell=1}^{n} \Psi_\ell J_{\ell,n}]_- + \sum_{\ell=1}^{n} \sum_{j=1}^{r} \Psi_{\ell,s} \nu_{\ell j} J_j^s - \sum_{\ell=1}^{n} \rho_\ell^s \vec{F}_{\ell,s} . \vec{v}^s$$

(4.13)

We shall assume that the potential energy is conserved in a chemical reaction:

$$\sum_{\ell=1}^{n} \Psi_{\ell,s} \nu_{\ell j} = 0 \qquad (4.14)$$

If the system is put in a gravitational field or in an electric field, this property is satisfied. Eq. (4.12) then reduces to

$$\rho^s \frac{d^s}{dt} \Psi^s = - \sum_{\ell=1}^{n} \Psi_{\ell,s} \vec{\nabla} . \vec{J}_\ell^s - [(v_n - v_n^s)\rho(\Psi - \Psi^s) + \sum_{\ell=1}^{n} \Psi_\ell J_{\ell,n}]_- - \sum_{\ell=1}^{n} \rho_\ell^s \vec{F}_{\ell,s} . \vec{v}^s$$

(4.15)

It follows from the above equations that neither the kinetic nor the potential energy are conserved quantities. As we will discuss below only the total energy is conserved.

5. CONSERVATION OF ENERGY

The total energy of the system is conserved. As a consequence one has for the specific energy density e

$$\frac{\partial}{\partial t} \rho e + \mathrm{div}(\rho e \vec{v} + \vec{J}_e) = 0 \qquad (5.1)$$

The total energy density can be written as the sum of the kinetic energy, the potential energy and the internal energy u:

$$e^\pm = \frac{1}{2}|\vec{v}^\pm|^2 + \Psi^\pm + u^\pm \qquad \text{for the bulk regions}$$

$$e^s = \frac{1}{2}|\vec{v}^s|^2 + \Psi^s + u^s \qquad \text{for the interface} \qquad (5.2)$$

The energy current is similarly the sum of a mechanical work term, a potential–energy flux due to diffusion and a heat flow $\vec{J}_q$

$$\vec{J}_e = \overrightarrow{\overrightarrow{P}}.\vec{v} + \sum_{\ell=1}^{n} \Psi_\ell \vec{J}_\ell + \vec{J}_q \tag{5.3}$$

In fact one may use the above equations as a definition of internal energy and of the heat current. Notice the fact that we have not included the convective contribution $\rho e\vec{v}$ in the energy current $\overrightarrow{\overrightarrow{J}}_e$ as is done in [De Groot 1962].

Using energy conservation (5.1) one immediately finds a balance equation for the interfacial energy density e^s from the general form (3.24) of such a balance equation

$$\rho^s \frac{d^s}{dt} e^s + \operatorname{div} \vec{J}_e^s + [(v_n - v_n^s)\rho(e - e^s) + J_{e,n}]_- = 0 \tag{5.4}$$

The excess energy defined above does not contain a convective contribution and is as a consequence directed along the interface

$$J_{e,n}^s = 0 \tag{5.5}$$

The definition of the internal energy given in eq.(5.2) may then be used to find a balance equation for the internal energy. Together with the balance equations for the interfacial kinetic and potential energy given in the previous section and the definition of the heat current (5.3) one then obtains

$$\rho^s \frac{d^s}{dt} u^s = -\operatorname{div} \vec{J}_q^s - \overrightarrow{\overrightarrow{P}}^s : \vec{\nabla} \vec{v}^s + \sum_{\ell=1}^{n} \vec{F}_{\ell,s} . \vec{J}_\ell^s - \{(v_n - v_n^s)\rho[u - u^s - \tfrac{1}{2}|\vec{v} - \vec{v}^s|^2]$$

$$+ J_{q,n} + [(v_n - v_n^s)\rho(\vec{v} - \vec{v}^s) + \vec{n}.\overrightarrow{\overrightarrow{P}}].(\vec{v} - \vec{v}^s)\}_- \tag{5.6}$$

Furthermore one finds that also the excess heat flow is directed along the dividing surface

$$J_{q,n} = 0 \tag{5.7}$$

The internal energy is not a conserved quantity, because of conversion of kinetic and potential energy into internal energy and vice versa. The above balance equation for the excess internal energy gives in fact the first law of thermodynamics for the interface.

6. ENTROPY BALANCE

In order to also formulate the second law of thermodynamics we now consider the balance equation for the entropy

$$\frac{\partial}{\partial t} \rho s + \mathrm{div}(\rho s \vec{v} + \vec{J}) = \sigma \tag{6.1}$$

Here s is the entropy density per unit of mass, $\vec{J}$ is the entropy current and σ the entropy source. Notice the fact that we do not use subscripts s to indicate that the current or the source are of the entropy. The reason for this is that this would lead to confusion with the subindex s which indicates that the value of a bulk variable should be taken at the dividing surface. For the interface the equation for the entropy balance becomes, cf again eq.(3.24),

$$\rho^s \frac{d^s}{dt} s^s = -\,\mathrm{div}\, \vec{J}^s - [(v_n - v_n^s)\rho(s - s^s) + J_n]_- + \sigma^s \tag{6.2}$$

If we compare this with the expression for the balance of entropy in the bulk regions [De Groot 1962]

$$\rho^{\pm} \frac{d^{\pm}}{dt} s^s = \rho^{\pm} \left(\frac{\partial}{\partial t} + \vec{v}^{\pm}.\vec{\nabla}\right) s^s = -\,\mathrm{div}\, \vec{J}^{\pm} + \sigma^{\pm} \tag{6.3}$$

one sees that the balance equation for the interfacial entropy density contains, in addition to the divergence of a current and a source term, a flow of entropy from the bulk phases into the dividing surface. It is this last term and similar terms in the balance of excess energy and mass which will be found to also lead to an excess of the entropy production. The origin of the usual boundary conditions, which will be discussed below, is related to the appearence of these terms.

In the description of systems in which there are two phases separated by an interface using nonequilibrium thermodynamics it is assumed that in volume elements small compared to the over all size of the system but large compared to the molecular size one may use the usual thermodynamic relations. Thus one uses that in each volume element the second law of thermodynamics applies and concludes that the entropy production is non–negative

$$\sigma^- \geq 0, \qquad \sigma^s \geq 0 \qquad \text{and} \qquad \sigma^+ \geq 0 \tag{6.4}$$

While this property is well known in the bulk phases it is less obviously true for the interface.

In Gibbs' discussion of of the equilibrium thermodynamics of interfaces the interface is treated as a separate thermodynamic system. In the present context one may clearly use that in volume elements located in the surface of discontinuity one may also use the usual thermodynamic relations and it then follows that the excess entropy production is non–negative. Notice in this context that the surface of discontinuity as defined by Gibbs and as used here has a finite thickness, this contrary to the dividing surface also introduced by Gibbs which is a mathematically sharp two dimensional surface [Gibbs 1906].

From thermodynamics we know that the entropy and any other thermodynamic function, such as the hydrostatic pressure, are well–defined functions of the thermodynamic parameters necessary to specify the macroscopic state of the system. For the system under consideration we use as parameters the internal energy, the specific volume ($1/\rho^{\pm}$) or the specific surface area ($1/\rho^{s}$), and the mass fractions. For the entropy we may thus write

$$s^{-} = s^{-}(u^{-}, 1/\rho^{-}, c_{\ell}^{-}), \quad s^{s} = s^{s}(u^{s}, 1/\rho^{s}, c_{\ell}^{s}) \quad \text{and} \quad s^{-} = s^{-}(u^{-}, 1/\rho^{-}, c_{\ell}^{-}) \tag{6.5}$$

We stress the important fact that the excess entropy only depends on u^{s}, $1/\rho^{s}$ and c_{ℓ}^{s} and not on the extrapolated values $u_{s}^{\pm}$, $1/\rho_{s}^{\pm}$ and $c_{\ell,s}^{\pm}$ of these variabes in the bulk as is sometimes done [Defay 1966]. Nowhere do we find that such a dependence on the extrapolated bulk values is needed. In the context of the above argument that volume elements in the surface of discontinuity can be treated as separate thermodynamic systems, this would imply a dependence on the variables in neighbouring volume elements; such a nonlocal dependence is clearly contrary to the usual assumptions made in the context of nonequilbrium thermodynamics. In Gibbs discussion of interfaces he also discusses a possible dependence of the excess thermodynamic variables on the curvature of the surface. As he argues such a dependence is small for most systems and may therefore usually be neglected. Notable exceptions are systems in which the surface tension is very low due to the presence of sufficient surfactants like for instance microemulsions. We will in this paper always neglect the dependence on the curvature.

At equilibrium the total differential of the entropy is given by the Gibbs relation. In the bulk regions this relation is given by

$$T^{\pm} ds^{\pm} = du^{\pm} + p^{\pm} d(1/\rho^{\pm}) - \sum_{\ell=1}^{n} \mu_{\ell}^{\pm} dc_{\ell}^{\pm} \tag{6.6}$$

where $T^{\pm}$ is the temperature and $\mu_{\ell}^{\pm}$ is the chemical potential of component ℓ in the bulk regions. For the interface one similarly has [Gibbs 1906]

$$T^s ds^s = du^s - \gamma\, d(1/\rho^s) - \sum_{\ell=1}^{n} \mu_\ell^s\, dc_\ell^s \tag{6.7}$$

where T^s is the interfacial temperature and μ_ℓ^s is the interfacial chemical potential of component ℓ. If one writes the surface tension as minus the hydrostatic surface pressure p^s the Gibbs relation for the interface is term by term analogous to the Gibbs relation in the bulk.

In nonequilibrium thermodynamics the whole system is not in equilibrium. One assumes, however, that each volume element which is large compared to molecular sizes but small compared to the overall size of the system is in equilibrium. This is the socalled assumption of local equilibrium. In each volume element one may therefore write the entropy and other thermodynamic quantities as a function of the energy, the specific volume (surface area) and the mass fractions in the same volume element. Thus one has

$$s^-(\vec{r},t) = s^-(u^-(\vec{r},t),\ 1/\rho^-(\vec{r},t),\ c_\ell^-(\vec{r},t))$$

$$s^s(\xi_2,\xi_3,t) = s^s(u^s(\xi_2,\xi_3,t),\ 1/\rho^s(\xi_2,\xi_3,t),\ c_\ell^s(\xi_2,\xi_3,t))$$

$$s^-(\vec{r},t) = s^-(u^-(\vec{r},t),\ 1/\rho^-(\vec{r},t),\ c_\ell^-(\vec{r},t)) \tag{6.8}$$

The local equilibrium assumption implies that the Gibbs relation remains valid in the frame moving with the center of mass. We therefore have

$$T^\pm \frac{d^\pm}{dt} s^\pm = \frac{d^\pm}{dt} u^\pm + p^\pm \frac{d^\pm}{dt}(1/\rho^\pm) - \sum_{\ell=1}^{n} \mu_\ell^\pm \frac{d^\pm}{dt} c_\ell^\pm \tag{6.9}$$

in the bulk phases and

$$T^s \frac{d^s}{dt} s^s = \frac{d^s}{dt} u^s - \gamma \frac{d^s}{dt}(1/\rho^s) - \sum_{\ell=1}^{n} \mu_\ell^s \frac{d^s}{dt} c_\ell^s \tag{6.10}$$

for the interface.

The next step in the analysis is to substitute the expressions for the barycentric derivatives of the internal energy, the specific volume (surface area) and the mass fractions in the above Gibbs relations. This leads in particular for the interface to rather complicated expressions for the barycentric derivative of the entropy density. Comparing this expression for the barycentric derivative of the entropy density with eq.(6.3) one may then identify the entropy current and the entropy production. In this way one obtains as entropy current in the bulk phases [De Groot 1962]

$$\vec{J}^{\pm} = (\vec{J}_q^{\pm} - \sum_{\ell=1}^{n} \mu_\ell^{\pm} \vec{J}_\ell^{\pm})/T^{\pm} \tag{6.11}$$

For the interfacial entropy current one finds in a similar way

$$\vec{J}^{s} = (\vec{J}_q^{s} - \sum_{\ell=1}^{n} \mu_\ell^{s} \vec{J}_\ell^{s})/T^{s} \tag{6.12}$$

This expression is in essence identical to the one in the bulk phases.

For the entropy production one finds in the bulk phases [De Groot 1962]

$$\sigma^{\pm} = - (T^{\pm})^{-2} \vec{J}_q^{\pm} \cdot \text{grad } T^{\pm} + \vec{J}_\ell^{\pm} \cdot [(\vec{F}_{\ell,s} /T^{\pm}) - \text{grad}(\mu_\ell^{\pm}/T^{\pm})]$$
$$- (T^{\pm})^{-1} \vec{\vec{\Pi}}^{\pm} : \text{grad } \vec{v}^{\pm} - (T^{\pm})^{-1} \sum_{j=1}^{r} J_j^{\pm} A_j^{\pm} \tag{6.13}$$

The affinities $A_j^{\pm}$ of the chemical reactions in the bulk phases and the affinities A_j^s at the interface are defined by

$$A_j^{\pm} \equiv \sum_{\ell=1}^{n} \mu_\ell^{\pm} \nu_{\ell j} \quad \text{and} \quad A_j^{s} \equiv \sum_{\ell=1}^{n} \mu_\ell^{s} \nu_{\ell j} \tag{6.14}$$

For the excess entropy production one finds in this way after some tedious algebra

$$\sigma^{s} = - (T^{s})^{-2} \vec{J}_q^{s} \cdot \text{grad } T^{s} + \sum_{\ell=1}^{n} \vec{J}_\ell^{s} \cdot [(\vec{F}_{\ell,s} /T^{s}) - \text{grad}(\mu_\ell^{s}/T^{s})]$$

$$- (T^{s})^{-1} \vec{\vec{\Pi}}^{s} : \text{grad } \vec{v}^{s} - (T^{s})^{-1} \sum_{j=1}^{r} J_j^{s} A_j^{s}$$

$$+ \{T[J_n + (v_n - v_n^{s})\rho s][(1/T) - (1/T^{s})]\}_- - (1/T^{s})\{[\vec{n}.\vec{\vec{\Pi}} + (v_n - v_n^{s})\rho\vec{v}] .(\vec{v} - \vec{v}^{s})\}_-$$

$$- (1/T^s) \sum_{\ell=1}^{n} \{[J_{\ell,n} + (v_n - v_n^s)\rho_\ell][\mu_\ell - \mu_\ell^s - \tfrac{1}{2}|\vec{v}|^2 + \tfrac{1}{2}|\vec{v}^s|^2]\}_- \tag{6.15}$$

The above expression was first given in [Bedeaux 1986]. The expression in that paper contained $J_{q,n}/T$ rather then J_n which is incorrect [Albano 1987]. The first four terms in the excess entropy production are similar to the contributions found in the bulk regions and are a consequence of transport processes along the interface. The other terms are new and are the result of transport processes from the bulk regions into and through the interface. One may rewrite the above form of σ^s in a somewhat different but equivalent forms. Our reason to choose the above form is the fact that, in the terms corresponding to the flow into and through the interface, one systematically finds as currents the sum of the diffusive and the convective part of the heat, momentum and mass flow in the comoving reference frame. This seems to be a reasonable choice and gives a relatively elegant form. In the treatment of membrane transport the question how the choice of the reference frame affected the entropy production gave rise to considerable discussion [Mikulecki 1966]. Of course one may recombine the various terms in different combinations. In practice it is usually good to think about this choice carefully for every particular problem under consideration. Some contributions are usually more important than others and this must motivate the choice. In the analysis below we will in fact rewrite the above expression and combine all the contributions due to the convective part of the currents. This combined term will in fact turn out to couple the mass current through and into the interface to pressure differences.

In the derivation of the above expressions for the entropy production and current we have used the following thermodynamic identities:

$$T^\pm s^\pm = u^\pm + p^\pm/\rho^\pm - \sum_{\ell=1}^{n} \mu_\ell^\pm c_\ell^\pm \tag{6.16}$$

for the bulk phases and

$$T^s s^s = u^s - \gamma/\rho^s - \sum_{\ell=1}^{n} \mu_\ell^s c_\ell^s \tag{6.17}$$

for the interface. In the expression for the interface we have not used the freedom to make one of the excess densities equal to zero by the appropriate choice of the location of the dividing surface. If one would use this freedom to take ρ^s equal to zero it is better to write the last equation in the alternative form

$$T^s(\rho s)^s = (\rho u)^s - \gamma - \sum_{\ell=1}^{n} \mu_\ell^s (\rho c_\ell)^s \tag{6.18}$$

It is important to realize that $\rho^s = 0$ does not imply that for instance $(\rho s)^s$ is also zero. The definition $s^s\rho^s \equiv (\rho s)^s$ is somewhat confusing in this case. If one prefers to avoid this problem one should rewrite the above analysis using densities per unit of surface area rather than using densities per unit of surface mass. The resulting expression for the entropy current and the entropy production are of course excacly the same. In a given problem certain excess densities or currents may be zero to a good approximation. In that case it is most convenient to eliminate these terms from the relevant balance equations and from the above expressions for the excess entropy production. In view of the large number of force–flux pairs in the excess entropy production, such a simplification is most conveniently made at that stage of the analysis.

The diffusion currents in the bulk regions as well as along the interface are not independent. In fact their sum is zero, cf eq.(3.16). Using this property one may eliminate one of these diffusion currents from the expression for the entropy production. In solutions it is most convenient to eliminate the diffusion current of the solvent. If one eliminates $\vec{J}_n^s$ from the excess entropy production one obtains

$$\sigma^s = -(T^s)^{-2}\,\vec{J}_q^s \cdot \text{grad}\,T^s + \sum_{\ell=1}^{n-1} \vec{J}_\ell^s \cdot [((\vec{F}_{\ell,s} - \vec{F}_{n,s})/T^s) - \text{grad}((\mu_\ell^s - \mu_n^s)/T^s)]$$

$$- (T^s)^{-1}\,\vec{\vec{\Pi}}^s : \text{grad}\,\vec{v}^s - (T^s)^{-1} \sum_{j=1}^{r} J_j^s A_j^s$$

$$+ \{T[J_n + (v_n - v_n^s)\rho s][(1/T) - (1/T^s)]\}_- - (1/T^s)\{[\vec{n}.\vec{\vec{\Pi}} + (v_n - v_n^s)\rho\vec{v}].(\vec{v} - \vec{v}^s)\}_-$$

$$- (1/T^s) \sum_{\ell=1}^{n-1} \{[J_{\ell,n} + (v_n - v_n^s)\rho_\ell][\mu_\ell - \mu_\ell^s - \mu_n + \mu_n^s]\}_- - (1/T^s)\{(v_n - v_n^s)\rho(\mu_n - \mu_n^s)]\}_- \tag{6.19}$$

It is now convenient to combine the various contributions proportional to $\rho(v_n - v_n^s)$ and to write TJ_n in terms of the heat current and the diffusion currents, cf eq.(6.11). This yields the following expression

$$\sigma^s = -(T^s)^{-2}\,\vec{J}^s_q \cdot \text{grad}\,T^s + \sum_{\ell=1}^{n-1} \vec{J}^s_\ell \cdot [((\vec{F}_{\ell,s} - \vec{F}_{n,s})/T^s) - \text{grad}((\mu^s_\ell - \mu^s_n)/T^s)]$$

$$- (1/T^s)\,\overset{\rightrightarrows}{\Pi}{}^s : \text{grad}\,\vec{v}^s - (1/T^s)\sum_{j=1}^{r} J^s_j\,A^s_j + \{J_{q,n}[(1/T) - (1/T^s)]\}_-$$

$$- (1/T^s)\{\,\vec{\Pi}_{n,\|} \cdot (\vec{v}_\| - \vec{v}^s_\|)\}_- - \sum_{\ell=1}^{n-1} \{\,J_{\ell,n}\,[(\mu_\ell/T) - (\mu^s_\ell/T^s) - (\mu_n/T) + (\mu^s_n/T^s)\,]\}_-$$

$$- (1/T^s)\{[\rho(v_n - v^s_n)][s(T - T^s) + \sum_{\ell=1}^{n} c_\ell(\mu_\ell - \mu^s_\ell) + (\Pi_{n,n}/\rho) + \tfrac{1}{2}|\vec{v} - \vec{v}^s|^2]\}_- \qquad (6.20)$$

The subindex $\|$ indicates the projection parallel to the dividing surface. Notice furthermore that the subindex n is usually used to indicate the normal component of a vector. If it appears as a subindex of the chemical potential, however, it indicates the n–th component.

In order to understand the driving force for the mass flow through and into the interface a little better we use the following thermodynamic relation for the bulk phases

$$(1/\rho^\pm)dp^\pm = s^\pm dT^\pm + \sum_{\ell=1}^{n} c^\pm_\ell\, d\mu^\pm_\ell \qquad (6.21)$$

A similar relation which we will not need, however, is valid for the interface

$$(1/\rho^s)d\gamma = -\,s^s dT^s - \sum_{\ell=1}^{n} c^s_\ell\, d\mu^s_\ell \qquad (6.22)$$

Now we use the fact that the system, in the context of nonequilibrium thermodynamics, is never very far from equilibrium. This implies in the bulk regions that the gradient of the temperature and the gradient of the chemical potentials are small. At the interface this similarly implies on the one hand that these gradients are also small along the interface and on the other hand that the temperature and chemical potential jumps are small. We may therefore write

$$\{[\rho(v_n - v^s_n)][s(T - T^s) + \sum_{\ell=1}^{n} c_\ell(\mu_\ell - \mu^s_\ell) + \Pi_{n,n} + \tfrac{1}{2}|\vec{v} - \vec{v}^s|^2]\}_-$$

$$= \{[\rho(v_n - v_n^s)][(p - p(T^s,\mu_\ell^s) + \Pi_{n,n})/\rho + \frac{1}{2}|\vec{v} - \vec{v}^s|^2]\}_- \qquad (6.23)$$

This expression implies that the mass flow through and into the interface couples directly to pressure differences. $p^{\pm}(T^s,\mu_\ell^s)$ is the pressure of the bulk phases at the temperature and chemical potentials of the interface.

7. THE PHENOMENOLOGICAL EQUATIONS

In equilibrium the entropy production is zero. This implies that all the thermodynamic forces and fluxes are zero. Close to equilibrium the fluxes are linear functions of the thermodynamic forces. It should be realized that the linear nature of this relation between the fluxes and forces does not imply that the resulting differential equations describing the time dependent state of the system are linear. These equations are non linear for a number of reasons. The first is the presence of convective contributions to all the currents. The second is the nonlinear nature of the equations of state relating the various thermodynamic quantities to each other. The third is the dependence of the linear constitutve coefficients on the variables. The only crucial limitation is the fact that these constitutive coefficients do not themselves depend on the thermodynamic forces. If one would contemplate taking such a dependence into account, one should realize that the assumption of local equilibrium is no longer correct in that case.

It is clear from the above expression for the excess entropy production there is a rather large number of force–flux pairs. In the most general case all Cartesian components of the fluxes couple to all the Cartesian components of the fluxes. This would lead to an immense number of different constitutive coefficients. For a fluid–fluid interface this situation is greatly simplified because of the symmetry of the problem. The fluid–fluid interface is invariant for translation along the interface and for rotation around the normal on the interface. One may therefore distinguish 2x2 tensorial force flux pairs, 2–dimensional vectorial force–flux pairs and scalar force–flux pairs. In view of the Curie symmetry principle [De Groot 1962] forces and fluxes of a different tensorial nature do not couple.

We need one more identity before we can identify all thermodynamic forces and fluxes at the interface. The jump of a product from one side of the interface to the other satisfies

$$(ab)_- = a_- b_+ + a_+ b_- \qquad (7.1)$$

Thus one has, for instance,

$$\{\vec{\Pi}_{n,\|} \cdot (\vec{v}_\| - \vec{v}_\|^s)\}_- = \vec{\Pi}_{n,\|,-} \cdot (\vec{v}_\| - \vec{v}_\|^s)_+ + \vec{\Pi}_{n,\|,+} \cdot (\vec{v}_\| - \vec{v}_\|^s)_-$$

$$= \vec{\Pi}_{n,\|,-} . (\vec{v}_{\|,+} - \vec{v}^{s}_{\|}) + \vec{\Pi}_{n,\|,+} . \vec{v}_{\|,-} \tag{7.2}$$

The first term contains the slip force when the interface is moved relative to the two bulk phases, and the second term contains the slip force when one bulk phase slides over the other.

In order to get an overview of the forces and fluxes of a different tensorial nature we will tabulate them for each tensorial character.

1. **Symmetric traceless tensor**

flux: $[\vec{\vec{\Pi}}^{s}_{\|,\|} - \frac{1}{2}(\mathrm{tr}\vec{\vec{\Pi}}^{s})(\vec{\vec{1}} - \vec{n}\vec{n})]_{symm}$ force: $[(\vec{\nabla}_{\|}\vec{v}^{s}_{\|}) - \frac{1}{2}(\mathrm{tr}(\vec{\nabla}\vec{v}^{s}))(\vec{\vec{1}} - \vec{n}\vec{n})]_{symm}$

The linear constitutive equation which is the result gives the symmetric traceless part of the excess viscous pressure tensor as a constant times the symmetric traceless gradient of the velocity of the excess mass along the interface. The proportionality constant may be interpreted as the interfacial shear viscosity.

2. **Antisymmetric traceless tensor**

flux: $[\vec{\vec{\Pi}}^{s}_{\|,\|}]_{antisymm}$ force: $(\vec{\nabla}_{\|}\vec{v}^{s}_{\|})_{antisymm}$

The linear constitutive equation which is the result in this case gives the antisymmetric traceless part of the excess viscous pressure tensor as a constant times the antisymmetric traceless gradient of the velocity of the excess mass along the interface. The proportionality constant may be interpreted as the interfacial rotational viscosity. For a fluid–fluid interface one may use the invariance for rotation around the normal on the interface to show that the antisymmetric part of the excess viscous pressure is zero [Waldmann 1967] if the molecules have no interfacial spin [De Groot 1962]. This term will therefore usually be unimportant.

3. **Vectors**

even

fluxes: $\vec{J}^{s}_{q,\|}$ forces: $\vec{\nabla}(1/T^{s})$

$\vec{J}^{s}_{\ell,\|}$ (for $\ell = 1, \dots, n-1$) $(\vec{F}_{\ell,s} - \vec{F}_{n,s}) - \vec{\nabla}[(\mu^{s}_{\ell} - \mu^{s}_{n})/T^{s}]$

odd

fluxes: $\vec{\Pi}_{n,\|,-}$ forces: $(\vec{v}_+ - \vec{v}^s)_\|$

$\vec{\Pi}_{n,\|,+}$ $(\vec{v}_- - \vec{v}^s)_\|$

$\vec{\Pi}^s_{\|,n}$ $\vec{\nabla}\, v^s_n$

Even and odd gives the behaviour of the corresponding quantities for time reversal. The (n+3) vectorial fluxes can be expressed in terms of the (n+3) vectorial forces using a (n+3)×(n+3) matrix of constitutive coefficients. The coefficients appearing in this matrix can be identified in terms of an interfacial heat conductivity, interfacial diffusion coefficients, interfacial Dufour coefficients, interfacial thermal diffusion coefficients, coefficients of sliding friction and thermal slip coefficients. The constitutive coefficients for a number of cross effects have no special name. The Onsager symmetry relations state that coefficients which couple terms with the same symmetry for time reversal will be symmetric while coefficients which couple terms with a different symmetry for time reversal are antisymmetric [De Groot 1962]. The number of independent constitutive coefficients thus reduces to $\frac{1}{2}$(n+3)(n+4).

4. Scalars

even

fluxes: J^s_j (for j = 1, ... , r) forces: A^s_j

$J_{q,n,-}$ $(1/T)_+ - (1/T^s)$

$J_{q,n,+}$ $(1/T)_-$

$J_{\ell,n,-}$ (for ℓ = 1, ... , n-1) $(\mu_\ell - \mu^s_\ell - \mu_n + \mu^s_n)_-$

$J_{\ell,n,+}$ (for ℓ = 1, ... , n-1) $(\mu_\ell - \mu_n)_+$

odd

fluxes: $\Pi^s \equiv \mathrm{tr}(\overleftrightarrow{\Pi}^s)$ forces: $\mathrm{div}\, \vec{v}^s$

$[\rho(v_n - v^s_n)]_-$ $[(p - p(T^s,\mu^s_\ell) + \Pi_{n,n})/\rho]_+ + \frac{1}{2}|\vec{v} - \vec{v}^s|^2_+$

$[\rho(v_n - v^s_n)]_+$ $[(p - p(T^s,\mu^s) + \Pi_{n,n})/\rho]_- + \frac{1}{2}|\vec{v} - \vec{v}^s|^2_-$

The (2n + r + 3) scalar fluxes can be expressed in terms of the (2n + r + 3) scalar forces using a (2n + r + 3)×(2n + r + 3) matrix of constitutive coefficients. Some of the coefficient can be interpreted in terms of chemical reaction rates for reactions which take place in the interfacial region and in terms of temperature jump coefficients. The Onsager symmetry relations again state that coefficients which couple terms with the same symmetry for time reversal will be symmetric while coefficients which couple terms with a different symmetry for time reversal are antisymmetric [De Groot 1962]. The number of independent constitutive coefficients thus reduces to $\frac{1}{2}(2n + r + 3)(2n + r + 4)$.

The total number of constitutive coefficients which gives a full description of all the possible processes at the interface is clearly prohibitively large. For a one component system in which there clearly are no reactions taking place one needs already 26 constitutive coefficients. For two components and one reaction one needs 52 constitutive coefficients. In practice most (if not all) these coefficients will be unknown. Furthermore it will almost be impossible to measure all of them. In a practical situation one must decide which of the above fluxes and forces contribute to the dynamical processes in a significant manner and neglect all the others.

8. DISCUSSION

The above analysis has resulted in explicit expressions for the excess entropy production. It then gives all the fluxes and conjugate thermodynamic forces which play a role for the dynamical processes which take place at an interface. Though many of these forces and fluxes may not play a role in a given situation, it is good to know which ones are neglected in a simplified description. One may otherwise too easily assume that conclusions derived using a simplified description are also valid for a more complicated situation. We will not proceed to discuss these simplifications in any detail. This would usually take another paper for each of these cases and would therefore take too much space. We will here merely mention some of these simplifications. The first is the application to problems where all velocities and gradients are normal to a planar dividing surface .In that case only the scalar fluxes and forces play a role. This situation is usually appropriate for transport through membranes. Furthermore, in view of the fact that fluid–fluid interfaces due to gravity form a horizontal plane, this case is also applicable to problems like evaporation and condensation [Bedeaux 1990]. Another typical simplification is that certain densities and currents have no excess at the surface. If one considers for instance a surface between oil and water one may usually neglect the excess densities of the oil and of the water. It is then clear that the interfacial chemical potentials of oil and water are no longer independent variables. In fact one should in that case take the interfacial chemical potential of these substances equal to the extrapolated value in the adjacent fluid. This not only reduces the number of independent variables describing the interface but also eliminates the corresponding fluxes and forces. A more difficult case is when the heat capacity of the interface may be neglected. Then the interfacial temperature becomes an average of the extrapolated bulk temperatures. The weights in the average depend on the specific heat of the adjacent phases. The most simple case is the interface between a gas and a liquid where one may in good approximation set the interfacial temperature equal to the extrapolated value in the liquid [Bedeaux 1990].

REFERENCES

Albano, A.M., Bedeaux, D., and Vlieger, J. 1979. On the description of interfacial properties using singular densities and currents at a dividing surface. Physica 99A: 293

Albano, A.M., Bedeaux, D., and Vlieger, J. 1980. On the description of interfacial electromagnetic properties using singular fields, charge density and currents at a dividing surface. Physica 102A: 105

Albano, A.M., Bedeaux, D. 1987. Nonequilibrium electro–thermodynamics of polarizable multicomponent fluids with an interface. Physica 147A: 407

Bedeaux, D., Albano, A.M., and Mazur, P. 1976. Irreversible thermodynamics of boundary conditions Physica 82A: 438

Bedeaux, D. 1986. Nonequilibrium Thermodynamics and Statistical Physics, in Advances in Chemical Physics, Vol. 64, p.47, Ed. I. Prigogine and Stuart A. Rice. John Wiley & Sons, New York.

Defay, R., and Prigigine, I. 1966. Surface Tension and Adsorption. Longmans and Green, London.

De Groot, S.R., and Mazur, P. 1962. Nonequilibrium Thermodynamics. North–Holland, Amsterdam. (reprinted by Dover publications, New York, 1984)

Gibbs, J.W. 1906. The Scientific Papers of J. Willard Gibbs. Vol. I. Longmans and Green, London. (reprinted by Dover publications, New York, 1961)

Kovac, J. 1977. Nonequilibrium thermodynamics of interfacial systems. Physica 86A: 1

Kovac, J. 1981. Nonequilibrium thermodynamics of interfacial spin systems. Physica 107A: 280

Mikulecky, D.C., and Caplan, S.R. 1966. The choice of reference frame in the treatment of membrane transport by nonequilibrium thermodynamics. J. Chem. Phys. 70: 3049

Morse, P.M., and Feshbach, H. 1953. Methods of Theoretical Physics, Vol I, McGraw – Hill, New York.

Ronis, D., Bedeaux, D., and Oppenheim, I. 1978. On the derivation of dynamical equations for a system with an interface I: general theory. Physica 90A: 487

Waldmann, L. 1967. Z. Naturforsch. Nonequilibrium thermodynamics of boundary conditions. 22A: 1269

Wolff, P.A., and Albano, A.M. 1979. Nonequilibrium thermodynamics of interfaces including electromagnetic effects. Physica 98A: 491

Method of Transfer Matrix for Calculation of Thermodynamics and Kinetics of Surface Reactions

A. V. Myshlyavtsev and G. S. Yablonskii
Tuvinian Complex Department of Sibiria Branch of the Academy of Sciences of the USSR
Lenin 30, Kyzyl, Tuva 667000, USSR

1. INTRODUCTION

At present for many chemical systems it's shown a posibility of complicated kinetic behaviour. This is a multiplicity of steady states, kinetic hysteresis, "chaotic" behaviour, selfoscillation and so on. In particular such effects are typical for simple molecule reactions on the d-metals surfaces (Nieuwenhuys, 1983). For theoretical describing of these reactions the Langmuir adsorption mechanism is usually used; this mechanism often supplemented by steps taking into account features of concrete reaction (Yablonskii at all, 1983; 1984). As a rule the hypothesis of the ideal adsorbed overlayer is used, in foundation of which lie the following assumption:

a) The equivalence of all surface sites and the constancy of chemisorption energy at various surface coverage by different adsorbents.

b) The constancy of catalysator and the invariability of its characteristic from composition of reactive mixture.

c) Boltzmann distribution of energy.

However, the detailed investigation made with using of modern physical methods have shown that these assumption are rather rough.

It was found that during the process of gas adsorption and reaction on the surface of single crystals there appear various ordering structures of chemisorption particles (Engel at all, 1979; Weinberg, 1983; Roelofs at all, 1983). The appearance of ordering structures is the evidence of a strong interaction between adsorbed particles. The existance of such interactions results in dependence of kinetic constants of coverage. The simplest model, which allows describe the appearance of ordering structures is a lattice gas model. In the framework of this model the influence of lateral interactions of adsorbed particles on kinetic konstants may be described.

The calculation of thermodynamic parameters within the framework lattice gas model is a very difficult problem to be exactly resoluble only in exeptional cases (Baxter, 1983). However, general equations for kinetic constants within the framework of the lattice gas model have rather simple form (Zhdanov, 1981; 1988; Reed, 1981).

The expressions for the rate of monomolecular desorption and adsorption may be written as

$$d\theta_A/dt = -(\nu_0 F_{A'}/F_A) \sum P_{A,i} \exp[-(E_d+\Delta\varepsilon_i)/T]\theta_A \quad (1.1)$$

$$d\theta_A/dt = (\nu_o F_{A'}/F^g_A) \sum P_{o,i} \exp[-(E_a+\varepsilon'_i)/T] N_g \quad (1.2)$$

Where θ_A - surface coverage; $F_{A'}$, F_A, F^g_A - non-configuration partition function of the activated complex, adsorbed particles and gas phase particles accordingly; $P_{A,i}$ is the probability that an adsorbed particle has the environment marked by index i; $P_{o,i}$ is the probability that an empty site has the environment marked by index i; E_d, E_a are activation energy for desorption and adsorption at low coverage; $\Delta\varepsilon_i = \varepsilon_i - \varepsilon'_i$, where ε_i, ε'_i are the lateral interaction of the adsorbed particle A and activated complex A' with its environment; N_g is concentration in gas phase. Unfortunately, formulae (1.1), (1.2) are practically the statement of the problem, as the main difficulty lies in the calculation of the various probabilities appearing in this formulae. Below for simplification (it is not a principle thing) we will assume that $\varepsilon'_i = 0$. In this case we can write more simple expressions for constants of desorption and adsorbtion.

$$k_d = k_d(0)\ (1-\theta_A)\ \exp(\ \mu_A/T)/\theta_A \quad (1.3)$$

$$k_a = k_a(0) = \text{const} \quad (1.4)$$

Where μ_A is chemical potential and Boltzmann constant is sent to unity. Also we write the expression for coefficiant of surface diffusion.

$$D(\theta) = \nu_{ef} a^2\ S\ \exp(\ \mu/T) \frac{1}{T} \frac{\partial\mu}{\partial\theta} \quad (1.5)$$

$$S = \sum_i P_{oo,i} \exp(\ -\varepsilon'_i/T) \quad (1.6)$$

$$\nu_{ef} = \nu\ \exp(\ -E_a/T) \quad (1.7)$$

Where $P_{oo,i}$ is a probability to find two neighboring empty sites having the environment marked by index i; a is lattice parameter. Assuming $\varepsilon'_i = 0$ we rewrite expression (1.5) as

$$D(\theta) = D(0)\ P_{oo}\ \exp(\mu/T)\ \frac{1}{T} \frac{\partial\mu}{\partial\theta} \quad (1.8)$$

From formulae (1.3), (1.4), (1.8) it is seen that for describing for desorption process and surface diffusion it is necessary to calculate the dependence of chemical potential on coverage. This dependence can be found from grand partition function with using standart expression.

$$\theta = -T\left(\frac{\partial \ln \Xi}{\partial \mu}\right)_T \qquad (1.9)$$

Where Ξ is grand partition function fitting one site. Investigating kinetics of elementary physico-chemical processes within the framework of the lattice gas model, it is used various methods. We shall characterize shortly the most used of these methods.

Mean-field, quasi-chemical and Bete-Payerls approximations

A finite cluster is considered in Bete-Payerls approximation where interaction between particles, placed into cluster is exactly taking into account. Interaction with particles placed out the cluster substitutes for "effective" interaction. Its value is chosen from self-sequence calculation. This calculation includes following:
a) Probability of occupying various sites into claster is calculated using grand partitional function.
b) Self-sequence conditions are the equivalence of these probabilities.

As a result it appears non-linear equations. These equations are bulky and their complication increases rapidly with growing size of cluster. At the same time precision of received results increases rather slowly (Smart, 1966). For this reason usually it used the simplest form of Bete-Payerls approximation - quasi-chemical approximation. The interaction between central partical and its nearest neighbours takes exactly into account in this method. Quasi-chemical approximation allows to obtain expressions for various probabilities in explicit form. Its precision gets much less, by accounting for many-body and more further than next-neighbour two-body interactions. This is connected that these interactions to take into account with using mean-field approximation. However even in case of interaction only nearest-neighbours, quasi-chemical and more complicated cluster approximations take into account correlations in position of adsorbed particles insufficiently. As it shown by the theoretical and numerical analysis, quasi-chemical approximation has rather satisfied precision at $T>1.3T_c$, where T_c is temperature of phase transition (Smart, 1966). At present quasi-chemical approximation (sometimes more complicated cluster approximations) is often used when calculating various kinetic parameters within the framework of the lattice gas model (Zhdanov, 1988). We will give some examples of using its.

The influence of lateral interactions upon pre-exponencial factor of monomolecular desorption was investigated by (Leuthauser, 1980; Zhdanov, 1981). Characteristic feature of received dependences is the break near $\theta = 0.5$ for square lattice with repulsion of nearest-neighbours. Thermal desorption spectra were calculated with using the quasi-chemical approximation by (Adams, 1974; Zhdanov, 1983). Existence of lateral interactions becomes by doubling of thermo-desorption maximum. It is necessary to mark, that for hexogonal lattice ($Z = 6$) the quasi-chemical approximation gives much worse results. This explains less studying of hexogonal lattice. The main short-coming of

Bete-Payerls approximation (in particular quasi-chemical one) is the inapplicability of it at $T<T_c$. At the same time as it has been said there are rather many real systems where T_c has value of several hundreds K. Let's examine another frequently used method. This is the method of mathematical simulation or Monte-Karlo method.

Monte-Karlo method

This method is often used in statistical physics (Binder,1982). Using the Monte-Karlo method we can investigate as properties of system at thermodinamical equilibrium as kinetics of approaching to 2-D systems the infinite lattice substitutes by finite lattice (usually within interval $40\times40 \div 200\times200$) with periodical boundary conditions. Monte-Karlo method is rather general and allows to examine many of interested problems. The main short-coming of this is high demands to computers (especially in the field of phase transitions where relaxation to real thermodynamical equilibrium taking place rather slowly). Recently this method is often used to investigate the kinetics various surface processes. This method was used to investigate surface diffusion by (Reed and all, 1981; Sadiq and all, 1983; Tringides and all, 1984). It was shown that the coefficient of surface diffusion has singularities in the field of existence of ordered and condentional phases. The help of Monte-Karlo method detailes investigations of thermo-desorption spectra for many models (including those studying of wich are practically impossible by other methods) was made. (Sales and all, 1987; Silverberg and all, 1985; Lombardo and all, 1988; Sales and all, 1989).

Finally we should mention that the effective methods of modern theoretical physics are activly used by investigation of various lattice models. They are the renormalization-group method and the transfer-matrix method. The transfer- -matrix will be spoken below. Unique advantage of these methods is their applicability both above and below critical temperature T_c. Renormalization-group (McMahon, 1988) method can be used in immediate proximity to the critical point.With their help phase diagrams of many complicated systems were made. However, neither this nor that method was used when studying the kinetics of surface processes (as it's known to authors). In this paper we present the using of the transfer-matrix method for these purposes and show its high effectivness. Now we state the essense of transfer-matrix method.

2. THE TRANSFER-MATRIX METHOD

The detailed description of this method and necessary proofs you can find in (Domb, 1960). The transfer-matrix method allows to recieve precise solutions of set problems for a 1-D chain, the every site of wich can be found in m states, where m is any finite integer number. We'll consider that only nearest-neighbours interact with energy depends on site-state. It is easy to see any classical system with a finite radius of lateral interactions to be given to this case.

Let's examine the semi-infinite chain the right end of which is fixed and is in the state marked by index i ($1 \leqslant i \leqslant m$). Designate a grand partitional function of this chain as Z_i. From all partitional function for i = 1,...,m we can form vector **Z**. Adding one more site to the right end of the chain we can form vector **Z'** by analogy with that one. Putting the transfer-matrix **T** we can write the following equation.

$$\mathbf{Z}' = \mathbf{T}\mathbf{Z} \qquad (2.1)$$

It is to be noted that the transfer-matrix is not unique. We'll use it in the form (Kinzel and all, 1981) providing symmetry. Evidently, that the addition of one more site to a semi-infinity chain doesn't change the distribution of probabilities for different states. Hence $\mathbf{Z}' = \lambda\mathbf{Z}$ and

$$\mathbf{T}\mathbf{Z} = \lambda\mathbf{Z} \qquad (2.2)$$

From (2.2) it follows that **Z** is eigen vector of matrix **T** and λ is eigenvalue of matrix **T**. It is easy to see that λ is the largest eigenvalue of matrix **T**. We can take a semi-infinite chain with the fixed left end and by analogy

$$\mathbf{Z}\mathbf{T} = \lambda\mathbf{Z} \qquad (2.3)$$

From (2.2) and (2.3) it follows that the largest eigen- value of transfer-matrix is equal to the grand partitional function one site of the chain. Let's write some simple expressions allowing to use transfer-matrix method (TMM) for the investigation of kinetics of processes on the surface. Probability of the site of the infinite chain is in the state marked by index i can be calculated as

$$P_i = \frac{Z_i^2}{\sum_j Z_j^2} \qquad (2.4)$$

Where Z_i is the component of the eigen vector of matrix **T** fitting to the largest eigenvalue. The generalization of (2.4) is the expression for probability of that n neighbours sites are in the states $i_1, i_2, \ldots, i_n$.

$$P_{i_1,\ldots,i_n} = \frac{\mathbf{T}_{i_1 i_2} Z_{i_2}^2 \, \mathbf{T}_{i_2 i_3} Z_{i_3}^2 \ldots \mathbf{T}_{i_{(n-1)} i_n} Z_{i_n}}{(\lambda_o \sum_j Z_j^2)^{(n-1)}} \qquad (2.5)$$

Thus for 1-D chain TMM allows to calculate exactly the dependence of desorption constant and coefficient of surface diffusion on coverage.

Let's examine the application of TMM for investigation of 2-D systems. For these systems in a general case TMM doesn't give exact solution. However this method can be used as a strong approximate one. Let's describe its application for 2-D systems according (Kinzel, 1981; Runnels and all, 1966; Runnels and all, 1967). Let's examine the simplest case of square lattice. We shall approach semi-infinite system with a width M site in the X direction and the infinite in both direction along Y axis. Periodical boundary conditions are put along X axis. As a result we have an infinite cylinder with M sites on perimeter. Only one particle is assumed to be on the lattice site and there are particles of one type othis lattice. It follows that a number of states of one site is equal two and a number of states of the ring of M sites is equal 2^M. Examing every ring as one new site wich can be in $m = 2^M$ states we shall receive 1-D system described above. We would remind that for this system the TMM allows to get exact solution set problem. Hence the grand partitional function fitting on one site of initial lattice can be calculated as

$$\ln \Xi_{(M)} = \frac{1}{M} \ln \lambda^{(M)} \qquad (2.6)$$

Dimension of transfer-matrix T as it follows from the previous describing is equal $2^M \times 2^M$. It's seen that dimension is growing exponentionally together with M. Details of forming of transfer-matrix and the calculation of its largest eigenvalue we can find in (Kinzel, 1981; Runnels and all, 1966;). Parameters which are interested us in the first time are kinetic constants. They are local in the main, i.e. they are determined by the immediate environment. It means that a short-range order in the arrangement of adsorbed particles is essential for them. As concrete calculations show with comporatively small $M = 4 \div 8$ the short-range order examined in the cylindrical semi-infinite system is similar to the short-range order in the initial 2-D system. This is not always true near the critical point. As detailed studies have shown for many thermodynamical parameters following asymptotical expression is correct.

$$a(M) = a(\infty) + bM^{-\gamma} \qquad (2.7)$$

Thus taking three rather large numbers M we can determine the value $a(\infty)$ which is the value of corresponding parameter in limit $M \to \infty$. When choising value M used for concrete calculation it is necessery to pay attention to compatibility of boundary conditions with symmetry of ordered phases which should appear in studying system. For square lattice even numbers M are usually suitable. For hexagonal lattice numbers divisible by 3 are usually suitable. As concrete calculations have shown the asymptotical expression (2.7) is rather accuracive for many systems starting with M = 6. For example we should present matrix elements of transfer-matrix for lattice gas on square lattice with interaction of the nearest neighbours.

$$T_{kl} = \exp[(0.5u_k + 0.5u_l + u_{kl})/T] \qquad (2.8)$$

$$u_k = -\varepsilon \sum_{i=1}^{M} n_{ij} n_{(i+1)j} + \mu \sum_{i=1}^{M} n_{ij} \qquad (2.9)$$

n_{ij}; $n_{(i+1)j}$ belongs to the ring K

$$u_{kl} = -\varepsilon \sum_{i=1}^{M} n_{ij} n_{i(j+1)} \qquad (2.10)$$

n_{ij} belongs to the ring K and $n_{i(j+1)}$ belongs to the ring L.

Where n_{ij} are occupation numbers are equal 0 if the site is empty and 1 if the site is occupied; ε is energy of lateral interactions of the nearest neighbours; μ is chemical potential.

The TMM is rather universal and let examine rather complicated systems. Its precision is limited only by computers. By restricting only local parameters which depend on short-range order as it has been seen above, we shall have rather accuracive describing for the whole interval of temperature and coverage, by not very large numbers M. The TMM prefers to cluster methods in the first time in the field of existence of ordering phases. Further development of computers will actually broad the sphere of application of this simple and effective method. Let's go on describing of results given by using of TMM. Further account will be based on the authors' works.

3. ISOTHERMS, THE APPARENT ARRHENIUS PARAMETERS FOR MONOMOMOLECULAR DESORPTION

Isotherms

Forming of lattice gas isotherms comes to the finding of dependence of chemical potential on coverage. Transfer-matrix method gives a solutions of this problem. In the first time the field of the existence of ordered phases will be interested us. Hamiltonian of studying system for a square lattice can be defined as

$$H_{eff} = \varepsilon_1 \sum_{\langle nn \rangle} n_i n_j + \varepsilon_2 \sum_{\langle nnn \rangle} n_i n_j + \varepsilon_t \sum n_i n_j n_k - \mu \sum_i n_i \quad (3.1)$$

and for hexogonal lattice as

$$H_{eff} = \varepsilon_1 \sum_{\langle nn \rangle} n_i n_j + \varepsilon_t \sum n_i n_j n_k - \mu \sum_i n_i \quad (3.2)$$

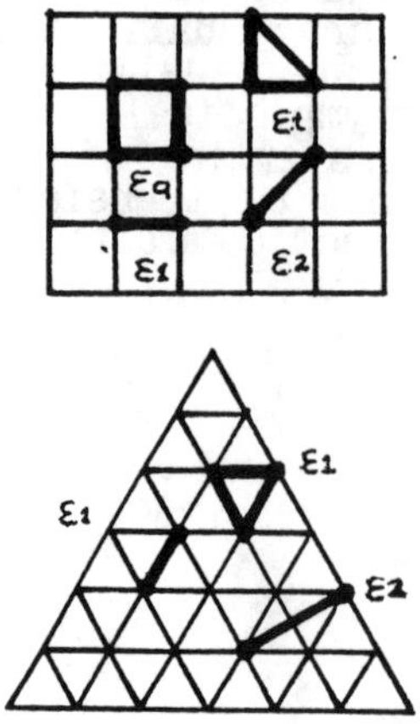

Figure 1. The types of lateral interactions.

The kinds of interactions to be taking into account show in Figure 1. The calculations were made of transfer- matrix method, by number M = 4,6,8 for square lattice and M = 6,9 for hexogonal lattice. As numerical results show the numbers M = 4 for square lattice and M = 6 for hexagonal lattice give rather high precision. In the first time we will examine the square lattice with pairwise <nn> and <nnn> interactions. In Figure 2 the isotherms with various types of lateral interactions are shown. When having only <nn> attraction at temperature below critical it appears the field of coexistence of two phase: lattice gas - LG and lattice liquid - LL. There appears the vertical part on isotherms. The close analogy may be with Van der Vaals gas isotherms. In real systems we can find repulsion of the nearest neighbors more frequently. Below we'll concentrate our attention on this case. Here at rather low temperature there appears ordered phase C(2×2) - "chess-board" in the interval of coverage $0.3<\theta<0.7$. Near $\theta = 0.5$ a "plateau" on the isotherms corresponds to the phase C(2×2).

Now let's examine more complicated system, in which the diagonal attraction besides <nn> repulsion presents. Here there are three phases C(2×2), LG + C(2×2), LL + C(2×2) at low temperature. Behaviour of isotherms correlates well with phase diagram. Vertical parts of the isotherms correspond to phases LG, LL + C(2×2), and horisontal part correspond to phase C(2×2). The existence of vertical part of isotherms corresponds to phase transition "gas-liquid" and can be a source of hysteresis in adsorption. Let's examine briefly the influence of three-body interactions. Naturally, their influence can be observed only with rather large lattice occupation and break the symmetry of isotherms concerning $\theta = 0.5$. The influence of these interactions on isotherms is rather obvious and we are not going to discuss it in this article. The isotherms for hexogonal lattice in the presence of <nn> repulsion and three-body attraction are given in Figure 3. It takes notice of the presence of two "plateau" with $\theta = 1/3$ and $\theta = 2/3$ on the isotherms, corresponding to ordered phases $(\sqrt{3}\times\sqrt{3})R30^\circ$ and $(\sqrt{3}\times\sqrt{3})R^*30^\circ$. In the presence of three-body attractions with $\theta>1/3$ on the isotherm there are appears a vertical part. It corresponds to the formation of condensed phase.

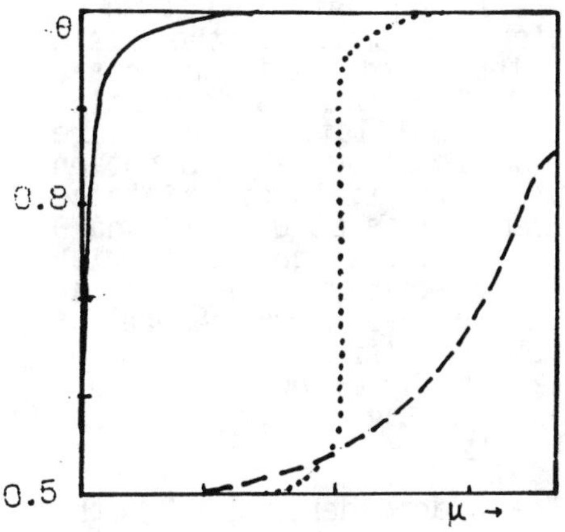

Figure 2. Isotherms.
—— <nn> attraction; - - - <nn> repulsion; ...... <nn> repulsion and <nnn> attraction. (Square lattice)

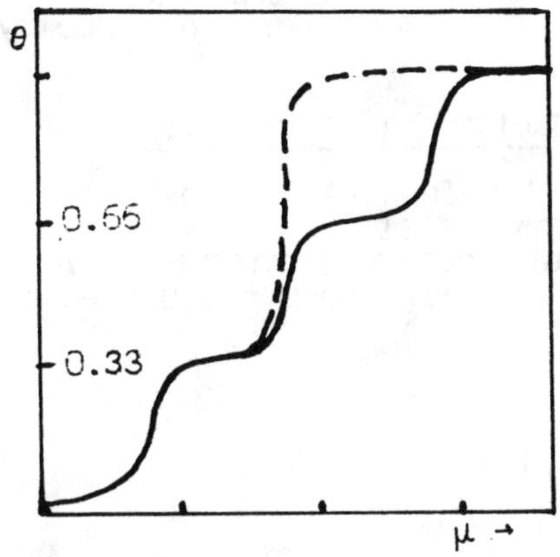

Figure 3 Isotherms.
—— <nn> repulsion; - - - <nn> repulsion and three-body attraction. (Hexogonal lattice)

The constant of a monomolecular desorption.

We shall restrict ourselves to a square lattice only. The dependences of the constant of a monomolecular desorption with <nn> and <nnn> interactions are given in Figure 4. We take the case which is the most interesting from physical and practical point of view - <nn> repulsion and <nnn> attraction. We can note two importent facts:

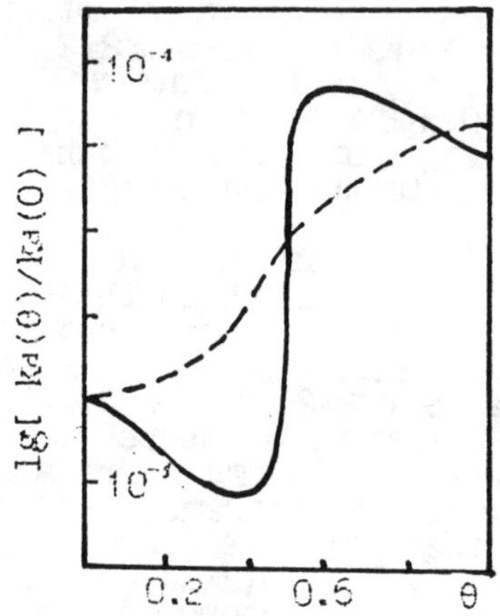

Figure 4. Desorption constant. —— <nn> repulsion and <nnn> attractions; - - - only <nn> repulsion.

a) There is a great advantage of desorption constant in the field of existence of ordered phase C(2×2) at θ = 0.5. Accordingly, for a hexogonal lattice there are two great advantage of desorption constant in the field of existence of ordered phases at θ = 1/3 and θ = 2/3. Calculations made with using quasi-chemical approximation give rather slow change of desorption constant near θ = 0.5, connected with effects of correlation taking into account incompletely.

b) The dependence of the desorption constant and the desorption rate upon coverage becomes non-monotonous, when taking into account simultaneously <nn> repulsion and <nnn> attraction. A complicated dependence of the desorption constant upon coverage can become apparent on kinetics of surface reactions very greatly. Three-body interactions on account of breaking of the symmetry "particles-holes", make the dependence of the desorption constant upon coverage non symmetry relative to θ = 0.5.

Activation energy and pre-exponetial factor for desorption.

As it's known, the parameters of the desorption constant are the activation energy and the pre-exponential factor. We assume this dependence of desorption constant upon temperature to have the Arrhenius form. The activation energy have been calculated by formula which is usually accepted.

$$E_d(\theta) = T^2 \, d \ln k_d(\theta)/dT \qquad (3.3)$$

and accordingly the pre-exponential factor

$$\nu(\theta) = k_d(\theta) \exp[E_d(\theta)/T] \qquad (3.4)$$

The dependence of the activation energy upon coverage with <nn> and <nnn> interactions is given in Figure 5. These dependences can be easily understood; they correlate very well with the phase diagram of lattice gas. In the case of <nn> attraction a long "plateau" with roughly permanent value of the activation energy corresponds to the phase LG + LL. On curves E(θ) with <nn> repulsion there are a region of a sharp decrease near θ = 0.5, more sharp than that one turned out by calculation with using quasi-chemical approximation. The region of sharp decrease of the activation energy corresponds to ordered phase C(2×2). In the presence <nn> repulsion and <nnn> attraction maximum of complicated form appeared on the curves E(θ). The primary rapid increase corresponds to the phase LG. The part of roughly permanent value E(θ) corresponds to the phase LG + C(2×2) and following rapid increase with sharp drop corresponds to the phase C(2×2). Curves E(θ) are antisymmetry relative to θ = 0.5. It is a result of existence of symmetry "particles-holes". Such kinds of dependence energy has been observed in the experiment for the system CO/Ru(100) (Pfnur and all, 1983). A traditional quasi-chemical approach cannot describe such dependence E(θ).

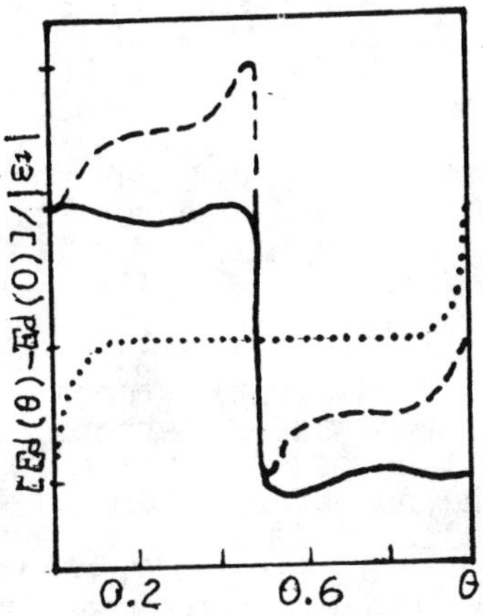

Figure 5. The activation energy .... <nn> attraction; —— <nn> repulsion without <nnn> interaction; ——— <nn> repulsion and <nnn> attraction.

The dependence of the pre-exponential factor upon coverage with <nn> repulsion and <nnn> attraction is given in Figure 6. Here by analogy with behaviour of activation energy we can see a complicated dependence similar to an experimental dependence for the system CO/Ru(100) (Pfnur and all, 1983). Calculation made by TMM show more better qualitative features, than the results received with using quasi-chemical approximation. In conclusion we can note that joint sharp decrease of pre-exponetial factor and activation energy for desorption at θ = 0.5 corresponds to ordered phase C(2×2).

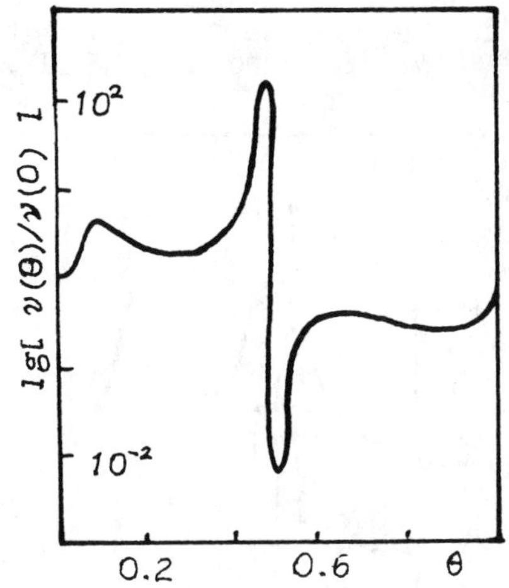

Figure 6. Pre-exponential factor

4. THERMAL DESORPTION SPECTRA

In general cases the kinetics of monomolecular desorption $A_s \rightarrow A_g$ is described as

$$d\theta/dt = -k_d\theta \qquad (4.1)$$

General expression for desorption constant within the framework of the lattice gas model can be written as (1.1). The desorption constant as stated above becomes complicated function depending on coverage and temperature when taking into account lateral interactions. When carrying out the thermal desorption experiment the temperature usually increases linearly.

$$T = T_o + \beta t \qquad (4.2)$$

Where T_0 is primary temperature; β is heating rate in K/s. Experimental thermal desorption spectra (TDS) can have one or several maximuma of desorption rate. As it was said above the lateral interactions in the case of square lattice can lead to the splitting of the single maximum into two maximuma. We shall use the TMM for calculation of TDS. This method allows to examine the cases more accurately, when a trajectory of the system in the thermal desorption process crosses the fields of ordered phase existence. We restrict ourselves to some general characteristics of TDS within the framework of lattice gas model, taking into account only pairwise lateral interactions. Usually the interaction between nearest neighbours is repulsion. If it is rather large there will be the splitting of TDS into two maximuma. This is connected with a sharp increase of desorption constant near $\theta = 0.5$. Taking into account another pairwise interactions, among which the diagonal interactions are the most important, the forms of maximuma usually change, while the number of maximuma is constant. The integrated intensity of both peaks is equal because the desorption constant makes a jump near $\theta = 0.5$. The high temperature peak is always lower and wider then the low temperature one. It is the second characteristic of these TDS. This is connected with the "particles-holes" symmetry. Let's examine TDS with <nn> and <nnn> interactions. The values of the energy of these interactions were chosen to be constant the whole decrease of activation energy.

$$E_d(0) - E_d(1) = 4\varepsilon_1 + 4\varepsilon_2 = 8 \text{ kcal/mol} \qquad (4.3)$$

The tipycal TDS, calculated with the help of different approximations are shown in Figure 7 (Myshlyavtsev and all, 1989).

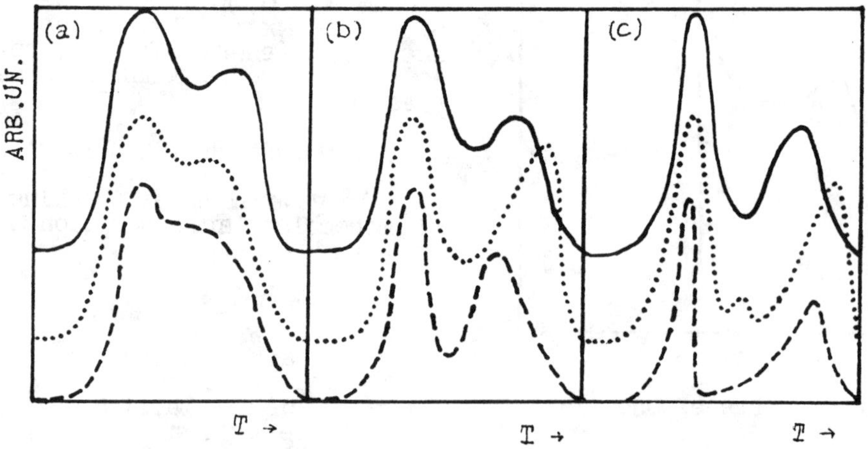

Figure 7. TDS for square lattice with <nn> and <nnn> interactions. ..... Monte-Karlo method. —— quasi-chemical approximation.- - - TMM; $E_d(0)$ = 35 kcal/mol; $\nu = 10^{13}$; β = 50 K/s; (a) ε_1 = 1.4, ε_2 = 0.6 kcal/mol (b) ε_1 = 2, ε_2 = 0 kcal/mol (c) ε_1 = 2.6, ε_2 = -0.6 kcal/mol.

When $T>T_c$ for the whole interval of coverage in the desorption process the spectra obtained with different methods are similar to each other. It is a case (a). A combination of the quasi-chemical and mean-field approximations underestimates a splitting of thermal desorption peaks in the cases (b) and (c). It is connected with thatthe trajectory of the system crosses the field of the existence of the ordered phases. The correlations in the arrangement of the particles in these phases are large. The quasi-chemical approximation does not take these correlations into account that leads to the smoothing of TDS. The results obtained with Monte-Karlo method and TMM are close to each other. However, in the case (c) the results obtained with Monte-Karlo method show a little intermediate peak. The existence of this peak is not confirmed by TMM. There are no phisical arguments in favour of the existance of this extremum. Apparently this peak is accounted by non-equilibrium effects connected with slow kinetics of ordered phase forming. These effects especially can be important in the case of the degeneration of the ordered phase. The phase C(2×2) is two fold degenerate. The influence of three-body interactions on TDS have been studied by (Myshlyavtsev and all, 1990). The comparison of the results obtained with Monte-Karlo method and TMM was made in this work. As it was mentioned above TDS submit to several qualitative rules, taking into account only pairwise lateral interactions. In particular, integrated intensity of peaks is identical. In spectra obtained experimentally this condition is rather frequently broken that makes many-body interactions to take into consideration. Another conditions are also broken in the same systems. For example: for system N_2/Ir(110) (Ibbotson and all, 1981) the low-temperature peak appears only at $\theta > 0.7$ and is lower than the high-temperature peak. The theoretical investigation of the influence of many-body interactions on monomolecular TDS have considerably less complete than for pairwise interactions. Three-body attraction increases the height of the first low-temperature peak while the three-body repulsion decreases its height. Integrated intensities of both peaks are prove neither equal, unlike the case of pairwise interactions (Myshlyavtsev and all, 1990). However, taking into account only three-body interactions as will be shown below it will turn out to be insuficient when interpreting experimental TDS. In particular, for square lattice four-body interactions having the symmetry of square are no less natural than three-body interactions. This interactions were not examined before. The Hamiltonian can be written as

$$H = \varepsilon_1 \sum n_i n_j + \varepsilon_t \sum n_i n_j n_k + \varepsilon_q \sum n_i n_j n_k n_l \qquad (4.4)$$

We should note one important circumstance. If we take "clear" three-body and four-body interactions, than the low-temperature peak for a four-body interaction is relatively lower than the low temperature peak for a three-body interaction. It is seen that the adding of any pairwise lateral interactions to "clear" many-body interactions will always lead to a relative increase for the low temperature peak. Thus, using only three-body interactions from all kinds of many-body interactions, we cannot make a relative height of low-temperature peak less than some value, i.e. the decrease of the relative height of the low-temperature peak under influence of the three-body interaction has its own limit. Therefore, if experimental TDS has a low-temperature peak less than some value, then it is necessery for its interpretation to take into account another many body interactions. Such example is the spectrum for above-mentioned system N_2/Ir(110). Experimental and theoretical spectra made by taking into account four-body interactions with using TMM and dependence activation energy upon coverage are given in Figure 8. We can see rather accuracy accordance between experiment and theory.

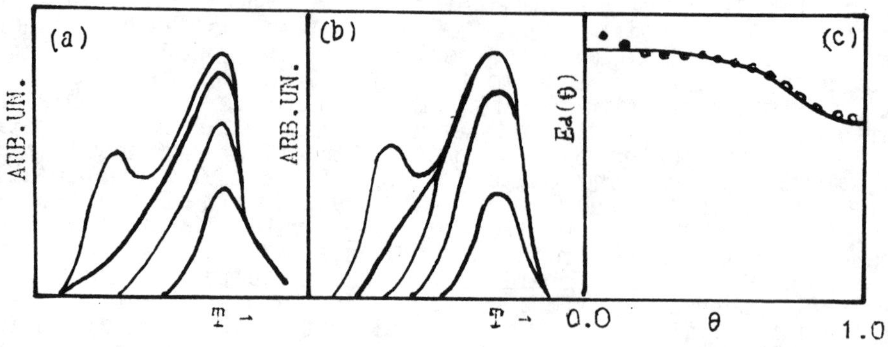

Figure 8. (a) - experimental spectra for N_2/Ir(100); (b) - theoretical spectra calculated with using of TMM; (c) - activation energy: —— theoretical value, ∘∘∘∘ experimental data.

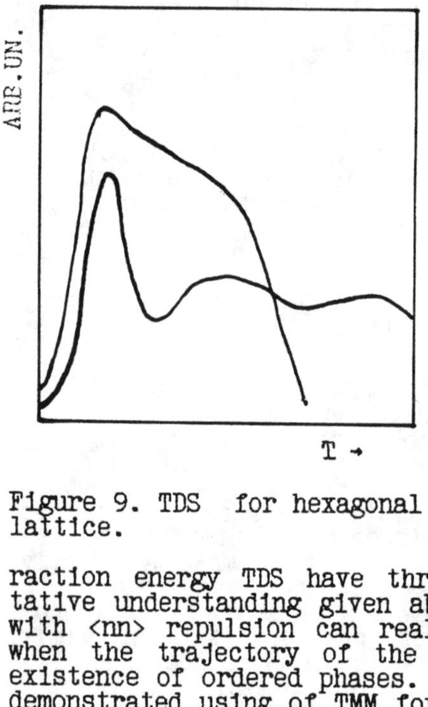

Figure 9. TDS for hexagonal lattice.

TMM allows to make TDS for hexogonal lattice also effectievly. Using of quasi-chemical approximation for this purpose is not possible in view of low precision. Therefore, TDS for hexagonal lattice were not practically examined. An analysis of the phase diagram of lattice gas alows to conclude that one can see three peaks in TDS spectra when taking into account only <nn> interactions. TDS at various values of the energy of <nn> repulsion are given in Figure 9. The calculation was made by TMM with M = 6. At rather low values of the interactions energy, TDS have one maximum and high-temperature "shoulder". The spectra forms are similar to the spectra forms for square lattice with diagonal repulsion. At rather large interaction energy TDS have three peaks in accordance with qualitative understanding given above. Thus TDS for hexagonal lattice with <nn> repulsion can really have three maximuma. It happens when the trajectory of the system crosses both fields of the existence of ordered phases. In this part of our article we have demonstrated using of TMM for calculating TDS as for square lattice as for hexagonal lattice. It is easy see that TMM more effective method than quasi-chemical approximation.

5. COEFFICIENT OF SURFACE DIFFUSION

Diffusion of the particles adsorbed on the plane of single crystals is very interesting as the example of diffusion in strong non-ideal systems. Besides, the surface diffusion often plays an important role in the kinetics of geterogenious chemical reactions. Phenomenologically this diffusion is described by well known Fick's laws.

$$J = D \nabla C; \qquad \partial C/\partial t = \nabla(D \nabla C) \qquad (5.1)$$

Where C is a concentration; J is a current; D is a diffusion coefficient. We shall assume a standart mechanism of diffusion within the framework of the lattice gas model. As experimental and theoretical results show, coefficient of surface diffusion most strongly depends upon coverage. This dependence is determined by lateral interactions and correlates well with phase diagram. Theoretical description of surface diffusion within the framework of the lattice gas model is a complicated problem. The most accuracive expression for the coefficient of the surface diffusion was given above (1.5) - (1.8). These formulae were derived under assumption that the diffusion is the result of activated jumps of the particles into the neighbouring unoccupied sites (Reed and all, 1981; Zhdanov, 1981). The coefficient of the surface diffusion has been calculated by quasi-chemical approximation (Zhdanov, 1988) and Monte-Karlo method (Reed and all, 1981).

The quasi-chemical approach is rather accuracive at high temperature ($T>1.3T_c$). In this field all three methods i.e. the quasi-chemical approuch, Monte-Karlo method and TMM give the same results. In the field of the ordered phase existence at $T<T_c$ the quasi-chemical approuch has insufficiently precision. In this field of parameters among the methods described above, Monte-Karlo method is the only accetable. At the same time as it was discussed above, TMM can be used both at high and at low temperature. The application of TMM for the calculation of the coefficient of surface diffusion are based on the formulae (1.8), (1.9), (2.5). TMM was not used for these purpose before.

The coefficient of surface diffusion as well as the constant of monomolecular desorption is a local parameter.Proceeding from this we can confirm, that TMM will have well precision in the field of the existence of ordered phases at comparatively little values M. We used numbers M = 4, 6 for concrete calculations. We shall study the dependence of coefficient of surface diffusion upon coverage on the square lattice with taking into account only pairwise interactions.

As noted above, lateral interactions of the nearest neighbours for chemisorbed particles are mostly repulsion. We shall concentrate our attention upon this case. Let's examine the system with <nn> repulsion on the square lattice. As it's known, the critical temperature of such system is determine by eqution $T_c = 0.567\varepsilon_1$. The dependences of the coefficient of surface diffusion upon coverage with $\varepsilon_1/T = 1$, i.e. at temperature much higher than critical and with $\varepsilon_1/T = 3$, i.e. at temperature much lower then critical are given in Figure 10. We can see a sharp maximum at θ =

0.5 at low temperature. We shall discuss the reason of the appearence of this maximum below. Let's examine what influence the diagonal attraction exerts on the dependence of the coefficient of surface diffusion upon coverage. The curves for various values of <nn> repulsion and <nnn> attraction energy are given in Figure 11.

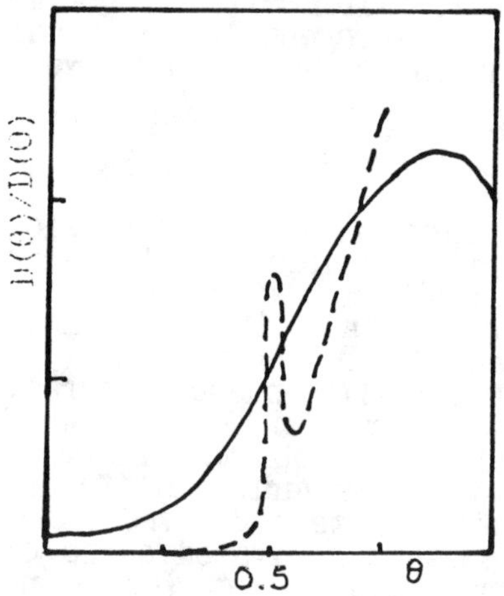

Figure 10. The coefficient of the surface diffusion with <nn> repulsion; —— $\varepsilon_1/T = 1$; - - - $\varepsilon_1/T = 3$.

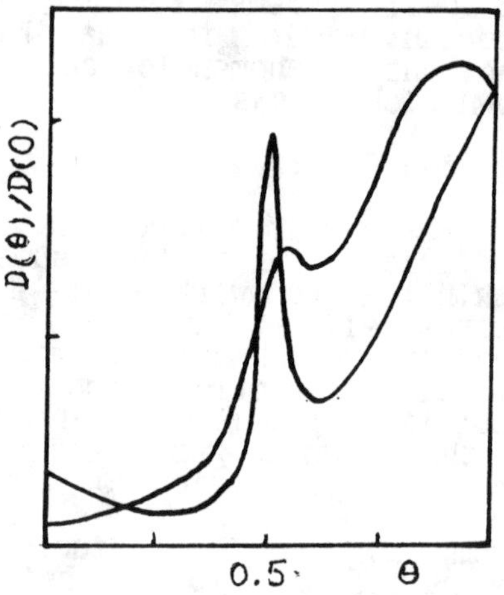

Figure 11. The coefficient of the surface diffusion with <nn> repulsion and <nnn> attraction.

The appearence of a sharp peak near $\theta = 0.5$ is a characteristic feature of certain curves. Its appearence is connected with the existence of ordered phase C(2×2) in these systems. Intervals of the decrease of the surface diffusion coefficient are determed by the existence of the phases LG, LL + C(2×2). In the case of <nnn> repulsion the dependences of surface diffusion coefficient are rather simple and have not sharp extrema. Let's examine the influence of three-body interactions on the diffusion coefficient. In general cases, the phase diagram for this system is rather complicated. However, in the field of parameters, most interested for our purpose, the phase diagram is considerably simplified and becomes similar to the phase diagram of the system without three-body interactions. As in case without three-body interactions, the existence of sharp peak at $\theta = 0.5$ is connected with the ordered phase C(2×2). In this system (with phase C(2×2)) the influence of three- body interactions becomes appreciable at $\theta>0.5$. At these values of θ, the behaviour of the curves depends strongly upon the sign of three-body interactions. Three-body attraction lead to a sharp decrease of diffusion coefficient at $\theta>0.5$. It is connected with greater easiness of formation of condensed phase LL + C(2×2). In the presence of three-body repulsion the sharp maximum is followed at once by so sharp "downfall" with the following fast, nearly vertical increase of diffusion coefficient. Strong three-body repulsion prevents the formation of the phase LL + C(2×2).

Characteristic features of many systems is the presence of sharp maximum at $\theta = 0.5$. As can be seen from given material the presence of such maximum is evidence of the existence of the ordered phase C(2×2). Let's examine the mechanism of this maximum formation. There are three factors depending on coverage in the expression for surface diffusion coefficient.

$$P_{oo},\ \exp(\ \mu/T\ ),\ \partial\mu/\partial\theta; \qquad (5.2)$$

The dependence of the product $P_{oo}\exp(\ \mu/T\ )$ upon coverage in temperature range according to the existence of sharp maximum of the diffusion coefficient is given in Figure 12. For simplicity we have examined the system having only <nn> repulsion. We can see that this product has not singularity at θ = 0.5, while diffusion coefficient has such one. Hence, the derivative $\partial\mu/\partial\theta$ at θ = 0.5 has singularity. In fact, isotherms corresponding to the presence of the phase C(2×2) have a long "plateau" of almost permanent value at θ = 0.5, i.e. a small change of coverage corresponds to a large change of chemical potential. Thus at θ = 0.5 the derivative $\partial\mu/\partial\theta$ has sharp maximum, which is a reason of maximum of diffusion coefficient. We can infer, that sharp peak of the coefficient of surface diffusion at θ = 0.5 is connected with presence of ordered structure C(2×2). The existence of ordered phases determines to the character of the dependence of the coefficient of the surface diffusion upon coverage. The influence of three-body interactions in the parameters range where there is a phase C(2×2), occurs only at θ>0.5. In absence of this phase three-body interactions have marked influence upon diffusion coefficient at rather small values θ = 0.2÷0.3. The appearance of maximum at θ<0.5 is connected with strong three-body attraction, which causes condensation of lattice gas and decreases the coefficient of surface diffusion. TMM showed its high effectivness for investigating surfase diffusion within the framework of lattice gas model in the field of ordered phases existence, where application of traditional methods is rather difficult. As marked above, TMM possesses rather high accuracy at comparatively small values M and allows to examine more complicated system than examined above. For example, the model of desorption with surface reconstruction is examined below.

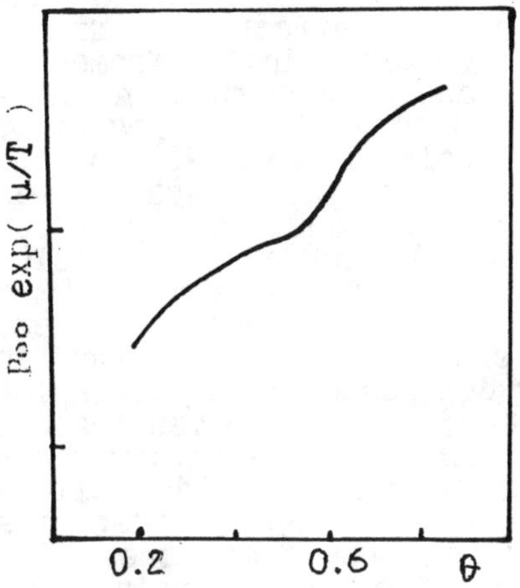

Figure 12. The dependence of product $P_{oo}\exp(\ \mu/T\ )$ upon coverage in the field of the existence of phase C(2×2).

6. DESORPTION FROM RECONSTRUCTED SURFACE

The compensation effect

The compensation effect essence, wich we are going to examine is that the activation energy and the pre-exponential factor decrease with increasing coverage, i.e. the pre-exponential factor change partly compensates the desorption constant increase. Data of the pre-exponential factor behaviour in more than fourty systems have been collected and analysed by (Seebaner and all, 1988). In about half of them the pre-exponential factor decreases more than 10^3 times. Frequently the compensation effect can be seen in systems with surface reconstruction. The H/W(100) system is a good example, where we can see both surface reconstruction and a strong (more than 10^5 times) compensaton effect. The influence of induced surface reconstruction on the Arrhenius parameters, supposing that there is the first order phase transition was investigated by (Zhdanov, 1985). It was shown that in this model there appears the compensation effect, but rather weak. More interesting results have been obtained in the framework of the model, that describes the surface reconstruction as a continuous phase transition. The model has been originally proposed by (Lau and all, 1980), on the basis of LEED data for the H/W(100) system. The total free energy of this system was represented as a sum of the phenomenological Landau's expression for the surface free energy, energy of the adsorbed particles calculated with using of the mean-field approximation and the additional interaction energy of the adsorbed particles and surface atoms. A strong compensation effect have been found in this model. The pre- exponential factor deacreased with increasing coverage more than 10^5 times. However, the phenomenological Landau's theory and the mean-field approximation are rather rough for describing of 2-D systems. It is connected the role of fluctuations in 2-D systems greater than in 3-D systems; the fluctuations do not take into account in these approaches. For this reason, it is of interest to analyse the problem using more accurate approximations.Below, for this purpose we use TMM.

Model of induced reconstraction

The model investigated us is a rather raugh approximation for describing of the H/W(100). We assume the continuous phase transitions to be in this system. At high temperature, tungsten atoms oscillate near the site of ideal square lattice occupy one of four equivalent places, while at low temperature these atoms do not oscillate. As the result of this localization the "zig-zag" structure is formed. Hydrogen atoms are in bridge positions between two tungsten atoms. Accurate analysis of such model is a difficult problem. For this reason we use a more simple model that contains the main characteristic of the initial system. Our model (Myshlyavtsev and all, 1990) contains two equilibrium position for surface atoms. In the simplest case of a hard localization of atoms we can describe their positions by spin variables $S_i = \pm 1$; $S_i = 1$ if a surface atom is located in the right position, and $S_i = -1$ if a surface atom is located in the left position. Adsorbed particles are are in bridge positions. This model is described by the Hamiltonian

$$H = \sum_{\langle nn \rangle} S_i S_j + \sum \varepsilon_{ij} n_i n_j + \lambda \sum_i n_i (S_{i+1} - S_i) \qquad (6.1)$$

The first term describes lateral interactions between surface atoms. We examined "antiferromagnetic" case (J>0). The second term describes lateral interactions between adsorbed particles. We assume that there are only two types of interactions: the interaction between nearest neighbours and the interactions between next nearest neighbours (<nn> and <nnn> interactions). Besides that we consider these interactions to be repulsive and equivalent by value. The third term represents interactions between adsorbed particles and adjacent surface atoms.

The influence of surface reconstruction on the desorption kinetics

At the first time, we have to calculate the phase diagram of this model. In general, it is a difficult problem. For our purpose it is enough to make qualitative picture, using the mean-field approximation. For clean lattice, we get the Ising model. For its we can write Onsager relation.

$$T_c(0) = 2.269J \quad (6.2)$$

In the mean-field approximation we can obtain the following equation for the critical temperature.

$$T_c(\theta)/T_c(0) = 0.5\ [1 + (1 + \lambda^2\ \theta\ (1 - \theta)/J^2)^{1/2}] \quad (6.3)$$

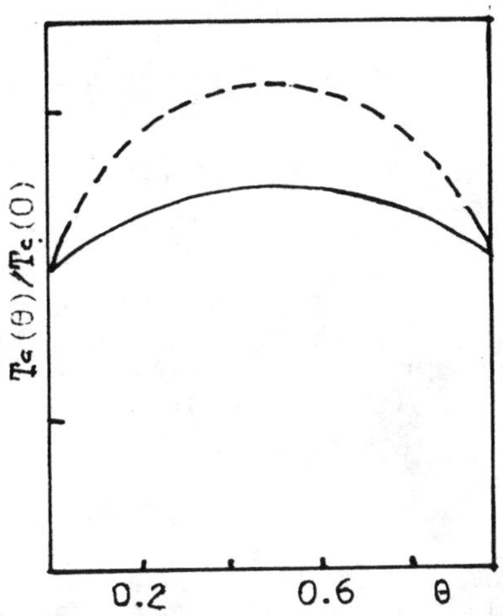

Figure 13. The phase diagram. - - - - $T_c(0)$ = 300K; ——— $T_c(0)$ = 500K.

It is necessary to note that this equation (6.3) does not contain lateral interaction energy. This is a consequence of our assumption that $\varepsilon_1 = \varepsilon_2$. The results of calculations are presented in Figure 13. Phase diagram is symmetrical conserning θ = 0.5, that is a consequence of special choise of lateral interactions, too. We use the TMM for investigation desorption as in the previous parts of this article. The only difference is that the site of the lattice may be in four states and transfer-matrix dimension is $4^M \times 4^M$. In our work we have used value M = 4. Precision of calculation is rather high as we interested in only local parameters: desorption constant, activation energy and pre-exponential factor. The desorpton constant have been calculated proceeding from general expression (1.1). The desorption have been studied near T = 400K. The lattice parameter J has been chosen so that $T_c(0)$ = 300K or 500K according to expression (6.2). The parameter of interaction energy between adsorbed particles and lattice - λ has been choosen to reproduce the dependence of the critical temperature upon coverage for the H/W(100) system. The phase diagram with $T_c(0)$ = 300K and λ = 1 kcal/mol is about the same that for the H/W(100) system at low coverage. The dependences of the activation energy and pre-exponential factor upon the coverage are given in Figure 14. We can see, that as a result of adsorbate-indused reconstruction there appears the compensation effect (the activation energy and the pre-exponential factor decrease with increasing coverage). We can see that $\nu(0)/\nu(1) > 10$. It is of interest that the compensation effect takes place both at $T<T_c(0)$ and at $T>T_c(0)$. In fact, the coverage dependence of the pre-exponential factor is stronger at $T>T_c(0)$ than at $T<T_c(0)$. This is explained by more strong changes in the short-range order with increasing coverage at $T>T_c(0)$. This effect, of course is absent if we use the phenomenolgical Landau's theory for describing reconstruction. In this case, the influence

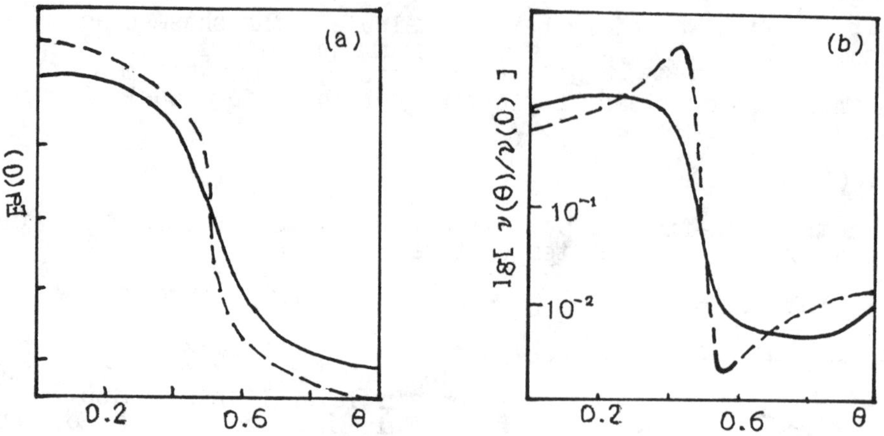

Figure 14. (a) The activation energy; (b) pre-exponential factor; $\varepsilon_1 = \varepsilon_2 = 0.5$ kcal/mol, $\lambda = 1$ kcal/mol, —— $T_c(0) = 300$K, – – — $T_c(0) = 500$K.

of reconstruction on the pre-exponential factor for desorption foils at $T>T_c(0)$ because the phenomenological theory does not take into account short-range order in the arrangement of particles. For associative desorption the expression for desorption constant is represented by (Zhdanov, 1988) as

$$k_d(\theta) = k_d(0)\ \exp(2\mu/T)P_{oo}/\theta^2 \qquad (6.4)$$

The analysis of (6.4) shows that the term $\exp(2\mu/T)$ makes the dominating contribution to nonideality of the kinetics of associative desorption. Thus, the influense of surface reconstruction on the Arrhenius parameters for associative desorption is doubled in comparison with monomolecular desorption, and the pre-exponential factor is changed more than 10^2 times. The experimental dependence of the pre-exponential factor upon the coyerage for H/W(100) give more strong decrease of it (more than 10^3 times). Apparently, it can be explained by more complicated type of reconstruction in H/W(100) system in comparison with investigating model.

The results of this part show the possibilities of TMM when investigating rather complicated systems.

7. CONCLUSION

As it is shown in the preceeding section, the TMM allows to calculate various coefficiens for rather complicated models. Let's examine some potencial possibilities of this method. The first application of its is an investigation of the kinetics of chemical reactions on plane of single crystal. In particular, the investigation of the influence of lateral interactions on bifurcation points of complicated systems is of a great interest. The second application is an examination of more complicated lattices with several kinds of centres, that will allow to simulate such phenomena as the "compression" structure. The TMM allows to study such interesting systems as a high-index plane of single crystals, which are regular step surfaces at equilibrium conditions. We consider that TMM can be used for investigations of many complicated problem which arise in the field of "surface science".

8. ACKNOWLEDGMENTS

We thanks Dr. V.P. Zhdanov for useful discussion.

Adams, D. L. 1974. Consequence of adsorbate-adsorbate interactions for thermal desorption and LEED measurements. Surf.Sci. 42: 12-36.

Baxter, R.J. 1982. Exactly solved models in statistical mechanics. New York. Academic Press.

Binder, K., Kinzel W., Landau D.P. Theoretical aspects of order-disorder transitions in adsorbed layer. Surf.Sci 117: 232-244.

Domb, C. 1960. On the theory of cooperative phenomena in crystals. Adv.Phys. 9: 149-244.

Engel, T., Ertl G. 1979. Elementary steps in the catalytic oxidation of the carbon monoxide on platinum metals. Adv.Catal. 28: 1-78.

Ibbotson, D. E., Wittrig T. S., Weinberg W. H. 1981. The chemisorption of N_2 on the (110) surface of iridium. Surf.Sci. 110: 313-328.

Kinzel, W., Schick M. 1981. Extent of exponent variation in a hard square lattice gas with second neighbour repulsion. Phys.Rev.B. 24: 324-330.

Lau, K. H., Ying S. C. 1980. Effect of H adsorption on the displasive transition of W(100) surface. Phys.Rev.Lett. 44: 1222-1225.

Leuthauser, U. 1980. Generalized quasichemical approximation for a lattice gas: application to CO on Ru. Z.Phys.B. 37: 65-67.

Lombardo, S. J., Bell A. 1988 T. A Monte-Karlo model for the simulation of temperature-programmed desorption spectra. Surf.Sci. 206: 101-123.

McMachon, P. D., Glandt E. D., Walker J. S. 1988. Renormalization group theory in solution thermodynamics. Chem.Eng.Sci. 43: 2561-2586.

Myshlyavtsev, A. V., Zhdanov V.P. 1989. The effect of nearest-neighbour and next-nearest-neighbour lateral interactions on thermal desorption spectra. Chem.Phys.Lett. 162: 43-46.

Myshlyavtsev, A. V., Sales J.L., Zgrablich G., Zhdanov V. P. 1990. The effect of three-body lateral interactions on thermal desorption spectra. J.Stat.Phys. 58: 1029-1039.

Myshlyavtsev, A. V., Zhdanov V. P. 1990. Effect of adsorbate-induced surface reconstruction on the apparent Arrhenius parameters for desorption. J.Chem.Phys.: in press.

Nieuwenhuys, B. E. 1983. Adsorption and reactions of CO, NO, H_2 and O_2 on group VIII metal surfaces. Surf.Sci. 126: 307-336.

Pfnur,H., Feulner P., Menzel D. 1983. The influence of adsorbate interactions on kinetics and equilibrium for CO on Ru(100). Desorption kinetics and equilibrium. J.Chem.Phys. 79: 4613-4623.

Reed, D. A., Ehrlich G. 1981. Surface diffusion, atomic jump rates and thermodynamics. Surf.Sci. 102: 588-609.

Reed, D. A., Ehrlich G. 1981. Surface diffusivity and the time correlation of concentration fluctuations. Surf.Sci. 105: 603-628.

Roelofs, L. D., Estrup P. J. 1983. Two-dimensional phases in chemisorption systems. Surf.Sci. 125: 51-73.

Runnels, L. K., Combs L. L. 1966. Exact finite method of lattice statistics. I. Square and triangular lattice gases of hard molecules. J.Chem.Phys. 45: 2482-2492.

Runnels, L. K., Combs L. L., Salvant J. P. 1967. Exact finite method of lattice statistics. II. Honeycomb-lattice gas of hard molecules. J.Chem.Phys. 47: 4015-4020.

Sadiq, A., Binder K. 1983. Diffusion of adsorbate atoms in ordered and disordered monolayers at surfaces. Surf.Sci. 128: 350-382.

Sales, J. L., Zgrablich G. 1987. Thermal desorption of interacting molecules from heterogeneous surfaces; application to CO desorption from MgO. Surf.Sci. 187: 1-20.

Sales, J. L., Zgrablich G., Zhdanov V. P. 1989. Lattice gas model for calculating thermal desorption spectra: comparison between analytical and Monte-Karlo results. Surf.Sci. 209: 208-214.

Seebaner, E. G., Kong A. C. F., Schmidt L. D. 1988. The coverage dependence of the pre-exponential factor for desorption. Surf.Sci. 193: 417-436.

Silverberg, M., Ben-Shaul A., Robentrost F. 1985. On the effect of adsorbate aggregation on the kinetics of surface reactions. J.Chem.Phys. 83: 6501-6513.

Smart, J. S. 1966. Effective field theories of magnetism. Philadelphia-London, W.B. Saunders company.

Tringides, M., Gomer R. 1984. A Monte-Karlo study of oxigen diffusion on the (110) plane of tungsten. Surf.Sci. 145: 121-144.

Weinberg, W.H. 1983. Order-disorder phase transitions in chemisorbed overlayers. Ann.Rev.Phys.Chem. 34: 217-243.

Yablonskii, G. S., Bykov V. I., Gorban' A. N. 1983. Kinetic models of catalytic reactions. Novosibirsk, Nauka.

Yablonskii, G. S., Bykov V. I., Elochin V. I. 1984. Kinetics of model reactions of heterogenious catalysis. Novosibirsk, Nauka. (in Russian)

Zhdanov, V. P. 1981. Effect of the lateral interaction of adsorbed molecules on pre-exponential factor of the desorption rate constant. Surf.Sci. 111: L662-666.

Zhdanov, V. P. 1983. Thermal desorption from adlayer of interacting particles. Surf.Sci. 133: 469-483.

Zhdanov, V. P. 1985. Simple model for adsorbate-induced surface phase transition. Surf.Sci. 164: L807-810.

Zhdanov, V. P. 1988. The elementary physico-chemical processes on surface. Novosibirsk, Nauka. (in Russian)

Glossary of Symbols

A Richardson-Dushman constant
A availability, affinity, extremized objective
A',A" partial values of Gibbs free energy for reactants and products, respectively
a_i activities, state variables, termal diffusivities
B magnetic field strength
B_a, B_e absorption and emission coefficients
B_k chemical component, transition probability
b_i state variables
C thermostatic capacity matrix
C heat capacity, capillary number, curvature
c molecular velocity
c,c_0 light speed and propagation speed, respectively
c_i molar concentration of i-th species
D diffusion coefficient
d thickness, diameter, density
E electric field
E energy, activation energy, energy density
e unit vector
e specific energy
e_p potential energy
F, **f** external forces
F free energy
F_A nonconfigurational partition function
F_i expected values, observables
F_s specific distribution function
f distribution function, activity coefficient, free energy
f_i microscopic variables, fluctuating variables
G Gibbs free energy, shear modulus, evolution operator
G_{jk} energy-momentum tensor
g_{jk} thermodynamic tensor
H Biot's heat vector
H magnetic field
H hamiltonian function, Boltzmann's H function
h specific enthalpy, heat exchange coefficient
h Planck constant
I action variable, electric current
I_v specific intensity (radiance) of energy radiation,
J, $\mathbf{J}_E$ total mass flux density and energy flux density, respectively
$\mathbf{J}_a$,$\mathbf{j}_a$ mass fluxes of a-th species
$\mathbf{J}_h$,$\mathbf{J}_s$ pure heat flux and entropy flux, respectively

J	lattice parameter
J_1, J_2	rates of elementary chemical steps
J_v	specific intensity (radiance) of entropy radiation
$\mathbf{K}$	momentum type variable, extra term of entropy flux
$\mathbf{K}, \mathbf{k}$	wave (propagation) vectors
K	Gaussian curvature, chemical equilibrium constant, memory kernel
k	Boltzmann constant
k	rate constant, thermal conductivity
$\mathbf{L}$	conductivity matrix
L	lagrangian, kinetic potential, total length, Onsager phenomenological coefficient
L_{ik}	Onsager's matrix
l	length, characteristic length, capillary length
M	molar mass, mass property, magnetization
Ma	Marangoni number
$\mathbf{m}$	unit direction vector
m	mass of a microobject
N	mole number, occupation number
Nu	Nusselt number
$\mathbf{n}$	vector of elementary extensities in Keizer theory, normal vector
n	dimensionality of state vector, number density, mole number
n_{ij}	occupation numbers
$\mathbf{P}$	pressure tensor, polarization vector
P	pressure, projection operator
Pr	Prandtl number
P_i, p_i	probabilities
$\mathbf{Q}$	energy flux density
Q	generalized coordinate, heat, electric charge
Q^*	heat of transport
$\mathbf{q}$	heat flux density
q_i	generalized coordinates, charges
$\mathbf{R}$	resistance matrix
R	universal gas constant, resistance, curvature radius
$\mathbf{r}$	radius vector
r	radius, chemical rate
S, s	total entropy and unit entropy, respectively
S_j	components of surface energy density vector, spin variables
S_{ij}	transition probability from state i to j, per unit time
s_{ij}	rate of deformation
$\mathbf{T}$	transfer matrix
T	absolute temperature
t	time, coordinate time

$t_{\mu\nu}$	energy-momentum tensor
U, u	internal energy and unit internal energy, respectively
u	hydrodynamic velocity, vector of transport potentials
u_i	displacement components
V	volume, scalar potential, interparticle potential, Liapunov function
v	speed, propagation speed
v	velocity, molecular velocity
W,w	probability density, work, velocity amplitude, volume energy density
w	velocity on a surface
X	vector of thermodynamic forces
X	state vector
X_i	macroscopic extensities, expectations, dimensionless coordinates
x	state vector, radius vector
x	mole fraction
Y	mass content function
Y_i	thermodynamic conjugates of variables X_i
y	thermodynamic conjugate of state x
Z	impedance matrix, eigenvector
Z	partition function
z	state vector
z^{α}	field potential, weight factor, fraction
α	absorption coefficient, thermal diffusivity
α_v	coefficient of isotropic scattering
β	nonequilibrium variable vector, vector Lagrangian multiplier
β	temperature reciprocal
Γ_i	coefficients of entropy flux, Planck potentials
Γ_{ij}	metric functions
γ	relativistic factor, coefficient of thermal inertia
δ	Dirac function, thickness, effective diameter
ε	density of energy, energy of lateral interactions
Φ	dissipation function, phase, Gibbs free energy, energy flux
ϕ	material action variable, flux dependent dissipation potential, electric potential
η	efficiency, shear viscosity
θ	inertial coefficient, characteristic function, amplitude, surface coverage
Λ	Lagrangian density of fluid

λ Lagrangian multiplier, thermal conductivity, wave length, weighting function, eigenvalue

μ chemical potential

ν frequency, stoichiometric coefficient, kinematic viscosity

Π Peltier coefficient

Π_{ij} stress tensor

ρ density, total mass density

ρ_{ea} energy density of a-th component

ρ_a mass density of a-th component

ρ_s total entropy density

Σ grand partition function

σ entropy source, hydrostatic part of stress, Stefan-Boltzmann constant

σ_u source of u-th variable

σ_{ik} stress tensor components

τ time constant, relaxation time, Thompson coefficient, stress

τ_h, τ_d, τ_m relaxation times for heat diffusion and momentum, respectively

τ_{ij} stress tensor components

Ω grand potential, statistical weight, solid angle, volume

$d\Omega$ element of solid angle

ω angular frequency

ψ external scalar field, force dependent dissipation potential

ζ_i state coordinates, progress variables, surface coordinates

Index

RETURN PHYSICS LIBRARY
TO 351 LeConte Hall 64